Lecture Notes Editorial Policies

Lecture Notes in Statistics provides a format for the informal and quick publication of monographs, case studies, and workshops of theoretical or applied importance. Thus, in some instances, proofs may be merely outlined and results presented which will later be published in a different form.

Publication of the Lecture Notes is intended as a service to the international statistical community, in that a commercial publisher, Springer-Verlag, can provide efficient distribution of documents that would otherwise have a restricted readership. Once published and copyrighted, they can be documented and discussed in the scientific literature.

Lecture Notes are reprinted photographically from the copy delivered in camera-ready form by the author or editor. Springer-Verlag provides technical instructions for the preparation of manuscripts. Volumes should be no less than 100 pages and preferably no more than 400 pages. A subject index is expected for authored but not edited volumes. Proposals for volumes should be sent to one of the series editors or addressed to "Statistics Editor" at Springer-Verlag in New York.

Authors of monographs receive 50 free copies of their book. Editors receive 50 free copies and are responsible for distributing them to contributors. Authors, editors, and contributors may purchase additional copies at the publisher's discount. No reprints of individual contributions will be supplied and no royalties are paid on Lecture Notes volumes. Springer-Verlag secures the copyright for each volume.

Series Editors:

Professor P. Bickel
Department of Statistics
University of California
Berkeley, California 94720
USA

Professor P. Diggle
Department of Mathematics
Lancaster University
Lancaster LA1 4YL
England

Professor S. Fienberg
Department of Statistics
Carnegie Mellon University
Pittsburgh, Pennsylvania 15213
USA

Professor K. Krickeberg
3 Rue de L'Estrapade
75005 Paris
France

Professor I. Olkin
Department of Statistics
Stanford University
Stanford, California 94305
USA

Professor N. Wermuth
Department of Psychology
Johannes Gutenberg University
Postfach 3980
D-6500 Mainz
Germany

Professor S. Zeger
Department of Biostatistics
The Johns Hopkins University
615 N. Wolfe Street
Baltimore, Maryland 21205-2103
USA

Lecture Notes in Statistics 171

Edited by P. Bickel, P. Diggle, S. Fienberg, K. Krickeberg,
I. Olkin, N. Wermuth, and S. Zeger

Springer-Verlag Berlin Heidelberg GmbH

David D. Denison
Mark H. Hansen
Christopher C. Holmes
Bani Mallick
Bin Yu (Editors)

Nonlinear Estimation and Classification

Springer

David D. Denison
Department of Mathematics
Imperial College
180 Queen's Gate
London, SW7 2BZ
UK
d.denison@ic.ac.uk

Christopher C. Holmes
Department of Mathematics
Imperial College
180 Queen's Gate
London, SW7 2BZ
UK
c.holmes@ic.ac.uk

Bin Yu
Department of Statistics
University of California, Berkeley
Berkeley, CA 94720-3860
USA
binyu@stat.berkeley.edu

Mark H. Hansen
Room 2C283
Bell Laboratories, Lucent Technologies
600 Mountain Avenue
Murray Hill, NJ 07974-0636
USA
cocteau@bell-labs.com

Bani Mallick
Statistical Department
Texas A&M University
College Station, TX 77843-3143
USA
bmallick@stat.tamu.edu

Library of Congress Cataloging-in-Publication Data
Nonlinear estimation and classification / editors, D.D. Denison ... [et al.]
        p. cm. — (Lecture notes in statistics ; 171)
    Includes bibliographical references and index.
    ISBN 978-0-387-95471-4
        1. Estimation theory.   2. Nonlinear theories.   I. Denison, D.D. (David D.)   II. Lecture
    notes in statistics (Springer-Verlag) ; v. 171.
    QA276.8 .N64   2002
    519.5'44—dc21   2002030566
    ISBN 978-0-387-95471-4   ISBN 978-0-387-21579-2 (eBook)   Printed on acid-free paper.
    DOI 10.1007/978-0-387-21579-2

9 8 7 6 5 4 3 2 1   SPIN 10874011

Typesetting: Pages created by the authors using a Springer TeX macro package.

Contents

Introduction

David D. Denison, Mark H. Hansen,
Christopher C. Holmes, Bani Mallick,
and Bin Yu

Background

Researchers in many disciplines now face the formidable task of processing massive amounts of high-dimensional and highly-structured data. Advances in data collection and information technologies have coupled with innovations in computing to make commonplace the task of learning from complex data. As a result, fundamental statistical research is being undertaken in a variety of different fields. Driven by the difficulty of these new problems, and fueled by the explosion of available computer power, highly adaptive, nonlinear procedures are now essential components of modern "data analysis," a term that we liberally interpret to include speech and pattern recognition, classification, data compression and image processing. The development of new, flexible methods combines advances from many sources, including approximation theory, numerical analysis, machine learning, signal processing and statistics. This volume collects papers from a unique workshop designed to promote communication between these different disciplines.

History

In 1999, Hansen and Yu were both Members of the Technical Staff at Bell Laboratories in Murray Hill, New Jersey. They were exploring the connections between information theory and statistics. At that time, Denison and Mallick were faculty members at Imperial College, London, researching

Bayesian methods for function estimation; and Holmes was a graduate student at Imperial, studying with Mallick. In the summer of 1999, Denison, Holmes and Mallick (together with Robert Kohn at the University of New South Wales, and Martin Tanner at Northwestern University), began to think about a conference to explore Bayesian approaches to classification and regression. Holmes spent part that summer visiting Hansen and Yu at Bell Labs, and invited them to join the organizing committee of his conference. Very quickly, the focus of the meeting expanded to include a broad range of ideas relating to modern computing, information technologies, and large-scale data analysis. The event took on a strong interdisciplinary flavor, and soon we had a list of invited speakers from machine learning, artificial intelligence, applied mathematics, image analysis, signal processing, information theory, and optimization. Within each broad area, we tried to emphasize complex applications like environmental modeling, network analysis, and bioinformatics.

In the fall of 1999, the Mathematical Sciences Research Institute (MSRI) in Berkeley, California agreed to sponsor the workshop. Given the size of the problem area and the diversity of disciplines represented, we planned on a two week affair. In addition to the invited speakers mentioned above, the program also included a series of excellent contributed talks. This volume contains several of the papers given at the workshop, organized roughly around the sessions in which they were presented (the heading "Longer papers" refers to chapters written by invited participants; while the "Shorter papers" were submitted through the contributed track). The MSRI video taped all of the talks given during the workshop, and have graciously made them available in streaming format on their Web site, http://www.msri.org. In addition to videos, the MSRI also distributes electronic versions of each speaker's talk materials, an invaluable resource for anyone interested in the areas represented in this volume.

In preparing these papers for publication, we are reminded of the excitement that came from the interplay between researchers in different fields. We hope that that spirit comes through in this book.

Thanks

First, the MSRI workshop was supported by the National Science Foundation under Grant No. DMS9810361, by the National Security Agency under grant MDA904-02-1-0003, and by the US Army Research Office under grant number 41900-MA-CF.

Next, we would like to thank Joe Buhler, (then) Deputy Director of MSRI, for his encouragement, guidance and seemingly tireless enthusiasm for this workshop. David Eisenbud, Director of MSRI was also instrumental in arranging funding and providing support for the event. Margaret Wright, (then) head of the Scientific Advisory Committee (SAC) for MSRI helped us navigate the workshop approval process; and Wing Wong, (then) Statistics representative on the SAC helped sharpen our initial concept. Perhaps the most important practical contributions to this workshop were made by the Program Coordinator Loa Nowina-Sapinski who managed virtually every detail, from funding applications, to local arrangements, visas and travel reimbursements, to advertising and administrative support. Dave Zetland, Assistant to the Director of MSRI, and Rachelle Summers, Head of Computing at MSRI, were responsible for collecting presentation materials and organizing the streaming videos we mentioned above.

Finally, we are grateful to everyone who took the time to participate in this workshop, and are especially thankful for the impressive list of researchers who contributed to this volume.

List of Participants

In This Volume	Participant and Affiliation

Barbara Bailey, *University of Illinois at Urbana-Champaign*
- Richard G. Baraniuk, *Rice University*
Andrew Barron, *Yale University*
- Gilles Blanchard, *École normale supérieure*
- Amy Braverman, *Jet Propulsion Laboratory and the California Institute of Technology*
Leo Breiman, *UC Berkeley*
- Robert Burbidge, *Imperial College*
Nelson Butuk, *Prairie View A & M University*
- William S. Cleveland, *Bell Laboratories, Lucent Technologies*
Nando DeFreitas, *University of British Columbia*
- Maria De Iorio, *University of Oxford*
- Vin de Silva, *Stanford University*
Ronald DeVore, *University of South Carolina*
- Tom Dietterich, *Oregon State University*
Minh Do, *University of Illinois at Urbana-Champaign*
David Donoho, *Stanford University*
- Steve Ellis, *New York State Psychiatric Institute and Columbia University*
- Mário A. T. Figueiredo, *Instituto Superior Técnico*
Ernest Fokoue, *Glasgow University*
Jerome Friedman, *Stanford University*
Ashis Gangopadhyay, *Boston University*
- Donald Geman, *University of Massachusetts at Amherst*
Edward I. George, *University of Pennsylvania*
- Servane Gey, *Université Paris XI*
- Robert Gray, *Stanford University*
- Maarten Jansen, *Rice University*
Michael Jordan, *UC Berkeley*
- Harri Kiiveri, *CSIRO*
- Eric D. Kolaczyk, *Boston University*
- Charles Kooperberg, *Fred Hutchison Cancer Research Center*
- Adam Krzyżak, *Concordia University*
Yann LeCun, *NEC Research Institute*
- Juan Lin, *Rutgers University*
Alexander Loguinov, *UC Berkeley*
Steven Marron, *University of North Carolina*
Jacqueline J. Meulman, *Leiden University*
David Mumford, *Brown University*
Kevin Murphy, *MIT*
Andrew Nobel, *University of North Carolina*
- Robert D. Nowak, *Rice University*
- John Rice, *UC Berkeley*
- Ingo Ruczinski, *Johns Hopkins University*

In This Volume	Participant and Affiliation

- Ludger Rüschendorf, *University of Freiburg*
  Jennifer Pittman, *Duke University*
- Tomaso Poggio, *MIT*
- Katherine Pollard, *UC Berkeley*
  Adrian Raftery, *University of Washington*
- Jörg Rahnenführer, *Heinrich-Heine-Universität Düsseldorf*
- Robert E. Schapire, *AT&T Labs − Research*
  David Scott, *Rice University*
  Rahul Shukla, *Swiss Federal Institute of Technology*
  Terry Speed, *UC Berkeley*
- Charles Stone, *UC Berkeley*
  Claudia Tebaldi, *Athene Software, Inc.*
  Alexandre Tsybakov, *Université Paris VI*
- Angelika van der Linde, *Universität Bremen*
  Martin Vetterli, *Swiss Federal Institute of Technology*
- Grace Wahba, *University of Wisconsin*
  Margaret H. Wright, *New York University*
  Sally Wood, *Northwestern University*
  Yuhong Yang, *Iowa State University*
- Heping Zhang, *Yale University School of Medicine*
  Linda Zhao, *University of Pennsylvania*

Part I

Longer Papers

1

Wavelet Statistical Models and Besov Spaces

Hyeokho Choi and Richard G. Baraniuk[1]

1.1 Introduction

1.1.1 Natural images models

Natural image models provide a foundation for framing numerous problems encountered in image processing, from compressing images to detecting tumors in medical scans. A good model must capture the key properties of the images of interest. A typical photograph of a natural scene consists of piecewise smooth or textured regions separated by step edge discontinuities along contours. Modeling both the smoothness and edge structure is essential for maximum processing performance.

Research to date in image modeling has been split into two fairly distinct camps, with one group focusing on *deterministic* models and the other pursuing *statistical* approaches. Deterministic models define a set or vector space that contains the images of interest and a metric or norm typically based on smoothness. Statistical models place a probability measure on images, making natural images more likely than unnatural ones.

1.1.2 Deterministic image modeling

Working under the assumption that natural images consist of smooth or textured regions delineated by step edges, we can assess images in terms of the number of derivatives we can compute. The *Sobolev space*[2] $W^\alpha(L_2)$ contains all images having α derivatives of finite energy. However, since

---

[1]Richard G. Baraniuk is Professor, Electrical and Computer Engineering, Rice University. Hyeokho Choi is Research Professor/Lecturer, Electrical and Computer Engineering, Rice University. This work was supported by NSF, ONR, AFOSR, DARPA, and Texas Instruments. The authors thank the anonymous reviewer for many useful comments.

[2]Sobolev and Besov spaces will be formally defined in Section 1.3.

natural images contain many step edges, they do not in general belong to any Sobolev space.

To overcome this difficulty, researchers have turned to *Besov spaces*, which can be tuned to measure the smoothness "between" the step edges. Roughly speaking, the Besov space $B_q^\alpha(L_p)$ contains signals having α derivatives between edges.

Besov norms are naturally computed in the wavelet domain, since the wavelet transform computes derivatives of the image at multiple scales and orientations. Besov space concepts have been applied to assess the performance of image estimation [1] and compression [2] algorithms. Indeed, estimation by wavelet thresholding [1] has been proven near optimal for representing and removing noise from Besov space images.

1.1.3 Statistical image modeling

Statistical models characterize natural images by assigning a probability measure to the set of images. Each image is considered as a random realization from the resulting "natural image random process." The smoothly varying spatial properties of images have motivated considerable research on Markov random field (MRF) models to characterize the statistical dependencies between image pixels. While conceptually appealing, MRFs have severe computational limitations.

As an alternative, statistical models that capture the statistics of the wavelet coefficients have been deployed in applications such as Bayesian estimation [3, 4, 5, 6], detection/classification [4], and segmentation [7]. We consider two wavelet-domain statistical models here: one that models the wavelet coefficients as independent generalized Gaussian random variables, and one that models the wavelet coefficients as an "infinite mixture" of Gaussian random variables correlated in variance.

1.1.4 Links between two paradigms

To date, the deterministic Besov space and wavelet statistical model frameworks have been developed by essentially two distinct communities, and few connections have been made between the two approaches. We summarize the state of affairs in Table 1.1. Only recently, Abramovich et al. showed that their independent wavelet-domain model is related to Besov space [5] (see also [8]). However, the general relationships between Besov spaces and wavelet-domain statistical models have never been clear.

In this chapter, we uncover several surprising relationships between these two seemingly different modeling frameworks that open up generalizations of the Besov space concept both for more accurate modeling and for practical applications. We take an information-theoretic view of the two frameworks and show that a Besov space deterministic model is in a sense equivalent to an independent generalized Gaussian statistical model.

Table 1.1. Comparison of deterministic vs. statistical image models.

deterministic model	statistical model
function space	probability density function
norm $\|\mathbf{x}\|$	likelihood $f(\mathbf{x})$

This connection allows us to see for the first time the equivalence between many processing algorithms proposed in one or the other framework. We also point out some of the shortcomings of Besov spaces as natural image models and use the statistical framework to suggest new, more realistic image models that generalize Besov spaces.

This chapter is organized as follows. In Section 1.2, we review the basic theory of the wavelet transform and the properties of natural images in the wavelet domain. Section 1.3 introduces the Besov space as a deterministic image model, and Section 1.4 presents two simple wavelet statistical image models. Section 1.5 then forges the links between Besov space and the statistical models. Finally, we propose some extensions to the Besov model inspired by advanced statistical models and conclude in Section 1.6.

1.2   Wavelet Transforms and Natural Images

1.2.1   Wavelet transform

The discrete wavelet transform (DWT) represents a one-dimensional (1-D) signal $z(t)$ in terms of shifted versions of a lowpass scaling function $\phi(t)$ and shifted and dilated versions of a prototype bandpass wavelet function $\psi(t)$ [9]. For special choices of $\phi(t)$ and $\psi(t)$, the functions $\psi_{j,k}(t) := 2^{j/2}\psi(2^j t - k)$ and $\phi_{j,k}(t) := 2^{j/2}\phi(2^j t - k)$ with $j, k \in \mathbb{Z}$ form an orthonormal basis, and we have the representation [9]

$$z = \sum_k u_{j_0,k}\, \phi_{j_0,k} + \sum_{j=j_0}^{\infty} \sum_k w_{j,k}\, \psi_{j,k} \qquad (1.1)$$

with $u_{j,k} := \int z(t)\, \phi_{j,k}(t)\, dt$ and $w_{j,k} := \int z(t)\, \psi_{j,k}(t)\, dt$.

The *wavelet coefficient* $w_{j,k}$ measures the signal content around time $2^{-j}k$ and frequency $2^j f_0$. The scaling coefficient $u_{j,k}$ measures the local mean around time $2^{-j}k$. The DWT (1.1) employs scaling coefficients only at scale j_0; wavelet coefficients at scales $j \geq j_0$ add higher resolution details to the signal.

We can construct 2-D wavelets from the 1-D ψ and ϕ by setting for $\mathbf{x} := (x, y) \in \mathbb{R}^2$, $\psi^{\mathrm{HL}}(x,y) := \psi(x)\phi(y), \psi^{\mathrm{LH}}(x,y) := \phi(x)\psi(y), \psi^{\mathrm{HH}}(x,y) := \psi(x)\psi(y)$, and $\phi(x,y) := \phi(x)\phi(y)$. If we let $\Psi := \{\psi^{\mathrm{HL}}, \psi^{\mathrm{LH}}, \psi^{\mathrm{HH}}\}$, then the set of functions $\{\psi_{j,\mathbf{k}} = 2^j \psi(2^j \mathbf{x} - \mathbf{k})\}_{\psi \in \Psi, j \in \mathbf{Z}, \mathbf{k} \in \mathbf{Z}^2}$ and $\{\phi_{j,\mathbf{k}} = 2^j \phi(2^j \mathbf{x} - \mathbf{k})\}_{j \in \mathbf{Z}, \mathbf{k} \in \mathbf{Z}^2}$ forms an orthonormal basis for $L_2(\mathbb{R}^2)$. That is, for every $z \in L_2(\mathbb{R}^2)$, we have

$$z = \sum_{\mathbf{k} \in \mathbf{Z}^2} u_{j_0,\mathbf{k}} \phi_{j_0,\mathbf{k}} + \sum_{j > j_0, \mathbf{k} \in \mathbf{Z}^2, \psi \in \Psi} w_{j,\mathbf{k},\psi} \psi_{j,\mathbf{k}} \qquad (1.2)$$

with $w_{j,\mathbf{k},\psi} := \int_{\mathbf{R}^2} z(\mathbf{x}) \psi_{j,\mathbf{k}}(\mathbf{x}) d\mathbf{x}$ and $u_{j_0,\mathbf{k}} := \int_{\mathbf{R}^2} z(\mathbf{x}) \phi_{j_0,\mathbf{k}}(\mathbf{x}) d\mathbf{x}$.

Given discrete samples of the continuous image $z(\mathbf{x})$ and proper prefiltering, we can approximate the discrete samples by the scaling coefficients of $z(\mathbf{x})$ at a certain scale J; that is, the sampled image $z(\mathbf{k}) = u_{J,\mathbf{k}}$. Equivalently, we can build a continuous-time image corresponding to $z(\mathbf{k})$ as

$$\widetilde{z} = \sum_{\mathbf{k} \in \mathbf{Z}^2} u_{J,\mathbf{k}} \phi_{J,\mathbf{k}} \qquad (1.3)$$

or, using the wavelet coefficients,

$$\widetilde{z} = \sum_{\mathbf{k} \in \mathbf{Z}^2} u_{j_0,\mathbf{k}} \phi_{j_0,\mathbf{k}} + \sum_{j_0 < j < J, \mathbf{k} \in \mathbf{Z}^2, \psi \in \Psi} w_{j,\mathbf{k},\psi} \psi_{j,\mathbf{k}}. \qquad (1.4)$$

The coefficients $u_{j_0,\mathbf{k}}$ and $w_{j,\mathbf{k},\psi}$ are closely approximated using 2-D discrete-time wavelet filters and decimators operating on the samples $z(\mathbf{k})$ [9].

1.2.2   Wavelet-domain properties of natural images

Completely specifying the form of the set of all natural images is an impossible task. To get an idea of the difficulty, note that an $N \times N$ natural image z is a vector in the space $\mathbb{R}^{N^2}$, an extremely large space for any reasonable N.[3] Two things are clear. First, the set of natural images is extremely small, since it is very unlikely that an arbitrary vector z' picked from $\mathbb{R}^{N^2}$ will resemble a natural image. Second, the tiny set of natural images is extremely complicated. It has been conjectured that natural images lie near a nonlinear manifold in $\mathbb{R}^{N^2}$ [11, 12]. If this is true, modeling natural images can be interpreted as trying to understand the properties of this manifold.

Since studying the natural image set directly is an ill-posed problem, image modeling researchers have typically studied the more gross, or aggregated, properties of images, such as their local smoothness, self-similarity, and so on. The wavelet transform has proven a useful tool in this regard.

The wavelet transform can be interpreted as a multiscale differentiator or edge detector that represents the singularity content of an image at

---

[3]1000×1000 digital photographs lie in $\mathbb{R}^{10^6}$! Moreover, the intuition that we humans build up from inhabiting $\mathbb{R}^3$ easily leads us astray in higher-dimensional spaces [10].

multiple scales and three different orientations — horizontal, vertical, and diagonal. Roughly speaking, each image singularity is represented by a cascade of large wavelet coefficients across scale [13]. If the singularity is within the support of a wavelet basis function, then the corresponding wavelet coefficient is large. Hence, the wavelet coefficients at the singularity location tend to be large. Likewise, a smooth image region is represented by a cascade of small wavelet coefficients across scale.

Several features of wavelet transforms make the wavelet domain well suited to constructing deterministic and statistical image models [13, 14]:

W1. Locality: Each wavelet coefficient represents the image content localized in spatial location and frequency/scale.

W2. Multiresolution: The wavelet transform analyzes the image at a nested set of scales.

W3. Energy Compaction: Wavelet transforms of natural images tend to be sparse. A wavelet coefficient is large only if singularities are present within the support of the corresponding wavelet basis element.

W4. Decorrelation: The wavelet coefficients of natural images tend to be approximately statistically decorrelated.

The Locality and Multiresolution properties (**W1, W2**) enable the wavelet transform to efficiently represent the local edge content of images with large coefficients, resulting in the Compaction property (**W3**), because only a small portion of a typical image contains edges. The Compaction and Decorrelation properties (**W3, W4**) simplify the statistical modeling of natural images in the wavelet domain as compared with a direct spatial-domain modeling.

The Compaction (**W3**) of signal energy in the wavelet domain results in a nonGaussian marginal probability density of the wavelet coefficients as shown in Figure 1.1(a):

W5. NonGaussianity: The wavelet coefficients of natural images have peaky, heavy-tailed, nonGaussian marginal statistics.

As we will see, the Decorrelation property (**W4**) inspires simple spatially localized modeling of the wavelet coefficients. Approaches include modeling each wavelet coefficient independently with a nonGaussian marginal distribution, such as a generalized Gaussian distribution [3, 15] or Gaussian mixture [4, 5].

The self-similarity of natural images [16] translates into a $1/f^\gamma$ Fourier spectrum behavior (see Figure 1.1(b)). This behavior translates into the wavelet domain as (see Figure 1.1(c)):

W6. Exponential decay across scale: The magnitudes of the wavelet coefficients of natural images tend to decay exponentially across scale.

The decay rate is directly related to the image smoothness [13].

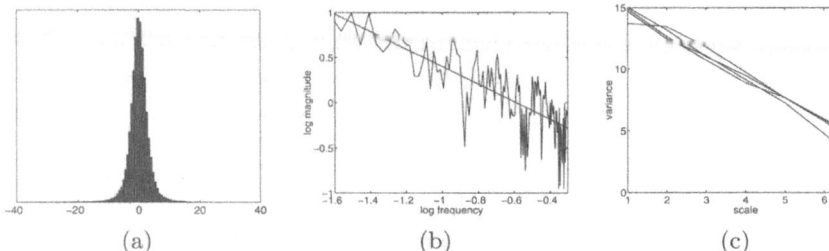

(a)   (b)   (c)

Figure 1.1. (a) Peaky and heavy-tailed marginal histogram of the finest scale wavelet coefficients of the Cameraman image. (b) Log-log magnitude plot of the Fourier spectrum of a 1-D slice of the Cameraman image. The spectral decay is close to linear, which implies the $1/f^\gamma$ property. (c) The $1/f^\gamma$ spectral decay induces an exponential decay of wavelet transform variance with scale, plotted here for the Lena, Cameraman, Boats, and Bridge test images.

1.3   Deterministic Image Models

In this section, we introduce Besov spaces as deterministic models for natural images. Since the Besov norm measures the smoothness of an image, it is no wonder that this norm can be related to the local derivatives computed by the wavelet transform. Indeed, wavelets provide the natural (unconditional) bases[4] for these spaces [9].

1.3.1   Besov function spaces

The theory of smoothness function spaces plays an ever more important rôle in signal and image processing. We shall consider the family of *Besov spaces* $B_q^\alpha(L_p(I))$ over a finite domain I, for example, the square $[0, 1]^2$, for $0 < \alpha < \infty$, $0 < p \le \infty$, and $0 < q \le \infty$. Images in these spaces have, roughly speaking, "α derivatives in $L_p(I)$"; the parameter q allows us to make finer distinctions in smoothness [2].

For $r > 0$ and $h \in \mathbb{R}^2$, define the r-th difference of a function z by

$$\Delta_h^{(r)} z(t) := \sum_{k=0}^{r} \binom{r}{k} (-1)^k z(t + kh) \tag{1.5}$$

for $t \in I_h^r := \{t \in I \mid t + rh \in I\}$. The $L_p(I)$-*modulus of smoothness* for $0 < p \le \infty$ is defined by

$$\omega_r(z, t)_p := \sup_{|h| \le t} \|\Delta_h^{(r)} z\|_{L_p(I_h^r)}. \tag{1.6}$$

---

[4]A basis $\{e_n\}$ for a set E is called *unconditional* if (i) for $\sum_n \mu_n e_n \in E$, $\sum_n |\mu_n| e_n \in E$, and (ii) for $\sum_n \mu_n e_n \in E$ and $\epsilon_n = \pm 1$ randomly chosen for every n, $\sum_n \epsilon_n \mu_n e_n \in E$.

The *Besov seminorm* of index (α, p, q) is defined for $r > \alpha$, $0 < p, q \leq \infty$ by

$$|z|_{B_q^\alpha(L_p(I))} := \left\{ \int_0^\infty \left(\frac{\omega_r(z,t)_p}{t^\alpha} \right)^q \frac{dt}{t} \right\}^{1/q}, \quad \text{if } 0 < q < \infty \qquad (1.7)$$

and by

$$|z|_{B_\infty^\alpha(L_p(I))} := \sup_{0 < t < \infty} \left\{ \frac{\omega_r(z,t)_p}{t^\alpha} \right\}. \qquad (1.8)$$

The *Besov norm* is then given by

$$\|z\|_{B_q^\alpha(L_p(I))} = \|z\|_{L_p(I)} + |z|_{B_q^\alpha(L_p(I))}. \qquad (1.9)$$

The Besov space $B_q^\alpha(L_p(I))$ is the class of functions $z : I \to \mathbb{R}$ satisfying $z \in L_p(I)$ and $|z|_{B_q^\alpha(L_p(I))} < \infty$. Various settings of the parameters yield more familiar spaces. For example, when $p = q = 2$, $B_2^\alpha(L_2(I))$ is the Sobolev space $W^\alpha(L_2(I))$, and when $\alpha < 1$, $1 \leq p \leq \infty$, $q = \infty$, $B_\infty^\alpha(L_p(I))$ is the Lipschitz space.

1.3.2 Wavelets and Besov space

Wavelets provide a simple characterization for the Besov spaces $B_q^\alpha(L_p(I))$. For analyzing ϕ and ψ possessing more than α vanishing moments [13], the Besov norm $\|z\|_{B_q^\alpha(L_p(I))}$ is equivalent[5] to a sequence norm of the wavelet coefficients [17]

$$\|z\|_{B_q^\alpha(L_p(I))} \asymp$$
$$\|u_{j_0}\|_p + \left(\sum_{j \geq j_0} 2^{jq(\alpha+1-2/p)} \left(\sum_{\mathbf{k}, \psi \in \Psi} |w_{j,\mathbf{k},\psi}|^p \right)^{q/p} \right)^{1/q}. \qquad (1.10)$$

The three hyper-parameters have natural interpretations: we take the ℓ_p norm of the wavelet coefficients within each scale (fixed j), weight these norms by an exponential factor of the smoothness parameter α, and then take the ℓ_q norm across scale j. We will take (1.10) as the definition of the Besov norm in the following.

In signal and image processing applications, we have three particularly simple cases of interest. First, when $p = q$, the Besov norm reduces to

$$\|z\|_{B_p^\alpha(L_p(I))} \asymp \|u_{j_0}\|_p + \left(\sum_{j \geq j_0, \mathbf{k}, \psi \in \Psi} 2^{j(\alpha p + p - 2)} |w_{j,\mathbf{k},\psi}|^p \right)^{1/p}, \qquad (1.11)$$

which is a weighted ℓ_p norm of the wavelet coefficients. The $B_1^1(L_1)$ and $B_\infty^1(L_1)$ spaces are interesting, because $B_1^1(L_1(I)) \subset \text{BV}(I) \subset B_\infty^1(L_1(I))$,

---

[5]Two norms $\|\cdot\|_a$ and $\|\cdot\|_b$ are *equivalent* and denoted $\|\cdot\|_a \asymp \|\cdot\|_b$ if and only if there exist positive constants m and M such that $m\|x\|_a \leq \|x\|_b \leq M\|x\|_a$ for all x.

with the set of bounded variation images BV(I) [18, 19] a popular image model in its own right. These Besov norms have a particularly simple structure:

$$\|z\|_{B_1^1(L_1(I))} \asymp \|u_{j_0}\|_1 + \sum_{j \geq j_0, \mathbf{k}, \psi \in \Psi} |w_{j,\mathbf{k},\psi}| \tag{1.12}$$

and

$$\|z\|_{B_\infty^1(L_1(I))} \asymp \|u_{j_0}\|_1 + \sup_{j \geq j_0} \sum_{\mathbf{k}, \psi \in \Psi} |w_{j,\mathbf{k},\psi}|. \tag{1.13}$$

1.3.3   Limitations of Besov space

While Besov spaces have proved enormously useful in a range of image processing applications, they are not without their shortcomings.

First is the proper interpretation of Besov norm. While it is clear that within the same Besov space $B_q^\alpha(L_p(I))$ images with small Besov norm are smoother, it is not immediate to interpret the Besov norm as an indicator of how much a given image "looks like" a natural image.

Second is the fact that Besov spaces contain continuous-variable functions, but, in practice, images are discretely sampled versions whose fine-scale information is truncated. Since the Besov norm depends greatly on the fine-scale wavelet coefficients, it is not straightforward to verify the smoothness of the original image given only its samples. As one extreme example, if we assume that all of the unavailable fine-scale wavelet coefficients are zero, then the continuous-space image corresponding to the sampled image is infinitely smooth in the sense of Besov smoothness and resides in every Besov space.

Third, as we readily see from the definition in (1.11), the Besov norm is invariant to coefficient *shuffles* (permutations) within scale j. Because the shuffling of wavelet coefficients destroys the edge structure of the image, a wavelet-shuffled image does not resemble a natural image (see Figure 1.2(b)). However, the wavelet-shuffled image belongs to the same Besov space (with identical Besov norm) as the original image. This implies that there exist many functions in a given Besov space that do not resemble natural images. This is equivalent to saying that the Besov norm lacks spatial localization of image smoothness. For more accurate image modeling, we must develop models that adapt to the local image characteristics.

Fourth, also from (1.11), we see that the Besov norm is invariant to *sign changes* of the wavelet coefficients. (Indeed, this follows directly from the fact that wavelets form an unconditional basis for Besov spaces [9].) As observed in Figure 1.2(c), the image resulting from random wavelet coefficient sign flips also does not resemble a natural image, while it belongs to the same Besov space as the original image with identical Besov norm.

Fifth, the set of natural images is not a linear space. Images are not formed by additive superpositions but rather by complicated nonlinear in-

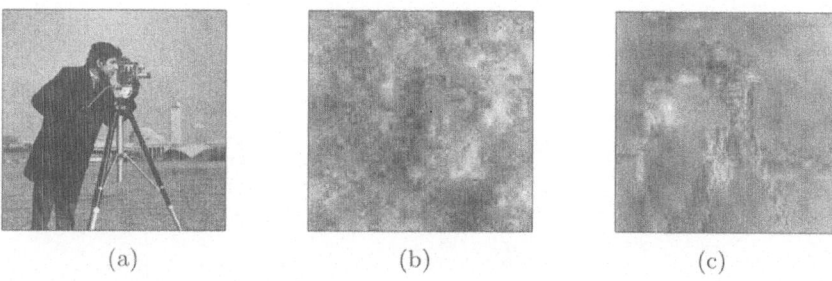

(a)   (b)   (c)

Figure 1.2. Limitations of Besov space and norm. (a) Image and inverse wavelet transforms of its wavelet coefficients after (b) random shuffles (permutations) within each scale band, and (c) random sign flips. All three images have identical Besov norm, indicating that the Besov norm is blind to much image structure.

Figure 1.3. Natural images do not form a linear space. For example, we would not form a new "natural image" by adding the cameraman and fruit images.

teractions like occlusions. See Figure 1.3, for example. The Besov spaces $B_q^\alpha(L_p(I))$ with $p < 1$ *are* non-convex, however, and so perhaps all is not lost.

For an extended discussion of additional Besov space shortcomings, including the difficulties encountered in estimating the Besov parameters p, q, α, see [20].

To overcome these limitations of Besov space as a characterization of natural images, it is clear that we must construct new spaces or sets based on more accurate image models. While little progress has been made to date in the deterministic setting, considerable progress has been made in the statistical setting, to which we now turn.

1.4   Statistical Image Models

Statistical image models specify the distribution of the random process generating the images of interest. In this section, we consider two common wavelet-domain statistical image models. We will discuss only models for the wavelet coefficients. The extension to the scaling coefficients is straightforward.

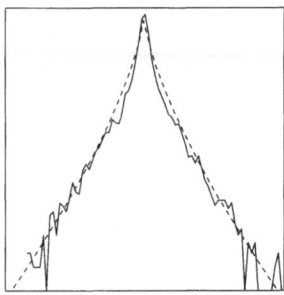

Figure 1.4. Fit (dashed) to the log of the wavelet coefficient histogram of Figure 1.1(a) (solid) by a generalized Gaussian distribution with parameter $\nu = 0.7$.

1.4.1   Independent generalized Gaussian model

The simplest wavelet-domain statistical models assume that the wavelet coefficients are statistically independent (extrapolating from the approximate decorrelation property (**W4**)). Under this assumption, modeling reduces to simply specifying the marginal distribution of each wavelet coefficient.

We begin with a model that aims to match the peaky, heavy-tailed marginal and exponential variance decay we saw in Figure 1.1. The ubiquitous Gaussian distribution is not a good choice here. A better choice is the zero-mean *generalized Gaussian distribution* (GGD) GGD($\sigma^2; \nu$) with variance σ^2 and shape parameter ν [15]. This probability density function (pdf) is defined as

$$f(x) := \frac{\nu \eta(\nu)}{2\Gamma(1/\nu)} \frac{1}{\sigma} \exp\left\{ -\left(\frac{\eta(\nu)|x|}{\sigma} \right)^\nu \right\}, \qquad (1.14)$$

with $\eta(\nu) = \sqrt{\frac{\Gamma(3/\nu)}{\Gamma(1/\nu)}}$. The GGD model contains the Gaussian and Laplacian distributions as special cases, using $\nu = 2$ and $\nu = 1$, respectively.

In an independent GGD wavelet model, each wavelet coefficient is generated independently according to a zero-mean GGD. For tractability, all wavelet coefficients at scale j are assumed to be independent and identically distributed (iid) with the same variance σ_j^2 and shape ν_j. Under this *iid in scale* model, we set $w_{j,\mathbf{k},\psi} \overset{\text{iid}}{\sim} \text{GGD}(\sigma_j^2; \nu_j)$. Thus, σ_j^2 and ν_j do not depend on the spatial location $\mathbf{k}$.

We choose the shape parameter ν_j to match the peakiness and heavy tail of the wavelet coefficient pdf. A large class of natural images share similar shape parameters [3]. In particular, the single fixed value $\nu_j \approx p = 0.7$ across all scales appears quite accurate for many natural images (see Figure 1.4).

The variance σ_j^2 represents the wavelet coefficient energy at scale j. It can be empirically estimated based on the given data [3], or it can be specified to decay exponentially (property **W6**, recall Figure 1.1(c)).

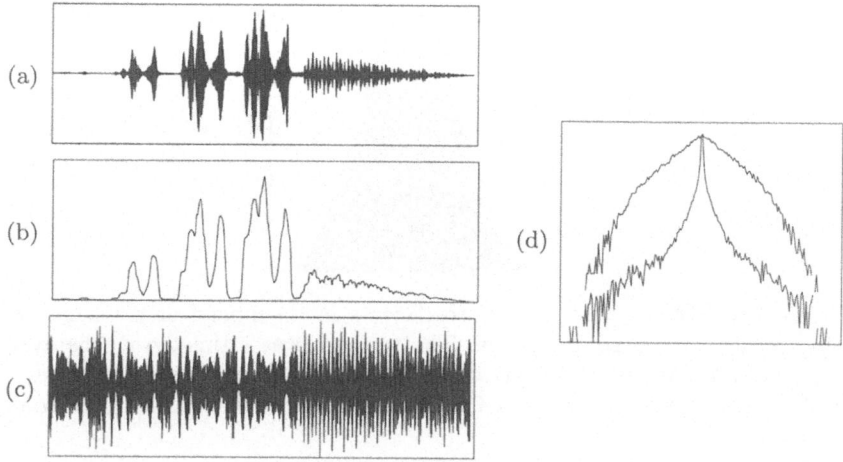

Figure 1.5. (a) Wavelet subband of a *cerulean warbler* bird song $w_{j_0,k}$, (b) its standard deviation $\hat{\sigma}_{j_0,k}$ computed in a local window, and (c) the rescaled subband $r_{j_0,k} = w_{j_0,k}/\hat{\sigma}_{j_0,k}$. Note the dramatic decrease in dynamic range. (d) Log histograms of $w_{j_0,k}$ (inner) and $r_{j_0,k}$ (outer). Note how the rescaling "Gaussianizes" the pdf (makes the log histogram closer to quadratic).

With fixed shape parameter p and fixed variance decay parameter β, the iid-in-scale GGD model Θ_p^β takes the form

$$\Theta_p^\beta : \quad w_{j,\mathbf{k},\psi} \overset{\text{iid}}{\sim} \text{GGD}(\sigma_j^2; p) \text{ with } \sigma_j = 2^{-j\beta}\sigma_0. \qquad (1.15)$$

The success of this simple GGD wavelet model in image denoising and texture classification problems [3, 6] is testimony to its accuracy for natural images.

1.4.2 Local Gaussian model

This model is inspired by how humans cope with the huge dynamic range in natural imagery and sounds — up to $1 : 10^5$ — that, if left untreated, would saturate the sensors in our eyes and ears. Humans depend on a form of *automatic gain control* encoded in the neural circuits of their sensory systems. The basic idea in the visual system is that the sensitivity of the eye decreases in bright regions and increases in dark regions. A roughly equivalent operation on an image is to renormalize each image pixel by some function of the energy of the pixels around it. Moreover, we can perform the same operation in the wavelet domain. Figure 1.5 illustrates the idea for a 1-D bird song signal with high dynamic range.

Local energy normalization has a calming effect on the marginal distribution of the wavelet coefficients, pulling up the values of the small coefficients and squashing down the large values. We see from Figures 1.5 and 1.6 that this is indeed the case. To compute the histogram in Figure 1.6, we renor-

Figure 1.6. Histogram of wavelet coefficients in a subband of the Lena image after local variance normalization. The variance was estimated as the average energy of the wavelet coefficients in a 3×3 window around each coefficient. The distribution is much closer to a Gaussian than the unnormalized histogram of Figure 1.1(a).

malized each wavelet coefficient $w_{j,\mathbf{k}}$ in a subband of the Lena image by dividing by the standard deviation of its eight surrounding neighbors in a 3×3 window. The resulting normalized coefficients are given by

$$\widetilde{w}_{j,\mathbf{k}} := \frac{w_{j,\mathbf{k}}}{\widehat{\sigma}_{j,\mathbf{k}}} \qquad (1.16)$$

with the local standard deviation computed from

$$\widehat{\sigma}_{j,\mathbf{k}}^2 := \frac{1}{L-1} \sum_{\mathbf{i} \in \mathcal{N}_{\mathbf{L}}(\mathbf{k})} w_{j,\mathbf{i}}^2 \qquad (1.17)$$

and $\mathcal{N}_L(\mathbf{k})$ an L-point neighborhood around the point $\mathbf{k}$. (In our example, $L = 8$.)

Figures 1.5 and 1.6 suggest a simple *locally Gaussian model* only slightly more complex than the iid GGD model from Section 1.4.1. We model the wavelet coefficients as independent Gaussian random variables but with a slowly varying variance field $\widehat{\sigma}_{j,\mathbf{k}}^2$ computed from (1.17)

$$\Omega_2 : \quad w_{j,\mathbf{k}} \sim \mathcal{N}(0, \widehat{\sigma}_{j,\mathbf{k}}^2). \qquad (1.18)$$

Although this model assumes that each coefficient is independent, the local variance estimation procedure implicitly captures dependencies between the sizes of neighboring coefficients. Thus, the model describes the wavelet coefficient joint statistics more accurately than the iid GGD model. Since each wavelet coefficient is allowed to have a different variance, the overall statistics of an entire subband of coefficients will be a Gaussian mixture distribution that approximates a GGD.

This locally Gaussian model is precisely the model behind the high-performance Estimation-Quantization (EQ) image coding algorithm [21,

22]. More general distributions than the Gaussian can also be chosen.[6] In the EQ coder, the window takes the form of a causal neighborhood following the standard scanning order of the wavelet coefficients both within and across scale. State-of-the-art compression performance and simplicity of implementation (no zero trees or other fancy appendages) results. An image denoising algorithm using similar methods also performs very well [23]. The size of the variance estimation window in [23] is determined by a bootstrap.

1.5   Links between Besov Space and Statistical Models

While seemingly quite disparate frameworks, there are deep commonalities between the above deterministic and statistical image models. In this section, we uncover a strong link between the Besov space model of Section 1.3 and the wavelet statistical image models of Section 1.4.

1.5.1   GGD realizations live in Besov space

The connection between wavelet-domain statistical models and Besov spaces was perhaps first noticed by Abramovich et al. [5] in their variation of the wavelet Gaussian mixture model. The decay of wavelet coefficients across scale determines the smoothness of the corresponding image; hence realizations of a statistical model with exponentially decaying variance will belong to certain Besov spaces. Interpreted in terms of the independent GGD model of (1.15), we have the following theorem.[7]

Theorem 1. *Suppose each wavelet coefficient $w_{j,\mathbf{k},\psi}$ is distributed according to the independent GGD model Θ_p^β from (1.15) with $\beta > 0$ and $\sigma_0 > 0$. Then, for $0 < p, q < \infty$, the image realizations from this model are almost surely in $B_q^\alpha(L_p(I))$ if and only if $\beta > \alpha + 1$.*

A very similar result applies for a large class of wavelet coefficient marginal pdf models, including the Gaussian mixture [5, 24] and even finite support pdf's [8]. It is also possible to impose further dependencies in the Gaussian mixture model to capture additional wavelet properties such as magnitude persistence across the scales of the wavelet tree [25].

---

[6]LoPresto et al. [22] argue that a GGD is more appropriate than a Gaussian distribution in the EQ algorithm only because in the compression case the variance must be estimated from *quantized* coefficients. If we are allowed to estimate the variance from the original coefficients, then the Gaussian becomes a better choice.

[7]The proof parallels the proof of Theorem 1 in [5].

Theorem 1 tells us that the class of realizations from the independent GGD statistical model is essentially a Besov space. But what of the Besov norm, which measures distance in this space?

1.5.2   Besov norm as a normalized likelihood

Consider normalizing the likelihood of a realization of a statistical model in terms of the maximum likelihood achievable. That is, define the *normalized likelihood function* by

$$f^N(x|\boldsymbol{\Theta}) := \frac{f(x|\boldsymbol{\Theta})}{\sup_x f(x|\boldsymbol{\Theta})}, \tag{1.19}$$

with the assumption that $0 < \sup_x f(x|\boldsymbol{\Theta}) < \infty$. Then, $f^N(x|\boldsymbol{\Theta}) \in [0,1]$, and we can tell that an observation x is "likely" if $f^N(x|\boldsymbol{\Theta})$ is close to 1 and "not likely" if it is close to 0. We can easily generalize the concept of normalized likelihood to finite random vectors using their joint pdfs.

For an infinite sequence of random variables, we can define the normalized likelihood as the limit (if it exists) of the normalized likelihood of its truncated subsequences. For the independent wavelet-domain models considered in Section 1.3, we can compute the limit as we move to finer scales, defining the normalized likelihood

$$f^N(\boldsymbol{w}) = \lim_{J \to \infty} \frac{\prod_{j=0}^J \prod_{\mathbf{k},\psi} f_{j,\mathbf{k},\psi}(w_{j,\mathbf{k},\psi})}{\sup \prod_{j=0}^J \prod_{\mathbf{k},\psi} f_{j,\mathbf{k},\psi}(w_{j,\mathbf{k},\psi})} \tag{1.20}$$

when $w_{j,\mathbf{k},\psi} \sim f_{j,\mathbf{k},\psi}(w_{j,\mathbf{k},\psi})$. For the independent GGD model with exponentially decaying variance $\boldsymbol{\Theta}_p^\beta$ from (1.15), the normalized likelihood is well defined, since the supremum is finite and the limit exists.

For the model $\boldsymbol{\Theta}_p^\beta$, the normalized likelihood computed using the coefficients between scales 0 and J equals

$$\begin{aligned}
f^N(\mathbf{w}_J) &= \prod_{j=0}^J \prod_{\mathbf{k},\psi} \exp\left\{-\left(\frac{\eta(p)|w_{j,\mathbf{k},\psi}|}{\sigma_j}\right)^p\right\} \\
&= \exp\left\{\sum_{0 \le j \le J,\mathbf{k},\psi} -\left(\frac{\eta(p)|w_{j,\mathbf{k},\psi}|}{\sigma_j}\right)^p\right\}. \tag{1.21}
\end{aligned}$$

Taking the negative log of the normalized likelihood function, we obtain

$$\begin{aligned}
-\log f^N(\boldsymbol{w}) &= \sum_{0 \le j,\mathbf{k},\psi} \left(\frac{\eta(p)|w_{j,\mathbf{k},\psi}|}{\sigma_j}\right)^p \\
&= \eta(p)^p \sum_{0 \le j,\mathbf{k},\psi} (2^{j\beta}|w_{j,\mathbf{k},\psi}|)^p, \tag{1.22}
\end{aligned}$$

which is equivalent to $\|z\|_{B_p^\alpha(L_p)}^p$, the Besov norm of the function z, when $\beta = \alpha + 2 - 2/p$.

With $p = 2$, we obtain an iid Gaussian model for the wavelet coefficients, and the corresponding normalized negative log likelihood is equivalent to the Sobolev norm of z. With $p = 0.7$ — corresponding to the $\nu = 0.7$ that closely matches the generalized Gaussian exponent of many real images (recall Figure 1.4) — we obtain a non-convex set $B_p^\alpha(L_p)$ (recall the discussion of Section 1.3.2).

In terms of the normalized likelihood function for the iid GGD model Θ_p^β, the Besov space $B_p^\alpha(L_p)$ can be equivalently defined as the set

$$\{w : f^N(w|\Theta_p^\beta) \neq 0\} \text{ with } \beta = \alpha + 2 - 2/p. \tag{1.23}$$

Thus, the signals in the Besov space $B_p^\alpha(L_p)$ are the "likely" signals under the statistical model Θ_p^β.

This link between Besov space and the GGD statistical model immediately unites many seemingly disparate algorithms. For instance, the deterministic variational problem of Chambolle et al. [26] corresponds to a Bayesian wavelet-based statistical estimator regularized by the likelihood of the approximant. The deterministic interpolation algorithm proposed by Choi et al. [27] becomes a maximum likelihood interpolator under the wavelet-domain GGD model.

1.5.3   Ties to information theory and coding

Much progress has been made on modeling image pdf's in the context of *image coding*. A typical wavelet-domain image compressor consists of two stages. First, the pdf of each wavelet coefficient is estimated as accurately as possible. (This step corresponds to confining the given image to a Besov-like set.) Then, the coefficient is quantized according to that pdf. (This corresponds to specifying the location of the image in the set.) For efficient compression, we must make the confining set as small as possible to reduce the number of code bits needed to specify the location within the set.

When images are described by a statistical model, the optimal compression performance is achieved by the Shannon source code for the model [28]. State-of-the-art image compression algorithms feature underlying statistical image models either implicitly [29] or explicitly [21]. Images conforming to the pdf model are well compressed, and thus the compressed codelength of an image in some sense measures how much it can be considered a "natural image" (at least in the eyes of the compression algorithm). The Shannon codelength enables an information-theoretic characterization of the natural image set.

All leading wavelet-domain image compression algorithms employ some type of spatially adaptive pdf estimation procedure, which is often hidden in the algorithm. To minimize the number of bits to describe the pdf prediction itself, a very simple parametric pdf form, such as a zero-mean generalized Gaussian or Laplacian, is typically assumed.

First, consider a simple compression algorithm based on the independent GGD model Θ_p^β from (1.15). Let $\boldsymbol{w}_J$ be the set of all the wavelet coefficients up to scale J. Under the Θ_p^β model, the pdf of $\boldsymbol{w}_J$ is given by

$$
\begin{aligned}
f(\boldsymbol{w}_J) &= C_J \prod_{j=0}^{J} \prod_{\mathbf{k},\psi} \exp\left\{-\left(\frac{\eta(p)|w_{j,\mathbf{k},\psi}|}{\sigma_j}\right)^p\right\} \\
&= C_J \exp\left\{\sum_{0\le j\le J,\mathbf{k},\psi} -\left(\frac{\eta(p)|w_{j,\mathbf{k},\psi}|}{\sigma_j}\right)^p\right\},
\end{aligned}
\tag{1.24}
$$

with

$$
C_J := \prod_{j=0}^{J} \prod_{\mathbf{k},\psi} \frac{\nu\eta(p)}{2\Gamma(1/p)} \frac{1}{\sigma_j}.
\tag{1.25}
$$

Because the exponential term equals 1 when all the wavelet coefficients are zero, C_J corresponds to the likelihood of an image having all zero wavelet coefficients, that is, a constant image. Using optimal source coding for this model, the Shannon codelength L_J of the wavelet coefficients up to scale J is given by

$$
L_J := -\log f(\mathrm{w}_J) = -\log C_J - \sum_{0\le j\le J,\mathbf{k},\psi} \left(\frac{\eta(p)|w_{j,\mathbf{k},\psi}|}{\sigma_j}\right)^p.
\tag{1.26}
$$

Recalling that $-\log C_J$ is the Shannon codelength of a constant image, we can write

$$
\widetilde{L}_J := L_J - (-\log C_J) = \sum_{0\le j\le J,\mathbf{k},\psi} \left(\frac{\eta(p)|w_{j,\mathbf{k},\psi}|}{\sigma_j}\right)^p.
\tag{1.27}
$$

If we set $\beta = \alpha+2-2/p$ and compute the limit of $\widetilde{L}_J$ as $J \to \infty$, we obtain the relation

$$
\lim_{J\to\infty} \widetilde{L}_J = \eta(\nu)^p \|z\|_{B_p^\alpha(L_p)}^p,
\tag{1.28}
$$

where $\alpha = \beta-2+2/p$. Because the constant image is the simplest image to code, we see that the Besov norm (ignoring the constant $\eta(p)^p$) measures how much more codelength is required to code the image z beyond that required for the constant image. In this sense, the Besov norm measures the coding complexity of the image under the iid GGD model.

The interpretation of the Besov norm as the Shannon codelength indicates that the Besov norm measures the coding complexity of images (see also [30]). Thus, the Besov regularization of Chambolle et al. [26] is equivalent to the minimum description length (MDL) denoising algorithm [31] under the independent GGD image prior. And the maximum smoothness interpolation algorithm of [27] simply finds the least complex signal among all signals satisfying the sampling constraints.

Unfortunately, the close equivalence between Besov spaces and the iid GGD model implies that the iid GGD model suffers from all of the same problems as Besov spaces, including shuffle and sign invariance (recall Figure 1.2).

1.5.4 Local Gaussian model = local Sobolev space

Under the locally Gaussian model Ω_2 from (1.18), the likelihood computation and definition of the normalized likelihood are straightforward. The likelihood of the entire set of wavelet coefficients is given by the product

$$f(\boldsymbol{w}) = \prod_{j,\mathbf{k}} g(w_{j,\mathbf{k}}), \tag{1.29}$$

where $g(w) := (2\pi\widehat{\sigma}_{j,\mathbf{k}}^2)^{-1/2} \exp(-\frac{w^2}{2\widehat{\sigma}_{j,\mathbf{k}}^2})$ is the zero-mean Gaussian density function with variance $\sigma_{j,\mathbf{k}}^2$.

From (1.29), the negative normalized log likelihood becomes

$$-\log f^N(\boldsymbol{w}) = \sum_{j,\mathbf{k}} \frac{w_{j,\mathbf{k}}^2}{2\widehat{\sigma}_{j,\mathbf{k}}^2}, \tag{1.30}$$

a weighted ℓ_2 "norm" of the wavelet coefficients that resembles a locally adapted Sobolev norm.[8] As a notation for the "norm" defined using this model, we use $\|\cdot\|_{IG}$. That is,

$$\|z\|_{IG}^2 := \sum_{j,\mathbf{k}} \frac{w_{j,\mathbf{k}}^2}{2\widehat{\sigma}_{j,\mathbf{k}}^2}. \tag{1.31}$$

The measure $\|z\|_{IG}^2$ is *not* shuffle invariant. However, since this spatially adapted model also uses zero-mean pdfs, the measure remains invariant to sign flips of the wavelet coefficients.

1.6 Conclusions and Beyond Besov Spaces

We have discovered several relationships between deterministic Besov spaces and wavelet-domain statistical image models. In particular, we have shown that the Besov norm is equivalent to a log likelihood or Shannon codelength under an independent GGD wavelet model.

While much progress has been made in both deterministic and statistical image modeling, we are still far from pinning down the set of natural images in the wavelet (or any other) domain. The most recent progress has been

---

[8]Note that (1.30) is not a valid norm, since it will not converge in general.

made in the statistical context, attacking the shuffle invariance and sign flip invariance of the Besov norm and the independent GGD wavelet model.

Jaffard's *oscillation spaces* are not invariant to wavelet-coefficient shuffling [8]. Likewise for the statistical *hidden Markov tree* (HMT) model, which fuses a Gaussian mixture marginal pdf with Markov dependencies through scale to capture the persistence of large and small wavelet coefficient values [4]. Encouragingly, the deterministic image model induced by the HMT is not a convex space. The promising complex wavelet HMT model [25] is neither shuffle invariant (thanks to an HMT model on the complex wavelet magnitude values) nor sign flip invariant (thanks to a model on the complex wavelet phase).

Future progress will require explicit attention to the fact that image edges lie along regular 1-D contours. The complex HMT model mentioned above has the capability to capture this property. Alternatively, we can move from the wavelet transform to other transforms — *curvelets* [32, 33] or *bandelets* [34], for example — that directly restructure the edge contours into a form that is easier to model.

Clearly the development of appropriate transforms and modeling frameworks for the transform coefficients of natural images will remain a challenging area of research for some time to come.

References

[1] D. Donoho and I. Johnstone, "Adapting to unknown smoothness via wavelet shrinkage," *J. Amer. Stat. Assoc.*, vol. 90, pp. 1200–1224, Dec. 1995.

[2] R. A. DeVore, B. Jawerth, and B. J. Lucier, "Image compression through wavelet transform coding," *IEEE Trans. on Information Theory*, vol. 38, no. 2, pp. 719–746, March 1992.

[3] P. Moulin and J. Liu, "Analysis of multiresolution image denoising schemes using generalized-Gaussian priors," in *Proc. IEEE-SP Int. Symp. Time-Freq. and Time-Scale Anal.*, Pittsburgh, PA, Oct. 6-9 1998, pp. 633–636.

[4] M. S. Crouse, R. D. Nowak, and R. G. Baraniuk, "Wavelet-based statistical signal processing using hidden Markov models," *IEEE Trans. Signal Proc.*, vol. 46, no. 4, pp. 886–902, April 1998.

[5] F. Abramovich, T. Sapatinas, and B. W. Silverman, "Wavelet thresholding via a Bayesian approach," *J. Roy Stat. Soc. Ser. B*, vol. 60, pp. 725–749, 1998.

[6] E. P. Simoncelli and E. H. Adelson, "Noise removal via Bayesian wavelet coring," in *Proc. IEEE Int. Conf. on Image Proc. — ICIP 1996*, Lausanne, Switzerland, Sept. 1996.

[7] H. Choi and R. G. Baraniuk, "Image segmentation using wavelet-domain classification," in *Proc. SPIE Conf. Math. Modeling, Bayesian Estimation, and Inverse Problems*, Denver, CO, July 1999, vol. 3816, pp. 306–320.

[8] S. Jaffard, "Beyond Besov spaces," *Preprint*, 2001.

[9] I. Daubechies, *Ten Lectures on Wavelets*, SIAM, New York, 1992.

[10] E. A. Abbot and A. Lightman, *Flatland: A Romance of many dimensions*, Penguin Classics, New York: NY, 1998.

[11] A. B. Lee, K. S. Pederson, and D. Mumford, "The complex statistics of high-contrast patches in natural images," in *Proc. 2nd International IEEE Workshop on Statistical and Computational Theories of Vision*, Vancouver, July 2001.

[12] M. T. Orchard, *Personal Communication*, 2000.

[13] S. Mallat, *A Wavelet Tour of Signal Processing*, Academic Press, San Diego, 1998.

[14] M. Vetterli and J. Kovačević, *Wavlets and Subband Coding*, Prentice Hall, Englewood Cliffs,NJ, 1995.

[15] S. Mallat, "A theory for multiresolution signal decomposition: The wavelet representation," *IEEE Transactions on Pattern Analysis and Machine Intelligence*, vol. 11, no. 7, pp. 674–693, July 1989.

[16] D. L. Ruderman and W. Bialek, "Statistics of natural images: Scaling in the woods," *Phys. Rev. Lett.*, vol. 73, no. 6, pp. 814–817, 1994.

[17] Y. Meyer, *Wavelets and Operators*, Cambridge University Press, Cambridge, 1992.

[18] L. C. Evans and R. F. Gariepy, *Measure Theory and Find Properties of Functions*, CRC Press, Boca Raton, FL, 1992.

[19] S. Osher L. I. Rudin and E. Fatemi, "Nonlinear total variation based noise removal algorithms," *Physica D*, vol. 60, pp. 259–268, 1992.

[20] C. S. Güntürk, "Harmonic analysis of two problems in signal quantization and compression," *Ph.D Thesis, Princeton University*, 2000.

[21] S. LoPresto, K. Ramchandran, and M. T. Orchard, "Image coding based on mixture modeling of wavelet coefficients and a fast estimation-quantization framework," in *Data Compression Conference*, Snowbird, Utah, 1997, pp. 221–230.

[22] S. LoPresto, K. Ramchandran, and M. T. Orchard, "Wavelet image coding via rate-distortion optimized adaptive classification," in *Proc. of NJIT Symposium on Wavelet, Subband and Block Transforms in Communications, New Jersey Institute of Technology*, 1997.

[23] M. K. Mihcak, I. Kozintsev, and K. Ramchandran, "Spatially adaptice statistical modeling of wavelet image coefficients and its application to denoising," in *Proc. IEEE Int. Conf. on Acoust., Speech, Signal Proc. — ICASSP '99*, Phoenix, AZ, March 1999.

[24] H. Choi and R. Baraniuk, "Wavelet-domain statistical models and Besov spaces," in *Proc. of SPIE Conf. Wavelet Applications in Signal Proc. VII*, Denver, July 1999, vol. 3813, pp. 489–501.

[25] J. Romberg, H. Choi, and R. G. Baraniuk, "Bayesian wavelet domain image modeling using hidden Markov models," *IEEE Trans. on Image Proc.*, vol. 10, pp. 1056–1068, July 2001.

[26] A. Chambolle, R. A. DeVore, N. Lee, and B. J. Lucier, "Nonlinear wavelet image processing: Variational problems, compression, and noise removal through wavelet shrinkage," *IEEE Trans. on Image Proc.*, vol. 7, pp. 319–355, March 1998.

[27] H. Choi and R. Baraniuk, "Interpolation and denoising of nonuniformly sampled data using wavelet domain processing," in *Proc. IEEE Int. Conf. on Acoust., Speech, Signal Proc. — ICASSP '99*, Phoenix, AZ, March 1999.

[28] T. M. Cover and J. A. Thomas, *Elements of Information Theory*, John Wiley & Sons, Inc., New York, 1991.

[29] J. Shapiro, "Embedded image coding using zerotrees of wavelet coefficients," *IEEE Trans. Signal Proc.*, vol. 41, no. 12, pp. 3445–3462, Dec. 1993.

[30] A. Cohen, I. Daubechies, O. G. Guleryuz, and M. T. Orchard, "On the importance of combining wavelet-based non-linear approximation with coding strategies," *IEEE Trans. on Information Theory*, Submitted.

[31] N. Saito, "Simultaneous noise supression and signal compression using a library of orthonormal bases and the MDL criterion," in *Wavelets in Geophysics*. 1994, pp. 299–324, New York: Academic Press, Editors: E. Foufoula-Georgiou and P. Kumar.

[32] E. Candès and D. Donoho, "Ridgelets: The key to high-dimensional intermittency?," *Phil. Trans. R. Soc. Lond. A.*, vol. 357, pp. 2495–2509, 1999.

[33] E. Candès and D. Donoho, "Curvelets: A surprisingly effective non-adaptive representation of objects with edges," in *Curves and Surface Fitting: Saint-Malo 1999*. 2000, Vanderbilt University Press, Nashville, TN, Editors A. Cohen, C. Rabut and L. L. Schumaker.

[34] E. L. Pennec and S. Mallat, "Image compression with geometrical wavelets," in *IEEE Int. Conf. on Image Proc. — ICIP '01*, Thessaloniki, Greece, Oct. 7–10 2001.

2

Coarse-to-Fine Classification and Scene Labeling

Donald Geman[1]

2.1 Introduction

The semantic interpretation of natural scenes, so effortless for humans, is perhaps the main challenge of artificial vision, having largely resisted any satisfying solution, at least in searching for multiple objects in real, cluttered scenes with arbitrary illumination. This problem is the motivation for the work in this chapter. Specifically, the models and algorithms presented here result from making computational efficiency the organizing principle for vision, a proposal recently explored in both theory [8, 2] and practice [1, 3, 4]; see also [9] and [6] for related examples of efficient visual search. "Theory" refers to analyzing the efficiency of coarse-to-fine (CTF) search under various statistical models and cost structures; summarized here are the general mathematical framework, including an abstract formulation of CTF classification based on multiresolution "tests," and some results about optimal testing strategies. "Practice" refers to designing computer algorithms for detecting objects in natural scenes; several such experiments on face detection are included together with a brief description of how the tests are realized as image functionals.

We model scene interpretation as a dynamic and adaptive process, generating a sequence of increasingly precise interpretations. At the beginning the labels are crude and too plentiful; there are confusions among objects of interest and between objects and clutter. Eventually, the labels become more precise, for instance object categories and presentations are refined, and confusions are removed. Certain fundamental tradeoffs then evolve – between invariance and discrimination, and between false positive error and computation. Similar themes are explored in [10] for visual processing in cortex.

---

[1]Donald Geman is Professor, Department of Mathematics and Statistics, University of Massachusetts at Amherst.

Figure 2.1. An example illustrating multifont optical character recognition where the objective is to identify the main symbols on the license plate.

For practical convenience, we separate the whole process of scene classification into two rough phases: non-contextual and contextual. (This distinction was previously explored in [1].) Noncontextual classification, or simply *detection*, comes first. The goal is to infer from the image data instances of highly visible objects of interest under the constraint that no objects be overlooked (no "missed detections"), but allowing for a limited number of false positives. *Detection is the focus in this chapter.* As conceived here, it is based on sparse representations and performed with very simple operations – essentially just counting local features. The result is a list of "classes" of objects and their "presentations" in the scene. The desired level of detail is application-dependent; for example, the class might be "face" rather than a specific individual and the presentation might be no more specific than a range of values for certain pose parameters (e.g., position, scale and orientation). Other aspects of the presentation might be of interest, such as the font of a character or the gender of a face.

An example of multifont optical character recognition is shown in Figure 2.1 where the objective is to identify the main symbols on the license plate. The detection phase is illustrated in Figure 2.2. For each of the six characters, there are multiple detections; some *but not all* of the detections "near" each symbol are erroneous, due to clutter or confusions. There are also false alarms away from the symbols of interest. This is ongoing work with Yali Amit and will be described elsewhere.

Figure 2.2. Characters detected during detection.

Contextual classification, not treated here, involves more intensive computation in the vicinity of detections in order to determine which of these are in fact objects of interest and to disambiguate among confusions, such as D's, 0's and O's detected at roughly the same location. The underlying process is again coarse-to-fine. Moreover, ultimately there is no way to avoid a fully contextual analysis in order to discover partially visible objects and other complex spatial arrangements. Processing which accounts for context and relationships is likely to require dense representations and be computationally intensive (e.g., involve online functional optimization). One proposal is "compositional vision" (cf. [7]).

If scene labeling is driven by computational efficiency, a natural and effective mechanism is CTF classification. It is certainly *one way* of gaining (online) efficiency and I would argue that any other way ultimately boils down to something similar. CTF classification depends on a CTF *representation* for the family of interpretations under investigation. In other words, the representation of objects and presentations must be structured to accommodate coarse-to-fine search. Thus, the events of interest must be characterized by "attributes" at many levels of resolution. CTF search then means investigating those attributes in a particular order, namely from coarse ones to fine ones. This is the way we play *Twenty Questions.*

We develop an abstract formulation of CTF classification. Roughly speaking, we consider a series of nested (hence increasingly fine) partitions of the set of possible explanations, and we define a binary "test" for each cell Λ of each partition. The test X_Λ associated with Λ must always respond "yes" to interpretations in Λ. The tests also have varying levels of "cost" and "discrimination" (statistical power); both increase as cell size decreases, and hence there is a tradeoff with invariance. The "detector" $\hat{Y}$ is a (set-valued) function of these tests; it consists of all interpretations which are confirmed at all levels of resolution, and is the primary object of our mathematical analysis. More specifically, we ask: *Which sequential (test by test) adaptive evaluation of $\hat{Y}$ minimizes average computation?* The answer is that under wide-ranging assumptions on the statistical distribution of the tests and how cost is measured, and among all testing strategies based on performing tests one at a time until a decision is reached (i.e., $\hat{Y}$ is determined), CTF questioning minimizes the *mean* of the sum of the costs of the all the tests which are utilized.

Further remarks about invariance and discrimination, and about parallel vs. serial processing, follow in Section 2.2. The abstract formulation is given in Section 2.3, where the statistical framework is laid out, including the definitions of cost, invariance and discrimination, and the definition of $\hat{Y}$. In Section 2.4 we introduce the family of possible evaluations of the detector and a model for measuring the computational efficiency of each candidate. Several results on optimal strategies are mentioned in Section 2.5 without proof and the error rates of the detector are specified in Section 2.6. In Section 2.7, we return to the scene interpretation problem and put everything in concrete terms, including how $\hat{Y}$ is constructed from image intensity data. Finally, some experiments on face detection and concluding remarks appear in Section 2.8.

2.2   Invariance vs. Discrimination

The rationale for CTF search is intuitive and transparent. Start with properties of objects and presentations which are simple and common, almost regardless of discriminating power; in other words, look for tests which *invariably* accept as many object/pose pairings as possible, even if many instances of clutter and non-targeted objects are found as well. Rejecting even a small percentage of background instances with cheap and universal tests is efficient. Then proceed to more discriminating properties, albeit more complex and specialized; whereas a greater number of tests must be designed or learned in order to "cover" all objects and poses, only relatively few of them will be needed during any given search due to pruning by coarser tests. Also the still significant false alarm rate is compensated by invariance (no missed detections) and low cost termination of the search. Finally, reserve computationally intensive, highly discriminating filters (basically, object-specific and pose-specific "template-matching") for the very end – for those inevitable and diabolical arrangements of clutter which "look" like objects in the eyes of the features.

This program amounts to creating "invariants" at many levels of power. But these are *not* the geometric and algebraic types sought after in continuum, shape-based approaches to object recognition. The invariants here are based on generic local features, not special points on curves, etc. And our requirements are more modest: Find binary image functionals which always respond positively for a given set of shapes but may respond positively to other shapes and image structures. It is only at the level of low invariance (specific poses) that we demand high discrimination. Consequently, during the course of processing there is then a steady progression from high invariance to low invariance and from low discrimination to high discrimination.

The image functionals we consider in Section 2.7 are of the form

$$X = \begin{cases} 1 & \text{if } \sum_{l \in L} \xi_l \geq t \\ 0 & \text{otherwise} \end{cases}$$

where each ξ_l is a local binary feature which signals an "edge" is present "near" location z_l and with orientation ϕ_l; L is a distinguished set of edges dedicated to a set of poses and t is an appropriate threshold. (How "near" depends on the desired level of invariance; see Section 2.7.) Thus, evaluating X consists of checking for at least t edges among a special ensemble which characterizes a particular set of shapes – certain types of objects at certain geometric poses. The complexity of such as test might simply be $|L|$. High discrimination and high complexity corresponds to "template-matching" and the set L might then provide a rather dense representation of the shapes. However, for such elementary tests, achieving high invariance (covering many poses) *and* high discrimination at the same time is likely to be impossible, regardless of cost.

It is clearly impossible to find common but localized attributes of two object presentations with significantly different (geometric) poses, say far apart in the scene. As a result, we use a simple, "divide-and-conquer" strategy based on object location. (Every object is assumed to have a visible, distinguished point.) One "base" detector, $\hat{Y}$, is designed to find all instances of objects with presentations in a "reference" cell, for example locations confined to a $k \times k$ block and scale confined to a $[\sigma_{min}, 2\sigma_{min}]$ where σ_{min} is the smallest scale entertained. The scene is partitioned into non-overlapping $k \times k$ blocks, and the detector $\hat{Y}$ is applied to the image data $I(z), z \in W$ in a window W centered at each such block; the dimension of W is sufficiently large to capture all objects at the given locations and scales. Objects at scales larger than $2\sigma_{min}$ are detected by repeatedly downsampling and parsing the scene in the same way.

In principle, the detector could be applied to each window simultaneously; this is the parallel component of the algorithm. The serial component – the CTF implementation of $\hat{Y}$ is each window – is the heart of the algorithm and the real source of efficient computation.

2.3 CTF Classification: Abstract Formulation

Let $\mathcal{S} = \{\lambda_1, \lambda_2, ...\}$ denote a set of *states* or *interpretations*. Each subset $\Lambda \subset \mathcal{S}$ will be called an *index*. In addition, fix a probability space $(\mathcal{I}, P)$ and suppose there is a *true index* $Y(I)$ for each $I \in \mathcal{I}$. Although we allow more than one true interpretation, we are primarily interested in the case in which either $Y = \{\lambda\}$ or $Y = \emptyset$.

In the application to detecting objects, λ is a pair (c, θ) where c is the "class" of an object and θ stands for the "presentation." Even one object class is challenging and frequently considered in computer vision. $\mathcal{I}$ is then the set of subimages $I = \{I(z), z \in W\}$ and P could be taken as an empirical measure; we shall be more specific about this later on. Finally, $Y(I)$ is the list of the objects and presentations appearing in I. In general, there is at most one object which is both visible and centered in a given W.

An important feature of the detection problem, and one that motivates an upcoming approximation of P, is that $P(Y = \emptyset) \gg P(Y = \Lambda)$ for any given Λ. We might even assume that $P(Y = \emptyset) \gg P(Y \neq \emptyset)$, so that the most likely interpretation is that there are no states "present" in I. Write P_λ for $P(.|\lambda \in Y)$ and P_0 for $P(.|Y = \emptyset)$.

Shortly we shall define a *detector* $\hat{Y}$ based on a family of functions X : $\mathcal{I} \longmapsto \{0, 1\}$ called *tests*. The basic constraint on $\hat{Y}$ is zero false negative error:

$$P(Y \subset \hat{Y}) = 1 \tag{2.1}$$

Equivalently,

$$P_\lambda(\lambda \in \hat{Y}) = 1, \quad \forall \lambda \in \mathcal{S}.$$

Assume each test has a *cost* or *complexity* $c(X)$ which represents the amount of online computation (or time) necessary to evaluate X and of course depends on how X is constructed. The *invariant set* for X is $\Lambda(X) = \{\lambda : P_\lambda(X = 1) = 1\}$. Finally, the *discrimination* or *power* of X is defined as $\beta(X) = P_0(X = 0)$. The tradeoff between cost and discrimination at different levels of invariance is shown in the lefthand panel of Figure 2.3; the righthand panel shows the tradeoff between invariance and discrimination at different costs.

Suppose we are given a family of tests $\mathbf{X} = \{X_\Lambda, \Lambda \in \mathbf{\Lambda}\}$ where the notation X_Λ means that $\Lambda = \Lambda(X)$. The reason for indexing the tests by their invariant sets is that we will build tests to a set of specifications. Basically, we *first* design a hierarchy of subsets of $\mathcal{S}$ and *then*, for each Λ in the hierarchy, we build a test X_Λ which is invariant with respect to the classes and poses in Λ.

Now define $\hat{Y}(I) \subset \mathcal{S}, I \in \mathcal{I}$, by

$$\hat{Y}(I) = \hat{Y}(\mathbf{X}(I)) = \{\lambda : X_\Lambda(I) = 1 \ \forall \lambda \in \Lambda\}. \tag{2.2}$$

The rationale is that we accept a state λ as part of our interpretation if and only if this state it is "verified" at all levels of resolution, in the sense that each test X which "covers" λ (meaning $\lambda \in \Lambda(X)$) responds positively.

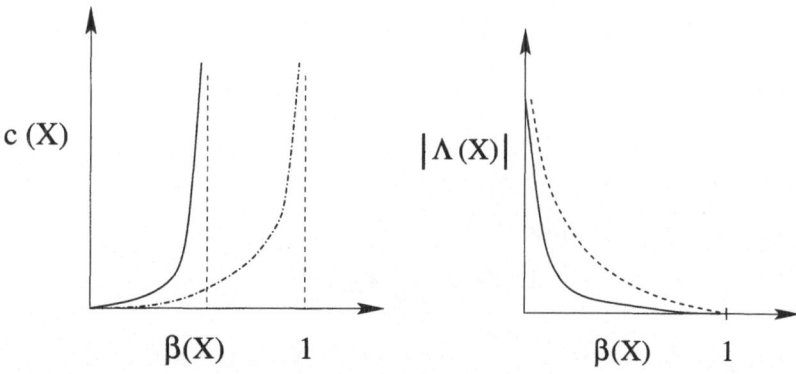

Figure 2.3. Left: The cost vs. discrimination tradeoff at two levels of invariance, "high" (solid line) and "low" (dashed line). Right: The invariance vs. discrimination tradeoff at two levels of cost, "low" (solid line) and "high" (dashed line).

2.4 Computation

Consider now adaptive (sequential) evaluations of $\hat{Y}$, i.e., tree-structured representations of $\hat{Y}$. Let $\mathcal{T}$ be the family of such evaluations. Each internal node of $T \in \mathcal{T}$ is labeled by a test X_Λ and each external node is labeled by an index – a subset of states. Our goal is to find the T which minimizes the mean cost of determining $\hat{Y}$ under assumptions on how $c(X)$ varies with $\beta(X)$, and how $\mathbf{X}$ is distributed under the "background model" P_0.

To illustrate such a computational procedure, take a simple example with $\mathcal{S} = \{\lambda_1, \lambda_2\}$ and three tests corresponding to $\Lambda_{1,2} = \{\lambda_1, \lambda_2\}$, $\Lambda_1 = \{\lambda_1\}$, $\Lambda_2 = \{\lambda_2\}$. One evaluation of $\hat{Y}$ is shown in Figure 2.4 where branching left means $X = 0$ and branching right means $X = 1$.

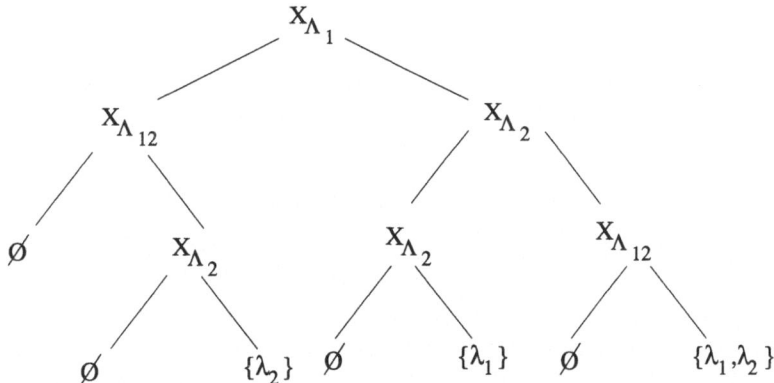

Figure 2.4. An example of a tree-structured evaluation of the detector $\hat{Y}$

Notice that $\hat{Y} = \emptyset$ if and only if there is a *null covering* of Λ: a subfamily $\{\Lambda_i\}$ such that $\bigcup_i \Lambda_i = \mathcal{S}$ and and $X_{\Lambda_i} = 0$ for each i. In Figure 2.4, one null covering corresponds to $\{X_{\Lambda_1} = 0, X_{\Lambda_2} = 0\}$ and one to $\{X_{\Lambda_{1,2}} = 0\}$.

2.4.1   Mean Cost

The *cost* of T is defined as

$$C(T) = \sum_{r \in T^\circ} \mathbf{1}_{H_r} c(X_r)$$

where T° is the set of internal nodes of T, X_r is the test at node r and H_r is the history of node r – the sequence of test results leading to r. (Equivalently, $C(T)$ is the aggregated cost of reaching the (unique) terminal node in T determined by $\mathbf{X}$.) Notice that $C(T)$ is a random variable. The *mean cost* $EC(T)$ is then

$$EC(T) \quad = \quad \sum_{r \in T^\circ} c(X_r) P(H_r) \qquad (2.3)$$

$$= \quad \sum_X c(X) P(X \text{ performed in } T) \qquad (2.4)$$

The second expression (2.4) is useful in proving the results mentioned in the following section.

Our optimization problem is then

$$\min_{T \in \mathcal{T}} EC(T) \qquad (2.5)$$

In other words, find the best testing strategy. As it turns out, the best strategies are often far more efficient than T's constructed with top-down, greedy procedures, such as those utilized in machine learning for building decision trees; see [5] for comparisons.

2.4.2   Hierarchical Tests

In order to rank computational strategies we must impose some structure on both Λ and the law of $\mathbf{X}$. From here on we consider the case of *nested binary partitions* of $\mathcal{S}$:

$$\Lambda = \{\Lambda_{m,j}, \ j = 1, ..., 2^m, \ m = 0, 1, ..., M\}.$$

Thus, $\Lambda_{0,1} = \Lambda_{1,1} \cup \Lambda_{1,2}$, $\Lambda_{1,1} = \Lambda_{2,1} \cup \Lambda_{2,2}$, etc. In Section 2.7, in experiments with a single object class, the hierarchy Λ is a "pose decomposition" wherein "cells" at level $m+1$ represent more constrained poses than cells at level m. As m increases, the level of invariance decreases and, in the models below, the level of discrimination increases. *This tree-structured hierarchy should not be confused with a tree-structured evaluation T of $\hat{Y}$.* The label $\hat{Y}$ at a terminal node of T depends on $\mathbf{X}$ and consists of all states in *any*

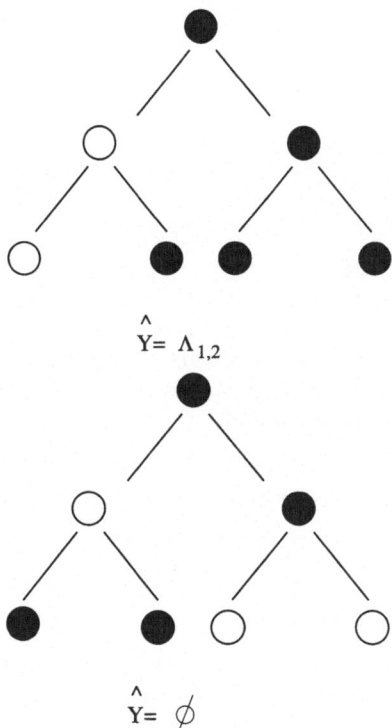

Figure 2.5. Examples of $\hat{Y}$ for two realizations of the tests in a nested hierarchy with three levels. Dark circles represent $X = 1$ and open circles represent $X = 0$. Top: There are two "chains of ones". Bottom: There is a null covering.

level M cell of Λ for which there is a "1-chain" back to $\Lambda_{0,1}$, i.e., $X_{\Lambda'} = 1$ for every $\Lambda' \supset \Lambda$.

As a simple illustration, consider the case $M = 2$, in which case there are exactly seven sets in the hierarchy. Notice that the most refined cells $\Lambda_{2,j}, j = 1, 2, 3, 4$, may each contain numerous states λ, i.e., may provide an interpretation at a level of resolution which is still rather "coarse." Figure 2.5 shows $\hat{Y}(\mathbf{X})$ for two realizations of $\mathbf{X}$. For the same hierarchy and realizations of $\mathbf{X}$, the lefthand panel of Figure 2.6 illustrates a "depth-first" CTF evaluation of $\hat{Y}$ and the righthand panel a "breadth-first" CTF evaluation.

Another way to interpret $\hat{Y}$ is the following: For each level m in the hierarchy define

$$\hat{Y}_m(\mathbf{X}) = \bigcup_{j \in J_m(\mathbf{X})} \Lambda_{m,j}$$

where $J_m = \{1 \le j \le 2^m : X_{m,j} = 1\}$. The detector $\hat{Y}$ is $\cap_m \hat{Y}_m$. We can think of $\hat{Y}_m$ as an estimate of Y at "resolution" m. Necessarily, $Y \subset \hat{Y}_m$

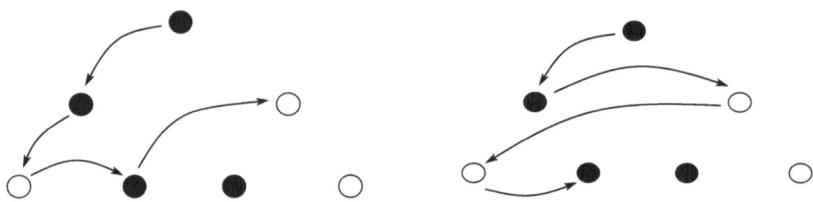

Figure 2.6. As in Figure 2.5, dark and light circles indicate positive and negative tests for a three level hierarchy. Left: A "depth-first" coarse-to-fine search. Right: A "breadth-first" coarse-to-fine search.

for each m, although in general it needn't happen that $\hat{Y}_{m+1} \subset \hat{Y}_m$ (or vice-versa).

2.4.3   An Approximation

We are going to assume that the mean cost is computed with respect to the background distribution P_0. This is motivated by the following argument. Recall that labeling the entire scene means applying the detector $\hat{Y}$ to subimages I centered on a sparse sublattice of the original scene. Whereas the likelihood of having some objects in the entire scene may be large, the *a priori* probability of the "null hypothesis" $Y = \emptyset$ is approximately one when evaluating $\hat{Y}$ for an arbitrary subimage at the scale of the objects. Therefore, the average amount of computation involved in executing our detection algorithm with a particular strategy T can be approximated by taking the expectation of $C(T)$ under P_0. Of course this approximation degrades along branches with many positive test responses, especially to discriminating tests associated with "small" class/pose cells, in which case the probability of an object being present may no longer be negligible. If one were to measure the mean computation under an appropriate (mixture) distribution, the conditional distribution of the tests under the various object hypotheses would come into play. Nonetheless, we continue to compute the likelihood of events under P_0 alone, thereby avoiding the need to model the behavior of the tests given objects are present.

2.5   Optimality Results

For simplicity, write $X_{m,j}$ for $X_{\Lambda_{m,j}}$. From here on, we make the following assumptions about the the distribution of **X**:

- $\{X_{m,j}\}$ are independent random variables under P_0;

- $\beta_{m,j} \doteq P_0(X_{m,j} = 0) = \beta_m,\ m = 0, 1, ..., M$;

- $\beta_0 \le \beta_1 \le \cdots \le \beta_M$;

- $c(X_{m,j}) = c_m,\ m = 0, 1, ..., M$;

- $c_m = \Phi(\beta_m)$ with $\Phi(0) = 0$ and Φ increasing.

The independence assumption is violated in practice, but we make it in order to facilitate a theoretical analysis. The other assumptions are realistic, partly by design, although the power of the tests may differ slightly within levels.

2.5.1   Testing $\hat{Y} = \emptyset$ vs. $\hat{Y} \ne \emptyset$

Consider first the problem of determining whether or not $\hat{Y} = \emptyset$, in other words, evaluating $\hat{Z} = \mathbf{1}_{\{\hat{Y} \ne \emptyset\}}$. The set-up is the same, except the terminal labels of T are simply "0" or "1." This problem was studied in [3] and [8] under the assumption that Φ is *convex*, i.e., cost is a convex, increasing function of power. One strategy is the depth-first CTF strategy, illustrated in Figure 2.7 for the case $M = 2$.

In Figure 2.8 two particular sample paths (branches) are depicted from a depth-first, CTF search of a five level hierarchy. The path on the left leads to the label $\hat{Z} = 0$ and the one on the right leads to $\hat{Z} = 1$.

The following result is proved in [8]; earlier, [3] had shown that the coarsest test $(X_{0,1})$ is necessarily at the root. The convexity assumption can be relaxed to supposing that $\frac{c_m}{\beta_m}$ is increasing, $m = 0, ..., M$.

Theorem: *If Φ is convex, and $P = P_0$, depth-first CTF search is the optimal strategy for evaluating $\hat{Z}$.*

2.5.2   Determining $\hat{Y}$

Here the objective is to determine *all* 1-chains instead of merely whether or not one such chain exists. For either CTF strategy, the expected cost under P_0 is

$$E_0 C(T) = c_0 + \sum_{m=1}^{M} c_m 2^m \prod_{j=0}^{m-1} (1 - \beta_j).$$

As it turns out, this is the smallest possible mean cost:

Theorem: *The CTF strategy is optimal for any increasing cost sequence if $\beta_0 > .5$*

The proof of this result, and others based on varying cost models and test hierarchies, will appear in [2].

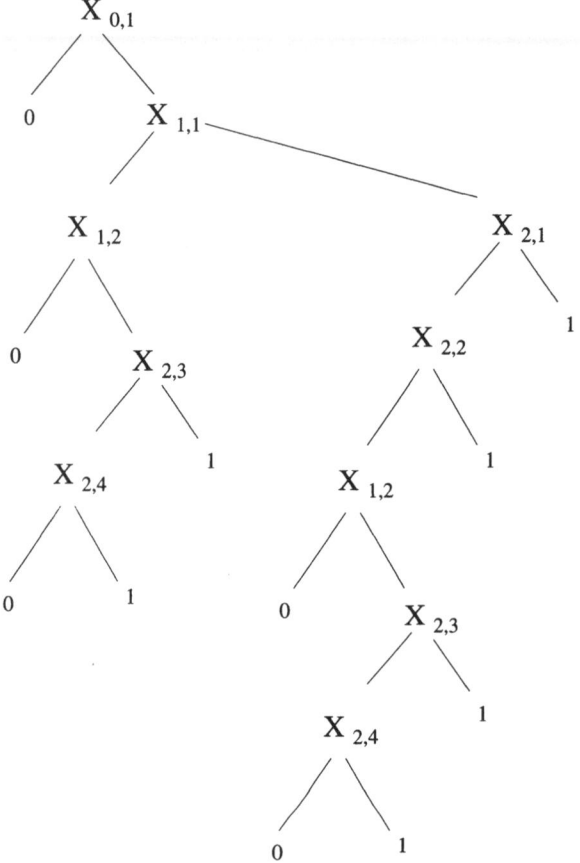

Figure 2.7. The CTF strategy tree for the case $M = 2$.

2.6   Error

We briefly discuss the theoretical error rates of the detector $\hat{Y}$ defined in (2.2).

In principle, the false negative rate is null. (Of course, in practice, this requires that X_Λ be invariant for Λ, which can be difficult to achieve.) The false positive error

$$\delta(\hat{Y}) \doteq P_0(\hat{Y} \neq \emptyset)$$

is determined by the (joint) distribution of $\{X_{m,j}\}$ under P_0, and depends on $\beta_0, ..., \beta_M$. The rate *per pixel* is then $\frac{\delta(\hat{Y})}{k^2}$; recall that the search for object locations is conducted in non-overlapping $k \times k$ blocks. A crude bound is to replace the probability of at least one 1-chain under P_0 by the

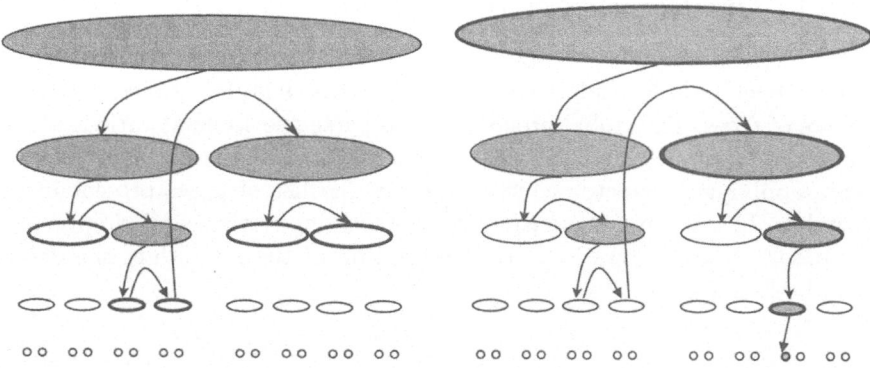

Figure 2.8. Two branches from a depth-first CTF search for a hierarchy with five levels. The tests are explored in the indicated order, with gray indicating a positive answer and white in bold outline indicating a negative answer; the other tests were not evaluated. Left: A null covering is encountered, ending the search. Right: A chain of ones is encountered (Gray in bold outline), ending the search when the goal is to determine if $\hat{Y} = \emptyset$.

sum of the probabilities, yielding

$$\delta(\hat{Y}) \leq 2^M \prod_{m=0}^{M} (1 - \beta_m)$$

$$\leq \frac{1}{2}\gamma^{M+1}, \ \gamma = 2(1 - \beta_0).$$

To calculate $\delta(\hat{Y})$ exactly, notice that there is a correspondence between realizations of a breath-first CTF evaluation of $\hat{Y}$ and realizations of a branching process with M generations starting from a single individual. Consequently,

Theorem: *The false positive error $\delta(\hat{Y})$ is the probability of no extinction for a non-homogeneous branching process with binomial family law $Bin(2, 1 - \beta_m)$ at generation $m = 0, ..., M$.*

2.7   Application to Face Detection

Recall that each $\lambda \in \mathcal{S}$ corresponds to the presentation or instantiation of an "object" in a predetermined library. The presentation includes information about the position, scale and other aspects of the geometric pose, and perhaps other properties of interest.

2.7.1   One Generic Class

Consider the simplified scenario of one generic object class and linear pose. More specifically, consider the problem of detecting all instances of frontal views of faces. The global procedure is to parse the scene at various scales, and at a sampling of locations, with a window of size 64 × 64, in each case applying a detector which computes the list of poses present in the window. In this case, the output $\hat{Y}$ of processing a window is simply a list of poses. What follows is a brief description of the algorithm; the details can be found in [4].

2.7.2   Pose Decomposition

The pose of a face is defined in the image plane, given by $\theta = (z, \sigma, \psi)$, where z is the center point between the eyes, σ is the distance (in pixels) between the eyes and ψ is the "tilt." The image is partitioned into non-overlapping 8 × 8 blocks, and the basic detector $\hat{Y}$ is applied to the image data in a window centered at each such block. These windows are the subimages in $\mathcal{I}$ in the previous sections. The detector $\hat{Y}$ produces a list of poses with $z \in [28, 36]^2, 8 \le \sigma \le 16, -20° \le \psi \le 20°$; of course in this case these are either false alarms or responses to the same face. Call this set of poses the "reference cell" $\Lambda_{0,1}$. Faces of scale $16 \le \sigma \le 32$ are found by downsampling the original scene and parsing again, etc.

The hierarchy $\{\Lambda_{m,j}\}$ of subsets of (reference) poses is constructed by recursively partitioning $\Lambda_{0,1}$ into a sequence of nested partitions; each cell Λ of each partition is a subset of poses which is included in exactly one of the cells in the preceding, coarser partition. At each level $m = 1, 2, ..., M$, one component of the pose θ is subdivided into equal parts – binary splits for scale and tilt and quaternary splits for position. Thus, for example, each cell $\Lambda_{1,j}, j = 1, 2, 3, 4$, of the the first partition corresponds constraining z to one of the four 4 × 4 subsquares of of the initial 8 × 8 square. There are two splits on z, two on σ and two on ψ, resulting in $M = 6$ levels (excluding the root).

2.7.3   Learning

Each test $X_{m,j} = X_{\Lambda_{m,j}}$ checks for a certain number of distinguished edge fragments, and hence is defined by a list $L_{m,j}$ of edges and a threshold (as described in Section 2.2) both determined during training. Consequently, the tests X are simply counting operators. Checking for an edge means evaluating a binary local feature ξ_l indexed by a position z_l and an orientation ϕ_l as described in Section 2.2. However, there is another, crucial, parameter – the "tolerance" of the edge – which allows one to achieve invariance to the poses in Λ. (The MAX filter in [10] has the same aim.) The tolerance η is the length of a strip of pixels centered at z_l and perpendicular to the direction of the edge. The feature $\xi_l = 1$ if there is an edge at the given orientation at *any* location in the strip. Thus, η controls the amount of ORing, which in turn depends on the desired degree of invariance. All ξ in the list have the same η. The tolerance η is "large" for the coarse cells $\Lambda_{m,j}$ and "small" for the fine cells. It controls the tradeoff between invariance and discrimination.

All the tests $X_{m,j}$ in the hierarchy are built with the same learning algorithm; the lists differ due to varying training sets, corresponding to varying constraints on the set of poses. The experiments shown here are based on the ORL database which contains 400 grayscale face pictures of size 112×96 pixels. For each cell $\Lambda_{m,j}$, a synthetic training set of 1600 face images is constructed whose poses are in $\Lambda_{m,j}$. Again, the details are in [4], including how the lists of edges $\xi_l, l \in L$ are chosen.

2.8 Experiments and Conclusions

The experiments use a breadth-first CTF evaluation of $\hat{Y}$. Nearby detections are clustered, resulting in one estimated pose per face, indicated by a triangle. The false negative rate is not null, and there are false positives, on the order of $1 - 10$ per scene on average. Thus, when computation is efficiently organized, one can use very simple components (such as the counters X described in the previous section) and still achieve reasonable error rates, in fact comparable to the best ones reported in the literature for high resolution images; see for example [11].In addition, detection is extremely fast, well under one second for a scene on the order of 400×400, which is faster than previously reported results. Two results are shown In Figures 2.9 and 2.11. As seen in Figure 2.12, coarse-to-fine processing leads to highly asymmetric scene processing in terms of spatial concentration, with orders of magnitude differences in the application of resources to different regions of the scene.

Recall that detection is far from a complete solution to scene interpretation. It leaves confusions unresolved and does not address occlusion and other complicating factors. Many examples of such specific class/pose confusions can be seen in a higher resolution rendering (not shown here) of the labeled detections in Figure 2.11. However, if highly visible objects are sure to be detected with a limited number of false alarms, it may then be computationally feasible to entertain very intense processing which is optimization-based but highly localized.

Finally, for detection, we can suppose that each test is constructed during an offline training phase, as in the previous section. Indeed, since we are not anticipating the specific confusions and occlusion patterns that might arise, we can afford to make a list of *all* the tests we wish to construct, learn them during training, and store all the instructions for execution. On the

Figure 2.9. An experiment from the CNN website.

Figure 2.10. An experiment with a group photograph.

contrary, during contextual classification we might be obliged to generate hypotheses and construct tests online, i.e., during scene parsing. Such ideas are currently being explored in the context of character recognition.

Figure 2.11. The detections for the group photo.

Figure 2.12. The coarse-to-fine nature of the algorithm is illustrated for the group photo by counting, for each pixel, the number of times the detector checks for the presence of an edge in its vicinity. The level of darkness is proportional to this count.

References

[1] Amit, Y. and D. Geman. 1999. A computational model for visual selection. *Neural Computation*, 11:1691-1715.

[2] Blanchard, G. and D. Geman. 2002. Computational models for coarse-to-fine search, in preparation.

[3] Fleuret, F. 2000. Detection hierarchique de visages par apprentissage statistique. Ph.D. thesis, University of Paris VI, Jussieu, France.

[4] Fleuret, F. and D. Geman. 2001. Coarse-to-fine face detection. *Inter. J. Computer Vision*, 41:85-107.

[5] Geman, D. and B. Jedynak. 2001. Model-based classification trees. *IEEE Trans. Info. Theory*, 47:1075-1082.

[6] Geman, S., K. Manbeck, and D. McClure. 1995. Coarse-to-fine search and rank-sum statistics in object recognition. Technical Report, Division of Applied Mathematics, Brown University.

[7] Geman, S., D. Potter, and Z. Chi. 2001. Composition systems. *Quarterly of Applied Mathematics*, to appear.

[8] Jung, F. 2001. Reconnaissance d'objects par focalisation et detection de changements. PhD thesis, Ecole Polytechnique, Paris, France.

[9] Lambdan, Y. and J.T. Schwartz and H.J. Wolfson. 1988. Object recognition by affine invariant matching. *Proc. IEEE Conf. on Computer Vision and Pattern Recognition*, 335-344.

[10] Riesenhuber, M. and T. Poggio. 1999. Hierarchical models of object recognition in cortex. *Nature Neuroscience*, 2:1019-1025.

[11] Rowley, A.R. 1999. Neural network-based face detection. PhD Thesis, Carnegie Mellon University.

3

Environmental Monitoring Using a Time Series of Satellite Images and Other Spatial Data Sets

Harri Kiiveri, Peter Caccetta,
Norm Campbell, Fiona Evans,
Suzanne Furby, and Jeremy Wallace[1]

3.1   Introduction

As a result of extensive farmland clearing over the last hundred years or so, dry-land salinity is a major problem in Western Australia. In fact, in some parts of the state, over 20 percent of Agricultural land is no longer productive. Prior to the work to be described in this chapter, no reliable large scale estimates of the extent or progression of salinity were available. This chapter describes a methodology for monitoring the historical extent of salinity, using a time series of satellite imagery, landform information derived from digital elevation models and ground truth data collected by experts with local knowledge. This work has served to highlight the salinity problem to decision makers in government and to provide input into the process of developing and applying remedial measures to arrest the spread of salinity.

Although the chapter primarily refers to salinity monitoring, the methodology is quite general and can be used in any situation in which the relationships amongst a number of images can be represented by a directed or undirected graph.

The chapter is structured as follows. The area monitored and the data used are described in Section 3.2. A short description of the data (pre)processing is given in Section 3.3. Whilst we don't not spend much

---

[1]The authors are all with the CSIRO Division of Mathematical and Information Sciences, Floreat Park, Western Australia.

time in Section 3.3, this part of the work is the most time consuming and critical to the success of the project. In Section 3.4 a model for integrating a time series of satellite images with other spatial data sets is constructed. Since the model contains latent (unobserved) images we consider the issue of identifiability of parameters in the model and also describe an EM algorithm for estimating the parameters. The model also explicitly allows for uncertainty in the inputs and outputs of the monitoring process. In Section 3.5, the chapter concludes with an example of the output products produced by the methodology.

3.2   Description of the data

The study area is covered by 15 Landsat TM scenes [18] and covers approximately 18 million hectares. This area, with satellite scenes overlaid, is shown in Figure 3.1. For each scene we have 6 satellite images, roughly every two years over the time period 1988 to 1999. Each satellite image has 8000 by 8000 pixels (picture elements). The images were re-sampled to give a pixel size of 25 metres and consist of six wavelength channels or bands. Hence for each date, at each pixel we have a data vector with six components. Previous work [10] has identified September/October to be the optimal time to discriminate salt from other ground classes and the images were acquired as close to this window as possible.

In addition, we have a landform map (image) derived from a digital elevation model for each scene. This is a five-class map identifying hilltops, valley floors and varying degrees of slopes.

Training data in the form of interpreted salinity maps over relatively small regions was available for each scene. These were supplied by local experts.

All in all for each scene we have approximately 4.5 gigabytes of data to store and process. Any methods we use to process this data need to take the large data volumes into consideration.

3.3   Data pre-processing

To produce input data for the model to be discussed in the next section required a large amount of data processing. Firstly, contour data was interpolated to produce grided digital elevation models. These were processed to produce water accumulation maps which were then converted into Landform maps, see [5].

Secondly, for each scene, the time series of satellite images needed to be accurately geo-coded or rectified. This was done carefully to ensure that

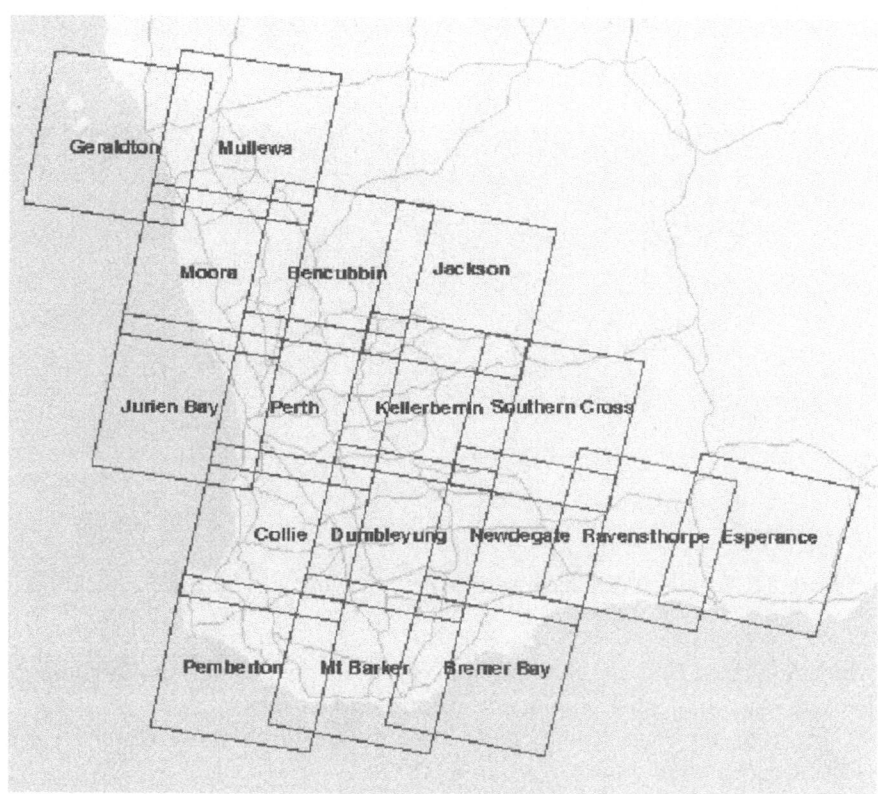

Figure 3.1. Location of satellite scenes over the study area

location errors were less than the pixel size [7]. This is essential to ensure that changes in ground cover are not confounded with locational errors.

Thirdly, images were calibrated to like values to ensure that areas which are not changing over time look the same in each image. To do this, linear transformations were defined using robust regression techniques to make the values of the images over a common set of (pseudo) invariant targets as close as possible [11].

Finally, each image was classified into approximately 12 ground cover classes, e.g., bush, water, agriculture, bare ground and different types of salt. The method used was Gaussian maximum likelihood [19] and this required the careful and time-consuming selection of training sites within each image. Using canonical variate analysis [6] the training site separation was studied and the training sites were eventually amalgamated into the final ground cover training classes. Figure 3.2 shows an idealised canonical variate plot and also illustrates the difficulty in using the satellite imagery to map salinity. The circles/ellipses in Figure 3.2 show the locations in canonical variate space of the cover types of interest. Unfortunately, there is an over lap with salt and bare, salt and bush and salt and agriculture

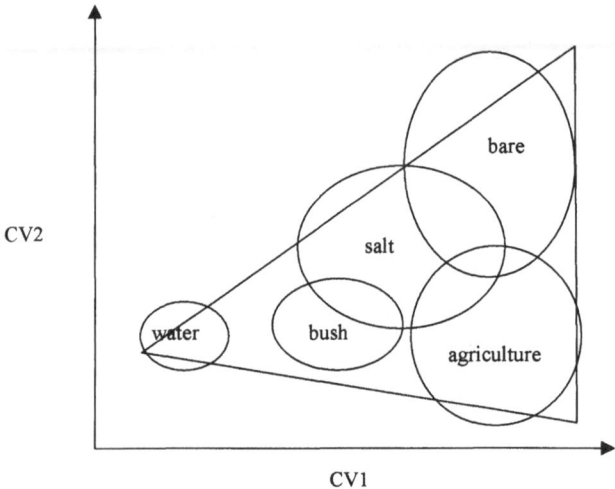

Figure 3.2. Idealised canonical variate plot showing ground cover separation

which indicated that the satellite imagery on its own could not discriminate between salt and other significant ground cover types.

Note that for each scene, the image for each date was classified i.e. 6 classifications were done.

Previous work had shown that the Gaussian assumption was reasonable for this problem. Although we used the Gaussian maximum likelihood classifier, any method which produces probabilities of class membership and has a notion of typicality could have been used.

3.4   A conditional probability network for image data

In the previous section we mentioned the difficulty in mapping salinity given a satellite image on one single date as the satellite sensor is unable to distinguish between salt and certain other ground cover classes. However, some simple prior knowledge about the process of salinisation suggests that a time series of images in conjunction with landform information can be used to effectively map salinity. It is known that salt is stable over time, so for example, paddocks which appear salt affected but in fact are bare, can be correctly mapped by noticing the temporal cover class sequence. If apparent salt is followed by healthy crop cover then we can deduce that the paddock is not salt affected. Similarly we know that salt tends to occur mainly in valley floors so that apparent salt in hill tops or upland regions

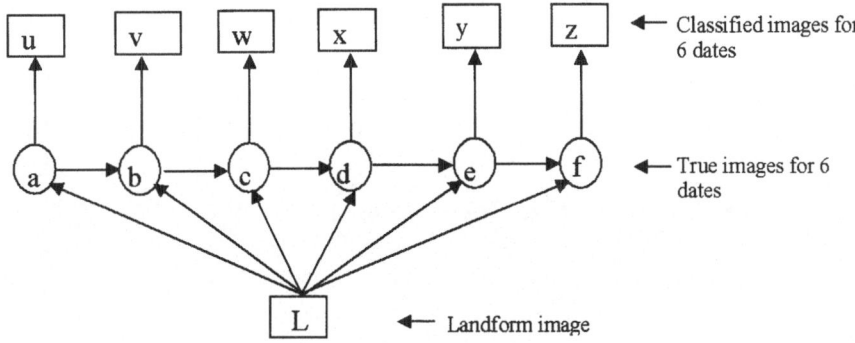

Figure 3.3. A conditional probability network (CPN) to integrate all the data for an area.

can also be down weighted. In this section we will develop a model for each scene which enables us to take this type of prior information into account.

3.4.1   The model

Conditional probability networks (CPNs) are typically represented by a directed acyclic graph, where the nodes represent variables and the edges define parent child relationships amongst the variables. Associated with the graph is a factorisation of the joint probability distribution of all the variables [12]. If we allow the nodes of the graph to represent images, then a model for a time series of images could be represented as follows in Figure 3.3.

In Figure 3.3, the six classified satellite images are denoted u, v, w, x, y, z, the true ground cover images are denoted by a, b, c, d, e, f and the landform image is denoted by L. The circles denote latent (unobserved) images and the squares denote observed images. By analogy with CPNs, the joint distribution of all the images would factorize as

$$p(u|a)\,p(v|b)\,p(w|c)\,p(x|d)\,p(y|e)\,p(z|f) \times$$
$$p(b|a, L)\,p(c|b, L)\,p(d|c, L)\,p(e|d, L)\,p(f|e, L)\,p(a|L)\,p(L) \tag{3.1}$$

Aside from the fact the the distributions are high dimensional e.g. u and a have 8000 by 8000 variables, this simply looks like a conditional probability network with latent variables.

To implement such a model we need to have tractable models for the probabilities in (3.1), determine the identifiability of the parameters in the model and have a method for estimating the parameters. Having fitted such a model, we then need to predict the unobserved true images given the observed images. Given the large volumes of data, we also need to have efficient algorithms for doing the calculations.

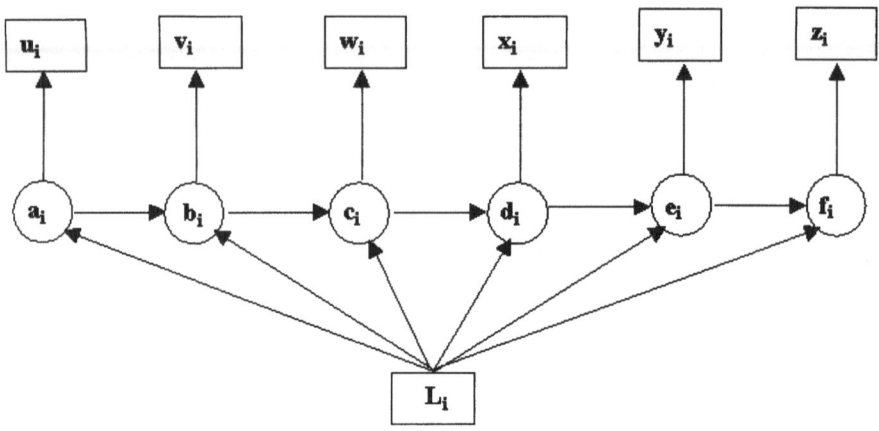

Figure 3.4. CPN model assuming independent pixels

3.4.2   Construction of a contextual "CPN" model for raster images

To build a suitable model we put together two pieces, the first piece will consider the relationship between images and the second piece will focus on the within image structure.

The first piece is a CPN for images assuming independent pixels i.e. the same CPN is applied to each pixel in the image data. Let u, v, w, x, y, z denotes the classified images for years 89, 90, 93, 94, 96, and 98 and a, b, c, d, e, f denotes the "true" images for the same years. Here, for example

$$u = (u_i, i = 1, \ldots, n),\ a = (a_i, i = 1, \ldots, n),\ z = (z_i, i = 1, \ldots, n)$$

and n is the number of pixels. L denotes the landform image. The probability density for all the images is

$$
\begin{aligned}
p &= p(a, b, c, d, e, f, u, v, w, x, y, z, L) \hspace{3cm} (3.2)\\
&= \prod_{i=1}^{n} p(u_i|a_i)p(a_i|L_i)p(v_i|b_i)p(b_i|a_i, L_i) \cdots p(z_i|f_i)p(f_i|e_i, L_i)p(L_i)
\end{aligned}
$$

The second piece we will use is a Markov Random Field model for images [2]. First define $n(i)$ to be the set of neighbours of pixel i and $r(i)$ the set of pixels excluding pixel i, i.e. the rest of the pixels. For concreteness we take the set of neighbours to be the eight nearest neighbours with appropriate modifications at the edges of the image. Writing $a_{n(i)}$ for the neighbouring set of image values and $a_{r(i)}$ for the image values at all pixels except pixel i, these models have the property that the conditional distribution of a_i given $a_{r(i)}$ depends only on $a_{n(i)}$ that is

$$p(a_i|a_{r(i)}) = p(a_i|a_{n(i)})$$

We use similar notation for the other unobserved images. We can make a hybrid model from the two components as follows

$$p^*(a, b, c, d, e, f, u, v, w, x, y, z, L) =$$

$$\exp\{\log p + \log p(a) + \cdots + \log p(f) \tag{3.3}$$

$$+ \log p(u) + \cdots + \log p(z) + \log p(L)\}/U$$

where p is the CPN model (3.2) for the images assuming independent pixels and the remaining terms in brackets correspond to Markov random field models for each of the images. The term U is simply a normalising constant. Note that (3.3) defines a Gibbs distribution.

To predict the unobserved true maps a, b, c, d, e, f we want to calculate the conditional probability

$$p^*(a, b, c, d, e, f | u, v, w, x, y, z, L) \tag{3.4}$$

and find the images a, b, c, d, e, f which maximise this probability. To see how to do this we first do a calculation. Writing

$$(a, b, c, d, e, f) = (a_i, b_i, c_i, d_i, e_i, f_i, a_{r(i)}, b_{r(i)}, c_{r(i)}, d_{r(i)}, e_{r(i)}, f_{r(i)})$$

and using factorisations for each term in (3.3) we get

$$p^*(a, b, c, d, e, f | u, v, w, x, y, z, L)$$

$$= p^*(a_i, b_i, c_i, d_i, e_i, f_i | a_{r(i)}, b_{r(i)}, c_{r(i)}, d_{r(i)}, e_{r(i)}, f_{r(i)}, u, v, w, x, y, z, L)$$

$$\times p^*(a_{r(i)}, b_{r(i)}, c_{r(i)}, d_{r(i)}, e_{r(i)}, f_{r(i)} | u, v, w, x, y, z, L)$$

$$= p^*(a_i, b_i, c_i, d_i, e_i, f_i | a_{n(i)}, b_{n(i)}, c_{n(i)}, d_{n(i)}, e_{n(i)}, f_{n(i)}, u, v, w, x, y, z, L)$$

$$\times p^*(a_{r(i)}, b_{r(i)}, c_{r(i)}, d_{r(i)}, e_{r(i)}, f_{r(i)} | u, v, w, x, y, z, L)$$

Given this result, a cyclic ascent algorithm for doing the maximisation can be constructed as follows:

1. Start with initial estimates of the unobserved images a, b, c, d, e, f e.g. by using the results for the independent pixels case

2. Visit each pixel i in turn keeping all labels fixed except a_i, b_i, c_i, d_i, e_i, f_i and compute

$$p^*(a, b, c, d, e, f | u, v, w, x, y, z, L)$$

$$= p^*(a_i, b_i, c_i, d_i, e_i, f_i | a_{n(i)}, b_{n(i)}, c_{n(i)}, d_{n(i)}, e_{n(i)}, f_{n(i)}, u, v, w, x, y, z, L)$$

$$\times p^*(a_{r(i)}, b_{r(i)}, c_{r(i)}, d_{r(i)}, e_{r(i)}, f_{r(i)} | u, v, w, x, y, z, L)$$

It can be shown [13] that

$$p^*(a_i, b_i, c_i, d_i, e_i, f_i | a_{n(i)}, b_{n(i)}, c_{n(i)}, d_{n(i)}, e_{n(i)}, f_{n(i)}, u, v, w, x, y, z, L)$$

factorises so that it can be computed from a CPN with a graph augmented with dummy neighbourhood nodes as in Figure 3.5. Choose

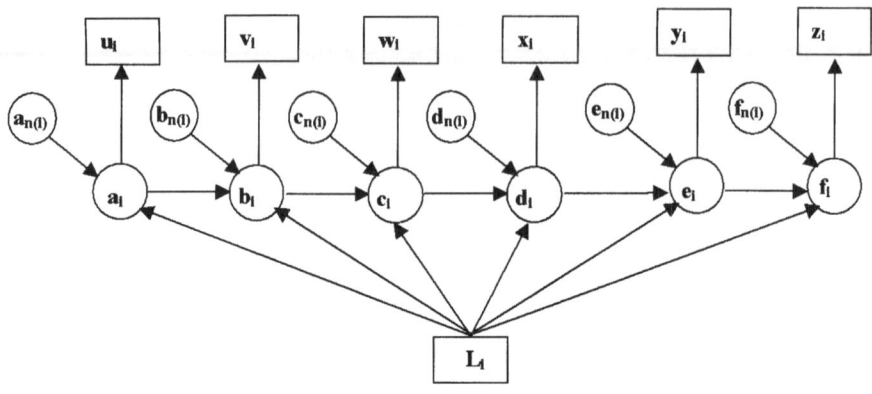

Figure 3.5. CPN augmented with neighbour information

map classes a_i, b_i, c_i, d_i, e_i and f_i to maximise (3.1). Another strategy at this point is to choose labels for the unobserved maps individually by maximising the marginal distributions

$$p^*(a_i|a_{n(i)}, b_{n(i)}, c_{n(i)}, d_{n(i)}, e_{n(i)}, f_{n(i)}, u, v, w, x, y, z, L)$$

and similarly for b_i, c_i, d_i, e_i, and f_i.

3. Continue cycling over all pixels until convergence. For the present application we only do a few iterations.

3.4.3 Estimation of parameters

To use the algorithm in practice requires the specification of parameters in the model. The specific Markov random field model we used for each unobserved image was

$$p(a_i|a_{n(i)}) = \exp\{\alpha + \beta N(a_i)\} \qquad (3.5)$$

where $N(a_i)$ is the number of 8 nearest neighbours of pixel i with label a_i, see for example [2]. For parameter values we used $\alpha = 0$ and $\beta = 1$. These parameters can be varied to change the relative weighting between image data and contextual information. The probabilities $p(u_i|a_i), \ldots, p(z_i|f_i)$ are assumed to be the same for each pixel (within a zone) and are estimated from error rates derived from the classification process. The remaining probabilities in (3.1) are estimated by the EM algorithm [9],[4] ignoring spatial dependence. Ignoring this dependence should not be too critical since we have large sample sizes. The transition probabilities for the ground cover classes are identifiable, see the Appendix. Alternative estimation procedures such as coding or pseudo likelihood could also be used [1, 2].

The EM algorithm is implemented as follows:

1. Guess initial values for all the unknown probabilities in (3.2), e.g., by random generation.

2. Perform the E step. For the model of Figure 3.4 this could be done by calculating the table of expected counts

$$m(a_i, b_i, c_i, d_i, e_i, f_i | u_i, v_i, w_i, x_i, y_i, z_i, L_i) =$$

$$\sum_{i=1}^{n} p(a_i, b_i, c_i, d_i, e_i, f_i | u_i, v_i, w_i, x_i, y_i, z_i, L_i) \quad (3.6)$$

where the conditional probability is computed using the model (3.2). Next, we could then calculate the expected marginal tables of counts defined by $m(a_i, L_i)$, $m(b_i, a_i, L_i)$, $m(c_i, b_i, L_i)$, $m(d_i, c_i, L_i)$, $m(e_i, d_i, L_i)$, $m(f_i, e_i, L_i)$, where for example arguments not appearing in $m(\cdot)$ are summed over in (3.6). However this can be done more efficiently, see [17].

3. Perform the M step. Estimate probabilities by calculating relative frequencies assuming the expected counts were observed data. For example $p(a_i|L_i) = m(a_i, L_i)/m(L_i)$ and $p(b_i|a_i, L_i) = m(b_i, a_i, L_i)/m(a_i, L_i)$.

4. Go to 2 until convergence.

3.4.4   A model for handling uncertainty in input class labels

When computing the probability of true class labels from the model in Figure 3.4, it is assumed that class labels are known. However, the Gaussian maximum likelihood classifier also produces posterior probabilities of class membership and we can use these in the mapping process. A graphical representation of a model for doing this is given in Figure 3.6.

In Figure 3.6 variable u_i^*, for example, refers to the six bands of the satellite image at pixel i on the first date in the series. Spatial context could also be included in the same manner as in Figure 3.5. Details about the modifications to the usual calculations are given in the Appendix.

3.5   An example

An example of a classification for a single date produced by using the methods described above is given in Figure 3.7. The figure is represented in gray scale, however colour would be more impactful.

To provide a (salinity) accuracy assessment for the classification, 124 validation sites were visited, and their salinity status noted. The results are given in Table 3.1 below. Of the 124 sites 6 were incorrectly labeled and 118 were correctly labeled, or in other words an overall salinity mapping

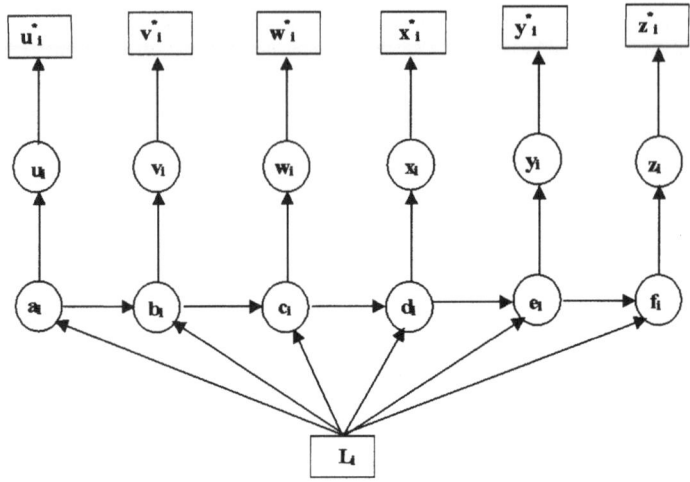

Figure 3.6. Modified CPN for handling uncertainty in input classification maps

Table 3.1. Accuracy assessment for example region

		Truth	
		Saline	non-saline
Map	Saline	40	5
	Non-saline	1	78

accuracy of approximately 95%. For the broader region, estimates of the percentage of this catchment affected by salinity range from 0 to 9.8%. Maps such as in Figure 3.7 are available for six dates for this area.

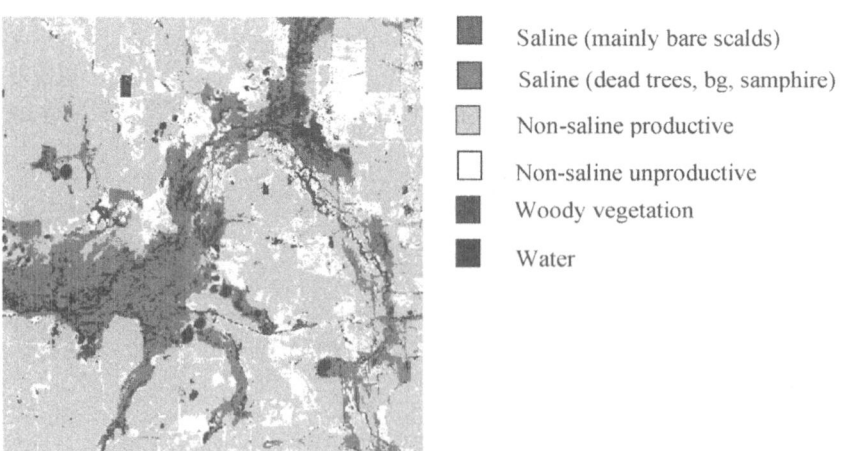

Saline (mainly bare scalds)

Saline (dead trees, bg, samphire)

Non-saline productive

Non-saline unproductive

Woody vegetation

Water

Figure 3.7. Example single date "true" map

There are of course significant issues in how to assess the accuracy of such large scale maps given limited resources, however we will not go into these here.

In conclusion, the methods discussed in this chapter seem promising for handling large scale environmental monitoring problems involving the use of time series of satellite imagery. A particularly desirable feature of the methods is the ability to propagate uncertainties in the inputs to the output products.

3.6   Appendix: Modifications to calculations when there is uncertainty in the classification input maps

We consider the independent pixels case and omit the subscript i on the variables. For generality and compactness we represent the classified satellite image class at pixel i at times $1, 2, 3, \ldots$ as $u_1, u_2, u_3, \ldots$ and the true classes as $a_1, a_2, a_3, \ldots$.

The joint distribution is

$$p(u_1, \ldots, u_p, a_1, \ldots, a_p, L) = \prod_{j=1}^{p} p(u_j^*|u_j)p(u_j|a_j)p(a_1, \ldots, a_p, L)$$

From this we can see that

$$p(a_1, \ldots, a_p|u_1^*, \ldots, u_p^*, L)$$
$$\propto \prod_{j=1}^{p} \left[\sum_{u_j} p(u_j^*|u_j)p(u_j|a_j) \right] p(a_1, \ldots, a_p, L) \qquad (3.7)$$

Writing

$$p(u_j|u_j^*) = \frac{p(u_j^*|u_j)}{\sum\limits_{u_j} p(u_j^*|u_j)}$$

for the posterior probability obtained from the Gaussian maximum likelihood classier using uniform prior probabilities (3.7) can be written as

$$p(a_1, \ldots, a_p|u_1^*, \ldots, u_p^*, L)$$
$$\propto \prod_{j=1}^{p} \left[\sum_{u_j} p(u_j|u_j^*)p(u_j|a_j) \right] p(a_1, \ldots, a_p, L) \qquad (3.8)$$

When $p(u_j|u_j^*)$ is an indicator vector i.e. zero everywhere except in one position (3.8) reduces to the usual formula for $p(a_1, \ldots, a_p|u_1, \ldots, u_p, L)$.

Note that all calculations can be done efficiently using the algorithm in [15]. See also [16]. We can also include spatial dependence as in Figure 3.5.

3.7   Appendix: Identifiability of the CPN model in the independent pixels case

Theorem 2. *If the G by G matrices A_i with elements defined by $p(u_i|a_i)$ for $u_i, a_i = 1, \ldots, G$ are known and full rank for all i then $p(a_1, \ldots, a_p, L)$ can be determined from the distribution $q(u_1, \ldots, u_p, L)$ of the observed variables.*

Proof. We use induction on p. First when $p = 1$ we have

$$\sum_{a_1} p(u_1|a_1)p(a_1, l) = q(u_1, l)$$

where $u_1 = 1, \ldots, G$ and $l = 1 \ldots, L$, say. This is a linear equation which can be solved uniquely for $p(a_1, l)$ since the matrix with elements $p(u_1|a_1)$ is full rank.

Next assume the result is true for p and we will demonstrate that it is true for $p + 1$. We have

$$\sum_{a_1, \ldots, a_{p+1}} \prod_{i=1}^{p+1} p(u_i|a_i)p(a_1, \ldots, a_{p+1}, l) = q(u_1, \ldots, u_{p+1}, l) \qquad (3.9)$$

Summing both sides of this equation over u_{p+1} gives

$$\sum_{a_1, \ldots, a_p} \prod_{i=1}^{p} p(u_i|a_i)p(a_1, \ldots, a_p, l) = q(u_1, \ldots, u_p, l) \qquad (3.10)$$

where we omit arguments of q which have been summed over. By the inductive hypothesis it follows that we can obtain $p(a_1, \ldots, a_p, l)$. It remains to show that we can obtain $p(a_p, a_{p+1}, l)$. This is sufficient since the joint distribution is defined by its marginals over the cliques, see [8] and Figure 3.4. Summing (3.9) over $u_1, \ldots, u_{p-1}$ gives

$$\sum_{a_p, a_{p+1}} p(u_{p+1}|a_{p+1})p(u_p|a_p)p(a_p, a_{p+1}, l) = q(u_p, u_{p+1}, l) \qquad (3.11)$$

Now writing $B(l)$ for the matrix with elements $p(a_p, a_{p+1}, l)$ and $Q(l)$ for the matrix $q(u_p, u_{p+1}, l)$ (3.11) can be written as the matrix equation

$$A_p B(l) A_{p+1} = Q(l)$$

Where A_p and A_{p+1} are square and full rank. Hence the result follows.   □

References

[1] Besag, J. E. Spatial interaction and the statistical analysis of lattice systems (with discussion). Journal of the Royal Statistical Society B, 36, 1974, pp. 192-326.

[2] Besag, J. E. On the statistical analysis of dirty pictures. Journal of the Royal Statistical Society B 48, 1986, pp. 259-302.

[3] Caccetta, P., Campbell, N., West, G., Kiiveri, H., and Gahegan, M. Aspects of reasoning with uncertainty in an agricultural GIS environment. The New Review of Applied Expert Systems 1, 1995, pp. 161-177.

[4] Caccetta, P. Remote Sensing, GIS and Bayesian Knowledge-based Methods for Monitoring Land Condition. PhD thesis, Department of Computer Science, Curtin University of Technology, Western Australia, 1997.

[5] Caccetta, P. C., Campbell, N. A. C., Evans, F., Furby, S. L., Kiiveri, H. T., and Wallace, J. F. (2000). Mapping and monitoring land use and condition change in the south west of Western Australia using remote sensing and other data. In Proceedings of the Europa 2000 Conference, Barcelona.

[6] Campbell, N. A. and Atchley, W. R. (1981), 'The geometry of canonical variate analysis', Syst. Zoology, Vol. 30, No. 3, pp. 268-280.

[7] Subpixel matching using cross correlation and second derivatives. Submitted to ISPRS Journal of Photogrammetry and Remote Sensing.

[8] Darroch, J. N., Lauritzen, S. L. and Speed, T. P. Log-linear models for contingency tables and Markov fields over graphs. Annals of Statistics 8, 1980, pp. 522-539.

[9] Dempster, A. P., Laird, N. M., and Rubin, D. B. Maximum likelihood from incomplete data via the EM algorithm. Journal of the Royal Statistical Society B 39, 1977, pp.1-21.

[10] Furby, S . L., (1994) Discriminating between pasture and barley grass and saltbush using multi-temporal imagery. CMIS technical report.

[11] Furby, S. L. and Campbell (2001), 'Calibrating images from different dates to like value digital counts', Remote Sensing of the Environment, 77, 186-196.

[12] Jensen, F. V. An Introduction to Bayesian Networks. Springer Verlag, New York, 1996.

[13] Kiiveri, H. T. and Caccetta, P. Data fusion, uncertainty and causal probabilistic networks for monitoring the salinisation of farmland. Digital signal processing, 8, 225-230

[14] Kiiveri, H. T. Some statistical models for remotely sensed data. In SISC96 Imaging Interface Workshop Proceedings, 1996.

[15] Lauritzen, S. L., and Spiegelhalter, D. Local computations with probabilities on graphical structures and their application to expert systems. Journal of the Royal Statistical Society B 50, 1988, pp. 157-224

[16] Lauritzen, S. L. Propagation of probabilities, means, and variances in mixed graphical association models. Journal of the American Statistical Association 87, 1992, pp. 1098-1108

[17] Lauritzen, S. L. (1995). 'The EM algorithm for graphical association models with missing data', Computational Statistics and Data Analysis, 19, pp. 191-201.

[18] NASA, (2001). Landsat 7 Science data users handbook. Available on line at http://ltpwww.gsfc.nasa.gov/IAS/handbook/handbook_toc.html.

[19] Rao, C. R. (1966),Linear statistical inference and its applications. Second Edition, Wiley, New York.

4

Traffic Flow on a Freeway Network

Peter Bickel, Chao Chen, Jaimyoung Kwon,
John Rice, Pravin Varaiya,
and Erik van Zwet[1]

4.1 Introduction

Traffic congestion is an unpleasant fact of modern life. Although difficult to quantify precisely, congestion must cost Californians millions of dollars per day. Since further extensive construction of freeways is unlikely, information technology is being increasingly looked to for amelioration by providing information allowing more efficient use of existing freeways. Statistics plays a major role in such efforts.

A large interdisciplinary team of faculty, postdocs, graduate students, and undergraduates at the University of California, Berkeley, has been working on a host of problems of this kind. Researchers come from Computer Science, Electrical Engineering, Statistics, and Transportation Engineering.

This chapter gives an overview of some of our activities, focusing on gathering statistics on traffic flow over the network of freeways in Los Angeles and on the prediction of travel times over this network. The chapter is organized as follows: The freeway system of Los Angeles is equipped with a densely deployed array of sensors, loop detectors, which we describe in the next section. Information from these sensors is captured in real time, displayed, and archived by the Freeway Performance Measurement System, as described in a Section 4.3. In Section 4.4 we describe briefly our attempts to globally model the evolution of the fascinating spatial-temporal field of traffic flow. Ultimately, however, rather than trying to fit and update

---

[1]Peter Bickel, Jaimyoung Kwon, John Rice and Erik van Zwet are with the Department of Statistics, University of California, Berkeley. Chao Chen and Pravin Varaiya are with the Department of Electrical Engineering and Computer Science, University of California, Berkeley.

Figure 4.1. A set of double loop recorders.

such comprehensive models, we found it preferable to use simpler, direct methods. These are described in Section 4.5 for the purpose of predicting the particular functional of interest, travel time. Section 4.6 contains final remarks.

4.2   Loop Detectors

Inductive loop detectors are the basic sensors monitoring the state of a freeway. A detector consists of a wire buried beneath the roadway. An alternating current generates an electromagnetic field, resulting in a change of inductance when an engine passes by on the surface. Such loops are located fairly densely on many freeway systems, with loops in each lane located in banks every half mile or so. Figure 4.1 shows a set of double loop recorders. Data from loops is usually sampled at rates ranging from 30 seconds to five minutes.

The fundamental variables that can be deduced from loops are flow (the number of vehicles per second) and occupancy (the percentage of time that vehicles are over the loops). The latter is essentially the density of vehicles. With assumptions about average vehicle length, these measurements can be converted to average velocity:

$$v(t) = g(t) \times \frac{c(t)}{o(t) \times T}. \tag{4.1}$$

Here $c(t)$ is the flow, $o(t)$ is the occupancy, and $g(t)$ is the effective vehicle length during a time period of duration T. The effective vehicle length depends upon the mix of traffic (trucks and cars) and thus upon the lane and the time of day an also on the electronics of an individual loop. If loops are spaced in nearby pairs, velocity can be measured directly, but single loop detectors are more common.

Because of transmission problems, data from banks of loops are often missing. Furthermore loops may malfunction due to a number of causes including stuck sensors, hanging (on or off), chattering, and cross-talk, the mutual coupling of magnetic fields of two or more neighboring detectors.

4.3   Freeway Performance Measurement

The Freeway Performance Measurement System (PeMS) is an experimental project conducted by researchers at the University of California at Berkeley, with the cooperation of California Department of Transportation. The intent of this project is to collect historical and real-time data from freeways in the State of California, in order to compute freeway performance measures, thus providing managers with a comprehensive assessment of freeway performance. It also provides a wide variety of tools for transportation researchers to examine historical loop detector data. PeMS receives 30 second loop data in real time from Caltrans districts in California; about 1 gigabyte per day comes in from District 7, Los Angeles. From flow and occupancy records of single loop detectors, velocity is derived by using an adaptive algorithm [3] to estimate the effective vehicle length in 4.1 for each loop. Los Angeles has 4,000 detector at 1,300 locations, on 400 conventional highway miles. 70% or more of the detectors are usually functional.

As an example of the kind of information available from PeMS, consider the following schematic representation of the classic traffic theory of the relationship of flow, density, and velocity, Figure 4.2. At low values of occupancy, traffic flows freely, as shown by the arrow emanating from the origin, and during this phase we have constant velocity (which is proportional to the slope of the line joining the point to the origin). Beyond some point of maximum efficiency, occupancy is sufficiently high so that the velocity decreases, and flow, the throughput of the system, decreases. Empirical versions of this diagram can be constructed from PeMS loop data as in Figure 4.2. Maximal efficiency of the system depends on keeping occupancy below the critical level, and this is the central aim of ramp metering.

PeMS can be interactively queried via a web browser.[2] A map of the entire freeway network can be displayed and updated every five minutes, Figure 4.3. These maps can also be played back in an animation, providing a vivid visualization of the propagation and dissipation of congestion.

In this chapter , we focus particularly on one aspect of PeMS: travel time prediction. We describe the statistical methodology underlying an interface

---

[2]http://transacct.eecs.berkeley.edu

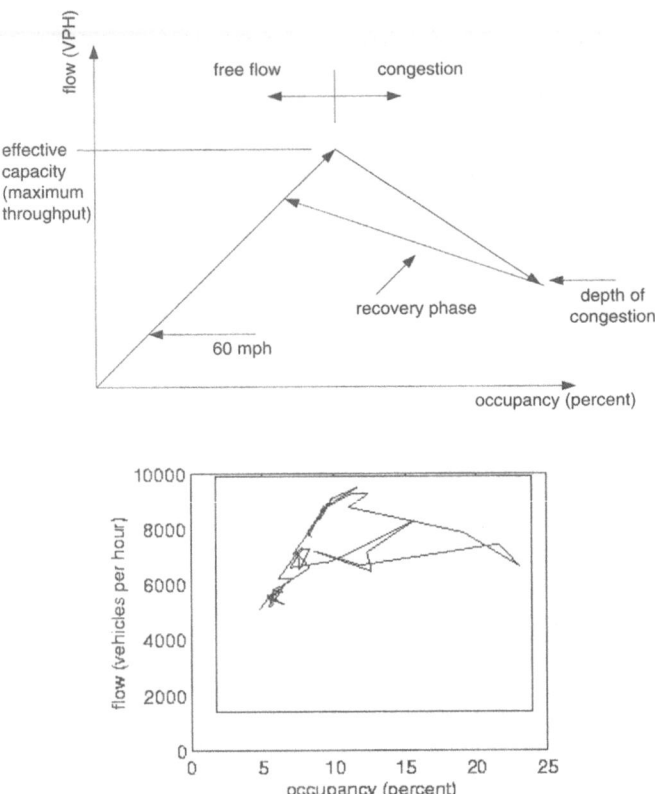

Figure 4.2. The relationship between flow an occupancy. Top: Classical flow-occupancy diagram. Bottom: Empirical values of flow and occupancy at a particular loop.

which allows a user to query the system for estimated travel times between locations selected by mouse-clicks, leaving at arbitrary times in the future. PeMS is also working on user interfaces via cell phones and via direct voice inquiry. In the not-too-distant future, continuously updated information on the state of the entire freeway system will be available to drivers as they negotiate it.

4.4   Global Models

Global models are inspired by the pattern of onset, propagation and dissipation of traffic congestion seen in Figures 4.4. Note in particular the characteristic wedge shapes reflecting the course of congestion from onset to dissipation. There are various theories [8] that try to explain such traffic

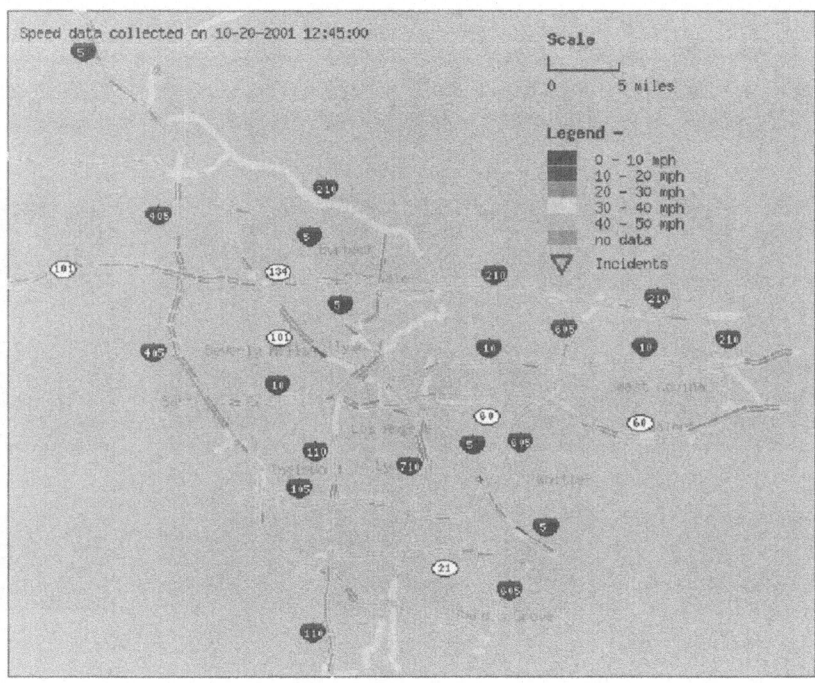

Figure 4.3. The Los Angeles freeways as currently displayed at the PeMS website.

dynamics, based on models of fluid flow, cellular automata, and microscopic computer simulations among others. Using such models for control and prediction is non-trivial, due to the complexity of the phenomenon.

4.4.1 A Coupled Hidden Markov Model

The model we consider, a coupled hidden Markov model (CHMM), is proposed as a phenomenological model for how the macroscopic dynamics of the freeway traffic arise from local interactions. This model views the velocity at each location as a noisy representation of the underlying binary traffic status (free flow or congestion) at that location and assumes the unobserved state vector is a Markov chain with a special structure of local dependencies. Such binary state assumption is justified from the clear distinction between free flow and congestion regime shown in Figure 4.5, although a richer state space can be incorporated.

Consider a fixed day, d. Given all other variables, the observed velocity $y_{l,t}$ (mph) at location l ($l = 1, ..., L$, from upstream to downstream) and time t ($= 1, ..., T$) has a distribution depends only on the underlying state variable $x_{l,t} \in \{0, 1\}$. The two states 0 and 1 correspond to 'congestion' and 'free flow' each, by setting $E(Y_{l,t}|X_{l,t} = 0) < E(Y_{l,t}|X_{l,t} = 1)$. In particular, assume that the observed velocity is Gaussian whose mean and

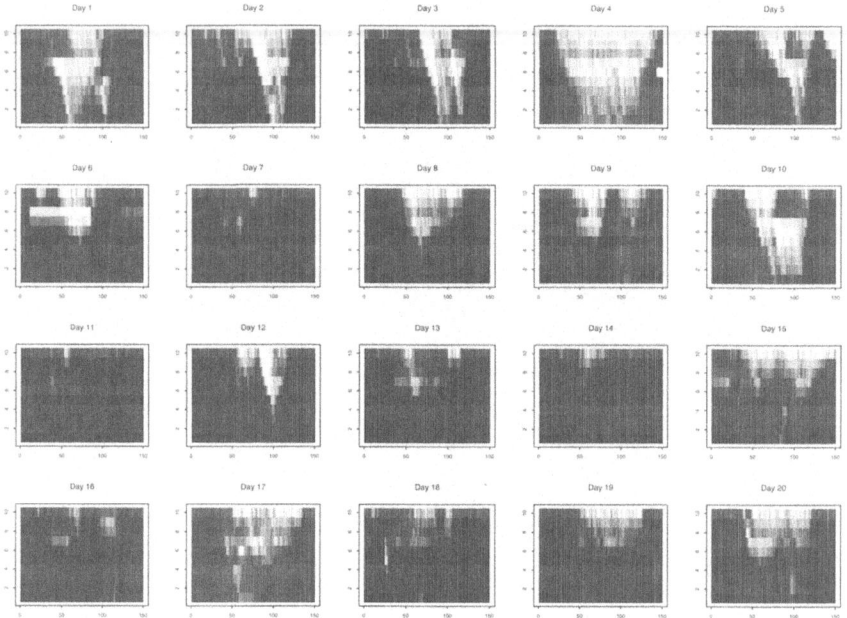

Figure 4.4. The velocity field for 20 weekdays, 2-7pm, between February 22 and March 19, 1993. The measurements come from 10 loop detectors (0.6 miles apart) in the middle lane of I-880 near Hayward, California, with 1 measurement per 2 minutes. The x-axis corresponds to time, and the y-axis to space. Vehicles travel upward in this diagram. The darkest gray-scale corresponds to the average velocity of 20 miles per hour (mph) and the lightest to 70 mph. The horizontally stretched bright blob on the left of the sixth day is due to a sensor failure.

variance depends on the underlying state at the location, or

$$P_\lambda(y_{l,t}|x_{l,t} = s) \sim N(\mu_l^{(s)}, \sigma_l^{(s)2}), \quad s = 0, 1, \tag{4.2}$$

where $\lambda = (\mu_l^{(s)}, \sigma_l^{(s)2}, s = 0, 1)$ are parameters for the emission probability.

We assume the hidden process of $x_t = (x_{1,t}, \cdots, x_{L,t}) \in \{0,1\}^{\otimes 2}$ is not only Markovian, i.e. $P(x_{t+1}|x_1, \cdots, X_t) = P(x_{t+1}|x_t)$, but also its transition probability allows the following decomposition.

$$P(x_{t+1}|x_t) = \prod_{l=1}^{L} P(x_{l,t+1}|x_t) = \prod_{l=1}^{L} P_\phi(x_{l,t+1}|x_{l-1,t}, x_{l,t}, x_{l+1,t}), \tag{4.3}$$

where ϕ specifies the transition probability and initial probability. This implies that the traffic condition at a location at time $t + 1$ is affected only by the conditions at the neighboring locations at time t. This local decomposability assumption in space-time should be reasonable for certain spatial and temporal scales.

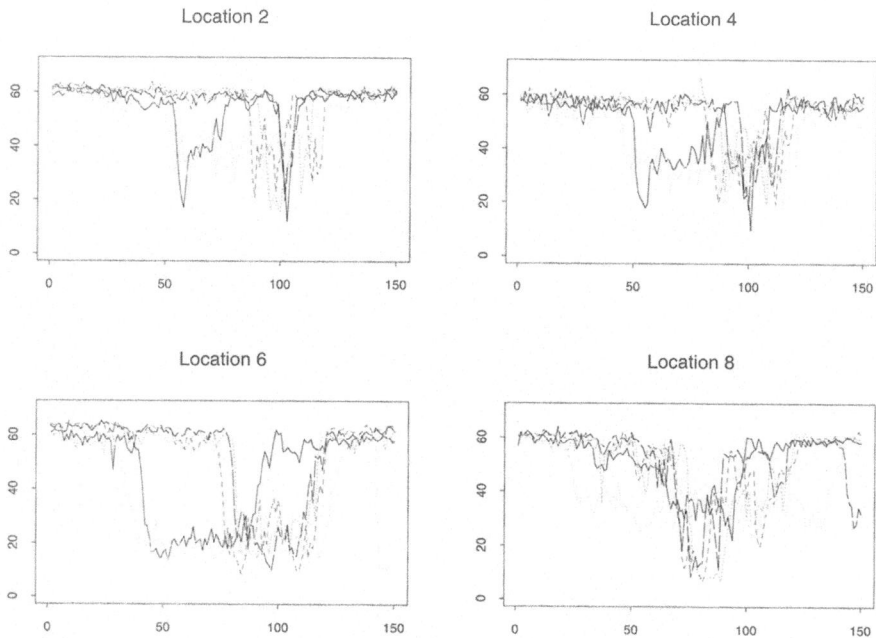

Figure 4.5. The velocity measurements for the first 5 days of Figure 4.4 at locations 2, 4, 6 and 8. The x-axis is in units of 2-minutes (0 to 150), the y-axis is mph and each line corresponds to a different day.

Then the complete likelihood of (x, y) is

$$P_\theta(x, y) = P_\phi(x)P_\lambda(y|x) = \prod_{i=1}^{n} P_\phi(x_{t+1}|x_t)P_\lambda(y_t|x_t).$$

where $\theta = (\phi, \lambda)$ is the parameter to estimate, and the likelihood of y is $P_\theta(y) = \sum_x P_\theta(x, y)$. This is the model for a single day, say d, and the extension to multiple days is apparent by viewing each day as an iid realization of this model. This is a hidden Markov model with a special structure and is called a *Coupled hidden Markov model* (CHMM; see [9]). See Figure 4.6 for the graphical model representation of the CHMM.

4.4.2 Computation

For hidden Markov models, the parameters are usually estimated by the maximum likelihood method via the expectation-maximization (EM) algorithm which tries to find a local maximum of the likelihood. Even though the CHMM has moderately large number (of order $O(L)$) parameters, the model is still hard to fit because of the dimensionality. In particular, the E-step is in general computationally intractable while the M-step is

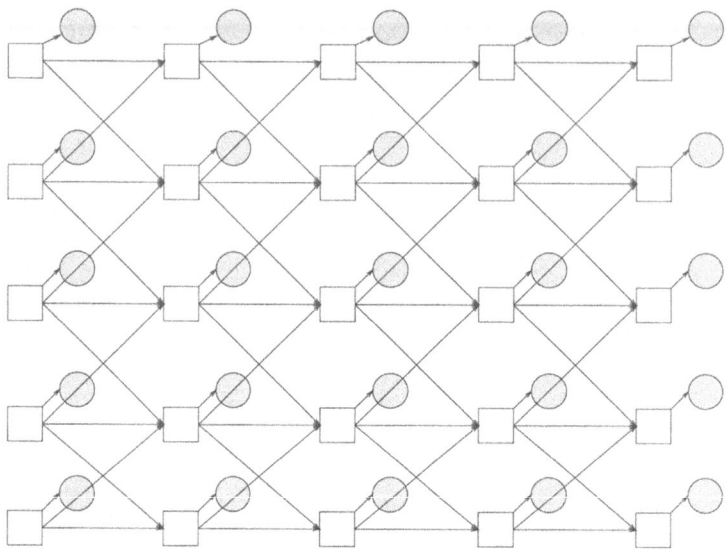

Figure 4.6. A coupled hidden Markov model represented as a dynamic Bayesian network. Square nodes represent discrete random variables (rv's) with multinomial distributions, round nodes represent continuous rv's with Gaussian distributions. Clear nodes are hidden, shaded nodes are observed. Here we show $L = 5$ chains and $T = 5$ timeslices.

straightforward. Sequential importance sampling with resampling (SISR; [6]) has been tried, leading to Monte Carlo EM (MC-EM) estimate of θ in which the exact E-step is replaced by an E-step approximated by the Monte Carlo sample from SISR. An alternative computational scheme, iterated conditional modes (ICM; [1]) has also been tried. For details about the computations, see [5] and [4].

4.4.3 Results and Conclusions

The approaches were applied to data collected from the I-880 freeway shown and explained in Figure 4.4. The ICM algorithm ran much faster than SISR (with 100 Monte Carlo samples), single iterations taking 1.63 and 417.73 seconds respectively on a 400MHz PC. Both algorithm seemed to stabilize after a few iterations, though the ICM algorithm, which is non-stochastic, seemed to fluctuate less. The latter however poses some conceptual difficulties.

Figure 4.7 shows the final estimates of μ and σ for the ten locations from the algorithms. Parameter estimates from the two algorithms are quite similar with each other. We can also observe that: (1)While mean free flow speed is similar (about 60 mph) for all locations, mean congestion speed

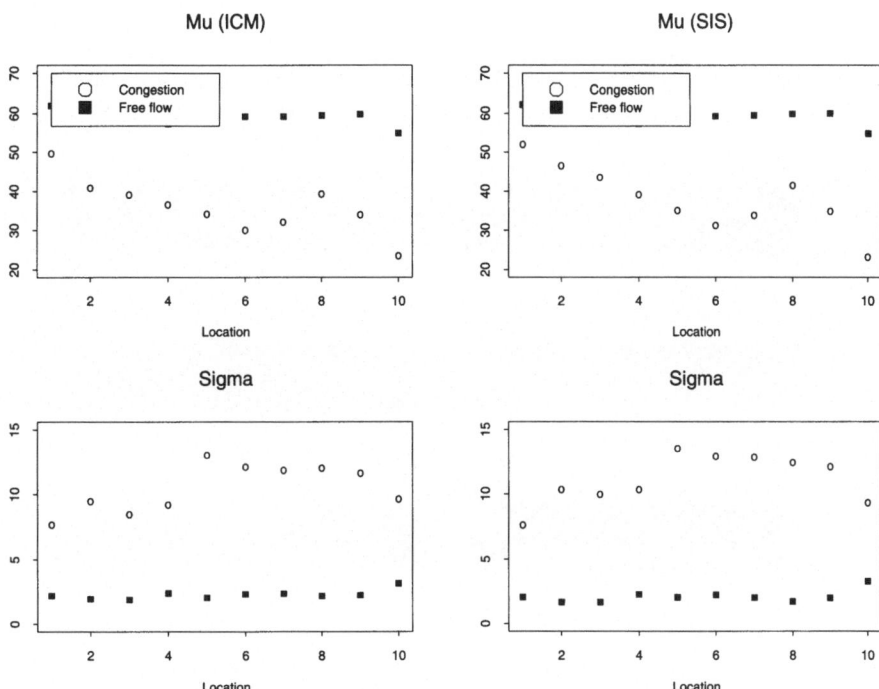

Figure 4.7. Conditional means and standard deviations of vehicle velocities for 10 locations estimated by the two algorithms.

varies greatly over locations, ranging from 20 to 50 mph. Location 8 where the mean congestion speed is the smallest, corresponds to the San Mateo Bridge, a notorious congestion spot, and (2) Standard errors of vehicle velocity are much larger for the congestion period than for the free flow period.

We simulated (binary) Markov chains using the parameters estimated by the algorithms to see qualitatively how well the fitted model reproduces the traffic dynamics. We can observe that inverted triangular regions of congestion are reproduced to some extent as bright patches in simulations from the SIS-estimated parameters, shown in Figure 4.8. Simulations using ICM results show similar behavior. Our initial hope for the CHMM model was to capture the behavior of the congestion and free flow regime and to reproduce pattern of propagation and dissipation of congestion. These goals have been achieved to a certain degree as illustrated above.

Given the success of such a global model, one might hope that it would prove useful in predicting future traffic patterns, including travel times which are only one of many possible functions of the predicted velocity vector. However, preliminary work on using the model for short term (5-30 minutes in the future) prediction of the velocity vector and travel time,

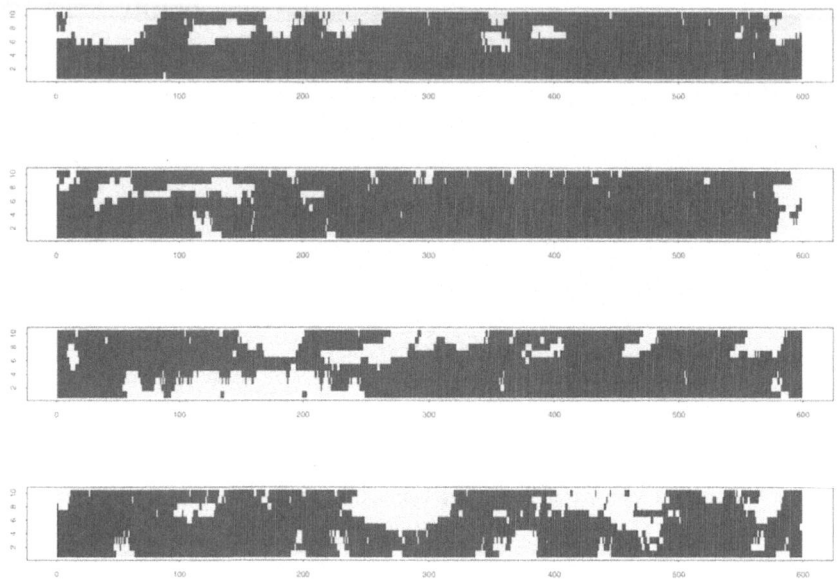

Figure 4.8. Four X fields simulated using the ϕ parameters estimated by the MC-EM algorithm using SIS. Light patches correspond to congestion and dark ones to free flow.

produced disappointing initial results, comparable only to a naive predictor like the current velocity vector/travel time. This performance for prediction may be due to many possible reasons including: (1) The model does not capture the dynamics of incident propagation/dissipation in full; (2) The model fails to accommodate certain important features of the dynamics like a time-of-day factor, which would requires an inhomogeneous hidden Markov chain, (3) Having too many parameters, the model is subject to overfit, and (4) The model may be inherently too complicated and general for the specific task of travel time prediction.

To sum up, even though the proposed global model captures some interesting characteristic of the macroscopic dynamics of the freeway traffic, it is not very useful for travel time prediction. Further work is required to show whether it can be improved to outperform naive predictors or whether such a limitation is inherent in this kind of a global model.

4.5   Travel Time Prediction

In this section we state the exact nature of our prediction problem. We then describe our prediction method and two alternative methods which will be used for comparison. This comparison is made in Section 4.5.4 with a collection of 34 days of traffic data from a 48 mile stretch of I-10 East in Los Angeles, CA. Finally, in Section 4.5.5, we summarize our conclusions, point out some practical observations and briefly discuss several extensions of our new method.

The data available for prediction can be represented as a matrix V with entries $V(d, l, t)$ $(d \in D,\ l \in L,\ t \in T)$ denoting the velocity that was measured on day d at loop l at time t. From V we can compute travel times $X_d(a, b, t)$, for all $d \in D$, $a, b \in L$ and $t \in T$. This travel time is to approximate the time it took to travel from loop a to loop b starting at time t on day d.

Suppose we have observed $V(d, l, t)$ for a number of days $d \in D$ in the past. Suppose a new day e has begun and we have observed $V(e, l, t)$ at times $t \leq \tau$. We call τ the 'current time'. Our aim is to predict $X_e(a, b, \tau+\delta)$ for a given (nonnegative) 'lag' δ. This is the time a trip that departs from a at time $\tau + \delta$ will take to reach b. Note that even for $\delta = 0$ this is not trivial. In forming the prediction we have historical data on travel times and for the trip to be predicted we have data up to time τ. The predictor is some function of $V(d, l, t)$, $t \leq \tau$: the problem is to select such a function from this very high dimensional space.

We can compute a proxy for these travel times which is defined by

$$X_d^*(a, b, t) = \sum_{i=a}^{b-1} \frac{2d_i}{V(d, i, t) + V(d, i+1, t)}, \qquad (4.4)$$

where d_i denotes the distance from loop i to loop $(i + 1)$. We call X^* the current status travel time (a.k.a. the snap-shot or frozen field travel time). It is the travel time that would have resulted from departure from loop a at time t on day d when no significant changes in traffic occurred until loop b was reached. It is important to notice that $X_d^*(a, b, t)$ can be computes at time t, whereas computation of $X_d(a, b, t)$ requires information of later times.

We fix an origin and destination of our travels and drop the arguments a and b from our notation. Define the historical mean travel time as

$$\mu(t) = \frac{1}{|D|} \sum_{d \in D} X_d(t). \qquad (4.5)$$

Two naive predictors of $X_e(\tau+\delta)$ are $X_e^*(\tau)$ and $\mu(\tau+\delta)$. We expect—and indeed this is confirmed by experiment—that $X_e^*(\tau)$ predicts well for small δ and $\mu(\tau+\delta)$ predicts better for large δ. We aim to improve on both these predictors for all δ.

4.5.1   Linear Regression with Time Varying Coefficients

Our main result is the discovery of an empirical fact: that there exist linear relationships between $X^*(t)$ and $X(t + \delta)$ for all t and δ. This empirical finding has held up in all of numerous freeway segments in California that we have examined. This relation is illustrated by Figure 4.9, which shows scatter plots of $X^*(t)$ versus $X(t + \delta)$ for a 48 mile stretch of I-10 East in Los Angeles. Note that the relation varies with the choice of t and δ. With this in mind we propose the following model

$$X(t + \delta) = \alpha(t, \delta) + \beta(t, \delta)X^*(t) + \varepsilon. \tag{4.6}$$

where ε is a zero mean random variable modeling random fluctuations and measurement errors. Note that the parameters α and β are allowed to vary with t and δ. Linear models with varying parameters are discussed in [2].

Fitting the model to our data is a familiar linear regression problem which we solve by weighted least squares. Define the pair $(\hat{\alpha}(t, \delta), (\hat{\beta}(t, \delta))$ to minimize

$$\sum_{\substack{d \in D \\ s \in T}} (X_d(s) - \alpha(t, \delta) - \beta(t, \delta)X_d^*(t))^2 K(t + \delta - s), \tag{4.7}$$

where K denotes the Gaussian density with mean zero and a certain variance which must be specified. The purpose of this weight function is to impose smoothness on α and β as functions of t and δ. We assume that α and β are smooth in t and δ because we expect that average properties of the traffic do not change abruptly. The actual prediction of $X_e(\tau + \delta)$ becomes

$$\widehat{X}_e^{\alpha\beta}(\tau + \delta) = \hat{\alpha}(\tau, \delta) + \hat{\beta}(\tau, \delta)X_e^*(\tau). \tag{4.8}$$

Writing $\alpha(t, \delta) = \alpha'(t, \delta) \times \mu(t + \delta)$ we see that (4.6) expresses a future travel time as a linear combination of the historical mean and the current status travel time—our two naive predictors. Hence our new predictor may be interpreted as the best linear combination of our naive predictors. From this point of view, we can expect our predictor to do better than both. In fact, it does, as is demonstrated in Section 4.5.4.

Another way to think about (4.6) is by remembering that the word "regression" arose from the phrase "regression to the mean." In our context, we would expect that if X^* is much larger than average—signifying severe congestion—then congestion will probably ease during the course of the trip. On the other hand, if X^* is much smaller than average, congestion is unusually light and the situation will probably worsen during the journey.

Besides comparing our predictor to the historical mean and the current status travel time, we subject it to a more competitive test. We consider two other predictors that may be expected to do well—one resulting from principal component analysis and one from the nearest neighbors principle. Next, we describe these two methods.

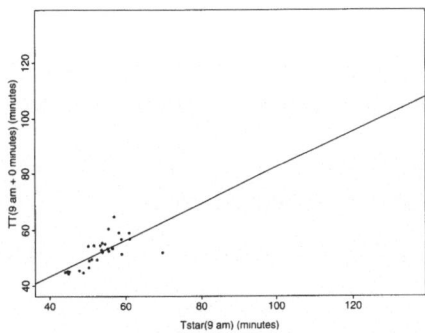 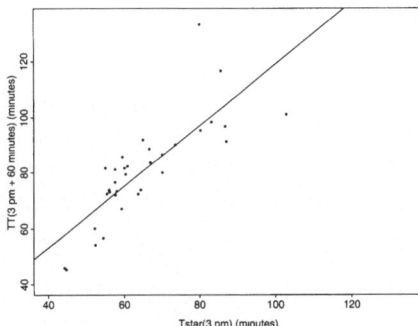

(a) $X(9$ am $+$ 0 min's$)$ vs. $X^*(9$ am$)$. Also shown is the regression line with slope α (9 am, 0 min)=0.65 and intercept β (9 am, 0 min)=17.3.

(b) $X(3$ pm $+$ 60 min$)$ vs. $X^*(3$ pm$)$). Also shown is the regression line with slope α (3 pm, 60 min)=1.1 and intercept β (3 pm, 60 min)=9.5.

Figure 4.9. Scatter plots of actual travel times, $X(t)$, versus "frozen field" travel times, $X^*(t)$.

4.5.2   Principal Components

Our predictor $\widehat{X}^{\alpha\beta}$ only uses information at one time point; the 'current time' τ. However, we do have information prior to that time. The following method attempts to exploit this by using the entire trajectories of X_e and X_e^* which are known at time τ.

Formally, let us assume that the travel times on different days are independently and identically distributed and that $\{X_d(t) : t \in T\}$ and $\{X_d^*(t) : t \in T\}$ are jointly multivariate normal. We estimate the covariance of this multivariate normal distribution by retaining only a few of the largest eigenvalues in the singular value decomposition of the empirical covariance of $\{(X_d(t), X_d^*(t)) : d \in D, \ t \in T\}$. We have experimented informally with the number of eigenvalues to retain, but one could also use cross-validation. Define τ' to be the largest t such that $t + X_e(t) \leq \tau$. That is, τ' is the latest trip that we have seen completed before time τ. With the estimated covariance we can now compute the conditional expectation of $X_e(\tau + \delta)$ given $\{X_e(t) : t \leq \tau'\}$ and $\{X_e^*(t) : t \leq \tau\}$. This is a standard computation which is described, for instance, in [7]. The resulting predictor is $\widehat{X}_e^{\mathrm{PC}}(\tau + \delta)$.

4.5.3   Nearest Neighbors

As an alternative to principal components, we now consider nearest neighbors, which is also an attempt to use information prior to the current time

τ. This method is nonlinear and makes fewer assumptions (such as joint normality) on the relation between X^* and X.

The method of nearest neighbors aims to find those days in the past which are most similar to the present day in some appropriate sense. The remainder of those past days beyond time τ are then used to form a predictor of the remainder of the present day.

The critical choice with nearest neighbors is in specifying a suitable distance m between days. We suggest two possible distances:

$$m(e, d) = \sum_{l \in L, \ t \leq \tau} |V(e, l, t) - V(d, l, t)| \tag{4.9}$$

and

$$m(e, d) = \left(\sum_{t \leq \tau} (X_e^*(t) - X_d^*(t))^2 \right)^{1/2}. \tag{4.10}$$

Now, if day d' minimizes the distance to e among all $d \in D$, our prediction is

$$\widehat{X}_e^{NN}(\tau + \delta) = X_{d'}(\tau + \delta). \tag{4.11}$$

Sensible modifications of the method are 'windowed' nearest neighbors and k-nearest neighbors. Windowed-NN recognizes that not all information prior to τ is equally relevant. Choosing a 'window size' w it takes the above summation to range over all t between $\tau - w$ and τ. So-called k-NN is basically a smoothing method, aimed at using more information than is present in just the single closest match. For some value of k, it finds the k closest days in D and bases a prediction on a (possibly weighted) combination of these. Alas, neither of these variants appear to significantly improve on the 'vanilla' $\widehat{X}^{NN}$.

4.5.4   Results

We gathered flow and occupancy data from 116 single loop detectors along 48 miles of I-10 East in Los Angeles. Measurements were done at 5 minute aggregation at times t ranging from 5 am to 9 pm for 34 weekdays between June 16 and September 8 2000. We used the method described in [3] to convert flow and occupancy to velocity. Fortunately, the quality of our I-10 data is quite good and we used simple interpolation to impute wrong or missing values. The resulting velocity field $V(d, l, t)$ is shown in Figure 4.10 where day d is June 16. The horizontal streaks typically indicate detector malfunction.

From the velocities we computed travel times for trips starting between 5 am and 8 pm. Figure 4.11 shows these $X_d(t)$ where time of day t is on the horizontal axis. Note the distinctive morning and afternoon congestions and the huge variability of travel times, especially during those periods.

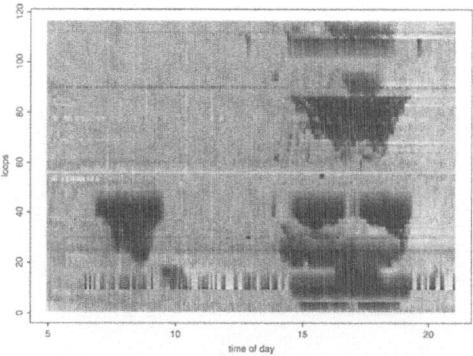

Figure 4.10. Velocity field $V(d, l, t)$ where day $d =$ June 16, 2000. Darker shades indicate lower speeds. Note the typical triangular shapes indicating the morning and afternoon congestions building and easing. The horizontal streaks are most likely due to detector malfunction.

During afternoon rush hour we find travel times ranging from 45 minutes to up to two hours. Included in the data are holidays July 3 and 4 which may readily be recognized by their very fast travel times.

We have estimated the root mean squared error of our various prediction methods for a number of 'current times' τ ($\tau =$6am, 7am,...,7pm) and lags δ ($\delta =$0 and 60 minutes). We evaluated the estimation error by leaving out one day at a time, performing the prediction for that day on the basis of the remaining other days and averaging the squared prediction errors.

The prediction methods all have parameters that must be specified. For the regression method we have chosen the standard deviation of the Gaussian kernel K to be 10 minutes. For the principal components method we have chosen the number of eigenvalues retained to be 4. For the nearest

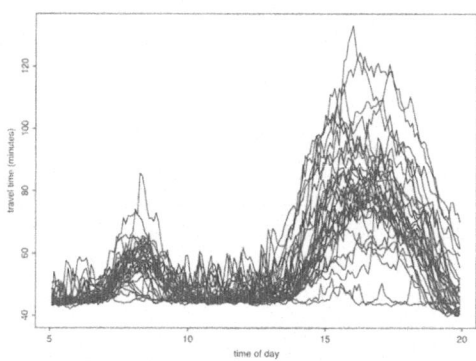

Figure 4.11. Travel Times $X_d(\cdot)$ for 34 days on a 48 mile stretch of I-10 East.

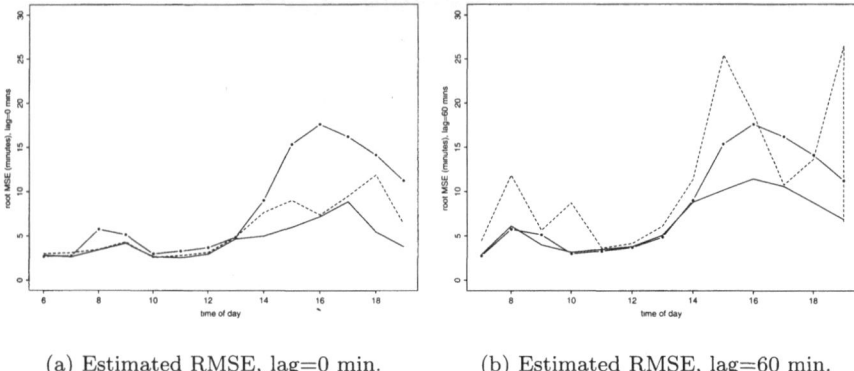

(a) Estimated RMSE, lag=0 min.   (b) Estimated RMSE, lag=60 min.

Figure 4.12. Comparison of Historical mean $(- \cdot -)$, current status $(- - -)$ and linear regression $(—)$.

neighbors method we have chosen distance function (4.10), a window w of 20 minutes and the number k of nearest neighbors to be 2.

Figure 4.12 shows the estimated root mean squared (RMS) prediction error of the historical mean $\mu(\tau + \delta)$, the current status predictor $X_e^*(\tau)$ and our regression predictor (4.8) for lag δ equal to 0 and 60 minutes, respectively. Note how $X_e^*(\tau)$ performs well for small δ $(\delta = 0)$ and how the historical mean does not become worse as δ increases. Most importantly, however, notice how the regression predictor beats both hands down, except during the time period of 10am-1pm.

Figure 4.13 again shows the RMS prediction error of the regression estimator. Here it is compared to the principal components predictor and the nearest neighbors predictor (4.11). Again, the regression predictor comes out on top, although the nearest neighbors predictor shows comparable performance.

The RMS error of the regression predictor stays below 10 minutes even when predicting an hour ahead. We feel that this is impressive for a trip of 48 miles right through the heart of L.A. during rush hour.

4.5.5   Conclusions and loose ends

We stated that the main contribution towards predicting travel times is the discovery of a linear relation between $X^*(t)$ and $X(t + \delta)$. But there is more. Comparison of the regression predictor to the principal components and nearest neighbors predictors unearthed another surprise. Given $X^*(\tau)$, there is not much information left in the earlier $X^*(t)$ $(t < \tau)$ that is useful for predicting $X(\tau + \delta)$. Some earlier attempts [5] at prediction using more complex models also turned out to be inferior to the regression method. In fact, we have come to believe that for the purpose of predicting travel

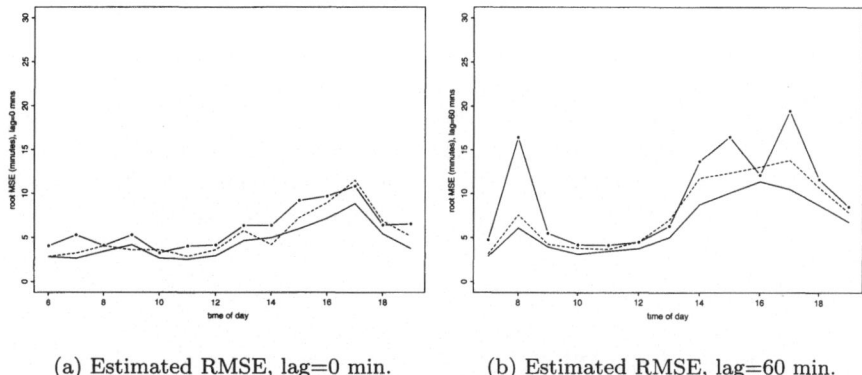

(a) Estimated RMSE, lag=0 min.   (b) Estimated RMSE, lag=60 min.

Figure 4.13. Comparison of Principal Components ($- \cdot -$), nearest neighbors (- - -) and linear regression (—).

times all the information in $\{V(l,t),\ l \in L,\ t \le \tau\}$ is well summarized by one single number: $X^*(\tau)$.

It is of practical importance to note that our prediction can be performed in real time. Computation of the parameters $\hat{\alpha}$ and $\hat{\beta}$ is time consuming but it can be done off-line in reasonable time. The actual prediction is trivial. In addition to making predictions available on the internet, it would also be possible to make them available for users of cellular telephones—and in fact we plan to do so in the near future.

It is also important to notice that our method does not rely on any particular form of data. In this chapter we have used single loop detectors, but probe vehicles or video data can be used in place of loops, since all the method requires is current measurements of X^* and historical measurements of X and X^*. It is straightforward to make the method robust to outliers by replacing the least squares refleast-squares criterion by a robust one.

We conclude this chapter by briefly pointing out two extensions of our prediction method.

1. For trips from a to c *via* b we have

$$X_d(a,c,t) = X_d(a,b,t) + X_d(b,c,t + X_d(a,b,t)). \qquad (4.12)$$

We have found that it is sometimes more practical or advantageous to predict the terms on the right hand side than to predict $X_d(a,c,t)$ directly. For instance, when predicting travel times across networks (graphs), we need only predict travel times for the edges and then use (4.12) to piece these together to obtain predictions for arbitrary routes. This is precisely what is done in forming a prediction for a trip over the LA freeway network. In this way, a complex non-linear predictor is formed by composition of simpler linear ones.

2. We regressed the travel time $X_d(t + \delta)$ on the current status $X_d^*(t)$, where $X_d(t+\delta)$ is the travel time departing at time $t+\delta$. Now, define $Y_d(t)$ to be the travel time *arriving* at time t on day d. Regressing $Y_d(t+\delta)$ on $X_d^*(t)$ will allow us to make predictions on the travel time subject to *arrival* at time $t + \delta$. The user can thus ask what time he or she should depart in order to reach his or her intended destination at a desired time.

4.6   Final Remarks

The complexity of traffic flow over the Los Angeles network is bewitching. In some intriguing fashion, microscopic interactions give rise to global patterns. Tempting as it is to model this process, it is not at all clear *a priori* that doing so is the best way to achieve accurate prediction of a particular functional, travel time in our case. Indeed, simpler, empirical methods such as those we have discussed in this chapter appear to be more effective.

Our results are hardly free of blemishes and warts. Probably the greatest problem we face is the variable quantity and quality of data from loop detectors. At any given time 20% or so of the loops may not report at all. Some of those that do report give erroneous results. This poses fascinating challenges for development of statistical methodology: stated abstractly, we have a large array of often faulty or non-reporting sensors and we wish to reconstruct the random field that is driving them. We are actively working on this problem, trying to take advantage of the high correlations between nearby sensors induced by the fact that they are measuring related aspects of a common random environment. A second problem we face is the lack of ground truth in District 12. We have tested our prediction method on a smaller, higher quality data set that includes dense coverage by probe vehicles, with good results [10].

The segmentation of trips into journeys over edges of the freeway network poses some interesting problems. First, there is the question of how to segment. We have taken taken the pragmatic approach of choosing to segment using nodes formed by the major freeway intersections, but there are certainly other possibilities, given that there are loops every half mile or so. Second, there are issue of internal consistency. Consider a trip from point A to point C passing through point B. We could use the regression method to predict the travel time from A to B and then the subsequent time from B to C, or we could use the method to directly predict the travel time from A to C. There is no guarantee that the two predictions will be identical. Adding to this the possibility of using the regression method to predict backwards in time, predicting when to leave A in order to arrive at C at a desired time, and there is no guarantee that we have not created wormholes in the space-time fabric of the Los Angeles freeway network.

Our group is engaged in other interesting activities. Foremost among these is a project in which an array of video cameras will survey a stretch of freeway near the San Francisco Bay Bridge in order to study the behavior of individual drivers and the microscopic causes of congestion. The first problem encountered in this effort is identifying vehicles on the videos in order to extract their trajectories—a non-trivial challenge in computer vision. Beyond that is the challenge of formulating effective statistical procedures to study the large quantity of data from measurements of this complex spatio-temporal stochastic process.

References

[1] J. Besag. Statistical analysis of non-lattice data. *The Statistician*, 24:179–195, 1975.

[2] T. Hastie and R. Tibshirani. Varying coefficient models. *Journal of the Royal Statistical Society Series B*, 55(4):757–796, 1993.

[3] Z. Jia, C. Chen, B. Coiffman, and P. Varaiya. The PeMS algorithms for accurate, real-time estimates of g-factors and speeds from single-loop detectors. In *Fourth International IEEE Conference on Intelligent Transportation Systems*.

[4] J. Kwon and K. Murphy. Modeling freeway traffic with coupled HMMs. 2000.

[5] Jaimyoung Kwon, B. Coifman, and Peter J. Bickel. Day-to-day travel time trends and travel time prediction from loop detector data. In *Transportation Research Record*, 2000. Accepted for publication.

[6] J. S. Liu and R. Chen. Sequential monte carlo methods for dynamic systems. *Journal of the American Statistical Association*, 93(443):1032–1044, 1998.

[7] K. V. Mardia, J. T. Kent, and S. M. Bibby. *Multivariate Analysis*. Academic Press, 1979.

[8] A. D. May. *Traffic Flow Fundamentals*. Prentice-Hall, 1990.

[9] L. K. Saul and M. Jordan. Boltzman chains and hidden markov models. In G. Tesauro, D. S. Touretzky, and T. Leen, editors, *Advances in Neural Information Processing Systems*. MIT Press, 1995.

[10] X. Zhang and J. Rice. Short term travel time prediction. *Transportation Research C*, 2002. to appear.

5

Internet Traffic Tends *Toward* Poisson and Independent as the Load Increases

Jin Cao, William S. Cleveland, Dong Lin, and Don X. Sun[1]

Summary

Network devices put packets on an Internet link, and multiplex, or superpose, the packets from different active connections.

Extensive empirical and theoretical studies of packet traffic variables — arrivals, sizes, and packet counts — demonstrate that the number of active connections has a dramatic effect on traffic characteristics. At low connection loads on an uncongested link — that is, with little or no queueing on the link-input router — the traffic variables are long-range dependent, creating burstiness: large variation in the traffic bit rate. As the load increases, the laws of superposition of marked point processes push the arrivals toward Poisson, the sizes toward independence, and reduces the variability of the counts relative to the mean. This begins a reduction in the burstiness; in network parlance, there are multiplexing gains.

Once the connection load is sufficiently large, the network begins pushing back on the attraction to Poisson and independence by causing queueing on the link-input router. But if the link speed is high enough, the traffic can get quite close to Poisson and independence before the push-back begins in force; while some of the statistical properties are changed in this high-speed case, the push-back does not resurrect the burstiness. These results reverse the commonly-held presumption that Internet traffic is everywhere bursty and that multiplexing gains do not occur.

Very simple statistical time series models — fractional sum-difference (FSD) models — describe the statistical variability of the traffic variables and their change toward Poisson and indepen-

---

[1] Jin Cao and Dong Lin are Members of the Technical Staff, Bell Laboratories, Lucent Technologies, Murray Hill, NJ. William S. Cleveland is Distinguished Member of the Technical Staff, Bell Laboratories, Lucent Technologies, Murray Hill, NJ. Don X. Sun is a Quantitative Strategist with Deephaven Capital Management.

dence before significant queueing sets in, and can be used to generate open-loop packet arrivals and sizes for simulation studies.

Both science and engineering are affected. The magnitude of multiplexing needs to become part of the fundamental scientific framework that guides the study of Internet traffic. The engineering of Internet devices and Internet networks needs to reflect the multiplexing gains.

5.1   Are There Multiplexing Gains?

When two hosts communicate over the Internet — for example, when a PC and a Web server communicate for the purpose of sending a Web page from the server to the PC — the two hosts set up a *connection*. One or more files are broken up into pieces, headers are added to the pieces to form packets, and the two hosts send packets to one another across the Internet. When the transfer is completed, the connection ends.

The headers, typically 40 bytes in size, contain much information about the packets such as their source, destination, size, etc. In addition there are 40-byte control packets, all header and no file data, that transfer information form one host to the other about the state of the connection. The maximum amount of file information allowed in a packet is 1460 bytes, so packets vary in size from 40 bytes to 1500 bytes.

Each packet travels across the Internet on a path made up of devices and transmission links between these devices. The devices are the two hosts at the ends and routers in-between. Each device sends the packet across a transmission link to the next device on the path. The physical medium, or the "wire", for a link might be a telephone wire from a home to a telephone company, or a coaxial cable in a university building, or a piece of fiber connecting two devices on the network of an Internet service provider. So each link has two devices, the sending device that puts the bits of the packet on the link, and the receiving device, which receives the bits. Each router serves as a receiving device for one or more input links and as the sending device for one or more output links; it receives the packet on an input link, reads the header to determine the output link, and sends bits of the packet.

Each link has a speed: the rate at which the bits are put on the wire by the sending device and received by the receiving device. Units are typically kilobits/sec (kbps), megabits/sec (mbps), or gigabits/sec (gbps). Typical speeds are 56 kbps, 1.5 mbps, 10 mbps, 100 mbps, 156 mbps, 622 mbps, 1 gbps, and 2.5 gbps. The *transmission time* of a packet on a link is the time it takes to put all of the bits of the packet on the link. For example, the transmission time for a 1500 byte (12000 bit) packet is 120 μs at 100 mbps and 12 μs at 1 gbps. So packets pass more quickly on a higher-speed link than on a lower-speed one.

The packet traffic on a link can be modeled as a marked point process. The arrival times of the process are the arrival times of the packets on the link; a packet arrives at the moment its first bit appears on the link. The marks of the process are the packet sizes. An Internet link typically carries the packets of many active connections between pairs of hosts. The packets of the different connections are intermingled on the link; for example, if there are three active connections, the arrival order of 10 consecutive packets by connection number might be 1, 1, 2, 3, 1, 1, 3, 3, 2, and 3. This intermingling is referred to as "statistical multiplexing" in the Internet engineering literature, and as "superposition" in the literature of point processes.

If a link's sending device cannot put a packet the link because it is busy with one or more other packets that arrived earlier, then the device puts the packet in a queue, physically, a buffer. Queueing on the device delays packets, and if it gets bad enough, and the buffer size is exceeded, packets are dropped. This reduces the quality of Internet applications such as Web page downloads and streaming video. Consider a specific link. Queueing of the packet in the buffer of the link's sending device is *upstream queueing*; so is queueing of the packet on sending devices that processed the packet earlier on its flight from sending host to receiving host. Queueing of the packet on the receiving device, as well as on devices further along on its path is *downstream* queueing.

All along the path from one host to another, the statistical characteristics of the packet arrivals and their sizes on each link affect the downstream queueing, particularly the queueing on the link receiving device. The most accommodating traffic would have arrivals and sizes on the link that result in a traffic rate in bits/sec that is constant; this would be achieved if the packet sizes were constant (which they are not) and if they arrived at equally spaced points in time (which they do not). In this case we would know exactly how to engineer a link of a certain speed; we would allow a traffic rate equal to the link speed. There would be no queueing and no device buffers. The utilization, the ratio of the traffic rate divided by the link speed, would be 100%, so the transmission resources would be used the most efficiently. If the link speed were 1.5 mbps, the traffic rate would be 1.5 mbps.

Suppose instead, that the traffic is stationary with Poisson arrivals and independent sizes. There would be queueing, so a buffer is needed. Here is how we would engineer the link to get good performance. Suppose the speed is 1.5 mbps. We choose a buffer size so that a packet arriving when the queue is nearly full would not have to wait more than about 500 ms; for 1.5 mbps this would be about 100 kilobytes, or 800 kilobits. An amount of traffic is allowed so that only a small percentage of packets are dropped, say 0.5%. For this Poisson and independent traffic, we could do this and and achieve a utilization of 95%, so the traffic rate would be 1.425 mbps.

Unfortunately, we do not get to choose the traffic characteristics. They are dictated by the engineering protocols that underlie the Internet. What can occur is far less accommodating than traffic that has a constant bit rate, or traffic that is Poisson and independent. The traffic can be very *bursty*. This means the following. The packet sizes and inter-arrival times are sequences that we can treat as time series. Both sequences can have persistent, long-range dependence; this means the autocorrelations are positive and fall of slowly with the lag k, for example, like $k^{-\alpha}$ where $0 < \alpha \leq 1$. Long-range dependent time series have long excursions above the mean and long excursions below the mean. Furthermore, for the sizes and inter-arrivals, the coefficient of variation, the standard deviation divided by the mean, can be large, so the excursions can be large in magnitude as well as long in time. The result is large downstream queue-height distributions with large variability. Now, when we engineer a link of 1.5 mbps, utilizations would be much lower, about 40%, which is a traffic rate of 0.6 mbps.

Before 2000, this long-range dependence had been established for links with relatively low link speeds and therefore low numbers of simultaneous active connections, or connection loads, and therefore low traffic rates. But beginning in 2000, studies were undertaken to determine if on links with higher speeds, and therefore greater connection loads, there were effects due to the increased statistical multiplexing. Suppose we start out with a small number of active connections. What happens to the statistical properties of the traffic as we increase the connection load? In other words, what is the effect of the increase in magnitude of the multiplexing? We would expect that the statistical properties change in profound ways, not just simply that the mean of the inter-arrivals decreases. Does the long-range dependence dissipate? Does the traffic tend toward Poisson and independent, as suggested by the superposition theory of marked point processes? This would mean that the link utilization resulting from the above engineering method increase. In network parlance, are there would be multiplexing gains.

In this chapter, we review the results of the new studies on the effect of increased statistical multiplexing on the statistical properties of packet traffic on an Internet link.

5.2   The View of the Internet Circa 2000

The study of Internet traffic beginning in the early 1990s resulted in extremely important discoveries in two pioneering articles [24, 30]: counts of packet arrivals in equally-spaced consecutive intervals of time are long-range dependent and have a large coefficient of variation (ratio of the standard deviation to the mean), and packet inter-arrivals have a marginal distribution that has a longer tail than the exponential. This means the arrivals are not a Poisson process because the counts of a Poisson are independent and the inter-arrivals are exponential. The title of the second article, "Wide-Area Traffic: The Failure of Poisson Modeling", sent a strong message that the old Poisson models for voice telephone networks would not do for the emerging Internet network. And because queue-height distributions for long-range dependent traffic relative to the average bit/rate are much greater than for Poisson processes, it sent a signal that Internet technology would have to be quite different from telephone network technology. The discovery of long-range dependence was confirmed in many other studies (e.g., [14, 18, 31]). The work on long-range dependence drew heavily on the brilliant work of Mandelbrot [26], both for basic concepts and for methodology.

Models of source traffic were put forward to explain the traffic characteristics [14, 19, 28, 33]. The sizes of transferred files utilizing a link vary immensely; to a good approximation, the upper tail of the file size distribution is Pareto with a shape parameter that is often between 1 and 2, so the mean exists but not the variance. A link sees the transfer of files whose sizes vary by many orders of magnitude. Modeling the link traffic began with an assumption of a collection of on-off traffic sources, each on (with a value of 1) when the source was transferring a file over the link, and off (with a value of 0) when not. Since the model has no concept of packets, just connections, multiplexing becomes summation; the link traffic is a sum, or aggregate, of source processes. Because of the heavy tail of the on process, the summation is long-range dependent, and for a small number of source processes, has a large coefficient of variation. We will refer to this as the *on-off aggregation theory*.

Before 2000, there was little empirical study of packet arrivals and sizes. Most of the intuition, theory, and empirical study of the Internet was based on a study of packet and byte counts. It took some time for articles to appear in the literature showing packet inter-arrivals and packet sizes are long-range dependent, although one might have guessed this from the results for counts. The first report in the literature of which we are aware appeared in 1999 [31]. The first articles of which we are aware that sizes are long-range dependent appeared in 2001 [3, 21].

While there was no comprehensive empirical study of the effect of multiplexing, before 2000 there were theoretical investigations. Some of the early, foundations-setting articles on Internet traffic contained conjectures that multiplexing gains did not occur. Leland *et al.* [24] wrote:

> We demonstrate that Ethernet LAN traffic is statistically *self-similar*, ... and that aggregating streams of such traffic typically intensifies the self-similarity ('burstiness') instead of smoothing it.

Crovella and Bestavros [14] wrote:

> One of the most important aspects of self-similar traffic is that there is is no characteristic size of a traffic burst; as a result, the aggregation or superposition of many such sources does not result in a smoother traffic pattern.

Further consideration and discussion however suggested that issues other than long-range dependence needed to be considered. Erramilli *et al.* [17] wrote

> ... the FBM [fractional Brownian motion] model does predict significant multiplexing gains when a large number of independent sources are multiplexed, the relative magnitude is reduced by $\sqrt{n}$ ....

Floyd and Paxson [19] wrote:

> ... we must note that it remains an open question whether in highly aggregated situations, such as on Internet backbone links, the correlations [of long-range dependent traffic], while present, have little actual effect because the variance of the packet arrival process is quite small.

In addition, there were theoretical discussions of the implications of increased multiplexing on queueing [1, 8, 16, 23, 32]. But the problem with such theoretical study is that results depend on the assumptions about the individual traffic sources being superposed, and different plausible assumptions lead to different results. Without empirical study, it was not possible to resolve the uncertainty about assumptions.

With no clear empirical study to guide judgment, many subscribed to a presumption that multiplexing gains did not occur, or were too small to be relevant. For example, Listani *et al.* [25] wrote:

> ... traffic on Internet networks exhibits the same characteristics regardless of the number of simultaneous sessions on a given physical link.

Internet service providers acted on this presumption in designing and provisioning networks, and equipment designers acted on it in designing devices.

5.3   Foundations: Theory and Empirical Study

Starting in 2000, a current of research was begun to determine the effect of increased multiplexing on the statistical properties of many Internet traffic variables, to determine if multiplexing gains occurred [3, 4, 6, 7, 11].

The empirical study of byte and packet counts of previous work was enlarged to include a study of arrivals and sizes. Of course, much can be learned from counts, but arrivals and sizes are the more fundamental traffic variables. It is arriving packets with varying sizes that network devices process, not aggregations of packets in fixed intervals, and packet and byte counts are derived from arrivals and sizes, but not conversely.

In keeping with a focus on arrivals and sizes, the superposition theory of marked point processes became a guiding theoretical framework, replacing the on-off aggregation theory that was applicable to counts but not arrivals and sizes [13, 15]. The two theories are quite different. For the on-off aggregation theory, one considers a sum of independent random variables, and a central limit theorem shows the limit is a normal distribution. For the superposition theory, in addition to the behavior of sums, one considers a superposition of independent marked point processes, and a central limit theorem shows the limit is a Poisson point process with independent marks, and quite importantly, *the theorem applies even when the inter-arrivals and marks of each superposed source point process are long-range dependent.*

The following discussion draws largely on the very detailed account in [4]. We will consider packet arrivals and sizes, and packet counts in fixed intervals. We omit the discussion of byte counts since their behavior is much like that of the packet counts.

5.4   Theory: Poisson and Independence

Let a_j, for $j = 1, 2, \ldots$ be the arrival times of packets on an Internet link where $j = 1$ is for the first packet, $j = 2$ is for the second packet, and so forth. Let $t_j = a_{j+1} - a_j$ be the inter-arrival times, and let q_j be the packet sizes. We treat a_j and q_j as a marked point process. a_j, t_j, and q_j are studied as time series in j. Suppose we divide time into equally-spaced intervals, $[\Delta i, \Delta(i + 1))$, for $i = 1, 2, \ldots$ where Δ might be 1 ms or 10 ms or 100 ms. Let p_i be the packet count, the number of arrivals in interval i. The p_i are studied as a time series in i.

Suppose the packet traffic is the result of multiplexing m traffic sources on the link. Each source has packet arrival times, packet sizes, and packet counts. The arrival times a_j and the sizes q_j of the superposition marked point process result from the superposing of the arrivals and sizes of the m source marked point processes. The packet count p_i of the superposition process in interval i results from summing the m packet counts for the m sources in interval i; theoretical considerations for the p_i are, of course, the same as those for the on-off aggregation theory described earlier.

Provided certain assumptions hold, the superposition theory of marked point processes prescribes certain behaviors for a_j, t_j, q_j, and p_i as m increases [15]. The arrivals a_j tend toward Poisson, which means the inter-arrivals t_j tend toward independent and their marginal distribution tends toward exponential. The sizes q_j tend toward independent, but there is no change in their marginal distribution. As discussed earlier, the t_j and q_j have been shown to be long-range dependent for small m. Thus the theory predicts that the long-range dependence of the t_j and the q_j dissipates. But the autocorrelation of the packet counts p_i does not change with m so its long-range dependence is stable. However, the standard deviation relative to the mean, the coefficient of variation, falls off like $1/\sqrt{m}$. This means that the burstiness of the counts dissipates as well; the durations of excursions of p_i above or below the mean, which are long because of the long-range dependence, do not change because the correlation stays the same, but the magnitudes of the excursions get smaller and smaller because the statistical variability decreases.

The following assumptions for the source packet processes lead to the above conclusions:

- Homogeneity: they have the same statistical properties.

- Stationarity: their statistical properties do not change through time.

- Independence: they are independent of one another and the size process of each is independent of the arrival process.

- Non-simultaneity: the probability of two or more packet arrivals for a source in an interval of length w is $o(w)$ where $o(w)/w$ tends to zero as w tends to zero.

We cannot take the source processes to be the individual connections; they are not stationary, but rather transient, that is, that have a start time and a finish time. Instead, we randomly assign each connection, to one of m source processes. Suppose the start times are a stationary point process, and let ρ be the arrival rate. Then the arrival rate for each source process is ρ/m. We let $\rho \to \infty$, keeping ρ/m fixed to a number sufficiently large that the source processes are stationary; so $m \to \infty$.

We refer to the formation of the source processes, the assumptions about them, and the implications, as the *superposition theory*. It is surely true that all we have done with this theory is to reduce our uncertainty about whether the superposition process is attracted to Poisson and independent with an uncertainty about whether the above construction creates source processes that satisfy the assumptions. But it is a least plausible, although by no means certain, that there are cases where the source process satisfies the above assumptions over a range of values of m. What we have done is to create a plausible hypothesis to be tested by empirical study which we describe shortly.

5.5   Theory: The Network Pushes Back

While we cannot verify the hypotheses of the superposition theory without empirical study, we can at least quite convincingly describe a way in which the network can push back and defeat assumptions. Once m is large enough, significant link-input queueing begins, and then grows as m gets larger still; at some point, the queueing will be large enough that the assumptions of independence of the different source processes and of independence of the inter-arrivals and the sizes of each source process, no longer serve as good approximations in describing the behavior of the source processes. (A small amount of queueing, which almost always occurs, does not invalidate the approximation.)

Consider two packets, $j = 19$ and $j = 20$. Suppose packet 20 waits in the queue for packet 19 to be transmitted. The two are back-to-back on the link, which means, because the arrival time is the first moment of transmission, that t_{19} is the time to put the bits of packet 19 on the link, which is equal to q_{19}/ℓ, where ℓ is the link speed. For example, at $\ell = 100$ mbps, the time for a 1500 byte (12000 bit) packet is 120 μ. So given q_{19} we know t_{19} exactly. Queueing can occur on routers further upstream than the link-input router and affect the assumptions as well.

The arrival times of the packets on the link, a_j, are the departure times of the packets from the queue. The departure times are the arrival times at the queue plus the time spent in the queue. If there are no other packets in the queue when packet j arrives, then a_j is also the arrival time at the queue. Suppose queueing is first-in-first-out. Then the order of the arriving packets at the queue is the same as the order of departing packets from the queue, so q_j is also the packet size process for the arrivals at the queue.

The effect of queueing on q_j is simple. Because the queueing does not alter the q_j, the statistical properties of the q_j are unaffected by the queueing; in particular, their limit of independence is not altered..

But statistical theory for the departure times from a queue is not developed well enough to provide much guidance for the affect of queueing on the statistical properties of t_j and p_i. However, the properties of the extreme case are clear. If m is so large that the queue never drains, then the t_j are equal to q_j/ℓ, so the t_j take on the statistical properties of q_j. Since the q_j tend to independence, the t_j eventually go to independence, so there is no long-range dependence. A Poisson process is a renewal process, a point process with independent inter-arrivals, with the added property that the marginal distribution of the inter-arrivals is exponential. The extreme t_j process is a renewal process but with a marginal distribution proportional to that of the packet sizes. The extreme p_i is the count process corresponding to the t_j renewal process; this implies the coefficient variation of p_i is a constant, so the decrease like $1/\sqrt{m}$ prescribed by the superposition theory is arrested, and it implies the p_i independent, so there is no long-range dependence. We do not expect to see the extreme case in our empirical study, but it does provide at least a point of attraction.

5.6   Empirical Study: Introduction

The superposition theory and the heuristic discussion of the effect of upstream queueing provide hypotheses about the statistical properties of the inter-arrivals t_j, the sizes q_j, and the counts p_i. We carried out extensive empirical studies to investigate the validity of the hypotheses [3, 4, 5].

In the early 1990s, Internet researchers put together a comprehensive measurement framework for studying the characteristics of packet traffic that allows not just statistical study of traffic, but performance studies of Internet engineering designs, protocols, and algorithms [9, 29]. The framework consists of capturing the headers of all packets arriving on a link and time-stamping the packet, that is, measuring the arrival time, a_j. The result of measuring over an interval of time is a *packet trace*. Packet trace collection today enjoys a very high degree of accuracy and effectiveness for traffic study [20, 27].

We put together a very large database of packet traces measuring many Internet links whose speeds range from 10 mbps to 2.5 gbps, and we built S-Net, a software system, based on the S language for graphics and data analysis, for analyzing very large packet header databases [6]. We put the database and S-Net work to study the multiplexing hypotheses.

For each studied trace, which covers a specific block of time on a link, we compute a_j, t_j, q_j, and 100-ms p_i. We also need a summary measure of the magnitude of multiplexing for the trace. At each point in time over the trace, the measure is the number of active connections. The summary measure, c, for the whole trace is the average number of active connections over all times in the trace.

Here, we describe some of the results of one of our empirical investigations in which we analyzed 2526 header packet traces, 5 min or 90 sec in duration, from 6 Internet monitors measuring 15 links ranging from 100 mbps to 622 mbps [4]. Table 5.1 shows information about the traces. Each row describes the traces for one link. The first column gives the trace group name: the trace length is a part of each name. Column 2 gives the number of traces. Column 3 gives the link speed. Column 4 gives the mean of the log base 2 of c for the traces of the link.

Consider each packet in a trace. Arriving after it is a back-to-back run of k packets, for $k = 0, 1, \ldots$; each packet in the run is back-to-back with its predecessor. If packet 19 has a back-to-back run of 3 packets, then packet 20 is back-to-back with 19, 21 is back-to-back with 20, 22 is back-to-back with 21, but 23 is not back-to-back with 22. The percent of packets with back-to-back runs of k or more is a measure of the amount of queueing on

Table 5.1. Trace Group: name including length of traces • Number: number of traces • Link: speed • log(c): log base 2 average number of active connections

Trace Group	Number	Link	log(c)
AIX1(90sec)	23	622mbps	13.09
AIX2(90sec)	23	622mbps	13.06
COS1(90sec)	90	156mbps	10.83
COS2(90sec)	90	156mbps	10.81
NZIX(5min)	100	100 mbps	10.75
NZIX7(5min)	100	100 mbps	9.60
NZIX5(5min)	100	100 mbps	8.66
NZIX6(5min)	100	100 mbps	7.85
NZIX2(5min)	100	100 mbps	7.32
NZIX4(5min)	100	100 mbps	7.17
BELL(5min)	500	100 mbps	6.97
NZIX3(5min)	100	100 mbps	6.54
BELL-IN(5min)	500	100 mbps	5.98
BELL-OUT(5min)	500	100 mbps	5.94
NZIX1(5min)	100	100 mbps	4.42

the link-input router. We studied this measure for many values of k. We need such study to indicate when the network is likely pushing back on the attraction to Poisson and independence.

Figure 5.1 graphs the percent of packets whose back-to-back runs are 3 or greater against log(c). Each point on the plot is one trace. Each of the 15 panels contains the points for one link. The panels are ordered, left to right and bottom to top, by the means of the log(c) for the 15 links, given in column 4 of Table 5.1.

Figure 5.1, and others like it for different values of k, show that only four links experience more than minor queueing — COS1, COS2, AIX1, and AIX2 — so we would not expect to see significant push-back except at these four. However, queueing further upstream than the link-input router can affect the traffic properties as well, but without creating back-to-back packets, so we reserve final judgment until we see the coming analyses.

Figure 5.1 also provides information about the values of c. Since the mean of log(c) increases left to right and bottom to top, the distribution shifts generally toward higher values in this order. The smallest c, which appears in the lower left panel, is 5.9 connections; the largest, which appears in the upper right panel, is 16164 connections.

5.7 Empirical Study: FSD and FSD-MA(1) Models

In this section we introduce two very simple classes of stationary time series models [5], one a subclass of the other, that we found provide excellent fits to the inter-arrivals t_j, the sizes q_j, and the counts p_i for the 2526 traces.

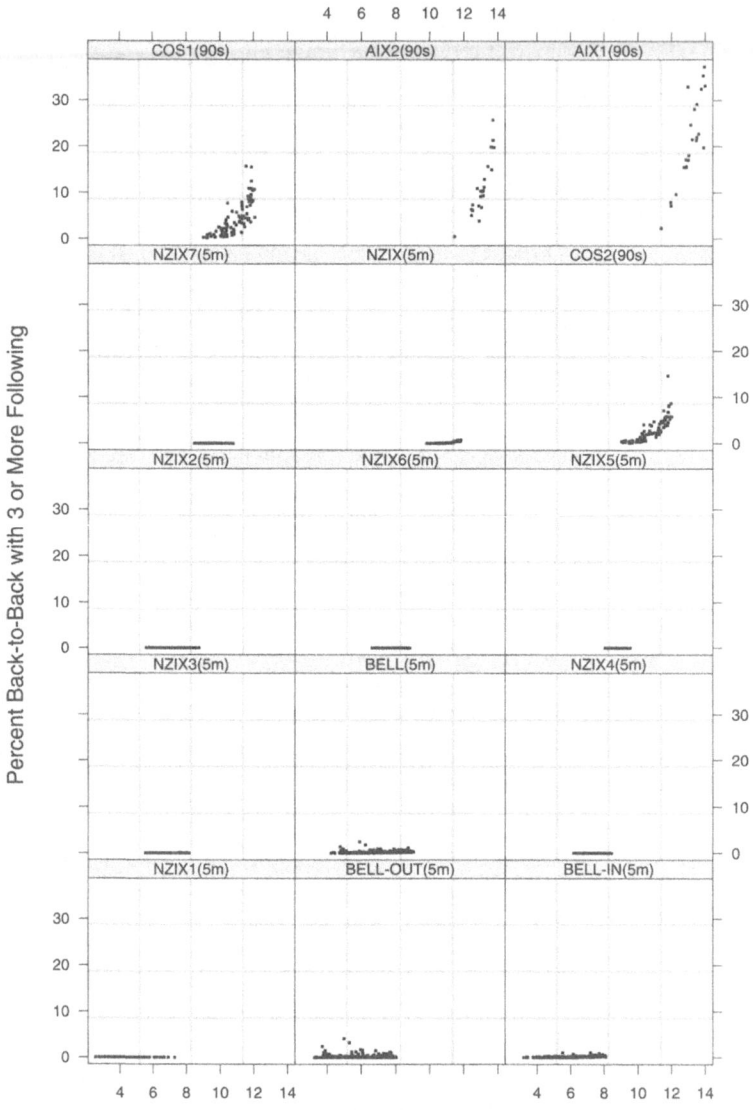

Figure 5.1. The percent of packets with back-to-back runs of 3 or greater is plotted against $\log(c)$.

The models are parametric. One of the parameters determines the amount of dependence. At low values of the parameter, the series has substantial autocorrelation and is long-range dependent. As the parameter increases, the amount of dependence decreases. At the largest value of the parameter, the series is independent. Other parameters determine the marginal distribution of the series and therefore the coefficient of variation. By fitting the models to each trace, we can study the multiplexing gains by studying the changing values of the parameters across the traces, and relating the changes to the average active connection load c of the traces.

The two model classes are fractional sum-difference (FSD) models and FSD-MA(1) models [5]. FSD models have two additive components: a simple fractional ARIMA and white noise. MA(1) refers to a first-order moving average [2]. FSD-MA(1) models replace the white noise of the FSD model with an MA(1). Since white noise is a special case of an MA(1), the FSD models are a subclass of the FSD-MA(1) models. As we will see, the names "transformation-Gaussian FSD models" and "transformation-Gaussian FSD-MA(1) models" would convey more information about the nature of the models, but for simplicity we will use the shorter names.

Suppose x_u for $u = 1, 2, \ldots$ is a stationary time series with a marginal cumulative distribution function $F(x; \phi)$ where ϕ is a vector of unknown parameters. For example, $F(x; \phi)$ might be log-normal or Weibull. Let $z_u = T(x_u; \phi)$ be a transformation of x_u such that the marginal distribution of z_u is normal with mean 0 and variance 1. If $G^{-1}(r)$ is the quantile with probability r of z_u, then $T(x_u; \phi) = G^{-1}(F(x_u; \phi))$. If x_u is log-normal and the vector ϕ consists of the mean μ and variance σ^2 on the log scale, then $T(x_u; \phi) = (\log(x_u) - \mu)/\sigma$.

Next we suppose that z_u is a Gaussian time series, that is, the joint distributions of all finite subsets of the time series are multivariate normal. Let

$$z_u = \sqrt{1 - \theta}\, s_u + \sqrt{\theta}\, n_u,$$

where s_u and n_u are independent of one another and each has mean 0 and variance 1. n_u is white noise, that is, an uncorrelated time series. s_u is a fractional ARIMA (FARIMA) model [22]

$$(I - B)^d s_u = \epsilon_u + \epsilon_{u-1}$$

where $Bs_u = s_{u-1}$, $0 < d < 0.5$, and ϵ_u is white noise with mean 0 and variance

$$\sigma_\epsilon^2 = \frac{(1 - d)\Gamma^2(1 - d)}{2\Gamma(1 - 2d)}$$

to make the variance of s_u equal to 1.

z_u is an FSD model. We coined this term because the model for z_u can be written as a combination of fractional and summation difference operators acting on z_u and on two white noise series:

$$(I - B)^d z_u = (I + B)\epsilon_u + (I - B)^d n_u.$$

These models are to FARIMA models what the very simple and widely applicable IMA(1,1) models are to ARIMA models [2]; the IMA(1,1) models can be written as

$$(I - B)z_u = (I + B)\epsilon_u + (I - B)n_u.$$

Generalizations of this latter model have been named *sum-difference models* [12].

The FSD-MA(1) model is

$$z_u = \sqrt{1 - \theta}\, s_u + \sqrt{\theta}\, n_u,$$

similar to the FSD model, but where n_u instead of white noise is a first order moving-average

$$n_u = \zeta_u + \beta\zeta_{u-1},$$

where ζ_u is Gaussian white noise with mean 0 and variance $(1 + \beta^2)^{-1}$, which makes the variance of n_u equal to 1. If $\beta = 0$, the moving-average component is white noise so the model is simply an FSD. We need the above restriction $d > 0$. If $d = 0$, the model is not identifiable because z_u, whose model has two parameters, is a first order moving average with variance 1, which has one parameter.

Suppose z_u is an FSD-MA(1) model. Let $r_z(k), r_s(k)$ and $r_n(k)$ be the autocorrelation functions of of z_u, s_u, and n_u, respectively, for lags $k = 0, 1, 2, \ldots$. Because $d > 0$, s_u is long-range dependent, and $r_s(k)$ falls off like k^{2d-1} and increases at all positive lags as d increases. $r_n(k) = \beta(1 + \beta^2)^{-1}\{k = 1\}$ where $\{k = 1\}$ is 1, if $k = 1$, and is 0 if $k > 1$. Thus

$$r_z(k) = (1 - \theta)r_s(k) + \theta\beta(1 + \beta^2)^{-1}\{k = 1\}.$$

As $\theta \to 1$, the long-range dependent component $\sqrt{1 - \theta}\, s_u$ contributes less and less variation to z_u. Finally, when $\theta = 1$, z_u is white noise if $\beta = 0$, and is a first-order moving average otherwise.

The power spectrum of z_u is

$$p_z(f) = (1 - \theta)\sigma_\epsilon^2 \frac{4\cos^2(\pi f)}{\left(4\sin^2(\pi f)\right)^d} + \theta\frac{1 + \beta^2 + 2\beta\cos(2\pi f)}{1 + \beta^2}$$

for $0 \leq f \leq 0.5$. The frequency f has units cycles/inter-arrival for t_j, cycles/packet for q_j, and cycles/interval-length for p_i. $p_z(f)$ decreases monotonically as f increases. Because $d > 0$, the term $\sin^{-2d}(\pi f)$ goes to infinity at $f = 0$, so if $\theta < 1$, no matter how close θ gets to 1, $p_z(f)$ gets arbitrarily large near $f = 0$, but its ascent begins closer and closer to 0 as θ gets closer to 1.

Figure 5.2 shows the power spectra for 16 FSD-MA(1) models. For each panel, the spectrum is evaluated at 100 frequencies, equally spaced on a log base 2 scale from -13 to -1. The value of d in all 16 cases is 0.41. θ varies from 0.39 to 0.99 by 0.2 as we go left to right through the columns. β varies from 0 to 0.3 by 0.1 as we go from bottom to top through the rows. So the bottom row shows spectra for the FSD model, while the other rows show the spectra for FSD-MA(1) models with positive β.

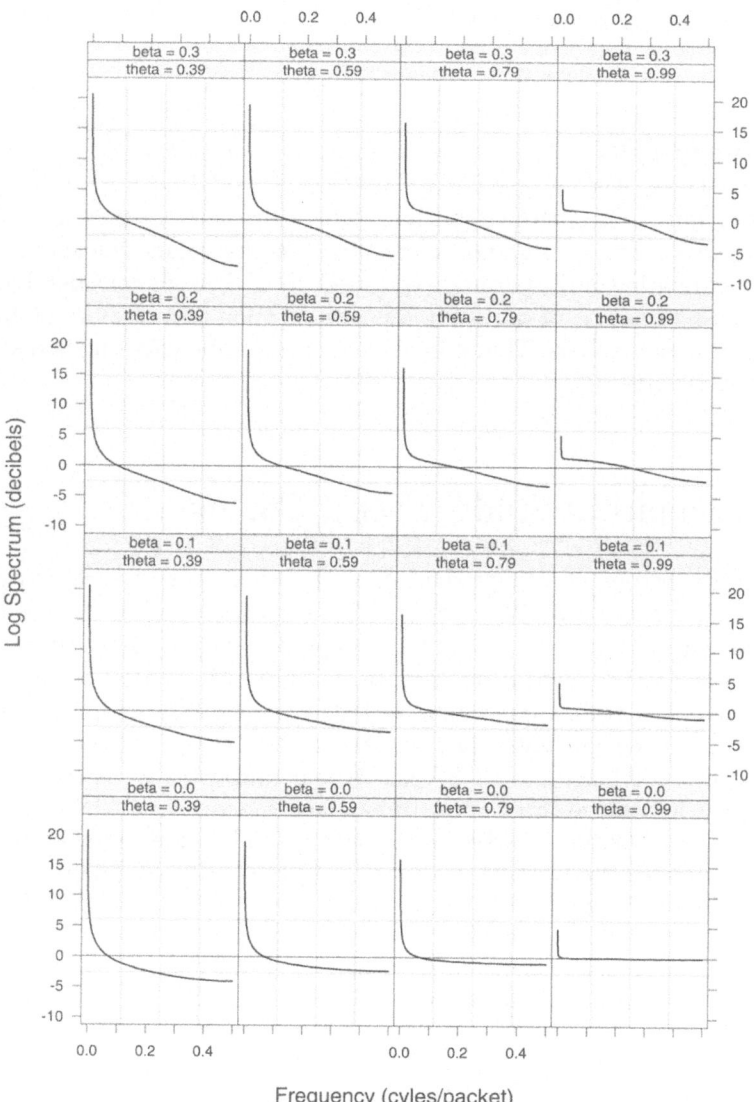

Figure 5.2. The log power spectrum of an FSD-MA(1) time series is plotted against frequency for different values of θ and β.

For all panels, there is a rapid rise as f tends to 0, and an overall monotone decrease in power as the frequency increases from 0 to 0.5. This is a result of the persistent long-range dependence. But for each row, as θ increases, the fraction of low-frequency power decreases, and the fraction of high-frequency power increases. In the bottom row, as θ increases, the spectrum at frequencies away from 0 shows a distinct flattening, tending toward the flat spectrum of white noise. In the remaining rows, the spectra, away from 0, tend toward that of a gently sloping curve, the spectrum of an MA(1).

We found that the 100-ms packet counts, p_i, and the packet sizes, q_j, for all but a few of the 2526 traces, are very well fitted by an FSD model. t_j is also typically well fitted by either an FSD model or an FSD-MA(1); for the traces of some links, an FSD-MA(1) model with a positive β is clearly required as c gets large, the result of the network pushing back on the attraction to Poisson and independence by upstream queueing.

The estimation of the parameters, especially of d, needs considerable care. But an essential part of the study was visualization tools that validated the resulting fitted models. The estimation and modeling checking is discussed in detail elsewhere [5].

5.8   Empirical Study: Packet Counts

The superposition theory predicts that the coefficient of variation of the p_i should decrease like $1/\sqrt{c}$. Figure 5.3 graphs the log of the coefficient against $\log(c)$. The theory predicts a slope of -0.5; the least squares line with slope -0.5 is shown on each panel. The rate of decline of the log coefficients is certainly consistent with a value of -0.5. At some sites, the decline is somewhat faster and at others, it is slower. Interestingly, the decline has not been altered by back-to-back occurrence, as predicted by the heuristics for the effect of upstream queueing, even for AIX1 and AIX2 which have the largest back-to-back percents. This presumably happened in part because the aggregation interval length is 100 ms; had we used a smaller interval, an effect might have been detected.

The p_i do not have a normal marginal until c gets large. A log-normal marginal does much better. Let p_i^* be $\log(1+p_i)$ normalized to have mean 0 and variance 1. An assumption of a Gaussian process for p_i^* is a reasonable approximation for much smaller c. We found that an FSD model fitted the p_i^* extremely well, except for a small fraction of intervals with low c where oscillatory effects of Internet transport protocols broke through and created spikes in the power spectrum. Even in these cases, the model serves as an excellent summary of the amount of long-range dependence in the correlation structure.

Estimates of d vary by a small amount across the traces and showed no dependence on c. The medians for the 15 links vary from 0.39 to 0.45 and their mean is 0.41. Estimates of θ also show no dependence on c; the mean of the 15 medians of θ is 0.53. Thus the p_i^* spectra look like the spectrum in column 2 and row 1 in Figure 5.2. The stability of the correlation structure of p_i^* is consistent with the superposition theory, which stipulates that the correlation structure of the p_i does not change with c. The heuristics for the effect of upstream queueing suggest that the autocorrelation should be

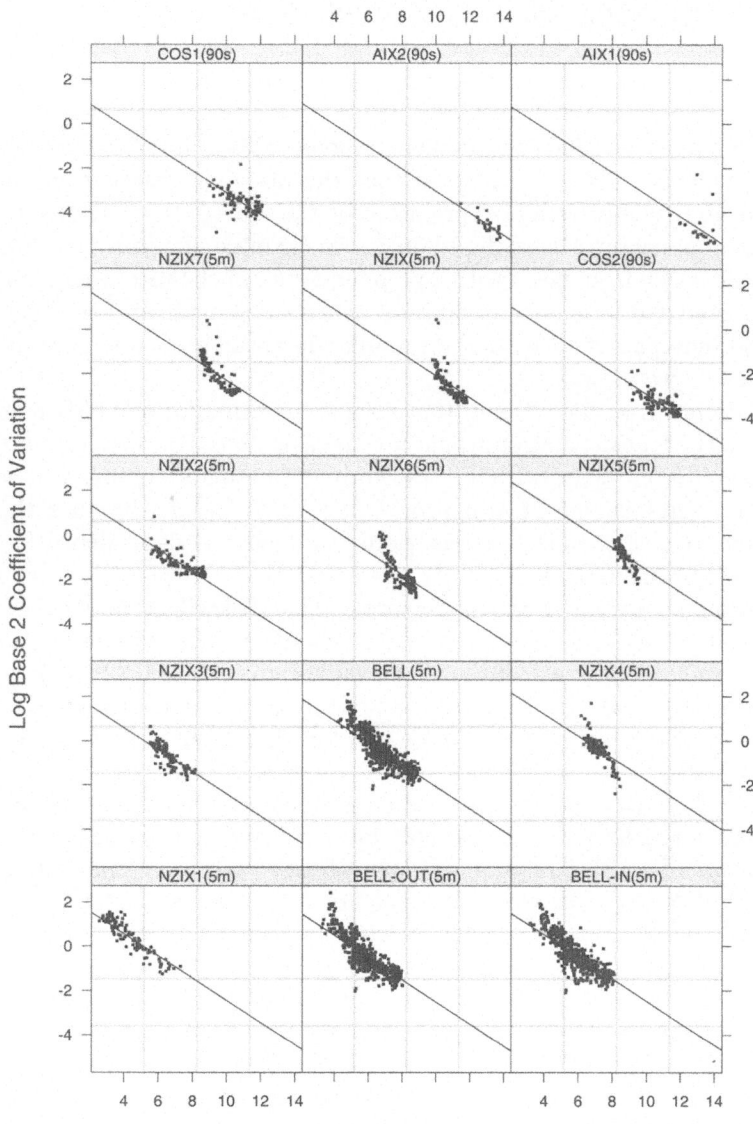

Figure 5.3. Log coefficient of variation of the 100-ms packet counts is plotted against $\log(c)$.

changed by a large amount of upstream queueing, but the effect does not appear to occur even for AIX1 or AIX2, where the occurrence of back-to-back packets is the greatest. As with the coefficient of variation, it is possible an effect would be seen for interval lengths less than 100 ms.

5.9   Empirical Study: Packet Sizes

A reasonable summary of the marginal distribution of the q_j is an atom at the minimum size of 40 bytes, an atom at the maximum size of 1500 bytes, an atom at 576 bytes, and continuous uniform from 40 bytes to 1500 bytes. Quantile plots [10] showed that the marginal distribution did not change appreciably with c, as predicted by the superposition theory, but did change appreciably across the 15 links. For example, if a link has traffic in a single direction from hosts with a preponderance of clients downloading web pages, then the frequency of 40 byte packets is greater and the frequency of 1500 byte packets less than for a link where the preponderance of hosts are serving web pages.

We do not transform the q_j to a normal marginal for our FSD modeling because the transformation would not be invertible. For analysis purposes, we treat the q_j as is, without transformation; this amounts to a second moment analysis, but it will provide adequate insight because the correlation coefficient is still a reasonable summary of dependence for such discrete-continuous data.

We found that an FSD model provided an excellent fit to the q_j. A combination of theory and empirical study show that d remains constant with c, and the estimate of the single value came out to 0.42, very close to the 0.41 for the p_i^*. For simplicity, we could not resist using a value of 0.41 for the q_j, the same as the estimate of d for the p_i. We fixed d to this value and estimated θ.

Figure 5.4 plots the estimates of θ against $\log(c)$. The smooth curve on each panel is a loess fit using robust locally linear fitting and a smoothing parameter of 1 [10]. Loess is a nonparametric procedure that puts curves through data by a moving local polynomial fitting procedure, the same in spirit, but not in detail, as a moving average smoothing a time series. The overall result in Figure 5.4 is that θ goes to 1 with c, so the spectrum changes as shown in the bottom row of Figure 5.2. Thus the q_j tend toward independence as prescribed by the the superposition theory. An increase in the percent of back-to-back packets with c for the COS1, COS2, AIX1, and AIX2 links does not alter the increase in θ, which is consistent with the heuristics for the effect of upstream queueing.

5.10   Empirical Study: Inter-Arrivals

We found, using Weibull quantile plots, that the marginal distribution of the inter-arrivals is well approximated by the Weibull distribution across all values of c. The back-to-back packets result in deviations from the Weibull, but because packet sizes vary by a factor of $1500/40 = 37.5$, the deviations are spread across the distribution, and overall the approximation remains

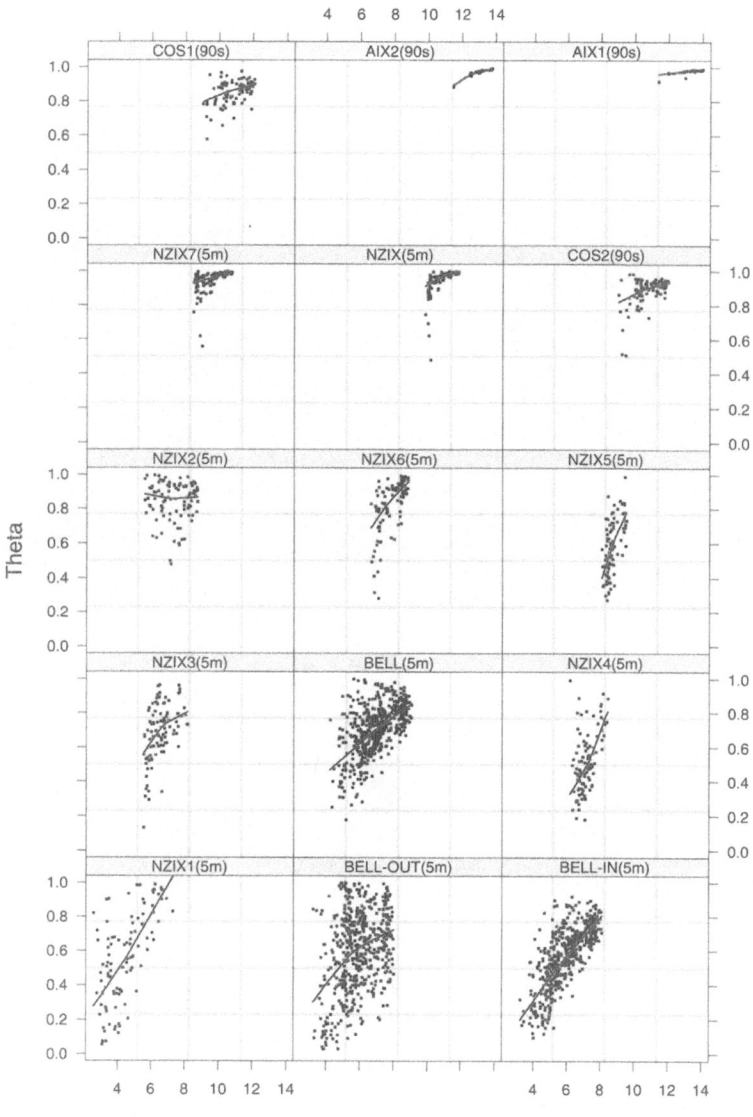

Figure 5.4. An estimate of θ for the packet sizes is plotted against $\log(c)$.

excellent, even for the traces with the largest occurrence of back-to-back packets. The Weibull has two parameters: α, a scale parameter, and λ a shape. When λ is 1, the Weibull is an exponential, the inter-arrival distribution of a Poisson process. When $\lambda < 1$, the tail is heavier than that of the exponential.

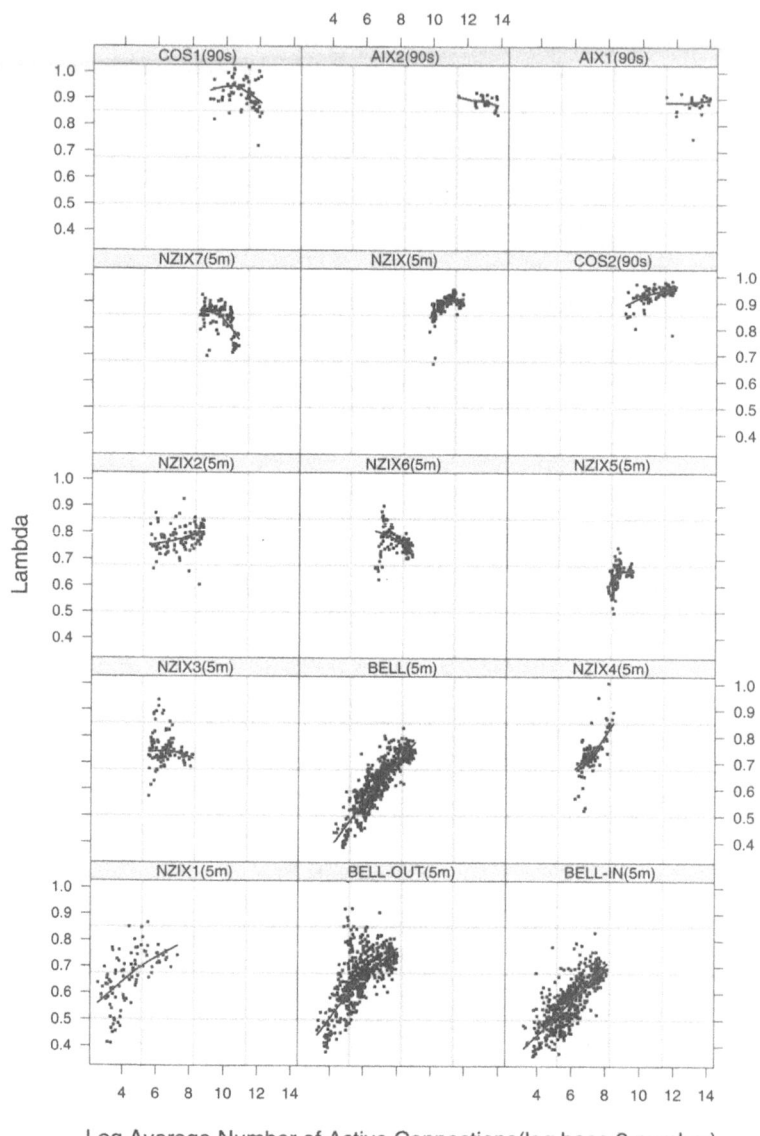

Figure 5.5. An estimate of λ for the inter-arrivals is plotted against $\log(c)$.

Figure 5.5 plots estimates of the Weibull shape parameter, λ, against $\log(c)$. The smooth curve on each panel is a loess fit with robust locally linear fitting and a smoothing parameter of 1. The overall result is that the shape estimates are less than 1, and as c increases, the shape tends toward 1. Consider the 5 links with the largest mean $\log(c)$ — NZIX, COS2, COS1, AIX2, and AIX1. Almost all of the values of c exceed 2^{10}, but few values for the remaining sites do so. For these top five, most estimates of λ exceed 0.9. For the remaining, most estimates are below 0.8. The top five appear to have a limit slightly less than 1; the back-to-back packets exert just enough influence to keep the estimates slightly below 1, but this is a small matter since a Weibull with shape of 0.95 is exceedingly close to exponential.

Because the t_j have a Weibull marginal, the transformation that takes them to normality is $T(t_j; \phi) = G^{-1}(F(t_j; \phi))$ where F is the Weibull cumulative distribution function and ϕ is the vector of parameters α and λ. Because λ changes, the transformation changes, but the change is not large and we found the transformations are well approximated by a single transformation, the sixth root of t_j. So for simplicity we used $t_j^* = t_j^{1/6}$.

We found that an FSD model or an FSD-MA(1) model provided an excellent fit to t_j^* except for a small fraction of intervals with low c where oscillatory effects of Internet transport protocols broke through and created spikes in the power spectrum. Even in these cases, the model serves as an excellent summary of the amount of long-range dependence in the correlation structure.

Theoretical results show that the value of d for the t_j, or for monotone transformations of t_j such as t_j^*, is the same as that for the p_i^*, so the estimate of d for the t_j^* was taken to be 0.41, that for p_i^*. This was done rather than estimating d from the t_j^* because, when θ gets close to 1, the long-range dependent component accounts for such a small fraction of the variation in the t_j^* that d is poorly estimated.

We fitted an FSD-MA(1) with $d = 0.41$ to the 2526 traces. Figure 5.6 graphs estimates of β against $\log(c)$. The smooth curve on each panel is a loess fit with robust locally linear fitting and a smoothing parameter of 1. 1.2% of the estimates are less than -0.4 and are not shown on the plot. Our model checking showed that the MA(1) component was important for producing a good fit for the largest values of c at NZIX7, NZIX, AIX1, and AIX2. The latter two sites show a large back-to-back occurrence, but not the first two. However, queueing upstream from the link-input router can affect the inter-arrivals without introducing back-to-back packets. In other words, our measure of back-to-back packets in Figure 5.1 does not tell the whole story of upstream queueing.

For $\beta \leq 0.1$, $n_u = \zeta_u + \zeta_{u-1}$ is nearly white noise. When $\beta = 0.1$, the variance of ζ_u is $1/(1+.1^2) = 0.990$, so n_u, whose variance is 1, is very close to white noise. But when $\beta = 0.3$, the variance of ζ_u is 0.917, so n_u contains significant correlated variation. It is only at NZIX7, NZIX, AIX1, and AIX2 that β is reliably above 0.1, getting as high as 0.3. At the other links, β is small enough, taking the greater variability of estimates as c decreases into account, that it is reasonable to omit the MA(1) component, that is, using just an FSD model. In particular, at COS1 and COS2, β is small.

Figure 5.7 graphs θ against $\log(c)$. The smooth curve on each panel is a loess fit with robust locally linear fitting and a smoothing parameter of 1.

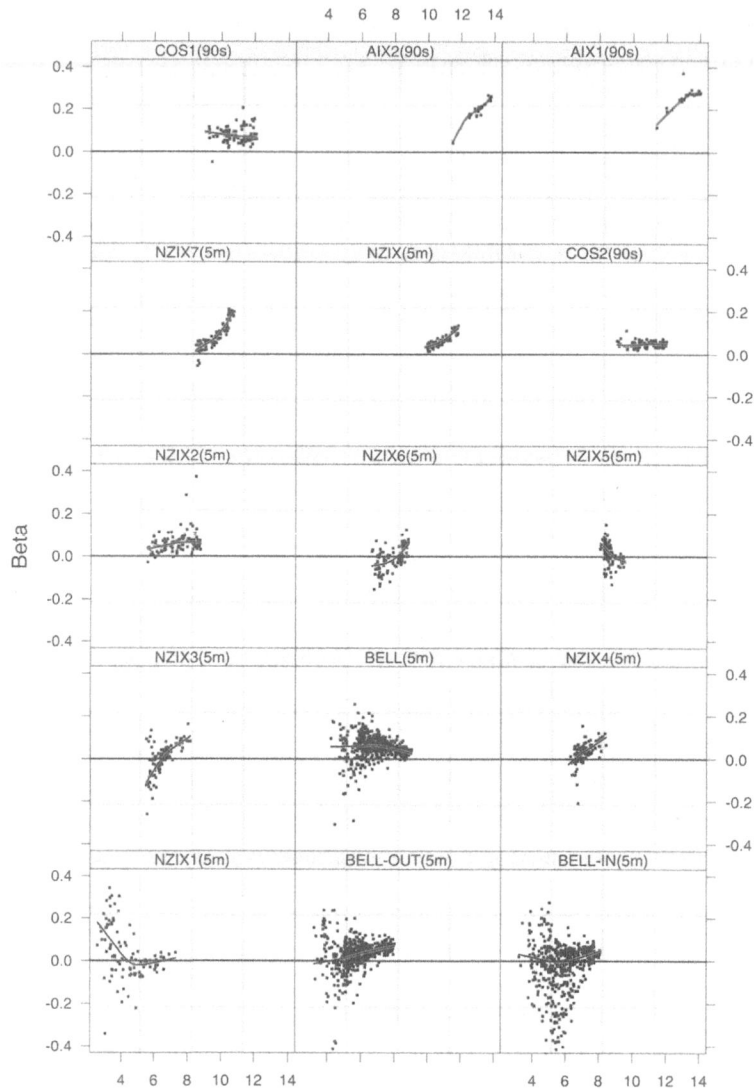

Figure 5.6. An estimate of β for the inter-arrivals is plotted against $\log(c)$.

The overall result is that θ goes to 1 with c. The long-range dependence of t_j^* dissipates, tending either to short-range dependence, an MA(1), or to independence. Thus all panels of Figure 5.2 convey the behaviors of the power spectra of the t_j^*.

These results for λ, β, and θ are consistent with the superposition theory and the heuristics for the effect of upstream queueing. Multiplexing creates an attraction to Poisson in the t_j; λ and θ tend toward 1 as the theory

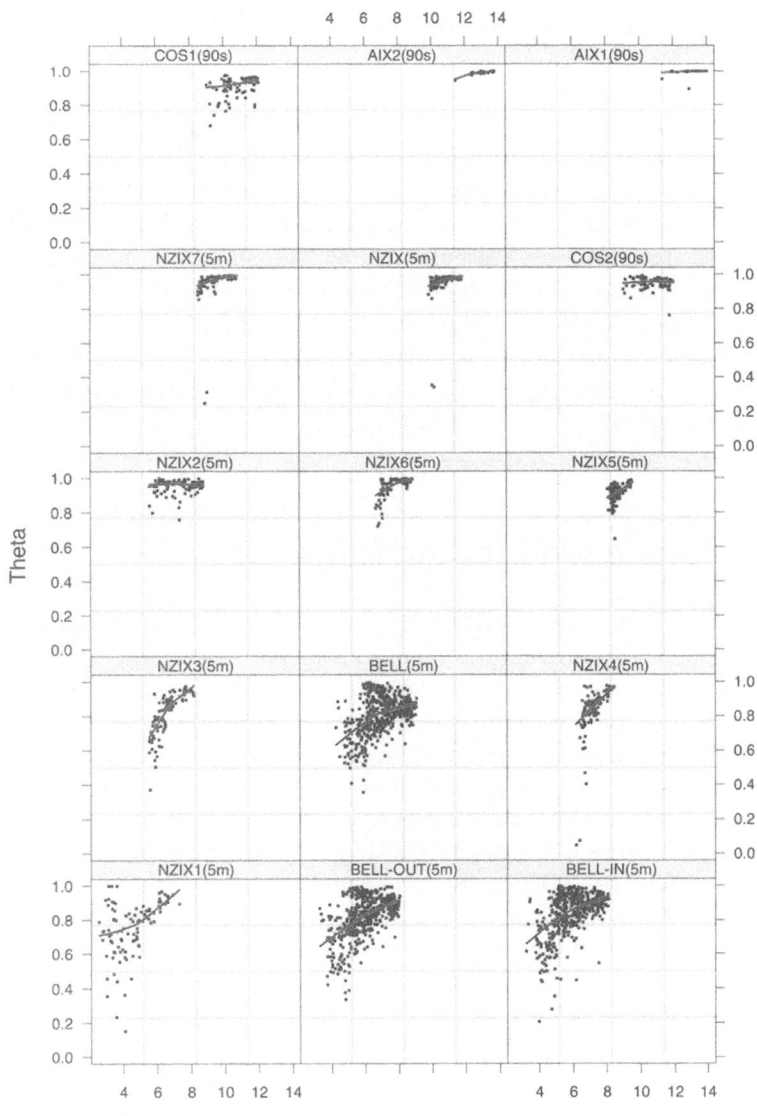

Figure 5.7. An estimate of θ for the inter-arrivals is plotted against $\log(c)$.

prescribes. But the network succeeds in pushing back in some cases, keeping λ slightly less than 1, and causing values of β for some links that indicate short-term dependence.

5.11   Open Loop Generation of Packet Traffic

The FSD models fitted to the sizes and inter-arrivals can be used for open-loop generation of synthetic traffic for simulation studies. The inter-arrival marginal is Weibull; the parameters are α and λ. The packet size marginal has atoms at specific packet sizes and has a continuous part that is uniform between 40 bytes and 1500 bytes; the parameters are the probabilities at the atoms. The inter-arrivals are generated by Gaussian FSD variables with d = 0.41 transformed to the Weibull marginal; the parameter is θ_t. The packet sizes are generated by Gaussian FSD variables with d = 0.41 transformed to the discrete-continuous marginal; the parameter is θ_q. α, λ, θ_t, and θ_q change with c according to certain models to reflect the multiplexing gains, so only c is specified to carry out generation.

5.12   There are Multiplexing Gains

The results here show that an increasing number of simultaneous active connections causes a dramatic change in the statistical properties of packet traffic on an Internet link. Starting at low connection loads on an uncongested link, packet arrivals tend toward Poisson and packet sizes tend toward independence as the load increases. A component of long-range dependence is retained in each of these variables, but the effect of the component gets increasingly small. Packet counts have a stable autocorrelation structure that does not change with the load, but the standard deviation of the counts relative to the mean gets small, so the counts become smooth. The network pushes back on this attraction to Poisson and independence through upstream queueing, which also increases with the connection load; very short term autocorrelation can develop in the inter-arrivals, and their marginal changes toward the distribution of packet sizes divided by the link speed. On a link with a sufficiently large speed that the increasing connection load can bring the traffic to Poisson and independence before substantial upstream queueing occurs, the onset of queueing does not resurrect the long-range dependence. All this means that the burstiness of traffic, once thought to pervade the whole Internet, dissipates with the connection load. There are multiplexing gains.

Inspired by these results on multiplexing gains, theoretical and empirical studies have now demonstrated that queueing on an Internet device tends to that of Poisson arrivals and independent sizes as the load increases, just as one would expect [3, 7]. This means that if a link speed is sufficiently large, queueing distributions relative to the bit/rate of the traffic get dramatically smaller.

The foundations of traffic analysis and modeling should reflect these results. The dramatic change in the statistical properties with the connection load makes clear that the load needs to play a central role in analysis and modeling. Theory must reflect the load. Empirical study must encompass a range of packet traces from small loads to large.

The results have important implications for Internet device engineering and Internet traffic engineering. On links with low speeds, at the edges of the Internet close to the user hosts, connection loads cannot get large, and traffic remains highly bursty. But on links with high speeds, toward the core of the Internet and carrying traffic made up of large numbers of connections, the traffic can be close to Poisson and independence, so the burstiness is gone. Engineering studies that are meant to apply to the Internet as a whole, and that use synthetic or live packet traffic to assess performance, need to consider packet traces varying across a wide range of link speeds and connection loads. Many issues of Internet engineering need to be revisited to determine how protocols, algorithms, device design, network design, and network provisioning should change to reflect the effect of the changing statistical properties of the traffic with the connection load.

References

[1] D. D. Botvich and N. G. Duffield. Large Deviations, the Shape of the Loss Curve, and Economies of Scale in Larger Multiplexers. *Queueing Systems*, 20:293–320, 1995.

[2] G.E.P. Box and G. M. Jenkins. *Time Series Analysis, Forecasting and Control*. Holden Day, San Francisco, 1970.

[3] J. Cao, W. S. Cleveland, D. Lin, and D. X. Sun. On the Nonstationarity of Internet Traffic. *ACM SIGMETRICS*, pages 102–112, 2001.

[4] J. Cao, W. S. Cleveland, D. Lin, and D. X. Sun. The Effect of Statistical Multiplexing on the Long-Range Dependence of Internet Packet Traffic. Technical report, Bell Labs, Murray Hill, NJ, 2002.

[5] J. Cao, W. S. Cleveland, and D. X. Sun. Fractional Sum-Difference Models for Open-Loop Generation of Internet Packet Traffic. Technical report, Bell Labs, Murray Hill, NJ, 2002.

[6] J. Cao, W. S. Cleveland, and D. X. Sun. S-Net: A Software System for Analyzing Packet Header Databases. In *Proceedings Passive and Active Measurement*, 2002.

[7] Jin Cao and Kavita Ramanan. A Poisson Limit for the Unfinished Work of Superposed Point Processes. In *Proceedings INFOCOMM*, 2002, to appear.

[8] G. L. Choudury, D.M. Lucantoni, and W. Whitt. Squeezing the Most Out of ATM. *IEEE Transactions on Communications*, 44(2):203–217, 1996.

[9] K. Claffy, H.-W. Braun, and G. Polyzos. A Parameterizable Methodology for Internet Traffic Flow Profiling. *IEEE Journal on Selected Areas in Communications*, 13:1481–1494, 1995.

[10] W. S. Cleveland. *Visualizing Data*. Hobart Press, Summit, New Jersey, U.S.A., 1993.

[11] W. S. Cleveland, D. Lin, and D. X. Sun. IP Packet Generation: Statistical Models for TCP Start Times Based on Connection-Rate Superposition. *ACM SIGMETRICS*, pages 166–177, 2000.

[12] W. S. Cleveland and C. Liu. Maximum Likelihood Estimation of Sum-Difference Time Series Models Using the EM Algorithm. Technical report, Bell Labs, 2002.

[13] D. R. Cox. *Renewal Theory*. Chapman and Hall, 1962.

[14] M. E. Crovella and A. Bestavros. Self-Similarity in World Wide Web Traffic: Evidence and Possible Causes. *ACM SIGMETRICS*, pages 160–169, 1996.

[15] D. J. Daley and D. Vere-Jones. *An Introduction to the Theory of Point Processes*. Springer-Verlag, New York, 1988.

[16] N. G. Duffield. Economies of Scale in Queues with Sources Having Power-Law Large Deviation Scaling. *Queueing Systems*, 33:840–857, 1996.

[17] Ashok Erramilli, Onuttom Narayan, and Walter Willinger. Experimental Queueing Analysis with Long-Range Dependent Packet Traffic. *IEEE/ACM Transactions on Networking*, 4:209–223, 1996.

[18] A. Feldman, A. A. Gilbert, and W. Willinger. Data Networks as Cascades: Explaining the Mulifractal Nature of Internet WAN Traffic. In *Proceedings ACM SIGCOMM*, pages 42–55, 1998.

[19] Sally Floyd and Vern Paxson. Why We Don't Know How to Simulate the Internet. Technical report, LBL Network Research Group, 1999.

[20] C. Fraleigh, C. Diot, B. Lyles, S. Moon, P. Owezarski, D. Papagiannaki, and F. Tobagi. Design and Deployment of a Passive Monitoring Infrastructure. In *Proceedings Passive and Active Measurement*, Amsterdam, 2001. Ripe NCC.

[21] J.B. Gao and I. Rubin. Multiplicative Multifractal Modeling of Long-Range-Dependent Network Traffic. *International Journal of Communications Systems*, 14:783–201, 2001.

[22] J. R. M. Hosking. Fractional Differencing. *Biometrika*, 68:165–176, 1981.

[23] K.R. Krishnan. A New Class of Performance Results for a Fractional Brownian Traffic Model. *Queueing Systems*, 22:277–285, 1996.

[24] W. Leland, M. Taqqu, W. Willinger, and D. Wilson. On the Self-Similar Nature of Ethernet Traffic. *IEEE/ACM Transactions on Networking*, 2:1–15, 1994.

[25] M. Listani, V. Eramo, and R. Sabella. Architectural and Technological Issues for Future Optical Internet Networks. *IEEE Communications Magazine*, September:82–86, 2000.

[26] B. B. Mandelbrot. Long-Run Linearity, Locally Gaussian Processes, H-Spectra and Infinite Variances. *International Economic Review*, 10:82–113, 1969.

[27] J. Micheel, S. Donnelly, and I. Graham. Timestamping Network Packets. In *Proceedings ACM SIGCOMM Internet Measurement Workshop*, San Francisco, 2001.

[28] K. Park, G. Kim, and M. Crovella. On the Relationship Between File Sizes, Transport Protocols, and Self-Similar Network Traffic. In *Proceedings of the IEEE International Conference on Network Protocols*, 1996.

[29] V. Paxson. End-to-End Internet Packet Dynamics. In *Proceedings ACM SIGCOMM*, pages 139–152, 1997.

[30] Vern Paxson and Sally Floyd. Wide-Area Traffic: The Failure of Poisson Modeling. *IEEE/ACM Transactions on Networking*, 3:226–244, 1995.

[31] Rudolf H. Riedi, Matthew S. Crouse, Vinay J. Ribeiro, and Richard G. Baraniuk. A Multifractal Wavelet Model with Application to Network Traffic. *IEEE Transactions on Information Theory*, 45(3):992–1019, 1999.

[32] I. Saniee, A. Neidhardt, O. Narayan, and A. Erramilli. Multi-scaling models of sub-frame VBR video and TCP/IP traffic. *KICS/IEEE Journal of Communication Networks*, 3(4):383–395, 2001.

[33] W. Willinger, M. S. Taqqu, R. Sherman, and D. V. Wilson. Self-Similarity Through High-Variability: Statistical Analysis of Ethernet LAN Traffic at the Source Level. *IEEE/ACM Transactions on Networking*, 5:71–86, 1997.

6

Regression and Classification with Regularization

Sayan Mukherjee, Ryan Rifkin, and Tomaso Poggio[1]

6.1 Introduction

The purpose of this chapter is to present a theoretical framework for the problem of learning from examples. Learning from examples can be regarded [13] as the problem of approximating a multivariate function from sparse data[2]. The function can be real valued as in regression or binary valued as in classification. The problem of approximating a function from sparse data is ill-posed and a classical solution is regularization theory [19].

Regularization theory, as we will consider here, formulates the regression problem as a variational problem of finding the function f that minimizes the functional

$$\min_{f \in \mathcal{H}} H[f] = \frac{1}{\ell} \sum_{i=1}^{\ell} V(y_i, f(\mathbf{x}_i)) + \lambda \|f\|_K^2 \qquad (6.1)$$

where $V(\cdot, \cdot)$ is a *loss function* (in the classical formulation the square loss was used), $\|f\|_K^2$ is a norm in a Reproducing Kernel Hilbert Space (RKHS) $\mathcal{H}$ defined by the positive definite function K, ℓ is the number of data points or examples (the ℓ training pairs $(\mathbf{x}_i, y_i)$) and λ is the regularization parameter. Under rather general conditions [14, 22, 17] the solution of

---

[1]Sayan Mukherjee is a PostDoctoral Fellow at the Center for Biological and Computational Learning (CBCL) at MIT and at the MIT/Whitehead Center for Genome Research; Ryan Rifkin recently received his PhD from CBCL at MIT, and now is at Honda Fundamental Research, Boston, USA. Tomaso Poggio is a Professor at in the Department of Brain Sciences at MIT and is affiliated with CBCL, the Artifical Intelligence Laboratory and the McGovern Institute for Brain Research. The authors would like to thank André Elisseeff and Ding Zhou for useful discussions.

[2]There is a large literature on the subject: useful reviews for this chapter are [9, 14, 10, 20] and references therein.

equation (6.1) is

$$f(\mathbf{x}) = \sum_{i=1}^{\ell} c_i K(\mathbf{x}, \mathbf{x}_i). \tag{6.2}$$

The accuracy of the approximated function is based upon its performance on future data, measured in terms of its *generalization error*. Given $\mathbf{x} \in \mathbb{R}^d$ and $y \in \mathbb{R}$ with underlying probability distribution $P(\mathbf{x}, y)$, the generalization error of a function f is

$$I_{exp}[f] \equiv \int V(y, f(\mathbf{x})) dP(\mathbf{x}, y). \tag{6.3}$$

Usually we do not know the distribution $P(\mathbf{x}, y)$. We have only the ℓ training pairs drawn from $P(\mathbf{x}, y)$ from which we can measure the *empirical error*

$$I_{emp}[f] \equiv \frac{1}{\ell} \sum_{i=1}^{\ell} V(y_i, f(\mathbf{x}_i)). \tag{6.4}$$

Regularization theory originates from Tikhonov's classical approach for solving ill posed problems. Existence, uniqueness and especially stability[3] can be restored via a regularizing operator defined through equation 6.1. The basic idea at the heart of the method – as in any approach to ill-posed problems – is to restrict appropriately the space of solutions f to an appropriately small hypothesis space[4] In the final section we will discuss in more detail Tikhonov's regularization approach and its three formulations. Within the universe of ill-posed problems, learning theory with its associated problem of function approximation from a finite set of sparse data has a specific requirement – the derivation of generalization performance and bounds on the performance. Two main approaches have emerged. The first approach – due to Vapnik – is "structural", starting from a bound on the complexity of the space of functions in which the solution is sought. The second, more recent approach starts instead from the stability property that is required of any algorithm solving an ill-posed problem. In the case of Tikhonov regularization the two approaches, as we will see in the last section, correspond to starting from two different but equivalent formulations of Tikhonov's regularization principle. The regularization parameter λ effectively controls both structural complexity and stability, since the two are of course directly related.

---

[3]Stability is defined as continuous dependence of the solution f on the data $(\mathbf{x}_i, y_i)$, e.g. the approximating function must vary little with small perturbations of training data.

[4]Tikhonov's method restricts the solution f to compact sets induced by $\|f\|_K^2 \leq A$ for any positive, finite A, see later.

In section 6.2 we state Tikhonov's regularization approach and some of its specific forms such as Support Vector Machines (SVM). In section 6.3 we outline the structural and the stability approaches to bounding the difference between the empirical and expected errors. In section 6.4 we describe the stability constraint in some detail and state deviation bounds for specific regularization algorithms. In sections 6.5 and 6.6 we use the structural criterion to state the deviation bounds. In section 6.5 we use covering numbers to bound the capacity of the class of functions; in section 6.6 we briefly state closely related results that use VC and V_γ dimension to measure the capacity. In section 6.7 we derive new relations between the two approaches and, in particular, we state that the specific stability property used here implies finite capacity but not vice versa.

6.2   Loss Functions and Regularization Algorithms

In this section we state some commonly used regression and classification algorithms and their associated regularization functional. All algorithms are based on the same regularization functional of equation (6.1) [9]:

$$\min_{f \in \mathcal{H}} H[f] = \frac{1}{\ell} \sum_{i=1}^{\ell} V(y_i, f(\mathbf{x}_i)) + \lambda \|f\|_K^2$$

but with different loss functions $V(\cdot, \cdot)$.

For the case of regression, the square loss corresponds to classical quadratic regularization networks of which splines and radial basis functions are specific cases. The ϵ-loss corresponds to Support Vector Machine (SVM) regression

- $V(y, f(\mathbf{x})) = (y - f(\mathbf{x}))^2$ for quadratic regularization
- $V(y, f(\mathbf{x})) = |y - f(\mathbf{x})|_\epsilon$ for SVM regression

where $|y - f(\mathbf{x})|_\epsilon = \max(0, |y - f(\mathbf{x})| - \epsilon)$.

For the case of binary classification (e.g. $y \in \{-1, 1\}$), SVM classification and variants correspond to the following loss functions

- $V(y, f(\mathbf{x})) = \Theta(-yf(\mathbf{x}))$ for classification with misclassification loss
- $V(y, f(\mathbf{x})) = (1 - yf(\mathbf{x}))_+$ for SVM classification with a soft margin
- $V(y, f(\mathbf{x})) = (f(\mathbf{x}) - y)^2$ for the quadratic classification loss
- $V(y, f(\mathbf{x})) = [yf(\mathbf{x}) - \log(1 + e^{f(\mathbf{x})})]$ for the Import Vector Machine (IVM) [24].

For all loss functions above the solution of equation (6.1) has the form

$$f(\mathbf{x}) = \sum_{i=1}^{\ell} c_i K(\mathbf{x}, \mathbf{x}_i).$$

In the case of SVM classification the label y is simply

$$y = \text{sign}(f(\mathbf{x}))$$

and in the case of the IVM

$$P(y = 1|\mathbf{x}) = \frac{e^{f(\mathbf{x})}}{1 + e^{f(\mathbf{x})}}.$$

6.3   Bounding the Generalization Error

In this section we first introduce the notation we will use and then sketch the two approaches to bound the generalization error.

Given an input space $\mathbf{x} \in \mathcal{X} \subseteq \mathbb{R}^d$ and an output space $y \in \mathcal{Y} \subseteq \mathbb{R}$, a training set

$$S = \{z_1 = (\mathbf{x}_1, y_1), ..., z_\ell = (\mathbf{x}_\ell, y_\ell)\},$$

of size ℓ in $\mathcal{Z} \in \mathcal{X} \times \mathcal{Y}$ is drawn i.i.d. from an unknown distribution D. We will refer to a set

$$S^i = \{z_1, ..., z_i', ..., z_\ell\},$$

where the point z_i in set S is replaced with z_i', a new point drawn i.i.d. from D. We will also refer to a leave-one-out set $S^{\backslash i}$ as the set

$$S^{\backslash i} = \{z_1, ..., z_{i-1}, z_{i+1}, ..., z_\ell\},$$

with the point z_i removed.

Given the training set S we estimate a function $f_S : \mathcal{X} \to \mathcal{Y}$, with $f_S \in \mathcal{H}$. The error of this function with respect to an example $z = (\mathbf{x}, y)$ is defined as

$$V(f_S, z) = V(f_S(\mathbf{x}), y).$$

Thus the empirical error of the function is

$$I_{emp}[f_S] = \frac{1}{\ell} \sum_{i=1}^{\ell} V(f_S, z_i)$$

and the expected or generalization error is

$$I_{exp}[f_S] = \int_{\mathcal{Z}} V(f_S, z)P(z)\ dz = \mathbb{E}_z[V(f_S, z)],$$

where $\mathbb{E}_z[\cdot]$ is the expectation for z sampled from the distribution D.

As we mentioned, to bound the deviation between the empirical error and the expected error two approaches can be used: the structural approach and the stability approach. The structural approach relies on a property called *uniform convergence in probability*: for any $\varepsilon > 0$ and for any set S, we seek bounds of the following form

$$\mathbb{P}_S \left\{ \sup_{f_S \in \mathcal{H}} |I_{emp}[f_S] - I_{exp}[f_S]| \geq \varepsilon \right\} \leq \delta, \qquad (6.5)$$

where the sample S of size ℓ is drawn according to D^ℓ. The capacity, $\mathcal{G}(\mathcal{H}, \varepsilon)$, of the function class $\mathcal{H}$ is used to bound the above probability

$$\delta = \mathcal{G}(\mathcal{H}, \varepsilon) \, e^{-\ell \varepsilon^2 / C}.$$

The capacity can be measured using either covering numbers (the minimal integer n such that there exist disks with radius ε covering $\mathcal{H}$)

$$\mathcal{G}(\mathcal{H}, \varepsilon) = \mathcal{N}(\mathcal{H}, \varepsilon),$$

or the V_γ dimension (of which VC is a special case), d_V [1]:

$$\mathcal{G}(\mathcal{H}, \varepsilon) = \left(\frac{e\ell}{d_V} \right)^{d_V}.$$

Note that the value of γ used in computing the V_γ dimension should be related to the level of precision required which is a function of ε. Using the bound in equation (6.5) one can state that with probability $1 - \delta$

$$I_{exp}[f_S] \leq I_{emp}[f_S] + O\left(\sqrt{\frac{1}{\ell}} \right).$$

In the stability approach a seemingly weaker requirement (but see discussion), *convergence in probability*, is needed to bound for any $\varepsilon > 0$ and for any set S

$$\mathbb{P}_S \left\{ |I_{emp}[f_S] - I_{exp}[f_S]| \geq \varepsilon + \beta \right\} \leq \delta$$

where the functions f_S are the output of an algorithm with the following stability property

$$\forall S, S^i \in Z^\ell \quad |V(f_S, \cdot) - V(f_S^i, \cdot)|_\infty \leq \beta, \qquad (6.6)$$

with $\beta = O(1/\ell)$ for the bounds to make sense. Throughout this chapter the term β-stability means stability with $\beta = O(1/\ell)$. The stability of algorithms is used to bound the above probability

$$\delta = \exp\left(-\frac{2\ell \varepsilon^2}{(4\ell\beta + 2C)^2} \right).$$

For $\beta = O(1/\ell)$ one can use the bound in equation (6.6) to state that with probability $1 - \delta$

$$I_{exp}[f_S] \leq I_{emp}[f_S] + O\left(\sqrt{\frac{1}{\ell}} \right).$$

6.4   Stability Based Bounds on Generalization Error

In this section, we outline the stability approach to prove generalization bounds for regularization algorithms. The approach and this exposition draw heavily on the work of Elisseef and Bousquet [4, 5].

The basic idea behind stability bounds is simple. Instead of bounding the complexity of a space of functions an algorithm is allowed to search, the goal is to test directly the stability of a function produced by a given algorithm, that the function will not change much when the training data is changed slightly. This leads to high-probability bounds on the difference between the generalization error and the training error for the function that the algorithm generates.

6.4.1   Notation and Preliminaries

The enabling mathematics for stability bounds are the so-called concentration inequalities, typified by McDiarmid's inequality [12]:

Theorem 6.4.1. (McDiarmid, 1989) *Given sets S and S^i as above, and $F : Z^m \to R$ a measurable function, assume there exist constants c_i $(i = 1, \ldots, \ell)$ such that*

$$\sup_{S, S^i \in Z^\ell} |F(S) - F(S^i)| \leq c_i.$$

Then, for any $\epsilon > 0$, the following bound holds

$$\mathbb{P}[F(S) - \mathbb{E}[F(S)] \geq \epsilon] \leq e^{-2\epsilon^2 / \sum_{i=1}^m c_i^2}.$$

McDiarmid's inequality can be viewed as a martingale version of the standard Hoeffding's inequality. Instead of bounding the range of a random variable, we bound the amount a function can change when we replace a data point.

An algorithm A is a procedure that takes a training set S and produces a hypothesis function A_S. The empirical error of A_S is

$$I_{emp}[A_S] = \frac{1}{\ell} \sum_{i=1}^{\ell} V(A_s, z_i).$$

The generalization error of A_S is defined to be

$$I_{exp}[A_S] = \mathbb{E}_z[V(A_s, z)].$$

It is this generalization error that we wish to control. However, since we do not know the underlying distribution D, we cannot measure it directly. We will see that for *stable* algorithms (defined below), we are able to bound the difference between the empirical and the generalization errors of A_S.

We will need several facts about convex functions. A *convex function F* is a function that satisfies

$$\lambda F(g) + (1 - \lambda)F(g') \geq F(\lambda g + (1 - \lambda)g')$$

for all g and g' and all $\lambda \in [0, 1]$. The *subgradient* of a convex function at a point g, denoted $\partial F(g)$, is the set a satisfying:

$$F(g') \geq F(g) + <g' - g, a>,$$

for all g'. If F is differentiable at g, the set $\partial F(g)$ is a singleton, and if g minimizes F, then $0 \in \partial F(g)$.

The *divergence* of F from g to g' is the set

$$d_F(g, g') = F(g) - F(g') - <g - g', \partial F(g') >$$

If F is differentiable at g', then $d_F(g, g')$ is a singleton, and if g' minimizes F, then $F(g) - F(g') \in d_F(g, g')$.

If $F = F_a + F_b$, where F_a and F_b are also convex functions, and F_a, F_b, and F are all differentiable at g', then

$$d_F(g, g') = d_{F_a}(g, g') + d_{F_b}(g, g')$$

If one or more of the functions are not differentiable at g', then, for any $d \in d_F(g, g')$, we can still *choose* $a \in d_{F_a}(g, g'), b \in d_{F_b}(g, g')$ such that $a + b = d$.

We also need the following fact about functions in a RKHS $\mathcal{H}$ defined by a kernel K [2, 9]:

$$\forall f \in \mathcal{H}, \forall x \in X, |f(x)| \leq \|f\|_K \sqrt{K(x, x)}. \tag{6.7}$$

6.4.2   Stability

Definition 6.4.1. *An algorithm A has* uniform stability β *with respect to a loss function V if*

$$\forall S, S^i \in \mathcal{Z}^\ell, \quad \forall i \in 1, \ldots, \ell \quad \|V(A_S, \cdot) - V(A_{S^i}, \cdot)\|_\infty \leq \beta$$

This key definition states that if an algorithm has uniform stability β, then for *any* training set S (which causes A to produce $f(A_S)$), the replacement of *any* point in S with *any* possible other point z' from the distribution D, generating a new training set S^i, gives $\|f(A_S) - f(A_{S^i})\|_\infty \leq \beta$. This may appear to be a very strong constraint. Not surprisingly, however, regularization algorithms exhibit uniform stability.

6.4.3   Generalization Bounds for Stable Algorithms

Using McDiarmid's Inequality and some probabilistic arguments, it is possible to derive the following [5]:

Theorem 6.4.2. (Elisseef and Bousquet, 2001) *Let A be an algorithm with uniform stability β with respect to a loss function V, satisfying $0 \leq V(A_S, z) \leq M$ for all $S, z \in \mathcal{Z}^{\ell+1}$. Then, for any $\ell \geq 1$, and any $\delta \in (0, 1)$, with probability at least $1 - \delta$ over the random draw of the sample S,*

$$I_{exp} \leq I_{emp} + 2\beta + (4\ell\beta + M)\sqrt{\frac{\ln 1/\delta}{2\ell}}$$

The theorem can be proved by applying McDiarmid's theorem to the random variable $I_{exp} - I_{emp}$.

6.4.4   Stability for Regularization Algorithms

In this section we will demonstrate that certain regularization algorithms exhibit uniform stability. A regularization algorithm looks for a minimizer such as

$$\min_f R_r(f) \equiv \frac{1}{\ell} \sum_{j=1}^{\ell} V(f, z_j) + \lambda N(f).$$

where N, called the stabilizer, is a functional from F to $\mathbb{R}^+$ which measures the complexity and typically also the smoothness of a function. Intuitively, regularization algorithms find a trade off between low empirical error (the first term) and a smooth or simple function (the second term increases with complexity and decreases with smoothness).

We will also make use of a truncated version of R_r,

$$R_{r\backslash i} \equiv \frac{1}{\ell} \sum_{j\neq i} V(f, z_j) + \lambda N(f).$$

We denote the minimizers of R_r and $R_{r\backslash i}$ as f, and $f^{\backslash i}$, respectively, and we define $\Delta f = f^{\backslash i} - f$. We will need a smoothness constraint on the loss function V as well, which states that the loss function does not vary too wildly as we vary its arguments:

Definition 6.4.2. *A loss function V is σ-admissible if it is convex w.r.t. its first argument and $\forall x \in \mathcal{X}, \forall S^1, S^2 \in \mathcal{Z}^\ell, \forall y' \in \mathcal{Y}$*

$$|V(f_{S^1}(x), y') - V(f_{S^2}(x), y')| \leq \sigma|f_{S^1}(x) - f_{S^2}(x)|.$$

If both the loss function V and the stabilizer N are convex and continuous, we have the following lemma:

Lemma 6.4.1. (Elisseef and Bousquet, 2001)

$$d_N(f, f^{\backslash i}) + d_N(f^{\backslash i}, f) \leq \frac{1}{\lambda\ell}(V(f^{\backslash i}, z_i) - V(f, z_i)) \leq \frac{\sigma}{\lambda\ell}|\Delta f(x_i)|.$$

PROOF:Assume that both V and N are differentiable. Using the fact that f and $f^{\backslash i}$ minimize R_r and $R_{r\backslash i}$, respectively, we have,

$$
\begin{aligned}
d_{R_r}(f^{\backslash i}, f) + d_{R_{r\backslash i}}(f, f^{\backslash i}) &= R_r(f^{\backslash i}) - R_r(f) + R_{r\backslash i}(f) - R_{r\backslash i}(f^{\backslash i}) \\
&\leq \frac{1}{\ell}V(f^{\backslash i}, z_i) - \frac{1}{\ell}V(f, z_i).
\end{aligned}
$$

Noting that $d_{R_r} = d_V + \lambda d_N$, the first inequality in the lemma follows. The second inequality then follows by the definition of σ-admissibility.

If the functions are not differentiable, the proof proceeds similarly, but we must choose appropriate subgradients of V and N that sum to zero subgradients of R_r and $R_{r\backslash i}$. $\square$

We are now ready to prove the key theorem that shows that regularization algorithms exhibit β stability.

Theorem 6.4.3. (Elisseef and Bousquet, 2001) *Let $\mathcal{H}$ be a RKHS with kernel K such that $K(x,x) \leq \kappa \leq \infty$ for all x. Let V be σ-admissible. The learning algorithm A_S which minimizes*

$$
\frac{1}{\ell}\sum_{i=1}^{\ell} V(g, z_i) + \lambda\|g\|_K^2
$$

has uniform stability β with

$$
\beta \leq \frac{\sigma^2\kappa^2}{2\lambda\ell}.
$$

PROOF:If the stabilizer $N(\cdot)$ is $\|\cdot\|_K^2$, then

$$
d_N(g, g') = \|g - g'\|_K^2.
$$

Applying lemma 6.4.1, the previous equation implies that

$$
2\|\Delta f\|_K^2 \leq \frac{\sigma}{\lambda\ell}|\Delta f(x_i)|.
$$

Since the functions f are in an RKHS,

$$
|\Delta f(x_i)| \leq \|\Delta f\|_K.
$$

Combining these last two inequalities and using equation 6.7, we obtain

$$
\|\Delta f\|_K \leq \frac{\kappa\sigma}{2\lambda\ell}.
$$

By the σ-admissibility of V and the RKHS property, we have

$$
|V(f,z) - V(f^{\backslash i}, z)| \leq \sigma\|f(.) - f^{\backslash i}(.)\|_\infty = \sigma\|\Delta f(x)\|_\infty \leq \sigma\kappa\|f\|_K,
$$

which combined with the previous equation yields the result. $\square$

6.5   Generalization Bounds Based on Covering Numbers

In this section we present the use of covering numbers or metric entropy to prove generalization bounds for regularization algorithms. Covering and entropy numbers have been used extensively in generalization bounds [7, 15]. The basic idea behind bounds based upon covering numbers is to first bound the difference between the generalization error and training error for a single function $f \in \mathcal{H}$. One then extends this bound to hold simultaneously for all functions in the class by first coarse-graining the space $\mathcal{H}$ into $\mathcal{N}$ prototype functions which cover the space and then applying the union bound to the bound for a single function. The covering number, $\mathcal{N}$, is a complexity measure of the function space $\mathcal{H}$. For regularization algorithms, recent results [6, 23, 18] demonstrate how to use classical results from approximation theory to compute covering numbers for functions belonging to a RKHS.

6.5.1   Bounds for a Single Function

We first bound the deviation between the empirical error and the expected error for a single function $f \in \mathcal{H}$.

Theorem 6.5.1. *Given a bounded loss function $0 \leq V(f(\mathbf{x}), y) \leq M$. Then for all $\epsilon > 0$*

$$\mathbb{P}_S \left\{ |I_{emp}[f] - I_{exp}[f]| \geq \epsilon \right\} \leq 2e^{-2\epsilon^2 \ell / M^2},$$

where ℓ is the number of samples in S.

The proof of the above theorem is an application of Hoeffding's inequality to the random variable $V(f(\mathbf{x}), y)$.

6.5.2   Bounds for a Class of Functions

We now bound the deviation between the empirical error and the expected error for the class of functions $\mathcal{H}$.

Let S be a metric space and $r > 0$. The covering number $\mathcal{N}(S, r)$ is the smallest $n \in \mathbb{N}$ such that there exist n disks of radius r that covers S. When S is compact (as induced by regularization in the case of RKHS [6]), the covering number is finite. The metric or epsilon entropy of a class of functions is defined as $\ln \mathcal{N}(S, r)$.

Theorem 6.5.2. *Given the square loss function bounded between $0 \leq V(f, x) \leq M$ and a compact $\mathcal{H}$. Then for all $\epsilon > 0$*

$$\mathbb{P}_S \left\{ \sup_{f \in \mathcal{H}} |I_{emp}[f] - I_{exp}[f]| \geq \epsilon \right\} \leq 2\mathcal{N} \left(\mathcal{H}, \frac{\epsilon}{8M} \right) e^{-\epsilon^2 \ell / 4M^2},$$

where ℓ is the number of samples in S.

The same result can be proved (in preparation) for any L_p or Lipschitz loss function except the constant 8 in $\frac{\epsilon}{8M}$ is replaced by a different constant. The proof consists of applying a union bound to theorem 6.5.1 over the covering number for the function class.

Theorem 6.5.3. *Given the square loss function bounded between $0 \leq V(f,x) \leq M$ and a compact $\mathcal{H}$. For all $\ell \geq \frac{4M^2}{\epsilon^2} \ln\left(\frac{\mathcal{N}}{\delta}\right)$ with probability $1 - \delta$*

$$I_{exp} \leq I_{emp} + \sqrt{\frac{4M^2}{\ell}\left[\ln\left(\frac{\mathcal{N}}{\delta}\right)\right]},$$

where $\mathcal{N}\left(\mathcal{H}, \frac{\epsilon}{8M}\right)$ is the covering number.

Again, the same result holds for any L_p or Lipschitz loss function with the constant 8 in $\frac{\epsilon}{8M}$ replaced by a different constant.

6.5.3   Computing the Covering Number

Measuring the size of a function class is a classic problem. The use of covering numbers as a measure was pioneered in [11]. When the RKHS is a finite dimensional space (the eigenvalue problem associated with the kernel has a finite number of strictly positive eigenvalues) one can compute covering numbers using classical results about sphere packing in Banach spaces. For the case of an infinite dimensional RKHS we use recent results [23, 18] to compute covering numbers for functions in a RKHS, $\mathcal{H}$, by embedding these function classes into a Sobolev space. Classic results exist for covering numbers of Sobolev spaces [8]. Inequalities (6.8), (6.9), (6.10), and (6.11) state bounds for the covering numbers generated by functions from an RKHS.

We first define the kernel function K that induces the RKHS, $\mathcal{H}$,

$$K : \mathcal{X} \times \mathcal{X} \to \mathbb{R}$$

to be continuous, symmetric and positive definite ($\mathcal{X} \subset \mathbb{R}^d$). The associated RKHS is a subset of $\mathcal{C}(\mathcal{X})$, the space of continuous functions on $\mathcal{X}$. The space of functions $\mathcal{H}$ can be embedded into $\mathcal{C}(\mathcal{X})$ with the inclusion denoted as $I_K : \mathcal{H} \to \mathcal{C}(\mathcal{X})$. For $R > 0$ B_R is the ball in $\mathcal{H}$ with radius R

$$B_R = \{f \in \mathcal{H} : \|f\|_K \leq R\}.$$

So $I_K(B_R)$ is a subset of $\mathcal{C}(\mathcal{X})$ and $\overline{I_K(B_R)}$, its closure in $\mathcal{C}(\mathcal{X})$, is a compact subset of $\mathcal{C}(\mathcal{X})$. We will bound the covering numbers of this space.

If the RKHS induced by the kernel is finite dimensional we can use sphere packing results to state the following bound

$$\ln \mathcal{N}(\overline{I_K(B_R)}, \epsilon) \leq N \ln \left(\frac{4R}{\epsilon} \right), \tag{6.8}$$

where N is the dimensionality of the RKHS.

For the case where the RKHS is infinite dimensional the following bounds hold. For the case where the kernel K is $\mathcal{C}^\infty$ and $\mathcal{X} \subset \mathbb{R}^d$ it was shown [6] that $\mathcal{H}$ can be embedded into the Sobolev space $L^{h,2}(\mathcal{X})$ for any $h > 0$. Using results in approximation theory for covering numbers of Sobolev spaces, Smale and Zhou then derived the following bound [18]:

$$\ln \mathcal{N}(\overline{I_K(B_R)}, \epsilon) \leq \left(\frac{R C_h}{\epsilon} \right)^{\frac{2d}{h}}, \tag{6.9}$$

where C_h is a constant independent of R and ϵ. This result was recently extended [23] to show that if K is $\mathcal{C}^s$, and s is odd, then $\mathcal{H}$ can be embedded into $\mathcal{C}^{s/2}(\mathcal{X})$. For the case where $\mathcal{X} = [0,1]^d$ and $s > 0$ it can be shown [23] that

$$\ln \mathcal{N}(\overline{I_K(B_R)}, \epsilon) \leq C \left(\frac{R}{\epsilon} \right)^{\frac{2d}{s}}, \tag{6.10}$$

where C is a constant independent of R and ϵ. If the kernel is an analytic function with exponential decay of its eigenvalues, for example the Gaussian, then the following bound holds [6, 18, 23]

$$\ln \mathcal{N}(\overline{I_K(B_R)}, \epsilon) \leq (d+1) \ln \left[C \ln \left(\frac{R}{\epsilon} \right) \right], \tag{6.11}$$

where C is a constant independent of R and ϵ. Note that for the covering number bounds to make sense R must be finite. This is always the case for functions in a RKHS.

6.6   Generalization Bounds Based on VC and V_γ Dimensions

In this section we summarize the use of two measures of the capacity of a function space VC [21] and V_γ [1] to prove generalization bounds for regularization algorithms. In the case of real valued functions finiteness of an extension of VC-dimension called V_γ-dimension is a *necessary and sufficient* condition for uniform convergence.

We now define V_γ-dimension for the case of real valued functions:

Definition 6.6.1. *Let* $0 \leq V(y, f(\mathbf{x})) \leq M$, $f \in \mathcal{H}$. *The* V_γ-*dimension of* $\mathcal{F}$ *(the set* $\{V(y, f(\mathbf{x})), \ f \in \mathcal{H}\}$) *is defined as the the maximum number* h *of vectors* $(\mathbf{x}_1, y_1) \ldots, (\mathbf{x}_h, y_h)$ *that can be separated into two classes in all* 2^h *possible ways using rules:*

$$\text{class 1 if: } V(y_i, f(x_i)) \geq s + \gamma$$
$$\text{class 0 if: } V(y_i, f(x_i)) \leq s - \gamma$$

for $f \in \mathcal{H}$ *and some* $s \geq 0$*. If, for any number* N*, it is possible to find* N *points* $(\mathbf{x}_1, y_1) \ldots, (\mathbf{x}_N, y_N)$ *that can be separated in all the* 2^N *possible ways, we will say that the* V_γ*-dimension of* $\mathcal{F}$ *is infinite.*

A uniform Glivenko-Cantelli class is a function class for which there exists uniform convergence in probability.

Theorem 6.6.1. (Alon et al, 1997) *Let* $\mathcal{F}$ *be a class of functions from* $\mathcal{X}$ *into* $[0, 1]$ *(that is the set comprised of the loss function* $V(\cdot)$ *and functions* $f \in \mathcal{H}$*).*
  1. The following are equivalent:
  (a) $\mathcal{F}$ *is a uniform Glivenko-Cantelli class.*
  (b) V_γ*-dim($\mathcal{F}$) is finite for all* $\gamma > 0$*.*

For the case of classification, $y \in \{-1, 1\}$, finite VC-dimension is a *necessary and sufficient* condition for uniform convergence. This is in contrast to regression where it is only a *sufficient* condition. So when $y \in \{-1, 1\}$, $V(y, f(\mathbf{x})) = \Theta(-yf(\mathbf{x}))$ is the misclassification error, and $\gamma = 0$, the V_γ-dimension reduces to the VC-dimension:

Definition 6.6.2. *The* VC*-dimension of a set* $\{\theta(f(\mathbf{x})), f \in \mathcal{H}\}$*, of indicator functions is the maximum number* h *of vectors* $\mathbf{x}_1, \ldots, \mathbf{x}_h$ *that can be separated into two classes in all* 2^h *possible ways using functions of the set. If, for any number* N*, it is possible to find* N *points* $\mathbf{x}_1, \ldots, \mathbf{x}_N$ *that can be separated in all the* 2^N *possible ways, we will say that the* VC*-dimension of the set is infinite.*

The uniform deviation between the empirical error and expected error for real valued functions can be bounded using the V_γ dimension. For the classification case the VC dimension can be used to bound the deviation:

Theorem 6.6.2. (Vapnik and Chervonenkis, 1979) *Let* $0 \leq V(y, f(\mathbf{x})) \leq 1$*,* $f \in \mathcal{H}$*,* $\mathcal{H}$ *be a set of bounded functions and* h *the VC-dimension of* $\mathcal{H}$*. Then, with probability at least* $1-\epsilon$*, the following inequality holds for all the elements* $f \in \mathcal{H}$*:*

$$I_{emp} \leq I_{emp} + \sqrt{\frac{h \ln \frac{2e\ell}{h} - \ln(\frac{\epsilon}{4})}{\ell}} \tag{6.12}$$

6.7 Generalization bounds: Stability and Capacity Control

We have described two approaches to generalization bounds, one based on the stability of an algorithm and the other based on the capacity of the underlying function class. In this section we discuss the relation between these two approaches.

Some historical perspective based on the classical approach to solving ill-posed problems is worthwhile. The basic idea in all approaches is to restore well-posedness – that is existence, uniqueness and stability of the solution – by restricting appropriately the data space and especially the solution space. Variational techniques achieve this goal in an implicit way. The so-called *generalized (or C-generalized) solutions* are sufficient to restore well-posedness of the learning problem (see [19, 3]). The basic idea behind *C*-generalized solutions is to restrict the solution space to a particular class of functions *C*. The induced continuous dependence of the solution on the data is not sufficient to be robust against noise, since the solution may be unstable for small changes in the initial data. In the context of learning this means that generalization is not automatically ensured. What is needed is sufficiently strong stability. The regularization method of Tikhonov was designed to deal with the noisiest or most difficult cases of ill-posed problems. This method also ensures a sufficiently strong stability to achieve good generalization bounds. Equation (6.1) defines a one-parameter family of solutions f_λ that approach the generalized solution of $f(\mathbf{x}_i) = y_i$ for $\lambda \to 0$. In general, optimization of the generalization error in learning theory requires an optimal value of the parameter λ. To gain some insight into the families of solutions corresponding to different λ, consider two other regularization methods which have been proposed and shown to be equivalent to Tikhonov's:

1. In Tikhonov's method:

$$\min_{f \in \mathcal{H}} \frac{1}{\ell} \sum_{i=1}^{\ell} V(y_i, f(\mathbf{x}_i)) + \lambda \|f\|_K^2$$

   for given $\{\mathbf{x}_i, y_i\}_{i=1}^{\ell}$ and λ;

2. In Phillips' method:

$$\min_{f \in \mathcal{H}} \|f\|_K^2$$

   subject to $\frac{1}{\ell} \sum_{i=1}^{\ell} V(y_i, f(\mathbf{x}_i)) \le \delta$ for given $\{\mathbf{x}_i, y_i\}_{i=1}^{\ell}$ and δ;

3. In Ivanov's method:

$$\min_{f \in \mathcal{H}} \frac{1}{\ell} \sum_{i=1}^{\ell} V(y_i, f(\mathbf{x}_i))$$

subject to $\|f\|_K^2 \leq R$ for given $\{\mathbf{x}_i, y_i\}_{i=1}^\ell$ and R.

The three techniques are equivalent in the sense that if a solution f_0 of problem 1. exists for $\lambda_0 > 0$, then f_0 is a solution of problem 2. with $\delta = \frac{1}{\ell} \sum_{i=1}^\ell V(y_i, f_0(\mathbf{x}_i))$ and is a solution of problem 3. with $R = \|f_0\|_K^2$. The converse is also true. If f_0 is a solution of problem 2. then a non-negative λ_0 exists such that f_0 is also a solution of problem 1 and $\frac{1}{\ell} \sum_{i=1}^\ell V(y_i, f_0(\mathbf{x}_i)) = \delta$. If f_0 is a solution of problem 3., then a non-negative λ_0 exists such that f_0 is also a solution of problem 1. with $\|f_0\|_K^2 = R$.

Though the three techniques are equivalent in the sense outlined above, they make explicit different viewpoints. Structural Risk Minimization and generalization bounds in terms of capacity constraints are most naturally incorporated in Technique 3 since the constraint on the function space is given explicitly in terms of R[5]. On the other hand generalization bounds in terms of stability can be easily derived directly for Technique 1. Thus, in a sense, stability offers a more natural route for regularization formulations such as equation (6.1).

The relation between the λ in the Tikhonov method and the bound on the RKHS norm R in the Ivanov method has been described for σ-admissible loss functions [5] and for square loss [6]. For the case of σ-admissible loss functions

$$R^2(\lambda) = \|f_S\|_K^2 \leq \frac{B_0}{\lambda} \tag{6.13}$$

where $B_0 = \sup_{y \in \mathcal{Y}} V(0, y)$. For the case of square loss if the target function g (the function to be approximated) is known then the Tikhonov regularization functional has the form

$$\min_{f \in \mathcal{H}} \|g - f\|^2 + \lambda \|A^{-1} g\|^2, \tag{6.14}$$

where A is the self-adjoint strictly positive compact operator associated with the kernel K, so $\|A^{-1} g\| = \|g\|_K$. The minimizer of (6.14) has the form $\hat{f}_\lambda = (I + \lambda A^{-2})g$. So we can write R as a function of λ

$$R(\lambda) = \|\hat{f}_\lambda\|_K. \tag{6.15}$$

Superficially, it seems that the conditions for applying covering number (or V_γ) bounds

$$\mathbb{P}_S \left\{ \sup_{f_S \in \mathcal{H}} |I_{emp}[f_S] - I_{exp}[f_S]| \geq \varepsilon \right\} \leq C_1 \left(\frac{e\ell}{V_\gamma} \right)^{V_\gamma} e^{-\ell \varepsilon^2 / C_2}$$

---

[5]The point of view given by Technique 1 can also be used but the reasoning is more involved, see [9].

may be stricter than those for stability bounds

$$\mathbb{P}_S\left\{|I_{emp}[f_S] - I_{exp}[f_S]| \geq \varepsilon + \frac{C_3}{\ell}\right\} \leq e^{-\ell\varepsilon^2/C_4}.$$

This is due to the fact that the covering number bounds have to hold simultaneously over a a class of functions rather than a particular function – the output of an algorithm – as is the case for stability bounds. However, closer inspection reveals that β-stability is a stricter requirement than finite covering number. To date, all β-stable algorithms are forms of Tikhonov regularization.

It will be shown that the "class of functions" output by a Tikhonov regularization algorithm has finite V_γ dimension but, empirical risk minimization using a hypothesis space of finite V_γ dimension is not necessarily β-stable.

For a Tikhonov regularization algorithm with σ-admissible loss functions the RKHS norm of the function output by the algorithm is bounded (6.13)

$$\|f_S\|_K \leq \sqrt{\frac{B_0}{\lambda}}.$$

We can then bound the V_γ dimension of the hypothesis space $\mathcal{H}$ of functions output by the regularization algorithm as

$$V_\gamma(\mathcal{H}(A), V) \leq \ln \mathcal{N}\left(\frac{1}{2}\sqrt{\frac{B_0}{\lambda}}, 2\gamma/\sigma\right),$$

where $V(\cdot)$ is a σ-admissible loss function, $\mathcal{H}(A)$ is the hypothesis space of functions the algorithm can output, $\mathcal{N}\left(\frac{1}{2}\sqrt{\frac{B_0}{\lambda}}, 2\gamma/\sigma\right)$ is the covering number (the number of disks of radius $2\gamma/\sigma$ that cover a ball of radius $\sqrt{B_0/\lambda}/2$). Using results from section 6.5 we get the following bounds on the V_γ dimension

$$N \text{ dimensional RKHS} \quad \Rightarrow \quad N \ln\left(\frac{\sqrt{B_0}\sigma}{\sqrt{\lambda}\gamma}\right)$$

$$\text{infinite dimensional RKHS, } K \text{ is } C^s \quad \Rightarrow \quad \ln\left(\frac{C\sqrt{B_0}\sigma}{4\gamma\sqrt{\lambda}}\right)^{2d/s}$$

$$\text{infinite dimensional RKHS, } K \text{ analytic} \quad \Rightarrow \quad (d+1)\ln\left[C\ln\left(\frac{\sqrt{B_0}\sigma}{4\gamma\sqrt{\lambda}}\right)\right]$$

where C is some constant and d is the dimensionality of $\mathcal{X}$. (For details about these results see [16].)

The fact that empirical risk minimization with functions that have finite V_γ dimension is not β-stable is illustrated in Figure (6.1). The loss function used was the quadratic loss $V(f(x), y) = (f(x) - y)^2$ and our hypothesis space consists of lines $f(x) = wx$. This corresponds to using linear least-

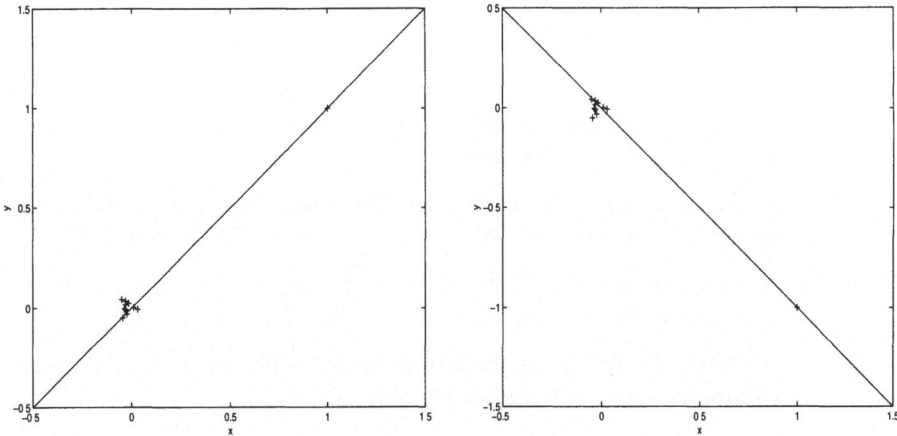

Figure 6.1. The two data sets and their respective solutions.

squares regression as our empirical risk minimization algorithm. The two pictures plot the functions obtained when this algorithm is run on two datasets that are identical at all points but one. Clearly, this algorithm is not β-stable with $\beta = O(1/\ell)$.

References

[1] N. Alon, S. Ben-David, N. Cesa-Bianchi, and D. Haussler. Scale-sensitive dimensions, uniform convergence, and learnability. *J. of the ACM*, 44(4):615–631, 1997.

[2] N. Aronszajn. Theory of reproducing kernels. *Trans. Amer. Math. Soc.*, 686:337–404, 1950.

[3] M. Bertero, T. Poggio, and V. Torre. Ill-posed problems in early vision. *Proceedings of the IEEE*, 76:869–889, 1988.

[4] O. Bousquet and A. Elisseeff. Algorithmic stability and generalization performance. In *Neural Information Processing Systems 14*, Denver, CO, 2000.

[5] O. Bousquet and A. Elisseeff. Stability and generalization. *Journal Machine Learning Research*, 2001. submitted.

[6] F. Cucker and S. Smale. On the mathematical foundations of learning. *Bulletin of AMS*, 2001. in press.

[7] R.M. Dudley, E. Gine, and J. Zinn. Uniform and universal glivenko-cantelli classes. *Journal of Theoretical Probability*, 4:485–510, 1991.

[8] D. Edmunds and H. Triebel. *Function Spaces, Entropy Numbers, Differential Operators.* Cambridge University Press, 1996.

[9] T. Evgeniou, M. Pontil, and T. Poggio. Regularization networks and support vector machines. *Advances in Computational Mathematics*, 13:1–50, 2000.

[10] F. Girosi, M. Jones, and T. Poggio. Regularization theory and neural networks architectures. *Neural Computation*, 7:219–269, 1995.

[11] A. Kolmogorov and V. Tikhomirov. ϵ-entropy and ϵ-capacity of sets in function spaces. *Transl. of the AMS*, 17:277–364, 1961.

[12] C. McDiarmid. On the method of bounded differences. *In Surveys in Combinatorics 1989*, pages 148–188, 1989.

[13] T. Poggio and F. Girosi. A theory of networks for approximation and learning. C.B.I.P. Memo No. 31, Center for Biological Information Processing, Whitaker College, 1989.

[14] T. Poggio and F. Girosi. Networks for Approximation and Learning. In C. Lau, editor, *Foundations of Neural Networks*, pages 91–106. IEEE Press, Piscataway, NJ, 1992.

[15] D. Pollard. *Convergence of stochastic processes.* Springer-Verlag, Berlin, 1984.

[16] R. Rifkin, S. Mukherjee, and T. Poggio. Stability, generalization, and uniform convergence. Ai memo, MIT, 2002. in press.

[17] B. Schölkopf, R. Herbrich, and A. Smola. A generalized representer theorem. In *Proc. of COLT*, p. 416–426. Springer Verlag, 2001.

[18] S. Smale and D. Zhou. Estimating the approxiation error in learning theory. 2001. preprint.

[19] A. N. Tikhonov and V. Y. Arsenin. *Solutions of Ill-posed Problems.* W. H. Winston, Washington, D.C., 1977.

[20] V. N. Vapnik. *Statistical Learning Theory.* Wiley, New York, 1998.

[21] V. N. Vapnik and A. Y. Chervonenkis. On the uniform convergence of relative frequences of events to their probabilities. *Th. Prob. and its Applications*, 17(2):264–280, 1971.

[22] G. Wahba. *Splines Models for Observational Data.* Series in Applied Mathematics, Vol. 59, SIAM, Philadelphia, 1990.

[23] D. Zhou. The regularity of reproducing kernel hilbert spaces in learning theory. 2001. preprint.

[24] J. Zhu and T. Hastie. Kernel logistic regression and the import vector machine. In *Proc. of Neural Information Processing Systems*, 2001. submitted.

7

Optimal Properties and Adaptive Tuning of Standard and Nonstandard Support Vector Machines

Grace Wahba, Yi Lin, Yoonkyung Lee, and Hao Zhang[1]

Summary

We review some of the basic ideas of Support Vector Machines (SVM's) for classification, with the goal of describing how these ideas can sit comfortably inside the statistical literature in decision theory and penalized likelihood regression. We review recent work on adaptive tuning of SVMs, discussing generalizations to the nonstandard case where the training set is not representative and misclassification costs are not equal. Mention is made of recent results in the multicategory case.

7.1   Introduction

This chapter is an expanded version of the talk given by one of the authors (GW) at the Mathematical Sciences Research Institute Berkeley Workshop on Nonlinear Estimation and Classification, March 20, 2001. In this chapter we review some of the basic ideas of Support Vector Machines(SVMs) with the goal of describing how these ideas can sit comfortably inside the statistical literature in decision theory and penalized likelihood regression, and we review some of our own related research.

---

[1]Grace Wahba is Bascom Professor of Statistics, Professor of Biostatistics and Medical Informatics, and Professor of Computer Sciences (by courtesy), University of Wisconsin-Madison. Yi Lin is Assistant Professor, Department of Statistics, University of Wisconsin-Madison. Yoonkyung Lee and Hao Zhang are doctoral candidates in the Department of Statistics, University of Wisconsin-Madison.

Support Vector Machines (SVM's) burst upon the classification scene in the early 90's, and soon became the method of choice for many researchers and practitioners involved in supervised machine learning. The talk of Tommi Poggio at the Berkeley workshop highlights some of the many interesting applications. The website http://kernel-machines.org is a popular repository for papers, tutorials, software, and links related to SVM's. A recent search in http://www.google.com for 'Support Vector Machines' leads to 'about 10,600' listings. Recent books on the topic include [24] [25] [5], and there is a section on SVM's in [10]. [5] has an incredible (for a technical book) ranking in amazon.com as one of the 4500 most popular books.

The first author became interested in SVM's at the AMS-IMS-SIAM Joint Summer Research Conference on Adaptive Selection of Models and Statistical Procedures, held at Mount Holyoke College in South Hadley MA in June 1996. There, Vladimir Vapnik, generally credited with the invention of SVM's, gave an interesting talk, and during the discussion after his talk it became evident that the SVM could be derived as the solution to an optimization problem in a Reproducing Kernel Hilbert Space (RKHS), [26], [30] [13], [28], thus bearing a resemblance to penalized likelihood and other regularization methods used in nonparametric regression. This served to link the rapidly developing SVM literature in supervised machine learning to the now obviously related statistics literature. Considering the relatively recent development of SVM's, compared to the 40 or so year history of other classification methods, it is of interest to question theoretically why SVM's work so well. This question was recently answered in [18], where it was shown that, provided a rich enough RKHS is used, the SVM is implementing the Bayes rule for classification. Convergence rates in some special cases can be found [19]. An examination of the form of the SVM shows that it is doing the implementation in a flexible and particularly efficient manner.

As with other regularization methods, there is always one, and sometimes several tuning parameters which must be chosen well in order to have efficient classification in nontrivial cases. Our own work has focused on the extension of the Generalized Approximate Cross Validation (GACV) [36] [17] [8] from penalized likelihood estimates to SVM's, see [21] [20] [33] [30]. At the Berkeley meeting, Bin Yu pointed GW to the $\xi\alpha$ method of Joachims [12], which turned out to be closely related to the GACV. Code for the $\xi\alpha$ estimate is available in SVM^{light} http://ais.gmd.de/~thorsten/svm_light/. At about this time there was a lot of activity in the development of tuning methods, and a number of them [27] [11] [23] [12] [2] turned out to be related under various circumstances.

We first review optimal classification in the two-category classification problem. We describe the standard case, where the training set is representative of the general population, and the cost of misclassification is the same for both categories, and then turn to the nonstandard case, where neither of these assumptions hold. We then describe the penalized likelihood estimate for Bernoulli data, and compare it with the standard SVM. Next we discuss how the SVM implements the Bayes rule for classification and then we turn to the GACV for tuning the standard SVM. Joachims' $\xi\alpha$ method is then described and the GACV and Joachims' $\xi\alpha$ method are compared. Next we turn to the nonstandard case. We describe the nonstandard SVM, and show how both the GACV and the $\xi\alpha$ method can be generalized in that case, from [32]. A modest simulation shows that they behave similarly. Finally, we briefly mention that we have generalized the (standard and nonstandard) SVM to the multicategory case [15].

7.2   Optimal Classification and Penalized Likelihood

Let $h_{\mathcal{A}}(\cdot), h_{\mathcal{B}}(\cdot)$ be densities of x for class $\mathcal{A}$ and class $\mathcal{B}$, and let $\pi_{\mathcal{A}} = $ probability the next observation (Y) is an $\mathcal{A}$, and let $\pi_{\mathcal{B}} = 1 - \pi_{\mathcal{A}} = $ probability that the next observation is a $\mathcal{B}$. Then $p(x) \equiv prob\{Y = \mathcal{A}|x\} = \frac{\pi_{\mathcal{A}} h_{\mathcal{A}}(x)}{\pi_{\mathcal{A}} h_{\mathcal{A}}(x) + \pi_{\mathcal{B}} h_{\mathcal{B}}(x)}$. Let $C_{\mathcal{A}} = $ cost to falsely call a $\mathcal{B}$ an $\mathcal{A}$ and $C_{\mathcal{B}} = $ cost to falsely call an $\mathcal{A}$ a $\mathcal{B}$. A classifier ϕ is a map $\phi(x) : x \rightarrow \{\mathcal{A}, \mathcal{B}\}$. The optimal (Bayes) classifier, which minimizes the expected cost is

$$\phi_{\text{OPT}}(x) = \begin{cases} \mathcal{A} & \text{if } \frac{p(x)}{1-p(x)} > \frac{C_{\mathcal{A}}}{C_{\mathcal{B}}}, \\ \mathcal{B} & \text{if } \frac{p(x)}{1-p(x)} < \frac{C_{\mathcal{A}}}{C_{\mathcal{B}}}. \end{cases} \tag{7.1}$$

To estimate $p(x)$, or, alternatively the logit $f(x) \equiv \log p(x)/(1 - p(x))$, we use a training set $\{y_i, x_i\}_{i=1}^n, y_i \in \{\mathcal{A}, \mathcal{B}\}, x_i \in \mathcal{T}$, where $\mathcal{T}$ is some index set. At first we assume that the relative frequency of $\mathcal{A}$'s in the training set is the same as in the general population. f can be estimated (nonparametrically) in various ways. If $C_{\mathcal{A}}/C_{\mathcal{B}} = 1$, and f is the logit, the optimal classifier is

$$f(x) > 0 \text{ (equivalently, } p(x) - \tfrac{1}{2} > 0) \rightarrow \mathcal{A}$$
$$f(x) < 0 \text{ (equivalently, } p(x) - \tfrac{1}{2} < 0) \rightarrow \mathcal{B}$$

In the usual penalized log likelihood estimation of f, the observations are coded as

$$y = \begin{cases} 1 & \text{if } \mathcal{A}, \\ 0 & \text{if } \mathcal{B}. \end{cases} \tag{7.2}$$

The probability distribution function for $y \,|\, p$ is then

$$\mathcal{L} = p^y(1-p)^{1-y} = \begin{cases} p & \text{if } y = 1 \\ (1-p) & \text{if } y = 0 \end{cases}.$$

Using $p = e^f/(1 + e^f)$ gives the negative log likelihood $-\log \mathcal{L} = -yf + \log(1 + e^f)$. For comparison with the support vector machine we will describe a somewhat special case (General cases are in [13], [17], [8], [35]). The penalized log likelihood estimate of f is obtained as the solution to the problem: Find $f(x) = b + h(x)$ with $h \in \mathcal{H}_K$ to minimize

$$\frac{1}{n} \sum_{i=1}^{n} \left[-y_i f(x_i) + \log(1 + e^{f(x_i)}) \right] + \lambda \|h\|_{\mathcal{H}_K}^2 \qquad (7.3)$$

where $\lambda > 0$, and $\mathcal{H}_K$ is the reproducing kernel Hilbert space (RKHS) with reproducing kernel

$$K(s, t), \quad s, t \in \mathcal{T}. \qquad (7.4)$$

For more on RKHS, see [1] [29]. RKHS may be tailored to many applications since any symmetric positive definite function on $\mathcal{T} \times \mathcal{T}$ has a unique RKHS associated with it.

Theorem: [13] f_λ, the minimizer of (7.3) has a representation of the form

$$f_\lambda(x) = b + \sum_{i=1}^{n} c_i K(x, x_i). \qquad (7.5)$$

It is a property of RKHS that

$$\|h\|_{\mathcal{H}_K}^2 \equiv \sum_{i,j=1}^{n} c_i c_j K(x_i, x_j). \qquad (7.6)$$

To obtain the estimate f_λ, (7.5) and (7.6) are substituted into (7.3), which is then minimized with respect to b and $c = (c_1, \ldots, c_n)$. Given positive λ, this is a strictly convex optimization problem with some nice features special to penalized likelihood for exponential families, provided that p is not too near 0 or 1. The smoothing parameter λ, and certain other parameters which may be inside K may be chosen by Generalized Approximate Cross Validation (GACV) for Bernoulli data, see ([17]) and references cited there. The target for GACV is to minimize the Comparative Kullback-Liebler (CKL) distance of the estimate from the true distribution:

$$CKL(\lambda) = E_{true} \frac{1}{n} \sum_{i=1}^{n} -y_{new.i} f_\lambda(x_i) + \log(1 + e^{f_\lambda(x_i)}), \qquad (7.7)$$

where $y_{new.i}$ is a new observation with attribute vector x_i.

7.3   Support Vector Machines (SVM's)

For SVM's, the data is coded differently:

$$y = \begin{cases} +1 & \text{if } \mathcal{A}, \\ -1 & \text{if } \mathcal{B}. \end{cases} \qquad (7.8)$$

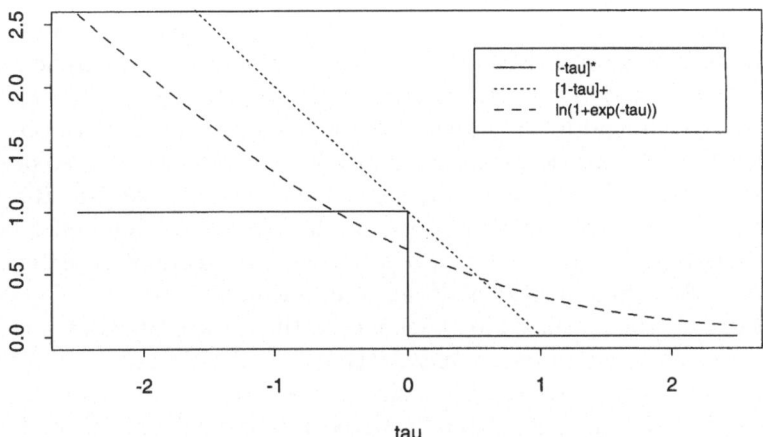

Figure 7.1. Adapted from [30]. Comparison of $[-\tau]_*, (1-\tau)_+$ and $\log_e(1 + e^{-\tau})$.

The support vector optimization problem is: Find $f(x) = b + h(x)$ with $h \in \mathcal{H}_K$ to minimize

$$\frac{1}{n}\sum_{i=1}^{n}(1 - y_i f(x_i))_+ + \lambda\|h\|_{\mathcal{H}_K}^2 \tag{7.9}$$

where $(\tau)_+ = \tau$, if $\tau > 0$, and 0 otherwise. The original support vector machine (see e. g. ([27]) was obtained from a different argument, but it is well known that it is equivalent to (7.9), see ([30], [26]). As before, the SVM f_λ has the representation (7.5). To obtain the classifier f_λ for a fixed $\lambda > 0$, (7.5) and (7.6) are substituted into (7.9) resulting in a mathematical programming problem to be solved numerically. The classifier is then $f_\lambda(x) > 0 \rightarrow \mathcal{A}, f_\lambda(x) < 0 \rightarrow \mathcal{B}$.

We may compare the penalized log likelihood estimate of the logit $\log p/(1 - p)$ and the SVM (the minimizer of (7.9)) by coding y in the likelihood as

$$\tilde{y} = \begin{cases} +1 & \text{if } \mathcal{A}, \\ -1 & \text{if } \mathcal{B}. \end{cases}$$

Then $-yf + \log(1 + e^f)$ becomes $\log(1 + e^{-\tilde{y}f})$, where f is the logit. Figure 7.1 compares $\log(1 + e^{-yf}), (1 - yf)_+$ and $[-yf]_*$ as functions of $\tau = yf$ where

$$[\tau]_* = \begin{cases} 1 & \text{if } \tau \geq 0, \\ 0 & \text{otherwise}. \end{cases}$$

Note that $[-yf]_*$ is 1 or 0 according as y and f have the same sign or not. Calling $[-yf]_*$ the misclassification counter, one might consider minimizing the misclassification count plus some (quadratic) penalty functional on f but this is a nonconvex problem and difficult to minimize numerically. Numerous authors have replaced the misclassification counter by some convex upper bound to it. The support vector, or ramp function $(1 - yf)_+$ is a convex upper bound to the misclassification counter, and Bin Yu observed that $\log_2(1+e^{-\tau})$ is also a convex upper bound. Of course it is also possible to use a penalized likelihood estimate for classification see [34]. However, the ramp function (modulo the slope) is the 'closest' convex upper bound to the misclassification counter, which provides one heuristic argument why SVM's work so well in the classification problem.

Recall that the penalized log likelihood estimate was tuned by a criteria which chose λ to minimize a proxy for the CKL of (7.7) conditional on the same x_i. By analogy, for the SVM classifier we were motivated in [20] [21] [30] [33] to say that it is optimally tuned if λ minimizes a proxy for the Generalized Comparative Kullback-Liebler distance (GCKL), defined as

$$GCKL(\lambda) = E_{true} \frac{1}{n} \sum_{i=1}^{n} (1 - y_{new \cdot i} f_\lambda(x_i))_+. \qquad (7.10)$$

That is, λ (and possibly other parameters in K) are chosen to minimize a proxy for an upper bound on the misclassification rate.

7.4   Why is the SVM so successful?

There is actually an important result which explains why the SVM is so successful: We have the Theorem:

Theorem [18]: The minimizer over f of $E_{true}(1 - y_{new} f(x))_+$ is sign $(p(x) - \frac{1}{2})$, which coincides with the sign of the logit.

As a consequence, if $\mathcal{H}_K$ is a sufficiently rich space, the minimizer of (7.9) where λ is chosen to minimize (a proxy for) $GCKL(\lambda)$, is estimating the sign of the logit. This is exactly what you need to implement the Bayes classifier! $E_{true}(1 - y_{new} f_\lambda)_+$ is given by

$E_{true}(1 - y_{new} f_\lambda)_+ =$

$$\left\{ \begin{array}{ll} p(1 - f_\lambda), & f_\lambda < -1 \\ p(1 - f_\lambda) + (1 - p)(1 + f_\lambda), & -1 < f_\lambda < +1 \\ (1 - p)(1 + f_\lambda), & f_\lambda > +1. \end{array} \right\} \qquad (7.11)$$

Since the true p is only known in a simulation experiment, $GCKL$ is also only known in experiments. The experiment to follow, which is reprinted

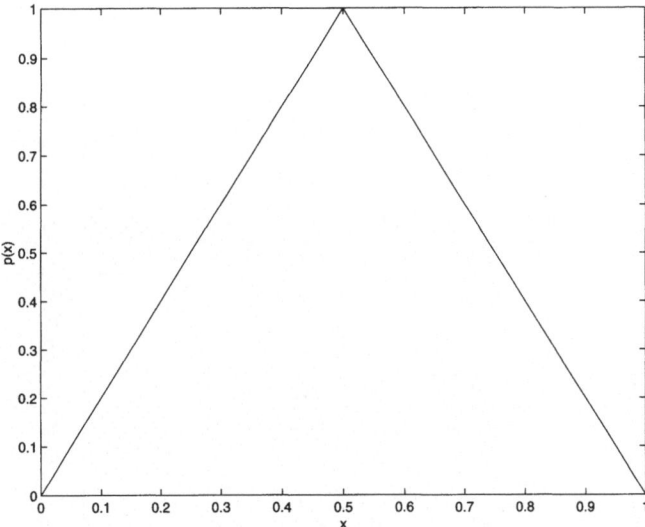

Figure 7.2. From [18]. The underlying conditional probability function $p(x) = Prob\{y = 1|x\}$ in the simulation.

from [18], demonstrates this theorem graphically. Figure 7.2 gives the underlying conditional probability function $p(x) = Prob\{y = 1|x\}$ used in the simulation. The function sign $(p(x) - 1/2)$ is 1, for $0.25 < x < 0.75; -1$ otherwise.

A training set sample of $n = 257$ observations were generated with the x_i equally spaced on $[0, 1]$, and p according to Figure 7.2. The SVM was computed and f is given in Figure 7.3 for $n\lambda = 2^{-1}, 2^{-2}, \ldots, 2^{-25}$, in the plots left to right starting with the top row and moving down. We see that solution f is close to sign $(p(x) - 1/2)$ when $n\lambda$ is in the neighborhood of 2^{-18}. 2^{-18} was the minimizer of the $GCKL$, suggesting that it is necessary to tune the SVM to estimate sign $(p(x) - 1/2)$ well.

7.5   The GACV for choosing λ (and other parameters in K)

In [30], [33], [20], [21] we developed and tested the GACV for tuning SVM's. In [30] a randomized version of GACV was obtained using a heuristic argument related to the derivation of the GCV [4], [9] for Gaussian observations and for the GACV for Bernoulli observations [36]. In [33], [20], [21] it was seen that a direct (non-randomized) version was readily available, easy to compute, and worked well. At about same time, there were several other tuning results [3] [11] [12] [23] [27] which are closely related to each other and to the GACV in one way or another. We will discuss these later. The

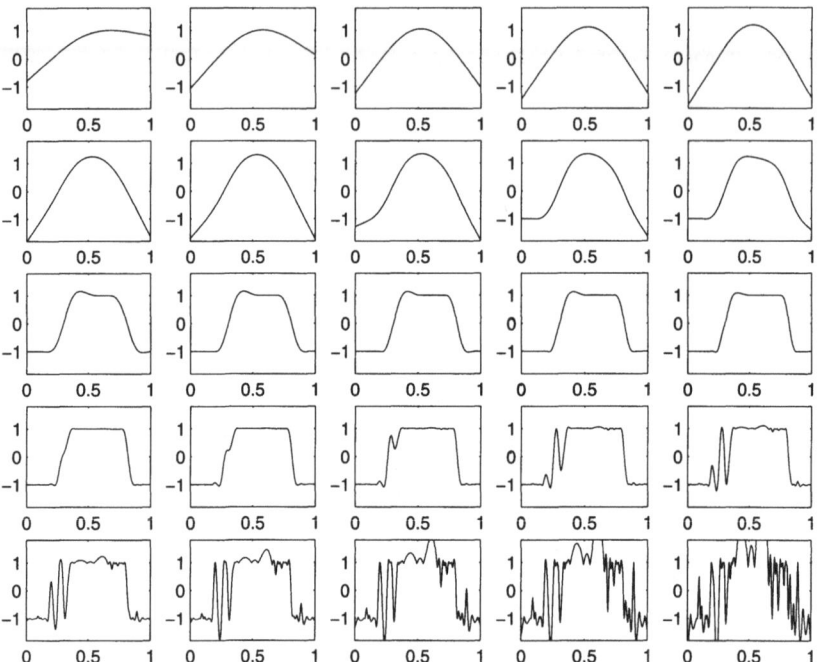

Figure 7.3. From [18]. Solutions to the SVM regularization for $n\lambda = 2^{-1}, 2^{-2}, \ldots, 2^{-25}$, left to right starting with the top row.

arguments below follow [33]. The goal here is to obtain a proxy for the (unobservable) $GCKL(\lambda)$ of (7.10). Let $f_\lambda^{[-k]}$ be the minimizer of the form $f = b + h$ with $h \in \mathcal{H}_K$ to minimize

$$\frac{1}{n} \sum_{\substack{i = 1 \\ i \neq k}} (1 - y_i f(x_i))_+ + \lambda\|h\|_K^2.$$

Let

$$V_0(\lambda) = \frac{1}{n} \sum_{k=1}^{n} (1 - y_k f_\lambda^{[-k]}(x_k))_+.$$

We write

$$V_0(\lambda) \equiv \text{OBS}(\lambda) + D(\lambda), \tag{7.12}$$

where

$$\text{OBS}(\lambda) = \frac{1}{n} \sum_{k=1}^{n} (1 - y_k f_\lambda(x_k))_+. \tag{7.13}$$

and

$$D(\lambda) = \frac{1}{n} \sum_{k=1}^{n} [(1 - y_k f_\lambda^{[-k]}(x_k))_+ - (1 - y_k f_\lambda(x_k))_+] \tag{7.14}$$

Using a rather crude argument, [33] showed that $D(\lambda) \approx \hat{D}(\lambda)$ where

$$\hat{D}(\lambda) = \frac{1}{n} \left[\sum_{y_i f_\lambda(x_i) < -1} 2\frac{\partial f_\lambda(x_i)}{\partial y_i} + \sum_{y_i f_\lambda(x_i) \in [-1,1]} \frac{\partial f_\lambda(x_i)}{\partial y_i} \right]. \tag{7.15}$$

In this argument, y_i is treated as though it is a continuous variate, and the lack of differentiability is ignored. Then

$$V_0(\lambda) \approx \mathrm{OBS}(\lambda) + \hat{D}(\lambda). \tag{7.16}$$

$\hat{D}(\lambda)$ may be compared to trace $A(\lambda)$ in GCV and unbiased risk estimates.

How shall we interpret $\frac{\partial f_\lambda(x_i)}{\partial y_i}$? Let $K_{n\times n} = \{K(x_i, x_j)\}$,

$$D_y = \begin{pmatrix} y_1 & & \\ & \ddots & \\ & & y_n \end{pmatrix}, \quad \begin{pmatrix} f_\lambda(x_1) \\ \vdots \\ f_\lambda(x_n) \end{pmatrix} = Kc + eb, \quad \text{and} \quad e = \begin{pmatrix} 1 \\ \vdots \\ 1 \end{pmatrix}.$$

We will examine the optimization problem for (7.9): Find (b, c) to minimize $\frac{1}{n}\sum_{i=1}^{n}(1 - y_i f_\lambda(x_i))_+ + \lambda c' K c$. The dual problem for (7.9) is known to be: Find $\alpha = (\alpha_1, \ldots, \alpha_n)'$ minimize $\frac{1}{2}\alpha' \left(\frac{1}{2n\lambda} D_y K D_y\right) \alpha - e'\alpha$ subject to

$$\begin{pmatrix} 0 \\ \vdots \\ 0 \end{pmatrix} \leq \begin{pmatrix} \alpha_1 \\ \vdots \\ \alpha_n \end{pmatrix} \leq \begin{pmatrix} 1 \\ \vdots \\ 1 \end{pmatrix}$$

and $y'\alpha = 0$, where $y = (y_1, \ldots, y_n)'$, and $c = \frac{1}{2n\lambda} D_y \alpha$. Then $(f_\lambda(x_1), \ldots, f_\lambda(x_n))' = \frac{1}{2n\lambda} K D_y \alpha + eb$, and we interpret $\frac{\partial f_\lambda(x_i)}{\partial y_i}$ as $\frac{\partial f_\lambda(x_i)}{\partial y_i} = \frac{1}{2n\lambda} K(x_i, x_i)\alpha_i$, resulting in

$$\hat{D}(\lambda) = \frac{1}{n} \left[2\sum_{y_i f_\lambda(x_i) < -1} \frac{\alpha_i}{2n\lambda} K(x_i, x_i) \right.$$
$$\left. + \sum_{y_i f_\lambda(x_i) \in [-1,1]} \frac{\alpha_i}{2n\lambda} K(x_i, x_i) \right] \tag{7.17}$$

and

$$GACV(\lambda) = OBS(\lambda) + \hat{D}(\lambda). \tag{7.18}$$

Let $\theta_k = \frac{\alpha_k}{2n\lambda} K(x_k, x_k)$, and note that if $y_k f_\lambda(x_k) > 1$, then $\alpha_k = 0$. If $\alpha_k = 0$, leaving out the kth data point does not change the solution. Otherwise, the expression for $\hat{D}(\lambda)$ in (7.17) is equivalent in a leaving-out-one argument, to approximating $[y_k f_\lambda(x_k) - y_k f_\lambda^{[-k]}(x_k)]$ by θ_k if $y_k f_\lambda(x_k) \in [-1, 1]$ and by $2\theta_k$ if $y_k f_\lambda(x_k) < -1$. Jaakkola and Haussler, [11] in the special case that b is taken as 0 proved that θ_k is an upper bound for $[y_k f_\lambda(x_k) - y_k f_\lambda^{[-k]}(x_k)]$ and Joachims [12] proved in the case considered here, that $[y_k f_\lambda(x_k) - y_k f_\lambda^{[-k]}(x_k)] \leq 2\theta_k$. Vapnik [27] in the case that b is set equal to 0, and $OBS = 0$, proposed choosing the parameters to minimize the so-called radius-margin bound. This works out to minimizing $\sum_i \theta_i$ when $K(x_i, x_i)$ is the same for all i. Chapelle and Vapnik [2] and Opper and Winther [23] have related proposals for choosing the tuning parameters. More details on some of these comparisons may be found in [3].

7.6   Comparing GACV and Joachims' $\xi\alpha$ method for choosing tuning parameters.

Let $\xi_i = (1 - y_i f_{\lambda i})_+$, and $K_{ij} = K(x_i, x_j)$. The GACV is then

$$GACV(\lambda) = \frac{1}{n}\left[\sum_{i=1}^n \xi_i + 2\sum_{y_i f_{\lambda i} < -1} \frac{\alpha_i}{2n\lambda} K_{ii} \right.$$
$$\left. + \sum_{y_i f_{\lambda i} \in [-1,1]} \frac{\alpha_i}{2n\lambda} K_{ii}\right]. \tag{7.19}$$

A more direct target than $GCKL(\lambda)$ is the misclassification rate, defined (conditional on the observed set of attribute variables) as

$$MISCLASS(\lambda) = E_{true}\frac{1}{n}\sum_{i=1}^n [-y_i f_{\lambda i}]_*$$
$$\equiv \frac{1}{n}\sum_{i=1}^n \{p_i[-f_{\lambda i}]_* + (1 - p_i)[f_{\lambda i}]_*\}. \tag{7.20}$$

Joachims [12], Equation (7) proposed the $\xi\alpha$ (to be called XA here) proxy for MISCLASS as:

$$XA(\lambda) = \frac{1}{n}\sum_{i=1}^n \left[\xi_i + \rho\frac{\alpha_i}{2n\lambda}K - 1\right]_*. \tag{7.21}$$

where $\rho = 2$ and here (with some abuse of notation) K is an upper bound on $K_{ii} - K_{ij}$. Letting $\theta_i = \rho\frac{\alpha_i}{2n\lambda}K$, it can be shown that the sum in $XA(\lambda)$ counts all of the samples for which $y_i f_{\lambda i} \leq \theta_i$. Since $y_i f_{\lambda i} > 1 \Rightarrow \alpha_i = 0$, XA may also be written

$$XA(\lambda) = \frac{1}{n}\left[\sum_{i=1}^n [-y_i f_{\lambda i}]_* + \sum_{y_i f_{\lambda i} \leq 1} I_{[\frac{\rho\alpha_i}{2n\lambda}K]}(y_i f_{\lambda i})\right], \tag{7.22}$$

where $I_{[\theta]}(\tau) = 1$ if $\tau \in (0, \theta]$ and 0 otherwise. Equivalently the sum in XA counts the misclassified cases in the training set plus all of the cases where $y_i f_{\lambda i} \in (0, \rho\frac{\alpha_i}{2n\lambda}K]$ (adopting the convention that if $f_{\lambda i}$ is exactly 0 then the example is considered misclassified). In some of his experiments

Joachims (empirically) set $\rho = 1$ because it achieved a better estimate of the misclassification rate than did the XA with $\rho = 2$. Let us go over how estimates of the difference between a target and its leaving out one version may be used to construct estimates when the 'fit' is not the same as the target - here the 'fit' is $(1 - y_i f_{\lambda i})_+$, while the 'target' for the XA is $[-y_i f_{\lambda i}]_*$. We will use the argument in the next section to generalize the XA to the nonstandard case in the same way that the GACV is generalized to its nonstandard version.

Let $f_{\lambda i}^{[-i]} = f_\lambda^{[-i]}(x_i)$. Suppose we have the approximation $y_i f_{\lambda i} \approx y_i f_{\lambda i}^{[-i]} + \theta_i$, with $\theta_i \geq 0$. A leaving out one estimate of the misclassification rate is given by $V_0(\lambda) = \frac{1}{n} \sum_{i=1}^n [-y_i f_{\lambda i}^{[-i]}]_*$. Now $V_0(\lambda) = \frac{1}{n} \sum_{i=1}^n [-y_i f_{\lambda i}]_* + D(\lambda)$ where

$$D(\lambda) = \frac{1}{n} \sum_{i=1}^n \{[-y_i f_{\lambda i}^{[-i]}]_* - [-y_i f_{\lambda i}]_*\}. \tag{7.23}$$

Now, the ith term in $D(\lambda) = 0$ unless $y_i f_{\lambda i}^{[-i]}$ and $y_i f_{\lambda i}$ have different signs. For $\theta_i > 0$ this can only happen if $y_i f_{\lambda i} \in (0, \theta_i]$. Assuming the approximation

$$y_i f_{\lambda i} \approx y_i f_{\lambda i}^{[-i]} + \frac{\alpha_i}{2n\lambda} K_{ii} \tag{7.24}$$

tells us that $\frac{1}{n} \sum_{y_i f_{\lambda i} \leq 1} I_{[\frac{\alpha_i}{2n\lambda} K_{ii}]}(y_i f_{\lambda i})$, can be taken as an approximation to $D(\lambda)$ of (7.23), resulting in (7.22). This provides an alternate derivation as well as an alternative interpretation of XA with $\rho = 1$, K replaced by K_{ii}.

7.7   The Nonstandard SVM and the Nonstandard GACV

We now review the nonstandard case, from [21]. Let π_A^s and π_B^s be the relative frequencies of the A and B classes in the training (sample) set. Recall that π_A and π_B are the relative frequencies of the two classes in the target population, C_A and C_B are the costs of falsely calling a B an A and falsely calling an A a B respectively, and $h_A(x)$ and $h_B(x)$ are the densities of x in the A and B classes, and that the probability that a subject from the target population with attribute x belongs to the A class is $p(x) = \frac{\pi_A h_A(x)}{\pi_A h_A(x) + \pi_B h_B(x)}$. However, the probability that a subject with attribute x chosen from a population with the same distribution as the training set, belongs to the A class, is $p_s(x) = \frac{\pi_A^s h_A(x)}{\pi_A^s h_A(x) + \pi_B^s h_B(x)}$. Letting $\phi(x)$ be the decision rule coded as a map from $x \in \mathcal{X}$ to $\{-1, 1\}$, where $1 \equiv A$ and $-1 \equiv B$, the expected cost, using $\phi(x)$ is $E_{x_{true}}\{C_B p(x)[-\phi(x)]_* + C_A(1 - p(x))[\phi(x)]_*\}$, where the expectation is taken over the distribution of x in the target population. The Bayes rule, which minimizes the expected cost is (from (7.1)) $\phi(x) = +1$ if $\frac{p(x)}{1-p(x)} > \frac{C_A}{C_B}$ and -1 otherwise. Since we don't observe a sample from the true distribution but only from the sampling distribution, we need to express the Bayes rule in terms of the sampling distribution p_s. It is shown in [21] that the Bayes rule can be written in terms of p_s as $\phi(x) = +1$ if $\frac{p_s(x)}{1-p_s(x)} > \frac{C_A}{C_B}\frac{\pi_A^s}{\pi_B^s}\frac{\pi_B}{\pi_A}$ and -1 otherwise. Let $L(-1) = C_A\pi_B/\pi_B^s$ and $L(1) = C_B\pi_A/\pi_A^s$. Then the Bayes rule can be expressed as $\phi(x) = sign\left[p_s(x) - \frac{L(-1)}{L(-1)+L(1)}\right]$. [21] proposed the nonstandard SVM to handle this nonstandard case as:

$$\min \frac{1}{n} \sum_{i=1}^{n} L(y_i)[(1 - y_i f(x_i))_+] + \lambda\|h\|_{H_K}^2 \tag{7.25}$$

over all the functions of the form $f(x) = b+h(x)$, with $h \in H_K$. This definition is justified there by showing that, if the RKHS is rich enough and λ is chosen suitably, the minimizer of (7.25) tends to sign $\left[p_s(x) - \frac{L(-1)}{L(-1)+L(1)}\right]$. In [7] and references cited there, the authors considered the nonstandard case and proposed a heuristic solution, which is different than the one discussed here.

The minimizer of (7.25) has same form as in (7.5). [20] show that the dual problem becomes minimize $\frac{1}{2}\alpha'\left(\frac{1}{2n\lambda}D_yKD_y\right)\alpha - e'\alpha$ subject to $0 \le \alpha_i \le L(y_i)$, $i = 1, 2, ..., n$, and $y'\alpha = 0$, and $c = \frac{1}{2n\lambda}D_y\alpha$. The GACV for nonstandard problems was proposed there, in an argument generalizing the standard case, as:

$$GACV(\lambda) = \frac{1}{n}\left[\sum_{i=1}^{n} L(y_i)\xi_i + 2\sum_{y_i f_{\lambda i} < -1} L(y_i)\frac{\alpha_i}{2n\lambda}K_{ii}\right.$$
$$\left. + \sum_{y_i f_{\lambda i} \in [-1,1]} L(y_i)\frac{\alpha_i}{2n\lambda}K_{ii}\right]. \tag{7.26}$$

It was shown to be a proxy for the nonstandard GCKL given by the nonstandard version of GCKL of (7.10), which can be written as:

$$GCKL(\lambda) = \frac{1}{n}\sum_{i=1}^{n}\{L(1)p_s(x_i)(1 - f_{\lambda i})_+ + L(-1)(1 - p_s(x_i))(1 + f_{\lambda i})_+\}.$$

(Compare (7.11).) We now propose a generalization, BRXA, of the XA as a computable proxy for the Bayes risk in the nonstandard case. Putting together the arguments which resulted in the GACV of (7.19), the XA in the form that it appears in (7.22) and the nonstandard GACV of (7.26), we obtain the BRXA:

$$BRXA(\lambda) = \frac{1}{n}[\sum_{i=1}^{n} L(y_i)[-y_i f_{\lambda i}]_*$$
$$+ \sum_{y_i f_{\lambda i} \le 1} L(y_i)I_{[\frac{\alpha_i}{2n\lambda}K_{ii}]}(y_i f_{\lambda i})]. \tag{7.27}$$

The BRXA is a proxy for BRMISCLASS, given by

$$BRMISCLASS(\lambda) =$$
$$\frac{1}{n}\sum_{i=1}^{n}\{L(1)p_s(x_i)[-f_{\lambda i}]_* + L(-1)(1-p_s(x_i))[f_{\lambda i}]_*\}. \tag{7.28}$$

A reviewer has asked if it is possible to construct uncertainty statements in the SVM context, as can be done in the logistic regression case. Since the SVM is not estimating a probability, a rigorous estimate may not be immediately available, but reasonable approximations or statements as to the confidence or strength in the classification may be available. This issue is mentioned in [22] and further discussion and details will be found in [14], where the idea of using appropriate costs to achieve a certain probability threshold is also mentioned.

7.8   Simulation Results and Conclusions

The two panels of Figure 7.4 show the same simulated training set. The sample proportions of the $\mathcal{A}$ (+) and $\mathcal{B}$ (o) classes are .4 and .6 respectively. The conditional distribution of x given that the sample is from the $\mathcal{A}$ class is bivariate Normal with mean (0,0) and covariance matrix diag (1,1). The distribution for x from the $\mathcal{B}$ class is bivariate Normal with mean (2,2) and covariance diag (2,1). The top panel in Figure 7.4 is for the standard case, assuming that misclassification costs are the same for both kinds of misclassification, and the target population has the same proportions of the $\mathcal{A}$ and $\mathcal{B}$ as the sample. For the bottom panel, we assume that the costs of the two types of errors are different, and that the target population has different relative frequencies than the training set. We took $C_\mathcal{A} = 1$ $C_\mathcal{B} = 2$, $\pi_\mathcal{A} = 0.1$, $\pi_\mathcal{B} = 0.9$. As before, $\pi_\mathcal{A}^s = 0.4$, and $\pi_\mathcal{B}^s = 0.6$, yielding $L(-1) = C_\mathcal{A}\pi_\mathcal{B}/\pi_\mathcal{B}^s = 1.5$, and $L(1) = C_\mathcal{B}\pi_\mathcal{A}/\pi_\mathcal{A}^s = 0.5$. Since the distributions generating the data and the distributions of the target populations are known and involve Gaussians, the theoretical best decision rules (for an infinite future population) are known, and are given by the curves marked 'true' in both panels.

The Gaussian kernel $K(x,x') = \exp\{-\|x-x'\|^2/2\sigma^2\}$ was used, where $x = (x_1, x_2)$, and σ is to be tuned along with λ. The curves selected by the GACV of (7.19) and the XA of (7.22) in the standard case are shown in the top panel, along with MISCLASS of (7.20), which is only known in a simulation experiment. The bottom panel gives the curves chosen by the nonstandard GACV of (7.26), the BRXA of (7.27) and the BRMISCLASS of (7.28). The optimal (λ,σ) pair in each case for the tuned curves was chosen by a global search. It can be seen from both panels in Figure 7.4 that the MISCLASS curve, which is based on the (finite) observed sample is quite close to the theoretical true curve (based on an infinite future population), we make this observation because it will be easier to compare

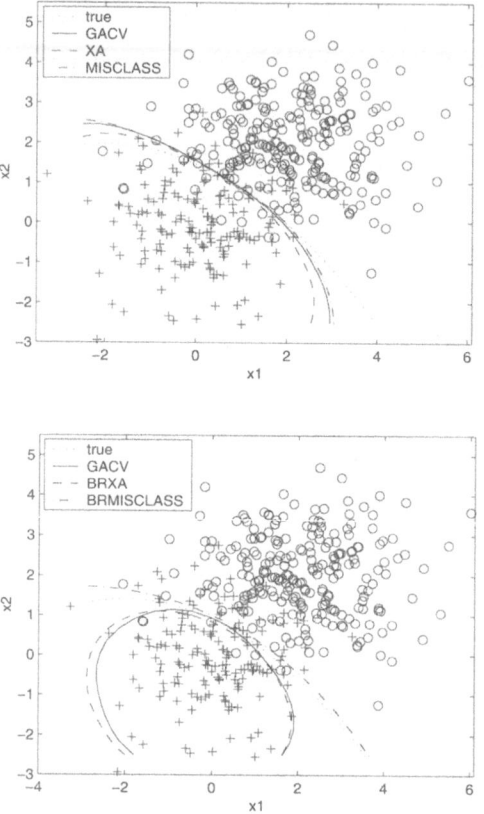

Figure 7.4. Observations, and true, GACV, XA and MISCLASS Decision Curves for the Standard Case (Top) and true, GACV, BRXA and BRMISCLASS Decision Curves for the Nonstandard Case (Bottom).

the GACV and the XA against MISCLASS than against the true, similarly for the BRMISCLASS curve. In both panels it can be seen that the decision curves determined by the GACV and the XA(BRXA) are very close.

We have computed the inefficiency of these estimates with respect to MISCLASS(BRMISCLASS), by inefficiency is meant the ratio of MIS-CLASS(BRMISCLASS) at the estimated (λ, σ) pair to its minimum value, a value of 1 means that the estimated pair is as accurate as possible, with respect to the (uncomputable) minimizer of MISCLASS(BRMISCLASS). The results for the standard case were: $GACV : 1.0064$, $XA : 1.0062 - 1.0094$ (due to multiple neighboring minima in the grid search, the 1.0062 case is in Figure 7.4); and for the nonstandard case: $GACV : 1.151, BRXA : 1.166$.

Figure 7.5 gives contour plots for GCKL, GACV, BRMISCLASS and BRXA as a function of λ and σ in the nonstandard case. It can be seen that

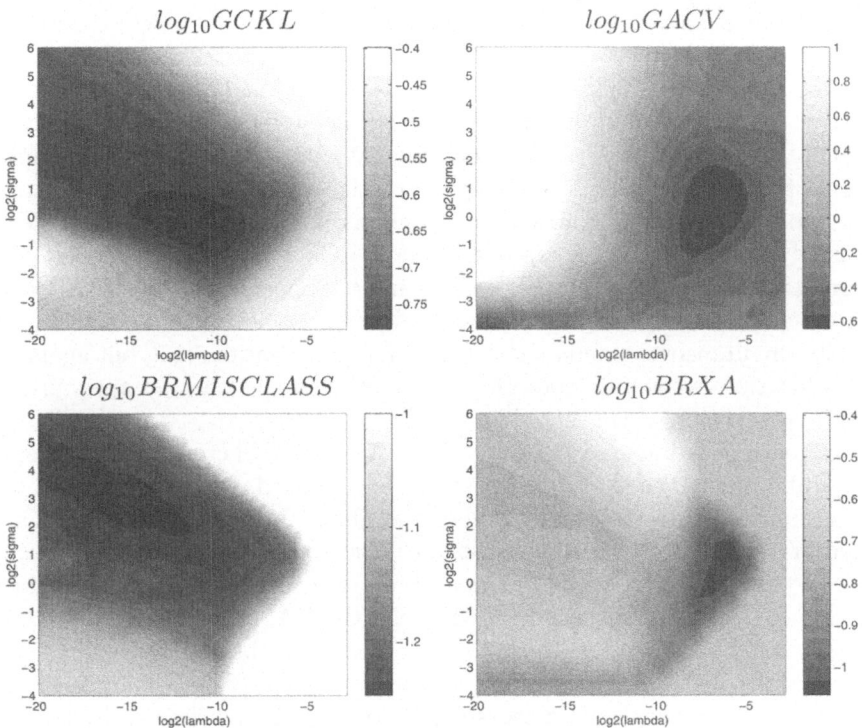

Figure 7.5. GCKL, GACV, BRMISCLASS, BRXA as functions of λ and σ^2, for the nonstandard example. Note different logarithmic scales in λ and σ.

the GACV and BRXA curves have nearly the same minima. The GCKL and BRMISCLASS curves both have long, shallow, tilted cigar-shaped minima, and the GACV and BRXA minima are near the lower right end. For the standard case (not shown) the minima are somewhat more pronounced and the GACV and XA minima are closer to the MISCLASS minimum, and this is reflected in inefficiencies nearer to 1. (BR)MISCLASS curves in other simulation studies we have done show this same behavior. We have observed (as did Joachims) that the value of XA in the standard case is a good estimate of the value of MISCLASS at its minimizer, only slightly pessimistic. The GACV at its minimizer is an estimate of twice the misclassification rate. The value of one half the GACV is somewhat more pessimistic. We note that once one obtains the solution to the problem the computation of both GACV and (BR)XA are equally trivial.

The GACV in (quadratically) penalized likelihood cases generally scatters about the minimizer of its target (analogous to GCKL)(see [31]) but here, both the GACV and the BRXA (along with the standard case) appear to be biased toward larger λ. The (BR)MISCLASS surfaces are so flat in λ in our examples this does not seem to be a serious problem (less so in the standard case).

Recently we have obtained a generalization of the SVM to the k category case, which solves a single optimization problem to obtain a vector $f_\lambda(x) = (f_{1\lambda}(x), \ldots, f_{k\lambda}(x))$ where the category classifier is the component of f that is largest, see [15]. Usual muticategory classification schemes do one-vs-many or $\binom{k}{2}$ pairwise comparisons, and the multicategory SVM has advantages in certain examples. The GACV has been extended to the nonstandard multicategory SVM case and it appears that the BRXA can also be extended. Penalized likelihood estimates which estimate a vector of logits simultaneously could also be used for classification, [16], but again, if classification is the only consideration, one can argue that an appropriate multicategory SVM is preferable.

Recently [6] compared the GACV, the XA, five-fold cross validation and several other methods for tuning, using the standard two-category SVM on four data sets with large validation sets available. It appears from the information given that the authors may not have always found the minimizing (λ, σ) pair. However, we note the authors' conclusions here. With regard to the comparison between the GACV and the XA, essentially similar conclusions were obtained as those here, namely that they behaved similarly, one slightly better on some examples the other slightly better on the other examples. However five-fold cross validation appeared to have a better accuracy record on three of the examples, and was tied with the GACV on the fourth. Several other methods were studied, none of which appeared to be related to any leaving out one argument, and those did not perform well. The five-fold cross validation will cost more in computer time, but with todays computing speeds, that is not a real consideration. In some of our own experiments we have found that the ten-fold cross validation beats or is tied with the GACV. It is of some theoretical interest to understand what appears to be a systematic overestimation of λ when using the Gaussian kernel and tuning σ^2 along with λ, by methods which are based on the leaving-out-one arguments around (7.24), especially since corresponding tuning parameter estimates in penalized likelihood estimation generally appear to be unbiased in numerical examples.

References

[1] N. Aronszajn. Theory of reproducing kernels. *Trans. Am. Math. Soc.*, 68:337–404, 1950.

[2] O. Chapelle and V. Vapnik. Model selection for support vector machines. In J. Cowan, G. Tesauro, and J. Alspector, editors, *Advances in Neural Information Processing Systems 12*, pages 230–237. MIT Press, 2000.

[3] O. Chapelle, V. Vapnik, O. Bousquet, and S. Mukherjee. Choosing multiple parameters for support vector machines. *Machine Learning*, 46(1):131–159, 2002.

[4] P. Craven and G. Wahba. Smoothing noisy data with spline functions: estimating the correct degree of smoothing by the method of generalized cross-validation. *Numer. Math.*, 31:377–403, 1979.

[5] N. Cristianini and J. Shawe-Taylor. *An Introduction to Support Vector Machines*. Cambridge University Press, 2000.

[6] K. Duan, S. Keerthi, and A. Poo. Evaluation of simple performance measures for tuning svm hyperparameters. Technical Report CD-01-11, Dept. of Mechanical Engineering, National University of Singapore, Singapore, 2001.

[7] T. Furey, N. Cristianini, N. Duffy, D Bednarski, M. Schummer, and D. Haussler. Support vector machine classification and validation of cancer tissue samples using microarray expression data. *Bioinformatics*, 16:906–914, 2001.

[8] F. Gao, G. Wahba, R. Klein, and B. Klein. Smoothing spline ANOVA for multivariate Bernoulli observations, with applications to ophthalmology data, with discussion. *J. Amer. Statist. Assoc.*, 96:127–160, 2001.

[9] G.H. Golub, M. Heath, and G. Wahba. Generalized cross validation as a method for choosing a good ridge parameter. *Technometrics*, 21:215–224, 1979.

[10] T. Hastie, R. Tibshirani, and J. Friedman. *The Elements of Statistical Learning*. Springer, 2001.

[11] T. Jaakkola and D. Haussler. Probabilistic kernel regression models. In *Proceedings of the 1999 Conference on AI and Statistics*, 1999.

[12] T. Joachims. Estimating the generalization performance of an SVM efficiently. In *Proceedings of the International Conference on Machine Learning*, San Francisco, 2000. Morgan Kaufman.

[13] G. Kimeldorf and G. Wahba. Some results on Tchebycheffian spline functions. *J. Math. Anal. Applic.*, 33:82–95, 1971.

[14] Y. Lee and C.-K. Lee. Classification of multiple cancer types by multicategory support vector machines using gene expression data. Technical Report Technical Report 1051, Department of Statistics, University of Wisconsin, Madison WI, 2002.

[15] Y. Lee, Y. Lin, and G. Wahba. Multicategory support vector machines. Technical Report 1043, Department of Statistics, University

of Wisconsin, Madison WI, 2001. To appear, The Interface Foundation, Computing Science and Statistics, v 33.

[16] X. Lin. Smoothing spline analysis of variance for polychotomous response data. Technical Report 1003, Department of Statistics, University of Wisconsin, Madison WI, 1998. Available via G. Wahba's website.

[17] X. Lin, G. Wahba, D. Xiang, F. Gao, R. Klein, and B. Klein. Smoothing spline ANOVA models for large data sets with Bernoulli observations and the randomized GACV. *Ann. Statist.*, 28:1570–1600, 2000.

[18] Y. Lin. Support vector machines and the Bayes rule in classification. Technical Report 1014, Department of Statistics, University of Wisconsin, Madison WI, to appear, *Data Mining and Knowledge Discovery*, 1999.

[19] Y. Lin. On the support vector machine. Technical Report 1029, Department of Statistics, University of Wisconsin, Madison WI, 2000.

[20] Y. Lin, Y. Lee, and G. Wahba. Support vector machines for classification in nonstandard situations. Technical Report 1016, Department of Statistics, University of Wisconsin, Madison WI, 2000. To appear, *Machine Learning*.

[21] Y. Lin, G. Wahba, H. Zhang, and Y. Lee. Statistical properties and adaptive tuning of support vector machines. Technical Report 1022, Department of Statistics, University of Wisconsin, Madison WI, 2000. To appear, *Machine Learning*.

[22] S. Mukherjee, P. Tamayo, D. Slonim, A. Verri, T. Golub, J. Mesirov, and T. Poggio. Support vector machine classification of microarray data. Technical Report 1677, AI Lab, Massachusetts Institute of Technology, 1999.

[23] M. Opper and O. Winther. Gaussian processes and svm: Mean field and leave-out-one. In A. Smola, P. Bartlett, B. Scholkopf, and D. Schuurmans, editors, *Advances in Large Margin Classifiers*, pages 311–326. MIT Press, 2000.

[24] B. Scholkopf, C. Burges, and A. Smola. *Advances in Kernel Methods-Support Vector Learning*. MIT Press, 1999.

[25] A. Smola, P. Bartlett, B. Scholkopf, and D. Schuurmans. *Advances in Large Marin Classifiers*. MIT Press, 1999.

[26] M. Pontil T. Evgeniou and T. Poggio. Regularization networks and support vector machines. *Advances in Computational Mathematics*, 13:1–50, 2000.

[27] V. Vapnik. *The Nature of Statistical Learning Theory.* Springer, 1995.

[28] G. Wahba. Estimating derivatives from outer space. Technical Report 989, Mathematics Research Center, 1969.

[29] G. Wahba. *Spline Models for Observational Data.* SIAM, 1990. CBMS-NSF Regional Conference Series in Applied Mathematics, v. 59.

[30] G. Wahba. Support vector machines, reproducing kernel Hilbert spaces and the randomized GACV. In B. Scholkopf, C. Burges, and A. Smola, editors, *Advances in Kernel Methods-Support Vector Learning*, pages 69–88. MIT Press, 1999.

[31] G. Wahba, X. Lin, F. Gao, D. Xiang, R. Klein, and B. Klein. The bias-variance tradeoff and the randomized GACV. In M. Kearns, S. Solla, and D. Cohn, editors, *Advances in Neural Information Processing Systems 11*, pages 620–626. MIT Press, 1999.

[32] G. Wahba, Y. Lin, Y. Lee, and H. Zhang. On the relation between the GACV and Joachims' $\xi\alpha$ method for tuning support vector machines, with extensions to the non-standard case. Technical Report 1039, Statistics Department University of Wisconsin, Madison WI, 2001.

[33] G. Wahba, Y. Lin, and H. Zhang. Generalized approximate cross validation for support vector machines. In A. Smola, P. Bartlett, B. Scholkopf, and D. Schuurmans, editors, *Advances in Large Margin Classifiers*, pages 297–311. MIT Press, 2000.

[34] G. Wahba, Y. Wang, C. Gu, R. Klein, and B. Klein. Structured machine learning for 'soft' classification with smoothing spline ANOVA and stacked tuning, testing and evaluation. In J. Cowan, G. Tesauro, and J. Alspector, editors, *Advances in Neural Information Processing Systems 6*, pages 415–422. Morgan Kauffman, 1994.

[35] G. Wahba, Y. Wang, C. Gu, R. Klein, and B. Klein. Smoothing spline ANOVA for exponential families, with application to the Wisconsin Epidemiological Study of Diabetic Retinopathy. *Ann. Statist.*, 23:1865–1895, 1995. Neyman Lecture.

[36] D. Xiang and G. Wahba. A generalized approximate cross validation for smoothing splines with non-Gaussian data. *Statistica Sinica*, 6:675–692, 1996.

8

The Boosting Approach to Machine Learning: An Overview

Robert E. Schapire[1]

Summary

Boosting is a general method for improving the accuracy of any given learning algorithm. Focusing primarily on the AdaBoost algorithm, this chapter overviews some of the recent work on boosting including analyses of AdaBoost's training error and generalization error; boosting's connection to game theory and linear programming; the relationship between boosting and logistic regression; extensions of AdaBoost for multiclass classification problems; methods of incorporating human knowledge into boosting; and experimental and applied work using boosting.

8.1 Introduction

Machine learning studies automatic techniques for learning to make accurate predictions based on past observations. For example, suppose that we would like to build an email filter that can distinguish spam (junk) email from non-spam. The machine-learning approach to this problem would be the following: Start by gathering as many examples as possible of both spam and non-spam emails. Next, feed these examples, together with labels indicating if they are spam or not, to your favorite machine-learning algorithm which will automatically produce a classification or prediction rule. Given a new, unlabeled email, such a rule attempts to predict if it is spam or not. The goal, of course, is to generate a rule that makes the most accurate predictions possible on new test examples.

---

[1]Robert E. Schapire is with AT&T Labs — Research, Shannon Laboratory, 180 Park Avenue, Florham Park, NJ 07932, USA (URL: www.research.att.com/~schapire).

Building a highly accurate prediction rule is certainly a difficult task. On the other hand, it is not hard at all to come up with very rough rules of thumb that are only moderately accurate. An example of such a rule is something like the following: "If the phrase 'buy now' occurs in the email, then predict it is spam." Such a rule will not even come close to covering all spam messages; for instance, it really says nothing about what to predict if 'buy now' does not occur in the message. On the other hand, this rule will make predictions that are significantly better than random guessing.

Boosting, the machine-learning method that is the subject of this chapter, is based on the observation that finding many rough rules of thumb can be a lot easier than finding a single, highly accurate prediction rule. To apply the boosting approach, we start with a method or algorithm for finding the rough rules of thumb. The boosting algorithm calls this "weak" or "base" learning algorithm repeatedly, each time feeding it a different subset of the training examples (or, to be more precise, a different distribution or weighting over the training examples[2]). Each time it is called, the base learning algorithm generates a new weak prediction rule, and after many rounds, the boosting algorithm must combine these weak rules into a single prediction rule that, hopefully, will be much more accurate than any one of the weak rules.

To make this approach work, there are two fundamental questions that must be answered: first, how should each distribution be chosen on each round, and second, how should the weak rules be combined into a single rule? Regarding the choice of distribution, the technique that we advocate is to place the most weight on the examples most often misclassified by the preceding weak rules; this has the effect of forcing the base learner to focus its attention on the "hardest" examples. As for combining the weak rules, simply taking a (weighted) majority vote of their predictions is natural and effective.

There is also the question of what to use for the base learning algorithm, but this question we purposely leave unanswered so that we will end up with a general boosting procedure that can be combined with any base learning algorithm.

Boosting refers to a general and provably effective method of producing a very accurate prediction rule by combining rough and moderately inaccurate rules of thumb in a manner similar to that suggested above. This chapter presents an overview of some of the recent work on boosting, focusing especially on the AdaBoost algorithm which has undergone intense theoretical study and empirical testing.

---

[2] A distribution over training examples can be used to generate a subset of the training examples simply by sampling repeatedly from the distribution.

Table 8.1. The boosting algorithm AdaBoost.

Given: $(x_1, y_1), \ldots, (x_m, y_m)$ where $x_i \in X$, $y_i \in Y = \{-1, +1\}$

Initialize $D_1(i) = 1/m$.

For $t = 1, \ldots, T$:

- Train base learner using distribution D_t.
- Get base classifier $h_t : X \to \mathbb{R}$.
- Choose $\alpha_t \in \mathbb{R}$.
- Update:

$$D_{t+1}(i) = \frac{D_t(i) \exp(-\alpha_t y_i h_t(x_i))}{Z_t}$$

where Z_t is a normalization factor (chosen so that D_{t+1} will be a distribution).

Output the final classifier:

$$H(x) = \text{sign} \left(\sum_{t=1}^{T} \alpha_t h_t(x) \right).$$

8.2   AdaBoost

Working in Valiant's PAC (probably approximately correct) learning model [75], Kearns and Valiant [41, 42] were the first to pose the question of whether a "weak" learning algorithm that performs just slightly better than random guessing can be "boosted" into an arbitrarily accurate "strong" learning algorithm. Schapire [66] came up with the first provable polynomial-time boosting algorithm in 1989. A year later, Freund [26] developed a much more efficient boosting algorithm which, although optimal in a certain sense, nevertheless suffered like Schapire's algorithm from certain practical drawbacks. The first experiments with these early boosting algorithms were carried out by Drucker, Schapire and Simard [22] on an OCR task.

The AdaBoost algorithm, introduced in 1995 by Freund and Schapire [32], solved many of the practical difficulties of the earlier boosting algorithms, and is the focus of this paper. Pseudocode for AdaBoost is given in Table 8.1 in the slightly generalized form given by Schapire and Singer [70]. The algorithm takes as input a training set $(x_1, y_1), \ldots, (x_m, y_m)$ where each x_i belongs to some *domain* or *instance space* X, and each *label* y_i is in some label set Y. For most of this paper, we assume $Y = \{-1, +1\}$; in Section 8.7, we discuss extensions to the multiclass case. AdaBoost calls a given *weak* or *base learning algorithm* repeatedly in a series of rounds $t = 1, \ldots, T$. One of the main ideas of the algorithm is to maintain a distribution or set of weights over the training set. The weight of this distribution on training example i on round t is denoted $D_t(i)$. Initially, all weights are set equally, but on each round, the weights of incorrectly classified examples are increased so that the base learner is forced to focus on the hard examples in the training set.

The base learner's job is to find a *base classifier* $h_t : X \rightarrow \mathbb{R}$ appropriate for the distribution D_t. (Base classifiers were also called rules of thumb or weak prediction rules in Section 8.1.) In the simplest case, the range of each h_t is binary, i.e., restricted to $\{-1, +1\}$; the base learner's job then is to minimize the *error*

$$\epsilon_t = \mathrm{Pr}_{i \sim D_t} \left[h_t(x_i) \neq y_i \right].$$

Once the base classifier h_t has been received, AdaBoost chooses a parameter $\alpha_t \in \mathbb{R}$ that intuitively measures the importance that it assigns to h_t. In the figure, we have deliberately left the choice of α_t unspecified. For binary h_t, we typically set

$$\alpha_t = \tfrac{1}{2} \ln \left(\frac{1 - \epsilon_t}{\epsilon_t} \right) \tag{8.1}$$

as in the original description of AdaBoost given by Freund and Schapire [32]. More on choosing α_t follows in Section 8.3. The distribution D_t is then updated using the rule shown in the figure. The *final* or *combined classifier H* is a weighted majority vote of the T base classifiers where α_t is the weight assigned to h_t.

8.3   Analyzing the training error

The most basic theoretical property of AdaBoost concerns its ability to reduce the training error, i.e., the fraction of mistakes on the training set. Specifically, Schapire and Singer [70], in generalizing a theorem of Freund and Schapire [32], show that the training error of the final classifier is bounded as follows:

$$\frac{1}{m} |\{i : H(x_i) \neq y_i\}| \leq \frac{1}{m} \sum_i \exp(-y_i f(x_i)) = \prod_t Z_t \tag{8.2}$$

where henceforth we define

$$f(x) = \sum_t \alpha_t h_t(x) \tag{8.3}$$

so that $H(x) = \mathrm{sign}(f(x))$. (For simplicity of notation, we write $\sum_i$ and $\sum_t$ as shorthand for $\sum_{i=1}^m$ and $\sum_{t=1}^T$, respectively.) The inequality follows from the fact that $e^{-y_i f(x_i)} \geq 1$ if $y_i \neq H(x_i)$. The equality can be proved straightforwardly by unraveling the recursive definition of D_t.

Eq. (8.2) suggests that the training error can be reduced most rapidly (in a greedy way) by choosing α_t and h_t on each round to minimize

$$Z_t = \sum_i D_t(i) \exp(-\alpha_t y_i h_t(x_i)). \tag{8.4}$$

In the case of binary classifiers, this leads to the choice of α_t given in Eq. (8.1) and gives a bound on the training error of

$$\prod_t Z_t = \prod_t \left[2\sqrt{\epsilon_t(1-\epsilon_t)} \right] = \prod_t \sqrt{1-4\gamma_t^2} \leq \exp\left(-2\sum_t \gamma_t^2\right) \quad (8.5)$$

where we define $\gamma_t = 1/2 - \epsilon_t$. This bound was first proved by Freund and Schapire [32]. Thus, if each base classifier is slightly better than random so that $\gamma_t \geq \gamma$ for some $\gamma > 0$, then the training error drops exponentially fast in T since the bound in (8.5) is at most $e^{-2T\gamma^2}$. This bound, combined with the bounds on generalization error given below prove that AdaBoost is indeed a boosting algorithm in the sense that it can efficiently convert a true weak learning algorithm (that can always generate a classifier with a weak edge for any distribution) into a strong learning algorithm (that can generate a classifier with an arbitrarily low error rate, given sufficient data).

Eq. (8.2) points to the fact that, at heart, AdaBoost is a procedure for finding a linear combination f of base classifiers which attempts to minimize

$$\sum_i \exp(-y_i f(x_i)) = \sum_i \exp\left(-y_i \sum_t \alpha_t h_t(x_i)\right). \quad (8.6)$$

Essentially, on each round, AdaBoost chooses h_t (by calling the base learner) and then sets α_t to add one more term to the accumulating weighted sum of base classifiers in such a way that the sum of exponentials above will be maximally reduced. In other words, AdaBoost is doing a kind of steepest descent search to minimize Eq. (8.6) where the search is constrained at each step to follow coordinate directions (where we identify coordinates with the weights assigned to base classifiers). This view of boosting and its generalization are examined in considerable detail by Duffy and Helmbold [23], Mason et al. [51, 52] and Friedman [35]. See also Section 8.6.

Schapire and Singer [70] discuss the choice of α_t and h_t in the case that h_t is real-valued (rather than binary). In this case, $h_t(x)$ can be interpreted as a "confidence-rated prediction" in which the sign of $h_t(x)$ is the predicted label, while the magnitude $|h_t(x)|$ gives a measure of confidence. Here, Schapire and Singer advocate choosing α_t and h_t so as to minimize Z_t ((8.4)) on each round.

8.4   Generalization error

In studying and designing learning algorithms, we are of course interested in performance on examples *not* seen during training, i.e., in the generalization error, the topic of this section. Unlike Section 8.3 where the training examples were arbitrary, here we assume that all examples (both train and test) are generated i.i.d. from some unknown distribution on $X \times Y$. The generalization error is the probability of misclassifying a new example, while the test error is the fraction of mistakes on a newly sampled test set (thus, generalization error is expected test error). Also, for simplicity, we restrict our attention to binary base classifiers.

Freund and Schapire [32] showed how to bound the generalization error of the final classifier in terms of its training error, the size m of the sample, the VC-dimension[3] d of the base classifier space and the number of rounds T of boosting. Specifically, they used techniques from Baum and Haussler [5] to show that the generalization error, with high probability, is at most[4]

$$\hat{\Pr}\left[H(x) \neq y\right] + \tilde{O}\left(\sqrt{\frac{Td}{m}}\right)$$

where $\hat{\Pr}[\cdot]$ denotes empirical probability on the training sample. This bound suggests that boosting will overfit if run for too many rounds, i.e., as T becomes large. In fact, this sometimes does happen. However, in early experiments, several authors [8, 21, 59] observed empirically that boosting often does *not* overfit, even when run for thousands of rounds. Moreover, it was observed that AdaBoost would sometimes continue to drive down the generalization error long after the training error had reached zero, clearly contradicting the spirit of the bound above. For instance, the left side of Fig. 8.1 shows the training and test curves of running boosting on top of Quinlan's C4.5 decision-tree learning algorithm [60] on the "letter" dataset.

In response to these empirical findings, Schapire et al. [69], following the work of Bartlett [3], gave an alternative analysis in terms of the *margins* of the training examples. The margin of example (x, y) is defined to be

$$\mathrm{margin}_f(x, y) = \frac{y f(x)}{\sum_t |\alpha_t|} = \frac{y \sum_t \alpha_t h_t(x)}{\sum_t |\alpha_t|}.$$

It is a number in $[-1, +1]$ and is positive if and only if H correctly classifies the example. Moreover, as before, the magnitude of the margin can be interpreted as a measure of confidence in the prediction. Schapire et al. proved that larger margins on the training set translate into a superior upper bound on the generalization error. Specifically, the generalization

---

[3]The Vapnik-Chervonenkis (VC) dimension is a standard measure of the "complexity" of a space of binary functions. See, for instance, refs. [6, 76] for its definition and relation to learning theory.

[4]The "soft-Oh" notation $\tilde{O}(\cdot)$, here used rather informally, is meant to hide all logarithmic and constant factors (in the same way that standard "big-Oh" notation hides only constant factors).

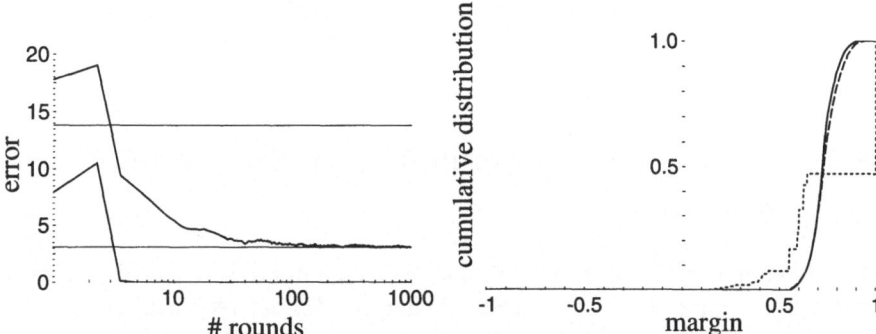

Figure 8.1. Error curves and the margin distribution graph for boosting C4.5 on the letter dataset as reported by Schapire et al. [69]. *Left*: the training and test error curves (lower and upper curves, respectively) of the combined classifier as a function of the number of rounds of boosting. The horizontal lines indicate the test error rate of the base classifier as well as the test error of the final combined classifier. *Right*: The cumulative distribution of margins of the training examples after 5, 100 and 1000 iterations, indicated by short-dashed, long-dashed (mostly hidden) and solid curves, respectively.

error is at most

$$\hat{\Pr}\left[\text{margin}_f(x, y) \leq \theta\right] + \tilde{O}\left(\sqrt{\frac{d}{m\theta^2}}\right)$$

for any $\theta > 0$ with high probability. Note that this bound is entirely independent of T, the number of rounds of boosting. In addition, Schapire et al. proved that boosting is particularly aggressive at reducing the margin (in a quantifiable sense) since it concentrates on the examples with the smallest margins (whether positive or negative). Boosting's effect on the margins can be seen empirically, for instance, on the right side of Fig. 8.1 which shows the cumulative distribution of margins of the training examples on the "letter" dataset. In this case, even after the training error reaches zero, boosting continues to increase the margins of the training examples effecting a corresponding drop in the test error.

Although the margins theory gives a qualitative explanation of the effectiveness of boosting, quantitatively, the bounds are rather weak. Breiman [9], for instance, shows empirically that one classifier can have a margin distribution that is uniformly better than that of another classifier, and yet be inferior in test accuracy. On the other hand, Koltchinskii, Panchenko and Lozano [44, 45, 46, 58] have recently proved new margin-theoretic bounds that are tight enough to give useful quantitative predictions.

Attempts (not always successful) to use the insights gleaned from the theory of margins have been made by several authors [9, 37, 50]. In addition, the margin theory points to a strong connection between boosting and the support-vector machines of Vapnik and others [7, 14, 77] which explicitly attempt to maximize the minimum margin.

8.5   A connection to game theory and linear programming

The behavior of AdaBoost can also be understood in a game-theoretic setting as explored by Freund and Schapire [31, 33] (see also Grove and Schuurmans [37] and Breiman [9]). In classical game theory, it is possible to put any two-person, zero-sum game in the form of a matrix $\mathbf{M}$. To play the game, one player chooses a row i and the other player chooses a column j. The loss to the row player (which is the same as the payoff to the column player) is $\mathbf{M}_{ij}$. More generally, the two sides may play randomly, choosing distributions $\mathbf{P}$ and $\mathbf{Q}$ over rows or columns, respectively. The expected loss then is $\mathbf{P}^{\mathrm{T}}\mathbf{M}\mathbf{Q}$.

Boosting can be viewed as repeated play of a particular game matrix. Assume that the base classifiers are binary, and let $\mathcal{H} = \{h_1, ..., h_n\}$ be the entire base classifier space (which we assume for now to be finite). For a fixed training set $(x_1, y_1), \ldots, (x_m, y_m)$, the game matrix $\mathbf{M}$ has m rows and n columns where

$$\mathbf{M}_{ij} = \begin{cases} 1 & \text{if } h_j(x_i) = y_i \\ 0 & \text{otherwise.} \end{cases}$$

The row player now is the boosting algorithm, and the column player is the base learner. The boosting algorithm's choice of a distribution D_t over training examples becomes a distribution $\mathbf{P}$ over rows of $\mathbf{M}$, while the base learner's choice of a base classifier h_t becomes the choice of a column j of $\mathbf{M}$.

As an example of the connection between boosting and game theory, consider von Neumann's famous minmax theorem which states that

$$\max_{\mathbf{Q}} \min_{\mathbf{P}} \mathbf{P}^{\mathrm{T}}\mathbf{M}\mathbf{Q} = \min_{\mathbf{P}} \max_{\mathbf{Q}} \mathbf{P}^{\mathrm{T}}\mathbf{M}\mathbf{Q}$$

for any matrix $\mathbf{M}$. When applied to the matrix just defined and reinterpreted in the boosting setting, this can be shown to have the following meaning: If, for any distribution over examples, there exists a base classifier with error at most $1/2 - \gamma$, then there exists a convex combination of base classifiers with a margin of at least 2γ on all training examples. AdaBoost seeks to find such a final classifier with high margin on all examples by combining many base classifiers; so in a sense, the minmax theorem tells us that AdaBoost at least has the potential for success since, given a "good" base learner, there must exist a good combination of base classifiers. Going much further, AdaBoost can be shown to be a special case of a more general algorithm for playing repeated games, or for approximately solving matrix games. This shows that, asymptotically, the distribution over training examples as well as the weights over base classifiers in the final classifier have game-theoretic intepretations as approximate minmax or maxmin strategies.

The problem of solving (finding optimal strategies for) a zero-sum game is well known to be solvable using linear programming. Thus, this formulation of the boosting problem as a game also connects boosting to linear, and more generally convex, programming. This connection has led to new algorithms and insights as explored by Rätsch et al. [62], Grove and Schuurmans [37] and Demiriz, Bennett and Shawe-Taylor [17].

In another direction, Schapire [68] describes and analyzes the generalization of both AdaBoost and Freund's earlier "boost-by-majority" algorithm [26] to a broader family of repeated games called "drifting games."

8.6   Boosting and logistic regression

Classification generally is the problem of predicting the label y of an example x with the intention of minimizing the probability of an incorrect prediction. However, it is often useful to estimate the *probability* of a particular label. Friedman, Hastie and Tibshirani [34] suggested a method for using the output of AdaBoost to make reasonable estimates of such probabilities. Specifically, they suggested using a logistic function, and estimating

$$\Pr_f\left[y = +1 \mid x\right] = \frac{e^{f(x)}}{e^{f(x)} + e^{-f(x)}} \tag{8.7}$$

where, as usual, $f(x)$ is the weighted average of base classifiers produced by AdaBoost ((8.3)). The rationale for this choice is the close connection between the log loss (negative log likelihood) of such a model, namely,

$$\sum_i \ln\left(1 + e^{-2y_i f(x_i)}\right) \tag{8.8}$$

and the function that, we have already noted, AdaBoost attempts to minimize:

$$\sum_i e^{-y_i f(x_i)}. \tag{8.9}$$

Specifically, it can be verified that Eq. (8.8) is upper bounded by Eq. (8.9). In addition, if we add the constant $1 - \ln 2$ to Eq. (8.8) (which does not affect its minimization), then it can be verified that the resulting function and the one in Eq. (8.9) have identical Taylor expansions around zero up to second order; thus, their behavior near zero is very similar. Finally, it can be shown that, for any distribution over pairs (x, y), the expectations

$$E\left[\ln\left(1 + e^{-2yf(x)}\right)\right]$$

and

$$E\left[e^{-yf(x)}\right]$$

are minimized by the same (unconstrained) function f, namely,

$$f(x) = \tfrac{1}{2}\ln\left(\frac{\Pr\left[y = +1 \mid x\right]}{\Pr\left[y = -1 \mid x\right]}\right).$$

Thus, for all these reasons, minimizing Eq. (8.9), as is done by AdaBoost, can be viewed as a method of approximately minimizing the negative log likelihood given in Eq. (8.8). Therefore, we may expect Eq. (8.7) to give a reasonable probability estimate.

Of course, as Friedman, Hastie and Tibshirani point out, rather than minimizing the exponential loss in (8.6), we could attempt instead to directly minimize the logistic loss in (8.8). To this end, they propose their LogitBoost algorithm. A different, more direct modification of AdaBoost for logistic loss was proposed by Collins, Schapire and Singer [13]. Following up on work by Kivinen and Warmuth [43] and Lafferty [47], they derive this algorithm using a unification of logistic regression and boosting based on Bregman distances. This work further connects boosting to the maximum-entropy literature, particularly the iterative-scaling family of algorithms [15, 16]. They also give unified proofs of convergence to optimality for a family of new and old algorithms, including AdaBoost, for both the exponential loss used by AdaBoost and the logistic loss used for logistic regression. See also the later work of Lebanon and Lafferty [48] who showed that logistic regression and boosting are in fact solving the same constrained optimization problem, except that in boosting, certain normalization constraints have been dropped.

For logistic regression, we attempt to minimize the loss function

$$\sum_i \ln\left(1 + e^{-y_i f(x_i)}\right) \tag{8.10}$$

which is the same as in (8.8) except for an inconsequential change of constants in the exponent. The modification of AdaBoost proposed by Collins, Schapire and Singer to handle this loss function is particularly simple. In AdaBoost, unraveling the definition of D_t given in Table 8.1 shows that $D_t(i)$ is proportional (i.e., equal up to normalization) to

$$\exp\left(-y_i f_{t-1}(x_i)\right)$$

where we define

$$f_t(x) = \sum_{t'=1}^{t} \alpha_{t'} h_{t'}(x).$$

To minimize the loss function in (8.10), the only necessary modification is to redefine $D_t(i)$ to be proportional to

$$\frac{1}{1 + \exp\left(y_i f_{t-1}(x_i)\right)}.$$

A very similar algorithm is described by Duffy and Helmbold [23]. Note that in each case, the weight on the examples, viewed as a vector, is proportional to the negative gradient of the respective loss function. This is because both algorithms are doing a kind of functional gradient descent, an observation that is spelled out and exploited by Breiman [9], Duffy and Helmbold [23], Mason et al. [51, 52] and Friedman [35].

Besides logistic regression, there have been a number of approaches taken to apply boosting to more general regression problems in which the labels y_i are real numbers and the goal is to produce real-valued predictions that are close to these labels. Some of these, such as those of Ridgeway [63] and Freund and Schapire [32], attempt to reduce the regression problem to a classification problem. Others, such as those of Friedman [35] and Duffy and Helmbold [24] use the functional gradient descent view of boosting to derive algorithms that directly minimize a loss function appropriate for regression. Another boosting-based approach to regression was proposed by Drucker [20].

8.7 Multiclass classification

There are several methods of extending AdaBoost to the multiclass case. The most straightforward generalization [32], called AdaBoost.M1, is adequate when the base learner is strong enough to achieve reasonably high accuracy, even on the hard distributions created by AdaBoost. However, this method fails if the base learner cannot achieve at least 50% accuracy when run on these hard distributions.

For the latter case, several more sophisticated methods have been developed. These generally work by reducing the multiclass problem to a larger binary problem. Schapire and Singer's [70] algorithm AdaBoost.MH works by creating a set of binary problems, for each example x and each possible label y, of the form: "For example x, is the correct label y or is it one of the other labels?" Freund and Schapire's [32] algorithm AdaBoost.M2 (which is a special case of Schapire and Singer's [70] AdaBoost.MR algorithm) instead creates binary problems, for each example x with correct label y and each *incorrect* label y' of the form: "For example x, is the correct label y or y'?"

These methods require additional effort in the design of the base learning algorithm. A different technique [67], which incorporates Dietterich and Bakiri's [19] method of error-correcting output codes, achieves similar provable bounds to those of AdaBoost.MH and AdaBoost.M2, but can be used with any base learner that can handle simple, binary labeled data. Schapire and Singer [70] and Allwein, Schapire and Singer [2] give yet another method of combining boosting with error-correcting output codes.

8.8   Incorporating human knowledge

Boosting, like many machine-learning methods, is entirely data-driven in the sense that the classifier it generates is derived exclusively from the evidence present in the training data itself. When data is abundant, this approach makes sense. However, in some applications, data may be severely limited, but there may be human knowledge that, in principle, might compensate for the lack of data.

In its standard form, boosting does not allow for the direct incorporation of such prior knowledge. Nevertheless, Rochery et al. [64, 65] describe a modification of boosting that combines and balances human expertise with available training data. The aim of the approach is to allow the human's rough judgments to be refined, reinforced and adjusted by the statistics of the training data, but in a manner that does not permit the data to entirely overwhelm human judgments.

The first step in this approach is for a human expert to construct by hand a rule p mapping each instance x to an estimated probability $p(x) \in [0, 1]$ that is interpreted as the guessed probability that instance x will appear with label $+1$. There are various methods for constructing such a function p, and the hope is that this difficult-to-build function need not be highly accurate for the approach to be effective.

Rochery et al.'s basic idea is to replace the logistic loss function in (8.10) with one that incorporates prior knowledge, namely,

$$\sum_i \ln\left(1 + e^{-y_i f(x_i)}\right) + \eta \sum_i \mathrm{RE}\left(p(x_i) \,\|\, \frac{1}{1 + e^{-f(x_i)}}\right)$$

where $\mathrm{RE}\,(p \,\|\, q) = p\ln(p/q) + (1 - p)\ln((1 - p)/(1 - q))$ is binary relative entropy. The first term is the same as that in (8.10). The second term gives a measure of the distance from the model built by boosting to the human's model. Thus, we balance the conditional likelihood of the data against the distance from our model to the human's model. The relative importance of the two terms is controlled by the parameter η.

8.9   Experiments and applications

Practically, AdaBoost has many advantages. It is fast, simple and easy to program. It has no parameters to tune (except for the number of round T). It requires no prior knowledge about the base learner and so can be

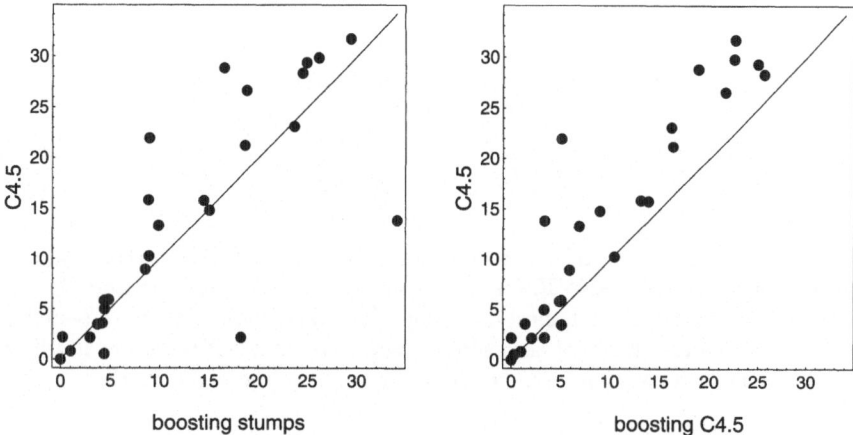

Figure 8.2. Comparison of C4.5 versus boosting stumps and boosting C4.5 on a set of 27 benchmark problems as reported by Freund and Schapire [30]. Each point in each scatterplot shows the test error rate of the two competing algorithms on a single benchmark. The y-coordinate of each point gives the test error rate (in percent) of C4.5 on the given benchmark, and the x-coordinate gives the error rate of boosting stumps (left plot) or boosting C4.5 (right plot). All error rates have been averaged over multiple runs.

flexibly combined with *any* method for finding base classifiers. Finally, it comes with a set of theoretical guarantees given sufficient data and a base learner that can reliably provide only moderately accurate base classifiers. This is a shift in mind set for the learning-system designer: instead of trying to design a learning algorithm that is accurate over the entire space, we can instead focus on finding base learning algorithms that only need to be better than random.

On the other hand, some caveats are certainly in order. The actual performance of boosting on a particular problem is clearly dependent on the data and the base learner. Consistent with theory, boosting can fail to perform well given insufficient data, overly complex base classifiers or base classifiers that are too weak. Boosting seems to be especially susceptible to noise [18] (more on this in Section sec:exps).

AdaBoost has been tested empirically by many researchers, including [4, 18, 21, 40, 49, 59, 73]. For instance, Freund and Schapire [30] tested AdaBoost on a set of UCI benchmark datasets [54] using C4.5 [60] as a base learning algorithm, as well as an algorithm that finds the best "decision stump" or single-test decision tree. Some of the results of these experiments are shown in Fig. 8.2. As can be seen from this figure, even boosting the weak decision stumps can usually give as good results as C4.5,

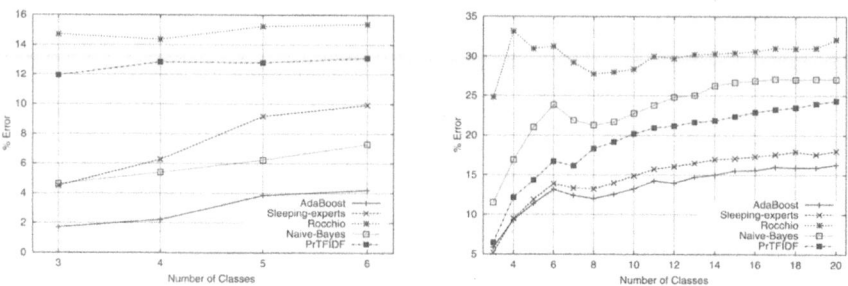

Figure 8.3. Comparison of error rates for AdaBoost and four other text categorization methods (naive Bayes, probabilistic TF-IDF, Rocchio and sleeping experts) as reported by Schapire and Singer [71]. The algorithms were tested on two text corpora — Reuters newswire articles (left) and AP newswire headlines (right) — and with varying numbers of class labels as indicated on the x-axis of each figure.

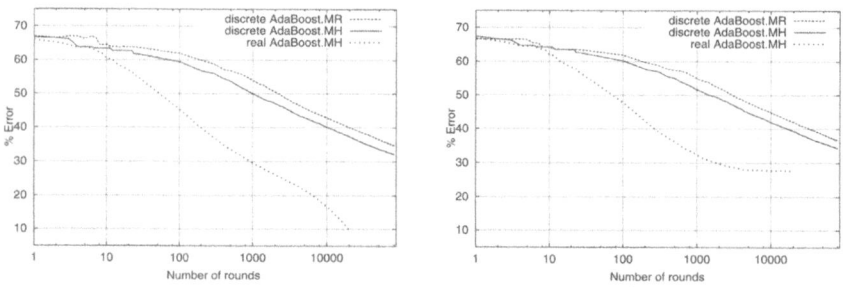

Figure 8.4. Comparison of the training (left) and test (right) error using three boosting methods on a six-class text classification problem from the TREC-AP collection, as reported by Schapire and Singer [70, 71]. Discrete AdaBoost.MH and discrete AdaBoost.MR are multiclass versions of AdaBoost that require binary ($\{-1, +1\}$-valued) base classifiers, while real AdaBoost.MH is a multiclass version that uses "confidence-rated" (i.e., real-valued) base classifiers.

while boosting C4.5 generally gives the decision-tree algorithm a significant improvement in performance.

In another set of experiments, Schapire and Singer [71] used boosting for text categorization tasks. For this work, base classifiers were used that test on the presence or absence of a word or phrase. Some results of these experiments comparing AdaBoost to four other methods are shown in Fig. 8.3. In nearly all of these experiments and for all of the performance measures tested, boosting performed as well or significantly better than the other methods tested. As shown in Fig. 8.4, these experiments also demonstrated the effectiveness of using confidence-rated predictions [70], mentioned in Section 8.3 as a means of speeding up boosting.

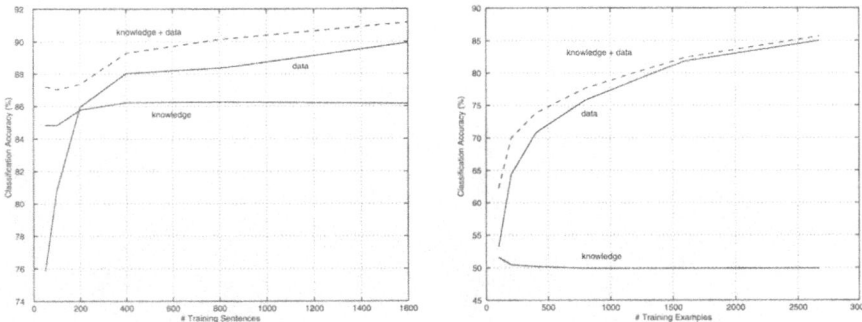

Figure 8.5. Comparison of percent classification accuracy on two spoken language tasks ("How may I help you" on the left and "Help desk" on the right) as a function of the number of training examples using data and knowledge separately or together, as reported by Rochery et al. [64, 65].

Boosting has also been applied to text filtering [72] and routing [39], "ranking" problems [28], learning problems arising in natural language processing [1, 12, 25, 38, 55, 78], image retrieval [74], medical diagnosis [53], and customer monitoring and segmentation [56, 57].

Rochery et al.'s [64, 65] method of incorporating human knowledge into boosting, described in Section 8.8, was applied to two speech categorization tasks. In this case, the prior knowledge took the form of a set of hand-built rules mapping keywords to predicted categories. The results are shown in Fig. 8.5.

The final classifier produced by AdaBoost when used, for instance, with a decision-tree base learning algorithm, can be extremely complex and difficult to comprehend. With greater care, a more human-understandable final classifier can be obtained using boosting. Cohen and Singer [11] showed how to design a base learning algorithm that, when combined with AdaBoost, results in a final classifier consisting of a relatively small set of rules similar to those generated by systems like RIPPER [10], IREP [36] and C4.5rules [60]. Cohen and Singer's system, called SLIPPER, is fast, accurate and produces quite compact rule sets. In other work, Freund and Mason [29] showed how to apply boosting to learn a generalization of decision trees called "alternating trees." Their algorithm produces a single alternating tree rather than an ensemble of trees as would be obtained by running AdaBoost on top of a decision-tree learning algorithm. On the other hand, their learning algorithm achieves error rates comparable to those of a whole ensemble of trees.

A nice property of AdaBoost is its ability to identify *outliers*, i.e., examples that are either mislabeled in the training data, or that are inherently ambiguous and hard to categorize. Because AdaBoost focuses its weight on the hardest examples, the examples with the highest weight often turn

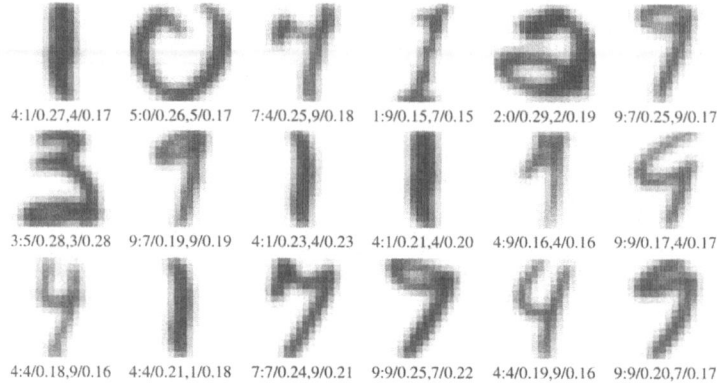

Figure 8.6. A sample of the examples that have the largest weight on an OCR task as reported by Freund and Schapire [30]. These examples were chosen after 4 rounds of boosting (top line), 12 rounds (middle) and 25 rounds (bottom). Underneath each image is a line of the form $d:\ell_1/w_1,\ell_2/w_2$, where d is the label of the example, ℓ_1 and ℓ_2 are the labels that get the highest and second highest vote from the combined classifier at that point in the run of the algorithm, and w_1, w_2 are the corresponding normalized scores.

out to be outliers. An example of this phenomenon can be seen in Fig. 8.6 taken from an OCR experiment conducted by Freund and Schapire [30].

When the number of outliers is very large, the emphasis placed on the hard examples can become detrimental to the performance of AdaBoost. This was demonstrated very convincingly by Dietterich [18]. Friedman, Hastie and Tibshirani [34] suggested a variant of AdaBoost, called "Gentle AdaBoost" that puts less emphasis on outliers. Rätsch, Onoda and Müller [61] show how to regularize AdaBoost to handle noisy data. Freund [27] suggested another algorithm, called "BrownBoost," that takes a more radical approach that de-emphasizes outliers when it seems clear that they are "too hard" to classify correctly. This algorithm, which is an adaptive version of Freund's [26] "boost-by-majority" algorithm, demonstrates an intriguing connection between boosting and Brownian motion.

8.10   Conclusion

In this overview, we have seen that there have emerged a great many views or interpretations of AdaBoost. First and foremost, AdaBoost is a genuine boosting algorithm: given access to a true weak learning algorithm that always performs a little bit better than random guessing on every distribution over the training set, we can prove arbitrarily good bounds on the training error and generalization error of AdaBoost.

Besides this original view, AdaBoost has been interpreted as a procedure based on functional gradient descent, as an approximation of logistic regression and as a repeated-game playing algorithm. AdaBoost has also been shown to be related to many other topics, such as game theory and linear programming, Bregman distances, support-vector machines, Brownian motion, logistic regression and maximum-entropy methods such as iterative scaling.

All of these connections and interpretations have greatly enhanced our understanding of boosting and contributed to its extension in ever more practical directions, such as to logistic regression and other loss-minimization problems, to multiclass problems, to incorporate regularization and to allow the integration of prior background knowledge.

We also have discussed a few of the growing number of applications of AdaBoost to practical machine learning problems, such as text and speech categorization.

References

[1] Steven Abney, Robert E. Schapire, and Yoram Singer. Boosting applied to tagging and PP attachment. In *Proceedings of the Joint SIGDAT Conference on Empirical Methods in Natural Language Processing and Very Large Corpora*, 1999.

[2] Erin L. Allwein, Robert E. Schapire, and Yoram Singer. Reducing multiclass to binary: A unifying approach for margin classifiers. *Journal of Machine Learning Research*, 1:113–141, 2000.

[3] Peter L. Bartlett. The sample complexity of pattern classification with neural networks: the size of the weights is more important than the size of the network. *IEEE Transactions on Information Theory*, 44(2):525–536, March 1998.

[4] Eric Bauer and Ron Kohavi. An empirical comparison of voting classification algorithms: Bagging, boosting, and variants. *Machine Learning*, 36(1/2):105–139, 1999.

[5] Eric B. Baum and David Haussler. What size net gives valid generalization? *Neural Computation*, 1(1):151–160, 1989.

[6] Anselm Blumer, Andrzej Ehrenfeucht, David Haussler, and Manfred K. Warmuth. Learnability and the Vapnik-Chervonenkis dimension. *Journal of the Association for Computing Machinery*, 36(4):929–965, October 1989.

[7] Bernhard E. Boser, Isabelle M. Guyon, and Vladimir N. Vapnik. A training algorithm for optimal margin classifiers. In *Proceedings*

of the Fifth Annual ACM Workshop on Computational Learning Theory, pages 144–152, 1992.

[8] Leo Breiman. Arcing classifiers. *The Annals of Statistics*, 26(3):801–849, 1998.

[9] Leo Breiman. Prediction games and arcing classifiers. *Neural Computation*, 11(7):1493–1517, 1999.

[10] William Cohen. Fast effective rule induction. In *Proceedings of the Twelfth International Conference on Machine Learning*, pages 115–123, 1995.

[11] William W. Cohen and Yoram Singer. A simple, fast, and effective rule learner. In *Proceedings of the Sixteenth National Conference on Artificial Intelligence*, 1999.

[12] Michael Collins. Discriminative reranking for natural language parsing. In *Proceedings of the Seventeenth International Conference on Machine Learning*, 2000.

[13] Michael Collins, Robert E. Schapire, and Yoram Singer. Logistic regression, AdaBoost and Bregman distances. *Machine Learning*, to appear.

[14] Corinna Cortes and Vladimir Vapnik. Support-vector networks. *Machine Learning*, 20(3):273–297, September 1995.

[15] J. N. Darroch and D. Ratcliff. Generalized iterative scaling for log-linear models. *The Annals of Mathematical Statistics*, 43(5):1470–1480, 1972.

[16] Stephen Della Pietra, Vincent Della Pietra, and John Lafferty. Inducing features of random fields. *IEEE Transactions Pattern Analysis and Machine Intelligence*, 19(4):1–13, April 1997.

[17] Ayhan Demiriz, Kristin P. Bennett, and John Shawe-Taylor. Linear programming boosting via column generation. *Machine Learning*, 46(1/2/3):225–254, 2002.

[18] Thomas G. Dietterich. An experimental comparison of three methods for constructing ensembles of decision trees: Bagging, boosting, and randomization. *Machine Learning*, 40(2):139–158, 2000.

[19] Thomas G. Dietterich and Ghulum Bakiri. Solving multiclass learning problems via error-correcting output codes. *Journal of Artificial Intelligence Research*, 2:263–286, January 1995.

[20] Harris Drucker. Improving regressors using boosting techniques. In *Machine Learning: Proceedings of the Fourteenth International Conference*, pages 107–115, 1997.

[21] Harris Drucker and Corinna Cortes. Boosting decision trees. In *Advances in Neural Information Processing Systems 8*, pages 479–485, 1996.

[22] Harris Drucker, Robert Schapire, and Patrice Simard. Boosting performance in neural networks. *International Journal of Pattern Recognition and Artificial Intelligence*, 7(4):705–719, 1993.

[23] Nigel Duffy and David Helmbold. Potential boosters? In *Advances in Neural Information Processing Systems 11*, 1999.

[24] Nigel Duffy and David Helmbold. Boosting methods for regression. *Machine Learning*, 49(2/3), 2002.

[25] Gerard Escudero, Lluís Màrquez, and German Rigau. Boosting applied to word sense disambiguation. In *Proceedings of the 12th European Conference on Machine Learning*, pages 129–141, 2000.

[26] Yoav Freund. Boosting a weak learning algorithm by majority. *Information and Computation*, 121(2):256–285, 1995.

[27] Yoav Freund. An adaptive version of the boost by majority algorithm. *Machine Learning*, 43(3):293–318, June 2001.

[28] Yoav Freund, Raj Iyer, Robert E. Schapire, and Yoram Singer. An efficient boosting algorithm for combining preferences. In *Machine Learning: Proceedings of the Fifteenth International Conference*, 1998.

[29] Yoav Freund and Llew Mason. The alternating decision tree learning algorithm. In *Machine Learning: Proceedings of the Sixteenth International Conference*, pages 124–133, 1999.

[30] Yoav Freund and Robert E. Schapire. Experiments with a new boosting algorithm. In *Machine Learning: Proceedings of the Thirteenth International Conference*, pages 148–156, 1996.

[31] Yoav Freund and Robert E. Schapire. Game theory, on-line prediction and boosting. In *Proceedings of the Ninth Annual Conference on Computational Learning Theory*, pages 325–332, 1996.

[32] Yoav Freund and Robert E. Schapire. A decision-theoretic generalization of on-line learning and an application to boosting. *Journal of Computer and System Sciences*, 55(1):119–139, August 1997.

[33] Yoav Freund and Robert E. Schapire. Adaptive game playing using multiplicative weights. *Games and Economic Behavior*, 29:79–103, 1999.

[34] Jerome Friedman, Trevor Hastie, and Robert Tibshirani. Additive logistic regression: A statistical view of boosting. *The Annals of Statistics*, 38(2):337–374, April 2000.

[35] Jerome H. Friedman. Greedy function approximation: A gradient boosting machine. *The Annals of Statistics*, 29(5), October 2001.

[36] Johannes Fürnkranz and Gerhard Widmer. Incremental reduced error pruning. In *Machine Learning: Proceedings of the Eleventh International Conference*, pages 70–77, 1994.

[37] Adam J. Grove and Dale Schuurmans. Boosting in the limit: Maximizing the margin of learned ensembles. In *Proceedings of the Fifteenth National Conference on Artificial Intelligence*, 1998.

[38] Masahiko Haruno, Satoshi Shirai, and Yoshifumi Ooyama. Using decision trees to construct a practical parser. *Machine Learning*, 34:131–149, 1999.

[39] Raj D. Iyer, David D. Lewis, Robert E. Schapire, Yoram Singer, and Amit Singhal. Boosting for document routing. In *Proceedings of the Ninth International Conference on Information and Knowledge Management*, 2000.

[40] Jeffrey C. Jackson and Mark W. Craven. Learning sparse perceptrons. In *Advances in Neural Information Processing Systems 8*, pages 654–660, 1996.

[41] Michael Kearns and Leslie G. Valiant. Learning Boolean formulae or finite automata is as hard as factoring. Technical Report TR-14-88, Harvard University Aiken Computation Laboratory, August 1988.

[42] Michael Kearns and Leslie G. Valiant. Cryptographic limitations on learning Boolean formulae and finite automata. *Journal of the Association for Computing Machinery*, 41(1):67–95, January 1994.

[43] Jyrki Kivinen and Manfred K. Warmuth. Boosting as entropy projection. In *Proceedings of the Twelfth Annual Conference on Computational Learning Theory*, pages 134–144, 1999.

[44] V. Koltchinskii and D. Panchenko. Empirical margin distributions and bounding the generalization error of combined classifiers. *The Annals of Statistics*, 30(1), February 2002.

[45] Vladimir Koltchinskii, Dmitriy Panchenko, and Fernando Lozano. Further explanation of the effectiveness of voting methods: The game between margins and weights. In *Proceedings 14th Annual Conference on Computational Learning Theory and 5th European Conference on Computational Learning Theory*, pages 241–255, 2001.

[46] Vladimir Koltchinskii, Dmitriy Panchenko, and Fernando Lozano. Some new bounds on the generalization error of combined classifiers. In *Advances in Neural Information Processing Systems 13*, 2001.

[47] John Lafferty. Additive models, boosting and inference for generalized divergences. In *Proceedings of the Twelfth Annual Conference on Computational Learning Theory*, pages 125–133, 1999.

[48] Guy Lebanon and John Lafferty. Boosting and maximum likelihood for exponential models. In *Advances in Neural Information Processing Systems 14*, 2002.

[49] Richard Maclin and David Opitz. An empirical evaluation of bagging and boosting. In *Proceedings of the Fourteenth National Conference on Artificial Intelligence*, pages 546–551, 1997.

[50] Llew Mason, Peter Bartlett, and Jonathan Baxter. Direct optimization of margins improves generalization in combined classifiers. In *Advances in Neural Information Processing Systems 12*, 2000.

[51] Llew Mason, Jonathan Baxter, Peter Bartlett, and Marcus Frean. Functional gradient techniques for combining hypotheses. In Alexander J. Smola, Peter J. Bartlett, Bernhard Schölkopf, and Dale Schuurmans, editors, *Advances in Large Margin Classifiers*. MIT Press, 1999.

[52] Llew Mason, Jonathan Baxter, Peter Bartlett, and Marcus Frean. Boosting algorithms as gradient descent. In *Advances in Neural Information Processing Systems 12*, 2000.

[53] Stefano Merler, Cesare Furlanello, Barbara Larcher, and Andrea Sboner. Tuning cost-sensitive boosting and its application to melanoma diagnosis. In *Multiple Classifier Systems: Proceedings of the 2nd International Workshop*, pages 32–42, 2001.

[54] C. J. Merz and P. M. Murphy. UCI repository of machine learning databases, 1999. www.ics.uci.edu/~mlearn/MLRepository.html.

[55] Pedro J. Moreno, Beth Logan, and Bhiksha Raj. A boosting approach for confidence scoring. In *Proceedings of the 7th European Conference on Speech Communication and Technology*, 2001.

[56] Michael C. Mozer, Richard Wolniewicz, David B. Grimes, Eric Johnson, and Howard Kaushansky. Predicting subscriber dissatisfaction and improving retention in the wireless telecommunications industry. *IEEE Transactions on Neural Networks*, 11:690–696, 2000.

[57] Takashi Onoda, Gunnar Rätsch, and Klaus-Robert Müller. Applying support vector machines and boosting to a non-intrusive monitoring system for household electric appliances with inverters. In *Proceedings of the Second ICSC Symposium on Neural Computation*, 2000.

[58] Dmitriy Panchenko. New zero-error bounds for voting algorithms. Unpublished manuscript, 2001.

[59] J. R. Quinlan. Bagging, boosting, and C4.5. In *Proceedings of the Thirteenth National Conference on Artificial Intelligence*, pages 725–730, 1996.

[60] J. Ross Quinlan. *C4.5: Programs for Machine Learning*. Morgan Kaufmann, 1993.

[61] G. Rätsch, T. Onoda, and K.-R. Müller. Soft margins for AdaBoost. *Machine Learning*, 42(3):287–320, 2001.

[62] Gunnar Rätsch, Manfred Warmuth, Sebastian Mika, Takashi Onoda, Steven Lemm, and Klaus-Robert Müller. Barrier boosting. In *Proceedings of the Thirteenth Annual Conference on Computational Learning Theory*, pages 170–179, 2000.

[63] Greg Ridgeway, David Madigan, and Thomas Richardson. Boosting methodology for regression problems. In *Proceedings of the International Workshop on AI and Statistics*, pages 152–161, 1999.

[64] M. Rochery, R. Schapire, M. Rahim, N. Gupta, G. Riccardi, S. Bangalore, H. Alshawi, and S. Douglas. Combining prior knowledge and boosting for call classification in spoken language dialogue. Unpublished manuscript, 2001.

[65] Marie Rochery, Robert Schapire, Mazin Rahim, and Narendra Gupta. BoosTexter for text categorization in spoken language dialogue. Unpublished manuscript, 2001.

[66] Robert E. Schapire. The strength of weak learnability. *Machine Learning*, 5(2):197–227, 1990.

[67] Robert E. Schapire. Using output codes to boost multiclass learning problems. In *Machine Learning: Proceedings of the Fourteenth International Conference*, pages 313–321, 1997.

[68] Robert E. Schapire. Drifting games. *Machine Learning*, 43(3):265–291, June 2001.

[69] Robert E. Schapire, Yoav Freund, Peter Bartlett, and Wee Sun Lee. Boosting the margin: A new explanation for the effectiveness of voting methods. *The Annals of Statistics*, 26(5):1651–1686, October 1998.

[70] Robert E. Schapire and Yoram Singer. Improved boosting algorithms using confidence-rated predictions. *Machine Learning*, 37(3):297–336, December 1999.

[71] Robert E. Schapire and Yoram Singer. BoosTexter: A boosting-based system for text categorization. *Machine Learning*, 39(2/3):135–168, May/June 2000.

[72] Robert E. Schapire, Yoram Singer, and Amit Singhal. Boosting and Rocchio applied to text filtering. In *Proceedings of the 21st Annual International Conference on Research and Development in Information Retrieval*, 1998.

[73] Holger Schwenk and Yoshua Bengio. Training methods for adaptive boosting of neural networks. In *Advances in Neural Information Processing Systems 10*, pages 647–653, 1998.

[74] Kinh Tieu and Paul Viola. Boosting image retrieval. In *Proceedings of the IEEE Conference on Computer Vision and Pattern Recognition*, 2000.

[75] L. G. Valiant. A theory of the learnable. *Communications of the ACM*, 27(11):1134–1142, November 1984.

[76] V. N. Vapnik and A. Ya. Chervonenkis. On the uniform convergence of relative frequencies of events to their probabilities. *Theory of Probability and its applications*, XVI(2):264–280, 1971.

[77] Vladimir N. Vapnik. *The Nature of Statistical Learning Theory*. Springer, 1995.

[78] Marilyn A. Walker, Owen Rambow, and Monica Rogati. SPoT: A trainable sentence planner. In *Proceedings of the 2nd Annual Meeting of the North American Chapter of the Associataion for Computational Linguistics*, 2001.

9

Improved Class Probability Estimates from Decision Tree Models

Dragos D. Margineantu and
Thomas G. Dietterich[1]

Summary

Decision tree models typically give good classification decisions but poor probability estimates. In many applications, it is important to have good probability estimates as well. This chapter introduces a new algorithm, Bagged Lazy Option Trees (B-LOTs), for constructing decision trees and compares it to an alternative, Bagged Probability Estimation Trees (B-PETs). The quality of the class probability estimates produced by the two methods is evaluated in two ways. First, we compare the ability of the two methods to make good classification decisions when the misclassification costs are asymmetric. Second, we compare the absolute accuracy of the estimates themselves. The experiments show that B-LOTs produce better decisions and more accurate probability estimates than B-PETs.

9.1   Introduction

The problem of statistical supervised learning is to construct a classifier f from a labeled training sample $\langle \mathbf{x}_i, y_i \rangle$, where each $\mathbf{x}_i$ is a vector of attributes (predictor variables), and $y_i \in \{1, \ldots, K\}$ is a class label. The classifier should accept new data points $\mathbf{x}$ and predict the corresponding class label $y = f(\mathbf{x})$. We assume that the training sample consists of independent and identically distributed draws from an unknown underlying distribution $\mathbf{P}(\mathbf{x}, y)$.

---

[1]Dragos D. Margineantu is with Adaptive Systems, Mathematics & Computing Technology, The Boeing Company. Thomas G. Dietterich is Professor, Department of Computer Science, Oregon State University.

There are two fundamental approaches to statistical learning of classifiers. One approach is based on constructing a probabilistic model $\hat{\mathbf{P}}(\mathbf{x}, y)$ of the data and then applying statistical inference methods to determine the conditional class probability distribution $\hat{\mathbf{P}}(y|\mathbf{x})$. The well-known Naive Bayes algorithm is an example of this approach, where the joint distribution $\hat{\mathbf{P}}(\mathbf{x}, y)$ is represented by a mixture model: $\hat{\mathbf{P}}(\mathbf{x}, y) = \sum_y \hat{\mathbf{P}}(y)\hat{\mathbf{P}}(\mathbf{x}|y)$ with one component for each class. Using such a mixture model, the class conditional distribution can be computed as

$$\hat{\mathbf{P}}(y|\mathbf{x}) = \frac{\hat{\mathbf{P}}(y)\hat{\mathbf{P}}(\mathbf{x}|y)}{\sum_{y'} \hat{\mathbf{P}}(y')\hat{\mathbf{P}}(y'|\mathbf{x})}.$$

The second approach to constructing classifiers is based on a direct representation of the classifier as a decision boundary. Decision tree methods [4, 24], for example, represent the decision boundary by a tree of tests on individual attribute values. For example, suppose that each $\mathbf{x}$ is a vector of real-valued attributes $\mathbf{x} = (x_1, \ldots, x_n)$ and $y \in \{1, 2\}$. Then a decision tree takes the form of a nested **if-then-else** statement such as the following:

if $x_2 > 3$ **then**
    if $x_1 > 4$ **then**
        $y = 1$
    else
        $y = 2$
    else if $x_3 > 2$ **then**
        $y = 2$
    else
        $y = 1$

Experience across many application problems suggests that the direct decision-boundary approach usually gives more accurate classifiers than the probability distribution approach. There seem to be two reasons for this.

First, the probability distribution approach requires that we make assumptions about the form of the probability distribution $\mathbf{P}(\mathbf{x}, y)$ that generated the data. These assumptions are almost always incorrect. Sometimes the assumptions are only slightly violated; in other cases, the assumptions introduce major errors.

Second, and perhaps more important, the probabilistic modeling approach does not adapt the complexity of the model to the amount and complexity of the data. In machine learning, there is always a tripartite tradeoff operating between the amount of data, the complexity of the models that are fit to the data, and the error rate of the resulting fitted model [17]. If the complexity of the models is held constant, then when there is insufficient data, the fitted model will exhibit high variance. If there is more than enough data (and the model assumptions are not completely correct), then the model will exhibit significant bias. Decision-boundary algorithms that dynamically adjust the complexity of the model to the amount and complexity of the data—such as decision trees—provide a better way of managing this three-way tradeoff.

However, decision boundary methods are designed only to make decisions, not to produce estimates of the class probabilities $\mathbf{P}(y|\mathbf{x})$. In many applications, it is essential to have these probability estimates. In particular, in cost-sensitive learning problems, such as medical diagnosis, fraud detection, or computer intrusion detection, the cost of making a false positive error may be substantially different from the cost of making a false negative error. Let $C(j, k)$ be the cost of predicting that $\mathbf{x}$ belongs to class j when in fact it belongs to class k. According to decision theory, we should choose the class to minimize the expected cost:

$$y = \operatorname*{argmin}_{j} \sum_{k} \mathbf{P}(k|\mathbf{x})C(j, k). \tag{9.1}$$

To carry out this computation, we need an estimate of the class probabilities $\mathbf{P}(k|\mathbf{x})$. How can we obtain such estimates from decision trees?

Existing decision tree algorithms estimate $\mathbf{P}(k|\mathbf{x})$ separately in each leaf of the decision tree by computing the proportion of training data points belonging to class k that reach that leaf. There are three difficulties with this. First, the decision tree was constructed using these same data points, so the estimates tend to be too extreme (i.e., close to 0 or 1).

Second, as one proceeds from the root of the decision tree down to the leaves, the training data is split into smaller and smaller subsets. The result is that each class probability estimate is based on a small number of data points, and hence, can be very inaccurate.

Third, each leaf of the decision tree corresponds to a region of the $\mathbf{x}$ space, and this procedure assigns the same probability estimate to all points in the region. To see why this is a problem, suppose we are estimating the probability of heart disease (the class, y) given blood pressure (bp). Imagine that there is a leaf of the tree that corresponds to $bp > 160$ and that out of the 100 examples that reach this leaf, 90 have heart disease. Then the class probability is estimated as $\mathbf{P}(\text{heart disease}|\mathbf{x}) = 0.9$. But suppose that some of these data points include a mix of patients, some of whom have blood pressure of 161 and others have blood pressure of 250. Surely we want to predict a lower probability of heart disease for the patient with 161 than for the patient with 250.

From this review, it is clear that there is a need to develop improved methods for estimating class probabilities from decision-boundary learning algorithms. This chapter presents and compares two methods for obtaining class probability estimates from decision trees. The first method, bagged probability estimation trees (B-PETs), was developed by Provost and Domingos [23]. The second, bagged lazy option trees (B-LOTs), was developed by the authors by extending previous work of Buntine [7], Breiman [5], Friedman, Kohavi, and Yun [16], and Kohavi and Kunz [21]. Other tree-based methods that can be extended to produce probability estimates include boosting [14, 15], random forests [6, 20, 12], and Bayesian CART [9, 11].

We evaluate the quality of the probability estimates produced by these two methods in two ways. First, we compare the predicted and true probabilities on a synthetic data set. Second, we compare the average misclassification costs of the two methods when applied to cost-sensitive learning problems. We employ a new bootstrap-based statistical test to determine which method (B-PET or B-LOT) gives lower expected misclassification cost across 16 data sets taken from the Irvine repository [2]. The results show that B-LOTs give more accurate probability estimates and lower expected misclassification costs. However, B-PETs appear to provide a better estimate of the *relative* class probabilities. Hence, if a way can be found to calibrate these relative rankings, B-PETs might also be made to provide good probability estimates. In either case, B-PETs and B-LOTs give much better performance than a single decision tree.

The remainder of this chapter is organized as follows. First, we review the bagged probability estimation trees introduced by Provost and Domingos. Next, we describe our new method, the bagged lazy option trees. Finally, we present our experimental evaluations of the probability estimates and discuss their significance.

9.2   Probability Estimation Trees

Provost and Domingos [23] have developed an interesting approach to obtaining class probabilities from decision trees. They start with Quinlan's popular C4.5 decision tree algorithm [24] and modify it in four ways.

First, they turn off the pruning phase of C4.5. Without pruning, C4.5 will continue growing the tree (and subdividing the training data) until each leaf is pure (i.e., contains examples from only one class) or contains only 2 examples.

Second, they turn off C4.5's collapsing mechanism. At each internal node, C4.5 invokes itself recursively to grow a decision subtree. When that recursive call returns, C4.5 compares the error rate of resulting subtree (on the training data) with the error rate that would be obtained by replacing that subtree with a single leaf node. Hence, the effect of these first two changes is to simplify C4.5 by removing all pruning procedures.

Third, they modify C4.5's probability estimation mechanism so that instead of simply estimating $\hat{\mathbf{P}}(k|\mathbf{x})$ by the proportion of examples in the leaf that belong to class k, it applies a Laplace correction as follows:

$$\hat{\mathbf{P}}(k|\mathbf{x}) = \frac{N_k + \lambda_k}{N + \sum_{k=1}^{K} \lambda_k} \qquad (9.2)$$

where N is the total number of training examples that reach the leaf, N_k is the number of examples from class k reaching the leaf, K is the number of classes, and λ_k is the prior for class y_k. In their experiments, they set λ_k uniformly to 1.0 for all $k = 1, \ldots, K$.

The effect of the Laplace correction [18, 8, 3] is to shrink the probability estimates toward $1/K$. For example, a leaf containing two examples (both from class $k = 1$) will estimate $\hat{\mathbf{P}}(1|\mathbf{x}) = (2+1)/(2+2) = 0.75$ when $K = 2$ instead of 1.0.

Finally, they apply an ensemble method known as Bagging [5]. Bagging constructs L decision trees and then averages their class probability estimates. Each decision tree is constructed by taking the original training data $S = \{(\mathbf{x}_i, y_i)\}_{i=1}^{N}$ and constructing a bootstrap replicate data set S_ℓ by drawing N data points uniformly with replacement from S [13]. This training set S_ℓ is then given to the modified C4.5 procedure to produce a decision tree T_ℓ and a probability estimator $\hat{\mathbf{P}}_\ell(y|\mathbf{x})$. The final probability estimates are computed as

$$\hat{\mathbf{P}}(y|\mathbf{x}) = \frac{1}{L} \sum_{\ell=1}^{L} \hat{\mathbf{P}}_\ell(y|\mathbf{x}).$$

These modifications to C4.5 address two of the three causes of poor probability estimates identified in the introduction. The Laplace correction addresses the problem that the estimates in each leaf are based on small amounts of data. The combination of no pruning and Bagging addresses the problem that the probability estimates are constant throughout the region of $\mathbf{x}$ space corresponding to a single leaf. The only problem unaddressed is that the probability estimates are still based on the training data, which was also employed to construct the decision trees.

We explored the possibility of using out-of-bag estimates to address this last problem. The idea is to compute the estimates $\hat{\mathbf{P}}_\ell(y|\mathbf{x})$ by using the data points that were *not* included in training set S_ℓ (i.e., the so-called "out-of-bag" data points). Unfortunately, this did not work well, perhaps because there are even fewer "out-of-bag" data points than there are "in-bag" data points.

9.3   Lazy Option Trees

The decision tree literature contains many methods for improving over the basic C4.5 system. We have combined two of these methods—lazy tree growing and options—to obtain further improvements over B-PETs.

Standard decision tree algorithms can be viewed as being "eager" algorithms—that is, once they have the training data, they grow the decision tree. "Lazy" algorithms, in contrast, wait until the unlabeled test data point $\mathbf{x}_u$ is available before growing the decision tree. In this, they are analogous to the nearest-neighbor classifier, which waits until $\mathbf{x}_u$ is available and then bases its classification decision on the data points that are nearest to $\mathbf{x}_u$ according to some distance measure [19, 1, 10, 25]

The advantage of lazy learning is that it can focus on the neighborhood around the test point $\mathbf{x}_u$. In the case of probability estimates, a lazy learning algorithm can base its estimate $\hat{\mathbf{P}}(y|\mathbf{x}_u)$ on the data points in the neighborhood of $\mathbf{x}_u$ and thereby side-step the problem that the probability estimates in decision trees are constant within each leaf. The disadvantage of lazy learning is that the amount of computer time required to classify each data point can be significantly larger than for eager algorithms. This makes lazy learning impractical for some applications.

Lazy learning in decision trees was explored by Friedman et al. [16]. In their approach, which we imitate, a lazy tree actually consists of only a single path from the root of the tree down to a leaf. This is the path that will be taken by the test data point $\mathbf{x}_u$. Since the tree is only constructed to classify this one point, there is no need to expand the other branches of the tree. In addition, the choice of which attribute and value to test at each node of the tree can be made focusing only on the one resulting branch. In standard decision tree algorithms, the choice of splitting attribute and value is based on the average improvement of all branches of the decision tree. In some cases, the accuracy of one branch may actually be decreased in order to obtain a large increase in the accuracy of the other branch. Lazy trees do not need to make this compromise.

Table 9.1. The lazy option tree learning algorithm.

LAZYOPTIONTREE(S,$\mathbf{x}_u$,$MaxTests$,$MinG$,$MinLeaf$)
Input: Training data $S = \{(\mathbf{x}_i, y_i)\}_{i=1}^{N}$, Unlabeled test example $\mathbf{x}_u$,
      Maximum number of tests in a node $MaxTests$,
      Minimum gain for a test $MinG$,
      Minimum number of examples in a leaf $MinLeaf$,
    for $k = 1$ **to** K **do**
        assign weights $\mathbf{w}$ such that $\sum_{i|y_i=k} w_i = 1$
    choose a set of T tests of the form $x_j > \theta$, $x_j < \theta$, or $x_j \neq \theta$ **such that**
        $T < MaxTests$
        if best test has gain g, each test has a gain $> MinG \cdot g$,
        at least $MinLeaf$ examples satisfy the test, and
        the set of tests contains no tests ruled out by the cousin rule
    if no such tests exist **then** terminate with leaf node
    else call LAZYOPTIONTREE on elements of S that satisfy the test
end LAZYOPTIONTREE

One risk of lazy trees is that the tests included in the decision tree are satisfied only by the test point $\mathbf{x}_u$. To reduce this risk, we restrict the form of the permitted tests. Let x_m be the value of the mth attribute of the test data point $\mathbf{x}_u$. We do not allow the test to directly refer to x_m. The test may have the form $x_j > \theta$, $x_j < \theta$, or $s_j \neq \theta$ as long as $\theta \neq x_m$.

To estimate the class probabilities at the leaf of the lazy tree, we do not apply a Laplace correction. Instead, we simply compute the proportion of training examples belonging to class k.

The other technique that we have included in our approach is options (see Buntine [7] and Kohavi and Kunz [21]). An option tree is a decision tree in which a given internal node in the tree is split in multiple ways (using multiple attribute-value tests, or "options"). The final probability estimate is the average of the estimates produced by each of these options.

Our lazy option tree algorithm selects in each node the best b_t tests with the highest accuracy (where accuracy is measured by the reduction in the estimated entropy of the class variable $H(y)$ from the parent to the child). To avoid redundancy in the tree, we check for and remove a particular pattern of tests (called the "cousin rule"). Let U and V be two tests (e.g., $x_j > \theta$ for some j and θ). Suppose that node A has two options. One option tests U and leads to child node B; the other option tests V and leads to child C. Now suppose that B chooses V as an option. Then we do not permit node C to choose U as an option, because this would create two identical paths (where U and V were true) in the option tree.

One final detail requires mention. Before choosing the options at a node, our algorithm assigns weights to the examples in each class in order to balance the weighted class frequencies.

Table 9.1 sketches the lazy option tree learning algorithm, and Figure 9.1 shows an example of a tree induced by the lazy option tree algorithm.

The options represent an alternative to the averaging mechanism of Bagging for improving the probability estimates of the decision trees. The advantage of the options mechanism is that tests having an information gain almost as high as the best test will be performed, while they might never be performed in decision trees, lazy trees, or even bagged trees. This increases the diversity of the ensemble and leads to better probability estimates.

Nonetheless, we found experimentally that the probability estimates can be further improved by employing bagging in addition to options. This completes our bagged lazy option trees (B-LOTs) algorithm.

9.4   Parameter Settings

To apply the B-PET method, the user must specify only two parameters: the minimum number of examples $MinLeaf$ permitted in each leaf of the tree, and the number of iterations L of bagging. To apply B-LOTs, the

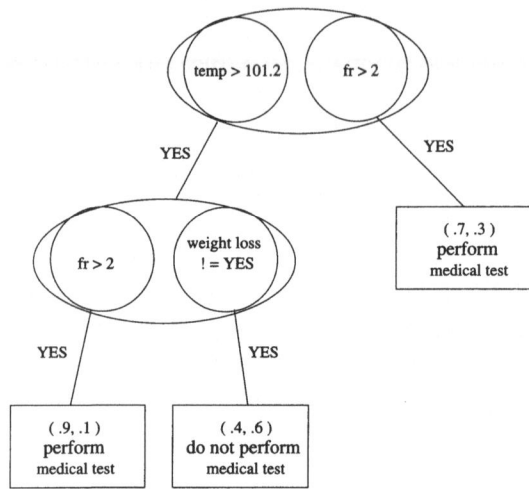

Figure 9.1. Example of a tree built by the lazy option tree algorithm for a specific unlabeled instance ⟨*fr* = 6, *weight loss* = *NO*, *blood pressure* = *HIGH*, *body temperature* = 101.5⟩. The task is to decide whether a patient is suspect of having AIDS, and therefore whether to perform an HIV test (*fr* stands for the frequency of recurrent infections per month, *temp* is the body temperature). The numbers in the leaves indicate the proportion of training examples that reach that leaf. Each leaf computes a probability estimate based on those numbers. These estimates are then averaged to produce the final estimates. The labels of the leaves indicate the decision in the case of 0/1 loss.

user must set *MinLeaf*, *L*, and also two more parameters: *MaxTests*—the maximum number of tests that are allowed in a node—and *MinG*—the gain proportion (0.0 < *MinG* < 1.0), a number indicating the minimum gain for a test in order to be selected. *MinG* is the proportion out of the gain of the best test achieved within that node (e.g., if the best test in a node has a gain on *g*, we will discard all tests that have a gain of less than *MinG* · *g*, within that same node).

9.5 Experiment 1: Probability Estimates on Synthetic Data

Figure 9.2 shows two one-dimensional gaussian distributions corresponding to $\mathbf{P}(\mathbf{x}|y = 1)$ and $\mathbf{P}(\mathbf{x}|y = 2)$. Let $\mathbf{P}(y = 1) = \mathbf{P}(y = 2) = 0.5$ be the probability of generating examples from classes 1 and 2, respectively. We constructed this synthetic problem so that we could measure the accuracy of the probability estimates produced by B-PETs and B-LOTs. We drew 1000 training examples from this joint distribution to form our training set S, which we then fed to the B-PET and B-LOT algorithms with the

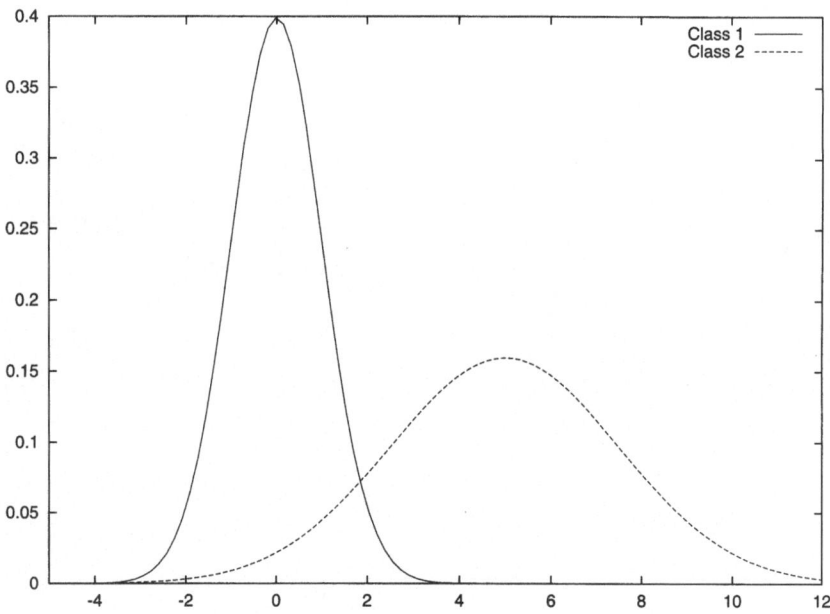

Figure 9.2. One-dimensional artificial data set

following parameter settings: $MinLeaf = 2$, $L = 30$, $MaxTests = 3$, and $MinG = 0.2$.

Figures 9.3 and 9.4 show scatter plots of the predicted and true probability values for the two methods. Perfect probability estimates would lie on the diagonal line shown in the plots. The scatter plot for the B-PETs shows that the points increase almost monotonically (with some errors), which indicates that the B-PETs have produced a good relative ranking of the test points. However, the absolute values of the probability estimates all lie in the range from 0.3 to 0.7, and the monotonic behavior is significantly non-linear. This behavior is partly a consequence of the Laplace shrinkage toward 0.5—however, if we remove the Laplace correction, the monotonic behavior of the curve is significantly degraded.

The plot for the B-LOTs, on the other hand, shows that the predicted probabilities cluster around the ideal diagonal and span the entire range from 0 to 1. However, individual points whose true probabilities are nearly identical can have predicted probabilities that differ by more than 0.2.

If this behavior holds in real domains, it would have two consequences. First, the B-PET estimates will not work well if they are plugged directly in to equation 9.1. This is because the given cost matrix C will determine an absolute probability threshold θ such that $\mathbf{x}_u$ should be classified as class 1 if $\hat{\mathbf{P}}(y|\mathbf{x}_u) > \theta$, and the absolute values of the B-PET predictions are poor. Consider, for example, if C dictates that $\theta = 0.2$. In such cases, all examples would be classified into class 1, with, presumably, high misclassification

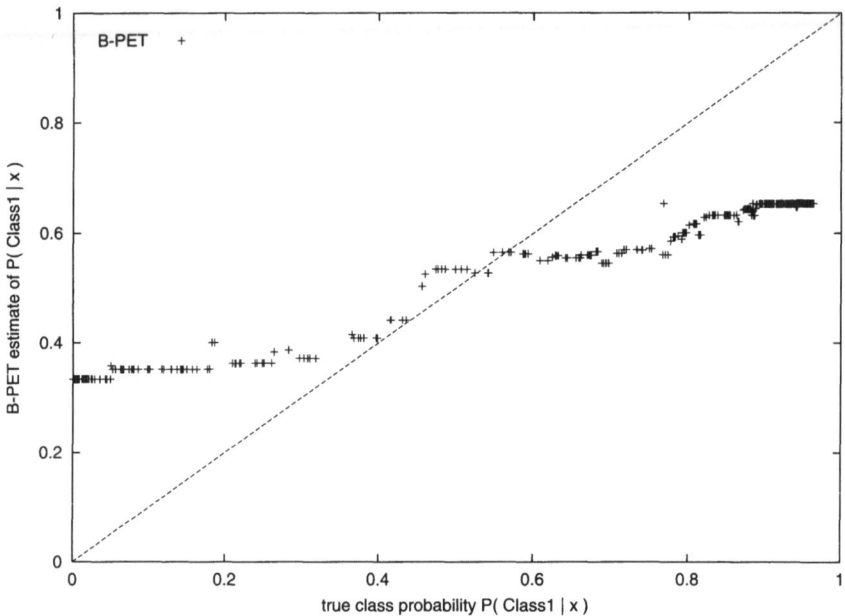

Figure 9.3. Scatterplot of B-PET probability estimates versus true probability values for 1000 test points

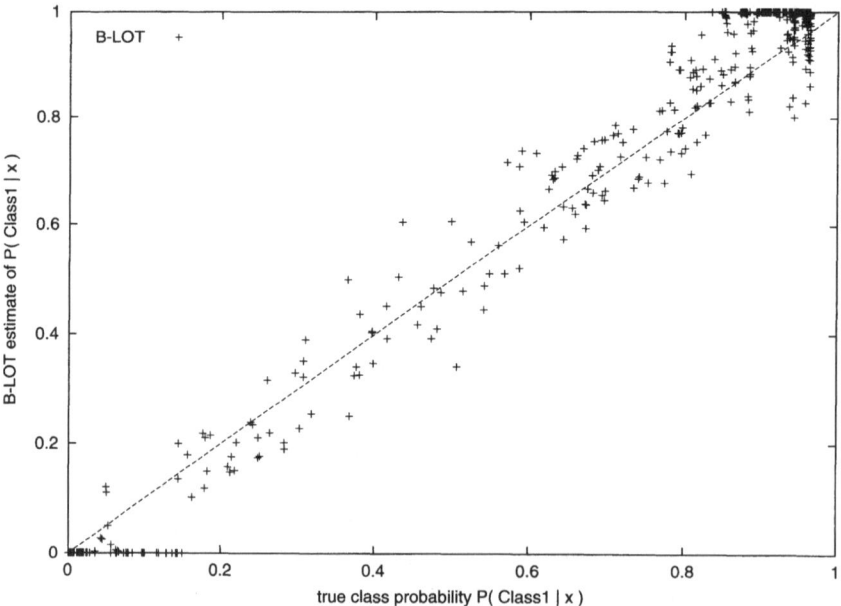

Figure 9.4. Scatterplot of B-LOT probability estimates versus true probability values for 1000 test points

costs. In contrast, the B-LOT estimates will work quite well in equation 9.1, because they are well-calibrated. However, there will be several misclassified points, because of the variance in the estimates.

In 2-class problems, it is not necessary to use equation 9.1. Instead, cross-validation methods can be employed to determine experimentally a threshold that minimizes the expected misclassification cost on hold-out data points. In such cases, the B-PET method may work well unless the threshold happens to land exactly on one of the values for which the plot in Figure 9.3 is flat. Note that there are several such values, particularly at the more extreme probability values (which could be important if the misclassification costs are highly asymmetrical). For problems with more than 2 classes, cross-validation is difficult to employ, because separate thresholds must be determined for each pair of classes. Setting so many thresholds by cross-validation requires very large hold-out sets.

9.6   Experiment 2: Misclassification Errors on Benchmark Data Sets

A more realistic way of comparing B-PETs and B-LOTs is to evaluate their performance on a range of benchmark data sets drawn from real problems. We chose sixteen data sets from the UC Irvine ML repository [2] for this study. Unfortunately, most of the UC Irvine data sets do not have associated cost matrices C. Hence, we decided to generate cost matrices at random according to some cost matrix models.

Table 9.2 describes eight cost models, M1 through M8. The second column of the table describes how the misclassification costs were generated for the off-diagonal elements of C. In most cases, the costs are drawn from a uniform distribution over some interval. However, for models M5 and M6, the costs are distributed according to the class frequencies. In model M5, rare classes are given higher misclassification costs. This is characteristic of medical and fraud detection problems where a failure to detect a rare event (a serious disease or a fraudulent transaction) has a much higher cost than the failure to recognize a common event. In model M6, rare classes are given lower costs, to see if this makes any difference. The third column of Table 9.2 specifies how the diagonal elements of C are generated. Only models M7 and M8 have non-zero diagonal values.

For each cost model we generated twenty-five cost matrices and performed ten-fold cross validation. This gives us 25 (matrices) × 8 (models) × 10 (folds) = 2000 runs of the B-PET and B-LOT procedures for each data set. For each of these 2000 runs, we applied a statistical test, BDELTA-COST, to determine whether the B-PETs and the B-LOTs had statistically significant different expected misclassification costs.

Table 9.2. The cost models used for the experiments. Unif[a, b] indicates a uniform distribution over the [a, b] interval. $P(i)$ represents the prior probability of class i.

Cost Model	$C(i,j)$ $i \neq j$	$C(i,i)$
M1	Unif[0, 10]	0
M2	Unif[0, 100]	0
M3	Unif[0, 1000]	0
M4	Unif[0, 10000]	0
M5	Unif[0, 1000 × $P(i)/P(j)$]	0
M6	Unif[0, 1000 × $P(j)/P(i)$]	0
M7	Unif[0, 10000]	Unif[0, 1000]
M8	Unif[0, 100]	Unif[0, 10]

BDELTACOST is a bootstrap test developed by the authors [22] for comparing the cost of two classifiers γ_1 and γ_2. It works by first constructing a generalized confusion matrix M. The contents of cell $M(i_1, i_2, j)$ is the number of (test set) examples for which γ_1 predicted that they belong to class i_1, γ_2 predicted that they belong to class i_2, and their true class was j. BDELTACOST then constructs a three-dimensional cost matrix Δ such that $\Delta(i_1, i_2, j) = C(i_1, j) - C(i_2, j)$. In other words, the value of $\Delta(i_1, i_2, j)$ is the amount by which the cost of classifier γ_1 is greater than the cost of classifier γ_2 when γ_1 predicts class i_1, γ_2 predicts class i_2, and the true class is j. Given M and Δ, the difference in the costs of γ_1 and γ_2 can be computed by taking the "dot product":

$$M \cdot \Delta = \sum_{i_1, i_2, j} M(i_1, i_2, j) \Delta(i_1, i_2, j).$$

BDELTACOST computes a confidence interval for $M \cdot \Delta$ and rejects the null hypothesis that $M \cdot \Delta = 0$ if this confidence interval does not include zero. The confidence interval is constructed as follows. Normalize M by dividing by the number of test examples N, so that M's elements sum to 1. Treating M as a multinomial distribution, draw 1000 simulated confusion matrices $\tilde{M}_u$ by drawing N triples (i_1, i_2, j) according to this distribution. Compute the cost $\tilde{c}_u$ of each simulated confusion matrix (by computing $\tilde{M}_u \cdot \Delta$), sort these costs, and choose the 25th and 976th elements to construct the confidence interval.

The barchart in Figure 9.5 shows the results of the 2000 test runs for each data set. It plots the number of wins for B-LOTs, ties, and wins for B-PETs computed by BDELTACOST with a significance level of $\alpha = 0.05$ (i.e., 95% confidence). For all tasks except the lung-cancer domain, the number of wins for the B-LOTs is noticeably larger than the number of B-PET wins. In the case of nine domains (audiology, abalone, iris, lymphography, segmentation, soybean, voting-records, wine and zoo), the B-LOTs were the clear winner. These results suggest that the B-LOTs give better probability estimates in these application domains.

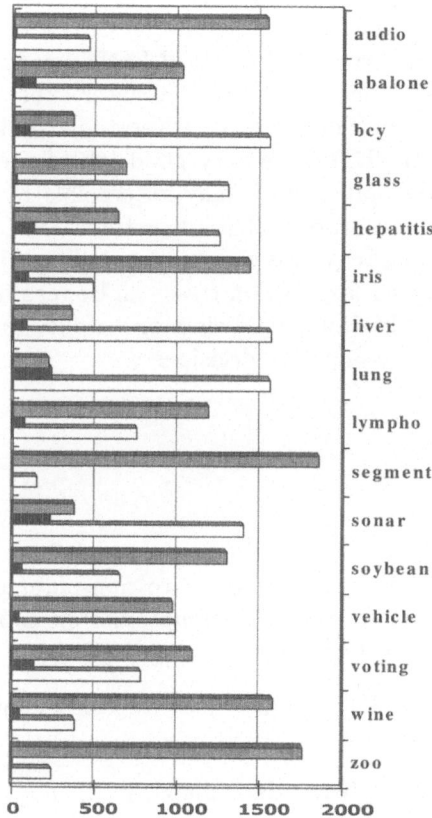

Figure 9.5. Barchart comparing the performance of B-LOTs and B-PETs for sixteen UCI data sets. White bars represent the number of ties, black bars represent the number of wins of B-PETs, and the gray bars represent the number of wins of B-LOTs computed by the BDELTACOST test.

9.7   Concluding Remarks

Decision-boundary methods—as exemplified by decision tree learning algorithms—have proven to be very robust in many applications of machine learning. However, unlike probabilistic models, they do not provide estimates of $\mathbf{P}(y|\mathbf{x})$. Such probability estimates are essential for applications that involve asymmetric loss functions. This chapter has addressed the question of whether techniques can be found to compute good class probability estimates from decision trees.

The chapter has introduced the bagged lazy option tree (B-LOT) method and compared it to the bagged probability estimate tree (B-PET) method. The results show experimentally that B-LOTs produce better-calibrated probability estimates and that these give lower expected misclassification cost across 16 benchmark datasets. However, the experiment with synthetic data showed that B-PETs do a very good job of predicting the relative probabilities of data points. So it may be possible to convert the B-PET estimates into more accurate probability estimates.

Based on the results from this chapter, for applications where lazy learning is practical, bagged lazy option trees can be recommended as the best known approach to obtaining good class probability estimates and making good cost-sensitive classification decisions.

References

[1] B. K. Bhattacharya, G. T. Toussaint, and R. S. Poulsen. The application of Voronoi diagrams to non-parametric decision rules. In *Proceedings of the 16th Symposium on Computer Science and Statistics: The Interface*, pages 97–108, 1987.

[2] C.L. Blake and C.J. Merz. UCI repository of machine learning databases, 1998.

[3] J. P. Bradford, C. Kunz, R. Kohavi, C. Brunk, and C. E. Brodley. Pruning decision trees with misclassification costs. In C. Nedellec and C. Rouveirol, editors, *Lecture Notes in Artificial Intelligence. Machine Learning: ECML-98, Tenth European Conference on Machine Learning*, volume 1398, pages 131–136, Berlin, 1998. Springer Verlag.

[4] L. Breiman, J. H. Friedman, R. A. Olshen, and C. J. Stone. *Classification and Regression Trees*. Wadsworth International Group, 1984.

[5] Leo Breiman. Bagging predictors. *Machine Learning*, 24(2):123–140, 1996.

[6] Leo Breiman. Random forests. Technical report, Department of Statistics, University of California, Berkeley, CA, 1999.

[7] W. L. Buntine. *A theory of learning classification rules*. PhD thesis, University of Technology, School of Computing Science, Sydney, Australia, 1990.

[8] B. Cestnik. Estimating probabilities: A crucial task in machine learning. In L. C. Aiello, editor, *Proceedings of the Ninthe European Conference on Artificial Intelligence*, pages 147–149. Pitman Publishing, 1990.

[9] H. Chipman, E. George, and R. McCulloch. Bayesian CART model search (with discussion). *Journal of the American Statistical Association*, 93:935–960, 1998.

[10] B. V. Dasarathy, editor. *Nearest neighbor (NN) norms: NN pattern classification techniques*. IEEE Computer Society Press, Los Alamitos, CA, 1991.

[11] D. G. T. Denison, B. K. Mallick, and A. F. M. Smith. A Bayesian CART algorithm. *Biometrika*, 85:363–377, 1998.

[12] Thomas G. Dietterich. An experimental comparison of three methods for constructing ensembles of decision trees: Bagging, boosting, and randomization. *Machine Learning*, 40(2):1, 2000.

[13] Bradley Efron and Robert J. Tibshirani. *An Introduction to the Bootstrap*. Chapman and Hall, New York, NY, 1993.

[14] Yoav Freund and Robert E. Schapire. Experiments with a new boosting algorithm. In *Proc. 13th International Conference on Machine Learning*, pages 148–146. Morgan Kaufmann, 1996.

[15] Jerome H. Friedman, Trevor Hastie, and Rob Tibshirani. Additive logistic regression: A statistical view of boosting. *Annals of Statistics*, 28(2):337–407, 2000.

[16] Jerome H. Friedman, Ron Kohavi, and Yeogirl Yun. Lazy decision trees. In *Proceedings of the Thirteenth National Conference on Artificial Intelligence*, pages 717–724, San Francisco, CA, 1996. AAAI Press/MIT Press.

[17] S. Geman, E. Bienenstock, and R. Doursat. Neural networks and the bias/variance dilemma. *Neural Computation*, 4(1):1–58, 1992.

[18] I. J. Good. *The estimation of probabilities: An essay on modern Bayesian methods*. MIT Press, Cambridge, MA, 1965.

[19] P. E. Hart. The condensed nearest neighbor rule. *IEEE Transactions on Information Theory*, 14:515–516, 1968.

[20] Tin Kam Ho. The random subspace method for constructing decision forests. *IEEE Transactions on Pattern Analysis and Machine Intelligence*, 20(8):832–844, 1998.

[21] Ron Kohavi and Clayton Kunz. Option decision trees with majority votes. In *Proc. 14th International Conference on Machine Learning*, pages 161–169. Morgan Kaufmann, 1997.

[22] D. D. Margineantu and T. G. Dietterich. Bootstrap methods for the cost-sensitive evaluation of classifiers. In *Proceedings of the Seventeenth International Conference on Machine Learning*, San Francisco, CA, 2000. Morgan Kaufmann.

[23] Foster Provost and Pedro Domingos. Well-trained PETs: Improving probability estimation trees. Technical Report IS-00-04, Stern School of Business, New York University, 2000.

[24] J. R. Quinlan. *C4.5: Programs for Empirical Learning*. Morgan Kaufmann, San Francisco, CA, 1993.

[25] Dietrich Wettschereck. *A Study of Distance-Based Machine Learning Algorithms*. PhD thesis, Department of Computer Science, Oregon State University, Corvallis, Oregon, 1994.

10

Gauss Mixture Quantization: Clustering Gauss Mixtures

Robert M. Gray[1]

Summary

Gauss mixtures are a popular class of models in statistics and statistical signal processing because Gauss mixtures can provide good fits to smooth densities, because they have a rich theory, because they can yield good results in applications such as classification and image segmentation, and because the can be well estimated by existing algorithms such as the EM algorithm. We here use high rate quantization theory to develop a variation of an information theoretic extremal property for Gaussian sources and its extension to Gauss mixtures. This extends a method originally used for LPC speech vector quantization to provide a clustering approach to the design of Gauss mixture models. The theory provides formulas relating minimum discrimination information (MDI) selection of Gaussian components of a Gauss mixture and the mean squared error resulting when the MDI criterion is used in an optimized robust classified vector quantizer. It also provides motivation for the use of Gauss mixture models for robust compression systems for random vectors with estimated second order moments but unknown distributions.

10.1   Introduction

Gauss mixtures have played an important role in modeling random processes for purposes of both theory and design. They have been used in statistics and statistical signal processing for many decades, but they have recently enjoyed renewed popularity in a variety of problems. Simply stated, a Gauss mixture probability density function (pdf) is a weighted sum of

---

[1]Robert M. Gray, is Professor, Information Systems Laboratory, Department of Electrical Engineering, Stanford University. (Email: rmgray@stanford.edu). This work was partially supported This work was partially supported by the National Science Foundation under Grant Number 0073050 and by the Hewlett Packard Corporation. The author would like to acknowledge the helpful comments and suggestions on the manuscript by Ms. Maya Gupta and by an anonymous referee.

a collection of distinct Gaussian pdf's. A single Gaussian pdf describing a
k-dimensional random variable X has the form

$$g(x) = \mathcal{N}(x, \mu, K) = \frac{1}{(2\pi)^{\frac{k}{2}}|K|^{\frac{k}{2}}} \exp\left(-\frac{1}{2}(x - \mu)^t K^{-1}(x - \mu)\right),$$

where $K = E[(X - EX)(X - EX)^t]$ is the $k \times k$ covariance matrix, $|K|$
the determinant of k, and $\mu = EX$ the mean vector. We assume that
the covariance is nonsingular and all moments are finite. A Gauss mixture
density has the form

$$\bar{g}(x) = \sum_i w_i g_i(x) = \sum_i w_i \mathcal{N}(x, \mu_i, K_i),$$

where

$$\sum_i w_i = 1$$

and we assume that there are only a finite number of components.

There are two distinct ways of interpreting the physical meaning of a mixture density which lead to different methods of estimating mixture densities
based on observed data. One view is to consider the weighted sum directly
as a single pdf with a particular structure and try to match such a pdf
directly to an observed sequence of sample vectors produced by the random phenomenon to be modeled. For example, the EM algorithm takes the
mixture model literally and tries to match a weighted sum of distributions
to observed data, generally attempting to approximate a maximum likelihood estimation assuming the components of the mixture to be Gaussian.
Relative entropy matching techniques also usually deal directly with the
weighted sum, yielding complicated formulas including sums of Gaussians
within logarithms [26]. An alternative viewpoint, which we adopt, is to instead view the mixture as a "doubly stochastic" phenomenon. First nature
selects which component of the mixture is to be used using the probability
mass function p. Once the component, say i, is chosen, the actual random
variable is then generated using the pdf g_i. The interpretation using the
doubly stochastic idea is that one is in fact observing a sample vector produced by one of the Gaussian components of the mixture, but one does
not know *which one*. If one is viewing a sequence of vectors from such a
source instead of a single vector (as is usually the case), the underlying
model is that of a composite or "switched"process, that is, the observed
vectors are produced by a random slowly varying switch that successively
connects distinct information sources to the output. The marginal pdf (for
an individual vector) for such a source is a mixture pdf. Composite source
models have been successful in the analysis of a variety of signal processing
and information theory problems and are explicit or implicit in the most
popular speech modeling algorithms.

The primary issue at hand is how to design a Gauss mixture considered as collection of Gaussian components together with a probability mass function providing their prior probabilities in the context of certain specific applications to be described. The answer provided is to use a clustering algorithm that attempts to match a collection of individual Gaussian components to a set of learning or training data. The underlying prior will follow from the relative frequencies at which these components are selected by the algorithm. The number of components will will follow as a natural byproduct of the clustering algorithm.

Gauss mixture models played a fundamental, although originally implicit, role in the development in the 1970s of linear predictive coded speech (LPC) modeling, which yielded the first truly low bit rate speech compression systems with reasonable quality and which still dominate speech coding and recognition applications. LPC was originally introduced as the partial correlation method (PARCOR) based on an approximately maximum likelihood fit of Gaussian components to nonoverlapping segments of speech by Itakura and Saito [18], It can be viewed as fitting Gauss mixture models to speech when the autoregressive (AR) models fit to segments of speech are excited by Gaussian residual processes. In this case the synthesized speech becomes a composite or Gaussian process with the property that at any given instant one is observing a single Gaussian component of the mixture.

The most popular means of fitting a Gauss mixture model to data is the EM algorithm, but similar ends can be achieved using Lloyd (or Steinhaus or k-means or principal points) clustering techniques [21] (see, e.g., [11] for a technical and historical survey) with suitable distortion measures between observed data and proposed models, a method described for low rate speech coding using the Itakura-Saito distortion in [16]. The Itakura-Saito distortion is an example of a minimum discrimination information (MDI) distortion, a measure based on model fitting techniques of Kullback using relative entropies (Kullback-Leibler numbers, cross entropies) [19]. Lloyd clustering techniques have several potential advantages over the EM algorithm. The first is the use of minimum distortion rules for selecting the "best" Gaussian component of a mixture to fit the current observed data, a form of nearest neighbor selection of a model for an observation. A second advantage is the existence of explicit formulas describing centroids with respect to the distortion measures, formulas which, when combined with quantization theory, provide quantitative relations between minimum discrimination information distortion measures and the performance of optimized robust compression systems. A final advantage is that the number of components in a Gauss mixture model is automatically determined by the clustering algorithm. Entropy constrained Lloyd clustering automatically eliminates unused clusters so as to improve the tradeoff between distortion and rate. When the algorithm converges, the number of cluster centers – which is also the number of components in the mixture model –

must be at least a locally optimum choice in terms of balancing rate and distortion.

One of the key properties of the Gaussian model is its role as a "worst case" in compression/source coding problems, a characterization most developed by Sakrison [24] and Lapidoth [20]. We describe a recent extension of this property to Gauss mixture models and explore applications to situations in which successive vectors produced by an information source are classified as belonging to one of a collection of Gaussian models which together form a Gauss mixture.

10.2   Preliminaries

The reader is referred to [11] for a summary, history, and extensive bibliography of quantization, including its intimate connections with clustering. We here recall several relevant definitions and results.

10.2.1   Information Sources

An information source is a stationary random process $\{X(n)\}$ where each $X(n)$ is a random vector taking values in $\Re^k$ according to a generic probability distribution P_X corresponding to a generic random vector X described by a smooth probability density function (pdf) f. Smoothness of the densities is required for the approximations we use which are based on Gersho's conjecture.

We focus on the generic random vector $X = (X_0, X_1, \ldots, X_{k-1})$ with the understanding that for any information theoretic problem in fact there will be a sequence of random vectors with the same distribution. For most information theory applications we do not need to assume that the successive vectors are independent, but this is often done because most coding structures code each vector in a manner independent of past and future, so that the average behavior is the same whether or not the vectors are independent. The vectors can be thought of rectangular blocks of pixels in an image, of transformed versions of those pixels using a Fourier or wavelet transform, or as features extracted from an image.

Notational problems arise when using ordinary vectors and matrices to model images, e.g., for some purposes it is more useful to think of an image as a raster or random field $X = \{X(i,j);\ i,j \in \mathcal{Z}_N\}$ rather than as a single integer-indexed vector. In the latter case the covariance matrix is easily described in vector notation as, but in the former case it is often more convenient to deal directly with the covariance function $\{K(i,j);\ i,j \in \mathcal{Z}_N^2\}$. For example, if an image is assumed to be spatially stationary, then the covariance function will be a Toeplitz operator, but if the raster is converted into a single indexed vector X to obtain the covariance matrix K, the matrix will not be Toeplitz. We will focus on the case where X is simply considered as a vector without regard to its origins, e.g., to how a raster image is scanned to convert it into a vector. It is worth pointing out that no matter how a raster X is "vectorized" into an integer-indexed vector, the resulting determinant will be unchanged since any vector versions of X will be related to others by a permutation matrix.

One of the useful properties of a Gaussian distribution is that it has a simple Shannon differential entropy:

$$h(g) = -\int dx f(x) \ln f(x) = \frac{1}{2} \ln(2\pi e)^k |K|. \tag{10.1}$$

Less well known outside of information theory is the fact that the Gaussian distribution plays an important extremal role in Shannon rate distortion theory, one version of which will be described in more detail shortly. Sakrison showed that of all distributions with a fixed mean and covariance, the Gaussian has the largest Shannon rate-distortion function [24], that is, the Gaussian requires the larges rate to meet a given fidelity requirement and hence has the worst performance over all sources with the same covariance. Lapidoth strengthened this result to show that for iid Gaussian sources, a code designed for a Gaussian source yields the same rate and distortion on an arbitrary source with the same second order moments [20]. This characterizes the Gaussian source as a "worst case" source for data compression and provides an approach to robust compression. A problem with this approach is that it can be too conservative, designing a code for a single Gaussian model may yield a trustworthy rate-distortion tradeoff, but it may yield far worse performance than that obtainable using a better source model. This motivates Gauss mixtures from a compression point of view: a Gauss mixture source may provide a "locally worst case" or "conditionally worst case" model if suitably used, yielding robust codes with better performance than a single Gaussian would provide.

10.2.2 Quantization

A vector quantizer Q (or Shannon lossy compression code or source code subject to a fidelity criterion) can be described by the following mappings and sets:

- an encoder $\alpha : \Omega \to \mathcal{I}$, where $\mathcal{I}$ is a countable index set, and an associated measurable partition $\mathcal{S} = \{S_i; \ i \in \mathcal{I}\}$ such that $\alpha(x) = i$ if $x \in S_i$,

- a decoder $\beta : \mathcal{I} \to \Omega$, and an associated reproduction codebook $\mathcal{C} = \{\beta(i); \ i \in \mathcal{I}\}$,

- an index coder $\psi : \mathcal{I} \to \{0, \ldots, D-1\}^*$, the space of all finite-length D-ary strings, and the associated length $L : \mathcal{I} \to \{1, 2, \ldots\}$ defined by $L(i) = \text{length}(\psi(i))$. ψ is assumed to be uniquely decodable (a lossless or noiseless code)[2].

- The overall quantizer is

$$Q(x) = \beta(\alpha(x)).\tag{10.2}$$

Without loss of generality we assume that the codevectors $\beta(i)$; $i \in \mathcal{I}$ are all distinct. Since the index coder must be uniquely decodable, the Kraft inequality (e.g., [6]) requires that the codelengths $L(i)$ satisfy

$$\sum_i D^{-L(i)} \leq 1.\tag{10.3}$$

It is convenient to measure lengths in a normalized fashion and hence we define the length function of the code in nats as $\ell(i) = L(i) \ln D$ so that Kraft's inequality becomes

$$\sum_i e^{-\ell(i)} \leq 1.\tag{10.4}$$

A set of codelengths $\ell(i)$ is said to be *admissible* if (10.4) holds.

As in Cover and Thomas [6], it is convenient to remove the restriction of integer D-ary codelengths and hence we define any collection of nonnegative real numbers $\ell(i); i \in \mathcal{I}$ to be an *admissible length function* if it satisfies (10.4). The primary reason for dropping the constraint is to provide a useful tool for proving results, but the general definition can be interpreted as an approximation since if $\ell(i)$ is an admissible length function, then for a code alphabet of size D the actual integer codelengths $L(i) = \lceil \frac{\ell(i)}{\ln D} \rceil$ will satisfy the Kraft inequality. Abbreviating $P_f(S_i)$ to p_i the average length (in nats) will satisfy

$$\sum_i p_i \ell(i) \leq (\ln D) \sum_i p_i L(i) < \sum_i p_i \ell(i) + \ln D.$$

If this is normalized by $1/k$, then the actual average length can be made arbitrarily close to the average length function by choosing a sufficiently large dimension. Let $\mathcal{A}$ denote the collection of all admissible length functions ℓ.

With a slight abuse of notation we will use the symbol Q to denote both the composite of encoder α and decoder β as in (10.2) and the complete quantizer comprising the triple (α, β, ℓ). The meaning should be clear from context.

The instantaneous rate of a quantizer is defined by $r(\alpha(x)) = \ell(\alpha(x))$. The average rate is

$$R_f(Q) = R_f(\alpha, \ell) = Er(\alpha(X)) = \sum_i p_i \ell(i).\tag{10.5}$$

---

[2]This is stronger than invertibility since it requires that sequences be invertible.

Given a quantizer Q, the Shannon entropy of the quantizer is defined in the usual fashion by

$$H_f(Q) = -\sum_i p_i \ln p_i$$

and we assume that $p_i > 0$ for all i.

For any admissible length function ℓ the divergence inequality [6] implies that

$$
\begin{aligned}
R_f(Q) - H_f(Q) &= \sum_i p_i \ln \frac{p_i}{e^{-\ell(i)}} \\
&\geq \sum_i p_i \ln \frac{p_i}{e^{-\ell(i)} / \sum_j e^{-\ell(j)}} \\
&\geq 0
\end{aligned}
$$

with equality if and only if $\ell(i) = -\ln p_i$. Thus in particular

$$H_f(Q) = \inf_{\ell \in \mathcal{A}} R_f(\alpha, \ell). \tag{10.6}$$

We assume a distortion measure $d(x, \hat{x}) \geq 0$ and measure performance by average distortion

$$
\begin{aligned}
D_f(Q) &= D_f(\alpha, \beta) \\
&= Ed(X, Q(X)) \\
&= Ed(X, \beta(\alpha(X))).
\end{aligned}
$$

For simplicity we assume squared error distortion with average

$$d(x, \hat{x}) = ||x - \hat{x}||^2 = \sum_{i=0}^{k-1} |x_i - \hat{x}_i|^2,$$

where $x = (x_0, \ldots, x_{k-1})$. The approach extends to more general measures (and will be explicitly considered for the minimum discrimination information (MDI) distortion measure for models or pdf's).

The optimal performance is the minimum distortion achievable for a given rate $R \geq 0$, the *operational distortion-rate function*

$$
\delta_f(R) = \inf_{Q:R_f(Q) \leq R} D_f(Q) \tag{10.7}
$$

$$
= \inf_{\alpha, \beta, \ell : R_f(\alpha, \ell) \leq R} D_f(\alpha, \beta). \tag{10.8}
$$

In order to describe necessary conditions for optimality of a code and provide a clustering algorithm for the design of codes having these properties, it is convenient to use a Lagrangian formulation of variable rate vector quantization [4]. Define for each value of a Lagrangian multiplier $\lambda > 0$ a Lagrangian distortion

$$
\begin{aligned}
\rho_\lambda(x, i) &= d(x, \beta(i)) + \lambda r(\alpha(i)) \\
&= d(x, \beta(i)) + \lambda \ell(i)
\end{aligned}
$$

a corresponding performance

$$\rho(f, \lambda, Q) = Ed(X, Q(X)) + \lambda E\ell(\alpha(X))$$
$$= D_f(Q) + \lambda R_f(Q),$$

and an optimal performance

$$\rho(f, \lambda) = \inf_Q \rho(f, \lambda, Q)$$

where the infimum is over all quantizers $Q = (\alpha, \beta, \ell)$, $\ell \in \mathcal{A}$. The Lagrangian formulation yields necessary conditions for optimality that generalize Lloyd's original formulation [21] of optimal scalar quantization. (See [11] for simple proofs and a history of these conditions and their intimate connections with k-means and other similar clustering techniques.) Intuitively the conditions simply capsulize the fact that each of the three components of the quantizer be optimal for the other two. In particular, for a given decoder β and length function ℓ, the optimal encoder is

$$\alpha(x) = \operatorname*{argmin}_i \left(d(x, \beta(i)) + \lambda\ell(i) \right)$$

(ties are broken arbitrarily). The optimal decoder for a given encoder and length function is the usual Lloyd centroid:

$$\beta(i) = \operatorname*{argmin}_y E(d(X, y) | \alpha(X) = i),$$

the conditional expectation in the case of squared error distortion. The optimal length function for the given encoder and decoder is, as we have seen, the negative log probabilities of the encoder output:

$$\ell(i) = -\ln p_i.$$

The Lloyd clustering algorithm iteratively applies these properties to improve a given code. The algorithm is well defined whenever both the minimum distortion rule and the centroid rule can be applied with reasonable complexity.

10.3   High-rate quantization theory

High-rate quantization theory treats the asymptotic behavior of the rate-distortion tradeoff as the rate becomes large and the distortion becomes small. Here we follow the approach to high rate analysis developed by Gersho [7] (see also [11]), which is an intuitive derivation of the rigorous results of Zador [28] (see also [13]) using the quantizer point density ideas of Lloyd.

Define the volume $V(S)$ of a set S as the integral over S with respect to Lebesgue measure, $V(S) = \int_S dx$. Assume that there are asymptotically many quantization levels with vanishingly small cell volume and that the density f is sufficiently well behaved. Assume also that $y_i \in S_i$, where as before $S_i = \{x : \alpha(x) = i\}$. In fact this is optimal for most common distortion measures. Under these assumptions, $f(x) \approx f(y_i)$; $x \in S_i$ From the mean value theorem of calculus

$$P_f(S_i) = \int_{S_i} f(x)\, dx \approx V(S_i)f(y_i)$$

and hence

$$f(y_i) \approx \frac{P_f(S_i)}{V(S_i)},$$

which implies that

$$D_f(Q) \approx \frac{1}{k} \sum_i P_f(S_i) \int_{S_i} \frac{\|x - y_i\|^2}{V(S_i)}\, dx.$$

For each i, let y_i be the Euclidean centroid of S_i and hence

$$\int_{S_i} \frac{\|x - y_i\|^2}{V(S_i)}\, dx$$

is the moment of inertia of the region S_i about its centroid.

It is convenient to use normalized moments of inertia so that they are invariant to scale: For any measurable set S define

$$M(S) = \frac{1}{kV(S)^{2/k}} \int_S \frac{\|x - y(S)\|^2}{V(S)}\, dx$$

where $y(S)$ denotes the centroid of S. If $c > 0$ and $cS = \{cs : s \in S\}$, then $M(S) = M(cS)$ and M depends only on shape and not upon scale. Thus

$$\begin{aligned}
D &\approx \sum_i P_f(S_i)M(S_i)V(S_i)^{2/k} \\
&= \sum_i f(y_i)M(S_i)V(S_i)^{1+2/k}
\end{aligned}$$

To be more precise, high rate quantization theory considers the behavior of a sequence of codes $\mathcal{C}_n$ or corresponding quantizers Q_n with increasing rate and decreasing distortion. In order to guarantee that distortion is decreasing, the reproduction vectors must become increasingly dense in the support set of the underlying pdf. We assume that as $n \to \infty$, the reproduction vectors $\mathcal{C}_n = \{y_{n,i};\ i = 1, \ldots, N\}$ have a smooth point density $\Lambda(x)$ in the sense that $\Lambda(x) \geq 0$ and

$$\# \text{ reproduction vectors of } \mathcal{C}_n \text{ in a set } S \to \int_S \Lambda(x)\, dx; \quad \text{all } S.$$

Note that a point density is like a mass density rather than a probability density since its integral need not be 1 (or even finite).

If we consider a tiny cell S containing a only a single reproduction vector, then from the mean value theorem

$$1 \approx \int_S \Lambda(x)\, dx \approx \Lambda(x)V(S); \quad \text{all } x \in S$$

and hence since $y_{n,i} \in S_{n,i}$,

$$V(S_{n,i}) \approx \frac{1}{\Lambda(y_{n,i})}.$$

Define for the sequence of codebooks $\mathcal{C}_n$; $n = 1, 2, \ldots$ the function $B_i^{(n)}(x)$ as the cell of the codebook $\mathcal{C}_n$ which contains x. Assume that as $n \to \infty$ and $B_i^{(n)}(x) \to x$, then

$$M(B_i^{(n)}(x)) \to m(x),$$

the so-called *inertial profile* of the sequence of codebooks [22]. This function is assumed to be smooth and occasionally it can be evaluated and usually bound above and below. We then have for large n that

$$
\begin{aligned}
D_f(Q_n) &\approx \sum_i f(y_{n,i})M(S_{n,i})V(S_{n,i})^{1+2/k} \\
&\approx \sum_i f(y_{n,i})m(y_{n,i})\frac{V(S_{n,i})}{\Lambda(y_{n,i})^{2/k}} \\
&\approx \int f(x)\frac{m(x)}{\Lambda(x)^{2/k}}\, dx \; = E\left[\frac{m(X)}{\Lambda(X)^{2/k}}\right]
\end{aligned}
$$

Gersho's conjecture [7] can be described as follows [11]. If $f(x)$ is smooth and the allowed rate rate $R_n = R_f(Q_n)$ of (10.5) is large, then, regardless of Λ, the optimal quantizer Q_n has cells $S_{n,i}$ that are (approximately) scaled, rotated, and translated copies of S^*, the convex polytope that tesselates $\mathcal{R}^k$ with minimum normalized moments of inertia $M(S)$, i.e.,

$$m^*(x) = c_k = \min_S M(S) = M(S^*).$$

In this case the average distortion becomes

$$D_f(Q_n) \approx c_k E_f\left(\left(\frac{1}{\Lambda(X)}\right)^{2/k}\right). \tag{10.9}$$

The minimum average rate of the quantizer given the encoder and decoder is given by the entropy of the quantizer:

$$E_f \ell(Q_n(X)) = H_f(Q_n) = -\sum_i P_f(S_{n,i}) \ln P_f(S_{n,i}).$$

and the approximations then imply that

$$P_f(S_{n,i}) \approx f(y_{n,i})/(\Lambda(y_{n,i})) = f(y_{n,i})V(S_{n,i})$$

so that

$$
\begin{aligned}
E_f \ell(\alpha(X)) &= H_f(Q_n) \\
&\approx -\sum_i \frac{f(y_{n,i})}{\Lambda(y_{n,i})} \ln \frac{f(y_{n,i})}{\Lambda(y_{n,i})} \\
&= -\sum_i V(S_{n,i}) f(y_{n,i}) \ln f(y_{n,i}) \\
&\quad\quad\quad\quad + \sum_i V(S_{n,i}) f(y_{n,i}) \ln(\Lambda(y_{n,i})) \\
&\approx -\int dy f(y) \ln f(y) + \int dy f(y) \ln(\Lambda(y)) \\
&\approx h(f) - E\left(\ln \frac{1}{\Lambda(X)}\right),
\end{aligned}
\tag{10.10}
$$

where $h(f)$ is the differential entropy. Thus for large N

$$
\begin{aligned}
H_f(Q_n) &\approx h(f) - E_f\left(\ln \frac{1}{\Lambda(X)}\right) \\
&= h(f) - \frac{k}{2} E\left(\ln\left(\frac{1}{\Lambda(X)}\right)^{2/k}\right).
\end{aligned}
\tag{10.11}
$$

From Jensen's inequality,

$$
\begin{aligned}
H_f(Q_n) &\geq h(f) - \frac{k}{2} \ln E_f\left(\left(\frac{1}{\Lambda(X)}\right)^{2/k}\right) \\
&\approx h(f) - \frac{k}{2} \ln c_k D_f(Q_n)
\end{aligned}
\tag{10.12}
$$

with equality iff $\Lambda(X)$ is a constant, in which case from (10.9) and (10.12) the constant is given by

$$
\Lambda(x) = \left(\frac{c_k}{D_f(Q_n)}\right)^{k/2} = e^{(H_f(Q_n)-h(f))}.
\tag{10.13}
$$

Thus for large n using the optimal point density

$$
D_f(Q_n) \approx c_k e^{\frac{2}{k}(h(f)-H_f(Q_n))}
\tag{10.14}
$$

The optimal performance is then

$$
\delta_f(R) = \inf_{Q:H_f(Q)\leq R} D_f(Q) \approx c_k e^{\frac{2}{k}(h(f)-R)}.
\tag{10.15}
$$

This approximate argument produces results that agree with the rigorous version [28, 13]. In particular, under suitable technical conditions the asympotic performance is characterized by

$$
\lim_{R\to\infty} 2^{\frac{2}{k}R} \delta_f(R) = b_k 2^{\frac{2}{k} h(f)}
\tag{10.16}
$$

where b_k is Zador's constant, which depends only on k and not f. If Gersho's conjecture is true, this identifies b_k with c_k.

The Lagrangian form follows a similar argument. Define

$$\rho(f, \lambda) = \inf_{\alpha, \beta, \ell} \left(E_f \|X - \beta(\alpha(X))\|^2 + \lambda E_f \ell(\alpha(X)) \right)$$

$$= \inf_{\alpha, \beta} \left(E_f \|X - \beta(\alpha(X))\|^2 + \lambda H_f(Q_n) \right)$$

and consider the behavior as $\lambda \to 0$. Since the limit corresponds to placing less cost on the entropy and more on the distortion, one would expect the entropy to grow and the distortion to shrink, which will happen under the same asymptotics as in the non-Lagrangian argument, that is, the rate will grow while the distortion will shrink so that the high rate approximations previously used will still hold, yielding for small λ the approximations

$$\frac{D_f(Q)}{\lambda} + H_f(Q) + \frac{k}{2} \ln \lambda$$

$$\approx \frac{c_k E_f((\frac{1}{\Lambda(X)})^{2/k})}{\lambda} + h(f) - \frac{k}{2} \ln E_f \left(\ln \left(\frac{1}{\lambda \Lambda(X)} \right)^{2/k} \right)$$

$$= \frac{k}{2} \left(E_f \left(\left(\frac{2c_k}{k \lambda \Lambda(X)} \right)^{2/k} \right) \right) - \ln E_f \left(\ln \left(\frac{2c_k}{k \lambda \Lambda(X)^{2/k}} \right) - 1 \right) \right)$$

$$+ h(f) + \frac{k}{2} \ln \frac{2ec_k}{k}$$

$$\geq h(f) + \frac{k}{2} \ln \frac{2ec_k}{k}$$

with equality iff

$$\Lambda(x) = (\frac{2c_k}{k\lambda})^{k/2}, \text{ all x}, \tag{10.17}$$

a constant point density function. The rigorous version of this result is proved in [13] and shows that under suitable conditions (including the assumptions that P_f is absolutely continous with respect to Lebesgue measure, $h(f)$ is finite, and $H_f(Q)$ is finite when Q corresponds to a partition of Euclidean space into cubes)

$$\lim_{\lambda \to 0} \left(\frac{\rho(f, \lambda)}{\lambda} + \frac{k}{2} \ln \lambda \right) = \frac{k}{2} \ln \frac{2eb_k}{k} + h(f) \tag{10.18}$$

where

$$\frac{k}{2} \ln \frac{2eb_k}{k} = \theta_k \triangleq \inf_{\lambda > 0} \left(\frac{\rho(u_1, \lambda)}{\lambda} + \frac{k}{2} \ln \lambda \right), \tag{10.19}$$

where u_1 is the uniform probability density on a k-dimensional unit cube.

A consequence of the result is that given a sequence λ_n decreasing to 0, there is a sequence of quantizers Q_n for which

$$\lim_{n \to \infty} \left(\frac{D_f(Q_n)}{\lambda} + E_f \ell_n(Q_n) + \frac{k}{2} \ln \lambda_n \right) = h(f) + \frac{k}{2} \ln \frac{2ec_k}{k}, \quad (10.20)$$

where ℓ is the optimal length function for the encoder/decoder (α_n, β_n) Such a sequence of quantizers is said to be *asymptotically optimal* for the density f.

10.4 The Gaussian Case

For a Gaussian pdf,

$$\delta_g(R) \approx b_k e^{\frac{2}{k} h(f)} e^{-2\frac{R}{k}} = b_k (2\pi e) e^{-\frac{2}{k}R} |K|^{\frac{1}{k}}$$

Combining this fact with the extremal property of the Gaussian pdf for differential entropy immediately provides a high-rate quantization variation on Sakrison's result:

$$\sup_{f : E_f[(X-EX)(X-EX)^t]=K} \delta_f(R) = \delta_g(R)$$

The Lagrangian formulation yields the same conclusion since from (10.20) the asymptotic performance depends on the source only through the differential entropy: the worst possible Lagrangian performance for a high rate quantizer applied to a source with known second order moments is achieved by the Gaussian source.

This property suggests a further extension. It is often the case that full knowledge of the covariance of a random vector is lacking, for example one might only have trustworthy estimates of covariance values for small lags, as in the case of low order correlations in LPC speech modeling. If the supremum above is instead taken over all f with only partial information, then the worst case will be achieved by the Gaussian pdf with the covariance consistent with the partial information and having the largest determinant. This is the famous "maximum determinant" or MAXDET problem [25] Given an index set $\overline{\mathcal{N}}$ and a partial covariance $\Sigma_{\overline{\mathcal{N}}} = \{\Sigma_{i,j}; \ (i,j) \in \overline{\mathcal{N}}\}$, find $\max_{K : K_{\overline{\mathcal{N}}} = \Sigma_{\overline{\mathcal{N}}}} |K|$. The K achieving the maximum (if it exists) is the MAXDET extension of $\Sigma_{\overline{\mathcal{N}}}$.

10.5 Quantizer Mismatch

Suppose now that Q is optimized for a Gaussian g, but applied to another source f. To quantify the performance change resulting from such a mismatch, we define the *relative entropy* (or Kullback-Leibler number or Kullback-Leibler I-divergence or directed divergence or discrimination). The reader is referred to [19, 23, 2, 3, 9] for thorough treatments of relative entropy and its properties. Given two pdf's f and g with the property that the induced measures P_f and P_g are such that P_f is absolutely continuous with respect to P_g, then following Csiszar's notation [2, 3] define

$$I(f\|g) = \int dx f(x) \ln \frac{f(x)}{g(x)}.$$

Consider the asymptotic approximations for a sequence of quantizers Q_n and assume that n is large. The distortion derivation mimics that of (10.9) and the point density is the optimal point density for g as given by (10.13) with g in place of f:

$$D_f(Q_n) \approx c_k E_f \left(\left(\frac{1}{\Lambda(X)} \right)^{2/k} \right) \approx D_g(Q_n). \qquad (10.21)$$

The average rate is found analogous to (10.10), but now the pdf being averaged over f differs from that used for the optimal length function, g:

$$
\begin{aligned}
E_f \ell(\alpha(X)) &\approx -\sum_i \frac{f(y_{n,i})}{\Lambda(y_{n,i})} \ln \frac{g(y_{n,i})}{\Lambda(y_{n,i})} \\
&= -\sum_i V(S_{n,i}) f(y_{n,i}) \ln g(y_{n,i}) \\
&\qquad\qquad + \sum_i V(S_{n,i}) f(y_{n,i}) \ln(\Lambda(y_{n,i})) \\
&\approx -\int dy f(y) \ln g(y) + \int dy f(y) \ln(\Lambda(y)) \\
&\approx H_g(Q_n) - h(g) - \int dy f(y) \ln g(y) \qquad (10.22)
\end{aligned}
$$

When g is Gaussian this becomes

$$
\begin{aligned}
R_f(Q_n) - R &\approx \frac{1}{2} \operatorname{Trace}(K_g^{-1} K_f) \\
&\quad + \frac{1}{2} (\mu_g - \mu_f)^t K_g^{-1} (\mu_g - \mu_f) - \frac{k}{2}.
\end{aligned}
$$

If $\mu_f = \mu_g$ and $K_f = K_g$, then $D_f(Q_n) \approx D_g(Q_n)$ and $R_f(Q_n) \approx R$, reminiscent of Lapidoth's fixed rate result for iid Gaussian processes: the performance predicted for the Gaussian is actually achieved for the non-Gaussian. The loss in performance here is that in theory one could do better if the code had been designed for f and not for g.

The mismatch result can be translated into the Lagrangian formulation, where the result provides particular insight for the applications to come. The basic high rate result of (10.20) implies that given a Gaussian source described by a pdf g there exists a sequence of Lagrange multipliers λ_n

decreasing to 0 and a corresponding asymptotically optimal sequence of codes $(\alpha_n, \beta_n, \ell_n)$ such that

$$\lim_{n\to\infty} \left(\frac{D_g(\alpha_n, \beta_n)}{\lambda_n} + E_g\ell(\alpha_n) + \frac{k}{2}\ln\lambda_n \right) = \frac{k}{2}\ln\frac{2e}{k}b_k + h(g), \quad (10.23)$$

the optimal value for g. Using the distortion approximation of (10.21) and the rate approximation of (10.22) with the quantizer point density function Λ_n defined by (10.17) we have for high rate that

$$\frac{D_f(Q_n)}{\lambda_n} + E_f(\ell_n) + \frac{k}{2}\ln\lambda_n$$

$$\approx \frac{k}{2}\left(\frac{E_f\left(c_k\frac{1}{\Lambda(X)}\right)^{2/k}}{\lambda_n} - E_f\left(\ln\frac{2ec_k}{k\lambda\Lambda(X)^{2/k}}\right) \right) - \int dy f(y)\ln g(y)$$

$$= -\int dy f(y)\ln g(y) + \frac{k}{2}\ln\left(\frac{2ec_k}{k}\right) \quad (10.24)$$

Thus applying the asymptotically optimal codes designed for g to the source f results in

$$\lim_{n\to\infty} \left(\frac{D_f(Q_n)}{\lambda_n} + E_f\ell(\alpha_n) + \frac{k}{2}\ln\lambda_n \right)$$

$$= -\int dy f(y)\ln g(y) + \frac{k}{2}\ln(\frac{2ec_k}{k})$$

$$= h(f) + I(f\|g) + \frac{k}{2}\ln(\frac{2ec_k}{k}).$$

Thus

$$\lim_{n\to\infty} \left(\frac{D_f(Q_n)}{\lambda_n} + E_f\ell(\alpha_n) + \frac{k}{2}\ln\lambda_n \right)$$

$$= \lim_{\lambda\to\infty}\inf_{\alpha,\beta,\ell} \left(\frac{D_f(\alpha, \beta)}{\lambda} + E_f\ell(\alpha) + \frac{k}{2}\ln\lambda \right) + I(f\|g). \quad (10.25)$$

In words, applying a sequence of codes which are asymptotically optimal for g to a mismatched source f yields a loss of performance from the optimal asymptotic performance for f of $I(f\|g)$. Thus $I(f\|g)$ can be viewed as a measure of the mismatch of two sources with respect to quantization or lossy compression performance. This is analogous to the similar role of discrete relative entropy for quantifying the mismatch in lossless coding. This fact will be used shortly to provide a distortion measure or cost function for clustering models or density functions. The reader is referred to [14] for a more thorough and precise development of the mismatch result.

10.6   Minimum Discrimination Information Quantization

Consider now the problem of fitting a Gaussian mixture model to observed data as given by a learning or training set. The primary motivation here is that Gaussian models will provide a worst case for the actual source data that is mapped into the model. Because there are many such Gaussian models which will be chosen according to the observed source data, the overall model is a composite Gaussian source or, confining attention to a single vector, a Gauss mixture. We follow Kullback's approach as applied to low rate speech coding in [16]. The method is simply an extension of the speech case to multiple dimensional sources such as images. As earlier discussed, we fit the mixture to the data by trying to map individual Gaussian components to each observed vector using a measure of the distortion or badness of approximation. This allows the use of Lloyd clustering ideas instead of the usual EM fitting. A philosophical difference is that we here admit the Gaussian models being fitted are just that, models. We are not claiming the actual data is itself Gaussian and we admit we do not know the true underlying densities, hence we cannot claim to using an ML method. Instead, as in speech, we argue that local Gaussian behavior provides a local worst case fit, and a distortion measure will be used to quantify the quality of that fit. In the case of speech, using Gaussian models to synthesize reproduced speech or to recognize words or phrases in speech can yield synthetic speech that sounds good and the best known recognizers. As a result, we can admit that in fact actual speech is highly nonGaussian, yet provide good signal processing by fitting Gaussian components to chunks of speech, providing overall an implicit Gauss mixture model.

Since each Gaussian model is described by its mean and covariance matrix, say (μ_l, K_l) for the lth model, the issue is how to measure the distortion between an observed vector x and each of the models in order to select the one with the smallest distortion. We assume that second order moments can be estimated from the observation x, that is, we have estimates $\hat{\mu}_x$ and $\hat{K}_x$. This effectively assumes that it is the second order characteristics which are important. Typically $\hat{\mu}_x$ is assumed to be a constant vector, as in the speech case where it is assumed to be 0. When a 0 mean assumption is not appropriate (as in untransformed image pixels), the mean can be estimated by a sample average. The covariance might be estimated, for example, by assuming that the underlying process is locally spatially stationary and computing scalar sample averages

$$\hat{K}_{x,m}(n) = \frac{\sum_{i,j:|i-j|=n}(x_i - m)(x_j - m)}{N(n)}; \ n \in \mathcal{Z}_N^2 \qquad (10.26)$$

where, e.g., one might choose $N(n) = \#\{i, j : |i-j| = n\}$. Choosing $m = \hat{\mu}_x$ in particular yields a covariance estimate. This is a notoriously bad estimate since some values are based on very few pixels, but the estimates will be smoothed when computing centroids in the Lloyd clustering. Alternatively, one might use sample averages only for small lags where they are reasonably trustworthy, e.g., only for adjacent pixels, and then find a "maximum entropy" extension if it exists, e.g., estimate the full $\hat{K}$ as that agreeing with the trusted value and having the maximum determinant $|K|$ (which means the maximum differential entropy over all pdf's with the known second order moments). This is an example of the MAXDET algorithm [25].

For a pdf estimate $\hat{f}$ consistent with the moment constraints the distortion from the input to g_l is given by the relative entropy $I(\hat{f}\|g_l) = \int \hat{f} \ln \hat{f}/g_l$. Choose the pdf $\hat{f}$ as the density consistent with the moment constraints which minimizes the relative entropy between $\hat{f}$ and the fixed g_l. This is the *minimum discrimination information (MDI) density estimate* of $\hat{f}$ given g_l and the second order constraints. If g is assumed to be Gaussian, then the minimizing $\hat{f}$ will also be Gaussian[19, 16] This follows in exactly the same manner as the proof that the Gaussian pdf yields the maximum differential entropy for a given covariance as proved, e.g., in [6]. Thus

$$
\begin{aligned}
d_{MDI}&(x, (\mu_l, K_l)) \\
&= I(\hat{f}\|g_l) \\
&= \frac{1}{2}[\ln \frac{|K_l|}{|\hat{K}_x|} + \mathrm{Tr}(\hat{K}_x K_l^{-1}) + (\hat{\mu}_x - \mu_l)^t K_l^{-1}(\hat{\mu}_x - \mu_l) - k].
\end{aligned}
$$

This can be rewritten by reverting from the matrix form to the raster form:

$$
\begin{aligned}
2\, d_{MDI}&(x, (\mu_l, K_l)) \\
&= \ln \frac{|K_l|}{|\hat{K}|} + \sum_{i,j\in\mathcal{Z}_N^2} \hat{K}_x(i,j) K_l^{-1}(i,j) \\
&\quad + \sum_{i,j\in\mathcal{Z}_N^2} (\hat{\mu}_x(i) - \mu_l(i))(\hat{\mu}_x(j) - \mu_l(j)) K_l^{-1}(i,j) - k \\
&= \ln \frac{|K_l|}{|\hat{K}_x|} \\
&\quad + \sum_{i,j\in\mathcal{Z}_N^2} K_l^{-1}(i,j)[\hat{K}_x(i,j) + (\hat{\mu}(i) - \mu_l(i))(\hat{\mu}(j) - \mu_l(j))] - k \\
&= \ln \frac{|K_l|}{|\hat{K}_x|} + \sum_{i,j\in\mathcal{Z}_N^2} K_l^{-1}(i,j) \hat{K}_{x,\mu_l}(i,j) - k
\end{aligned}
$$

Itakura and Saito [18] originally derived their "error matching measure" by an approximate maximum likelihood argument. A similar informal argument can be used here to argue that minimizing the above MDI distortion measure is approximately equivalent to a maximum likelihood selection of which Gaussian component is in effect based on observed data.

10.7   MDI Centroids

As in the analogous speech case [16], the MDI distortion measure is amenable to the Lloyd clustering algorithm, i.e., there is a well defined minimum distortion encoder using d_{MDI} and the distortion has well defined Lloyd centroids. In particular, the centroids μ_l and K_l must minimize the conditional expected distortion.

$$E[d_{\text{MDI}}(X, g_l) \mid \alpha(X) = l] = \frac{1}{2} E \left[\ln \frac{|K_l|}{|\hat{K}_X|} + \text{Tr}(\hat{K}_X K_l^{-1}) \right.$$

$$+ \left. (\hat{\mu}_X - \mu_l)^t K_l^{-1} (\hat{\mu}_X - \mu_l) - k \mid \alpha(X) = l \right]$$

where $\hat{\mu}_X$ and $\hat{K}_X$ are the mean and the covariance estimates for observation X. The mean centroids are given by $\mu_l = E[\hat{\mu}_X \mid \alpha(X) = l]$ regardless of K_l since this choice minimizes the quadratic term in the mean as 0 (the centroid with respect to a weighted quadratic measure is the mean). With this choice of μ_l we need K_l to minimize

$$E \left[\ln \frac{|K_l|}{|\hat{K}_X|} + \text{Trace}(\hat{K}_X K_l^{-1}) - k \mid \alpha(X) = l \right]$$

$$= \ln \frac{|K_l|}{|\overline{K}_l|} + \text{Trace}(\overline{K}_l K_l^{-1}) - k + E \left[\ln \frac{|\overline{K}_l|}{|\hat{K}_X|} \mid \alpha(X) = l \right]$$

$$\geq E \left[\ln \frac{|\overline{K}_l|}{|\hat{K}_X|} \mid \alpha(X) = l \right]$$

where $\overline{K}_l = E[(X - \mu_l)(X - \mu_l)^t]$ since the first three terms are just the relative entropy between two Gaussian distributions with the given covariances and 0 means, and this is nonnegative from the divergence inequality (see, e.g., [6]). Equality holds if $K_l = \overline{K}_l$.

10.8   MDI Clustering

Application of the Lloyd algorithm to the MDI distortion measures yields a Gaussian model VQ, a mapping of input vectors X (e.g., image blocks) into a Gaussian model chosen from a codebook of such models. The collection of models together with the probability of their occurrence when the quantizer is used on the training data provides a complete Gauss mixture model. Under reasonably general conditions, the Lloyd algorithm converges. If the algorithm converges to a stationary point, the centroid formulas yield the resulting MDI distortion in terms of the model covariances of the codebook and their probabilities of occurrence, i.e., in terms of the Gauss mixture model produced by the Lloyd algorithm.

Since we are considering variable rate systems, it is natural to consider an entropy constrained VQ for the models as well: $d_{\mathrm{ECMDI}}(x, g_l) = d_{\mathrm{MDI}}(x, g_l) - \mu \ln w_l$. Applying the MDI centroid expression provides a simple formula for the average ECMDI distortion:

$$D_{\mathrm{ECMDI}} = \frac{1}{2} \sum_l w_l \ln |K_l| - E[\ln |\hat{K}_X|] + \mu H(p) \qquad (10.27)$$

In Lloyd clustering with variable rate codes, standard practice is to begin with a large number of codewords. As the algorithm runs it will eliminate unused codewords, so that the total is monotonically decreasing. If the algorithm is initiated with too few codewords, then the number will usually not reduce with successive iterations and the usual practice is to restart the algorithm with a larger number. This practice is reinforced by theoretical results to the effect that under suitable conditions the optimal codes will have a finite number of reproduction vectors. "Suitable conditions" include the case of squared error distortion and pdf's with fine support and Gaussian pdf's [5]

10.9 Gauss Mixture VQ

We now consider an application where the MDI clustering approach to Gauss mixture modeling arises naturally. Suppose that X has mixture pdf f generated by classifying X into classes $l = 1, 2, \ldots$. For the moment the classifier is arbitrary, but obviously the clustering algorithm of the previous section could be used. Let $L = L(X)$ denote the integer-valued class. Then $\{f_{X|L}(x|l), w_l = \Pr(L = l)\}$ is a mixture model for f. A separate VQ can then be designed for each class, i.e., for each Gaussian source with a specified mean and covariance, yielding a classified VQ structure. For each class l one can estimate a conditional mean μ_l and covariance K_l, possibly only partially. The worst case source for quantizing this source is then the MAXDET Gaussian. Design an optimal code Q_l for each Gaussian component g_l, yielding a (D_l, R_l) distortion/rate pair with performance that can be approximated using the high rate formulas. This yields a two-step classified VQ: First classify X into Gaussian model g_l described by (μ_l, K_l), then quantize using optimal quantizer Q_l for g_l.

The idea for classified vector quantization dates back to switched quantizers and was extended to vector quantization for image coding by Gersho and Ramamurthi [8] in 1982. The approach has been highly successful for image coding applications by adapting the quantization or compression algorithms to local statistical behavior within an image. For a modern survey of such methods incorporating wavelet transforms and scalar quantizers, the reader is referred to Yoo, Ortega, and Yu [27]. Our treatment here, however, emphasizes arbitrary vector sources and vector quantizers and the theoretical performance when the bit rate is large.

On the average the total information rate for the lth component is $R_l - \ln w_l$, the number of nats needed to specify quantizer used plus the encoder output for that quantizer. The overall average distortion is then $D(Q) = \sum_l D_l w_l$ and the overall average rate $R(Q) = \sum_l R_l w_l + H(p)$. From high rate theory $D_l \approx \delta(R_l) \approx b_k e^{\frac{2}{k}(h_l - R_l)}$ whence $D(Q) = b_k \sum_l e^{\frac{2}{k}(h_l - R_l)} w_l$. The optimal rate allocation $\{R_l\}$ minimizing $b_k \sum_l e^{\frac{2}{k}(h_l - R_l)} w_l$ subject to $\sum_l R_l w_l + H(p) \leq R$ is readily solved by Lagrangian methods or directly using convexity arguments: $R_l = h_l + R - H(p) - \overline{h}$ where

$$\overline{h} = \sum_l h_l w_l = h(X|L) \ , \ \ H(p) = H(L)$$

The Lagrangian multiplier for the modified distortion for each quantizer is the same: $\lambda_l = (2b_k/k)e^{-\frac{2}{k}(R - H(p) - \overline{h})}$.

With this assignment it turns out that the optimum quantizer point density for all the quantizers is the same, $\Lambda(x) = e^{R - \overline{h} - H(p)}$, and that the conditional average distortion for each component is the same, $D_l = b_k e^{-\frac{2}{k}(R - \overline{h} - H(p))}$ so that $D = b_k e^{-\frac{2}{k}(R - \overline{h} - H(p))}$. Plugging in for the Gaussian case

$$D_{\mathrm{MSE}} = b_k(2\pi e)e^{\frac{2}{k}(\frac{1}{2}\sum_l w_l \ln |K_l| + H(p) - R)} \tag{10.28}$$

From the robustness property, *this formula also gives the performance for nonGaussian source classified into a mixture with* $\{(\mu_l, K_l, w_l)\}$!

If one designs a classified VQ using the MDI as described earlier, and then optimally designs VQs to minimize mean squared error for the resulting Gaussian models, and then applies the code by first classifying the input and then applying the optimal code for the class chosen, then the average distortion (assuming high rate and optimal bit allocation across the classes) is

$$D_{\mathrm{MSE}} = b_k(2\pi e)e^{\frac{2}{k}(D_{\mathrm{ECMDI}} + E[\ln |\hat{K}_X|])} \tag{10.29}$$

where b_k is a constant depending only on the dimension and not on the underlying pdf's and the MDI Lagrangian is chosen as $\mu = 1$. This relates the MSE in the resulting classified VQ to the ECMDI distortion used to design the classifier, providing a new relation between modeling accuracy and the resulting performance in a quantizer based on the model.

The relation shows that designing a Gauss mixture model using the MDI clustering algorithm described is equivalent to designing a robust classified vector quantizer with the overall goal of minimizing the reproduction mean squared error. The rightmost term above does not depend on the code, only on the estimation technique used to estimate the covariance.

Preliminary simulation results on various image databases may be found in [1, 15] and work is proceeding on applications to image classification and segmentation.

10.10   Parting Thoughts

The approach here described provides a hindsight interpretation as to why modern code excited linear prediction (CELP) techniques work as well as they do. They can be viewed as fitting a reproduction to an observed waveform by first selecting a Gaussian AR model (by LPC analysis and simple quantization), then populating a custom codebook for this model by driving the implied inverse filter by a codebook of vectors randomly generated from a memoryless Gaussian process, and then finding the closest resulting vector in the resulting output codebook to the original input vector. High rate quantization theory implies that driving the AR filters with independent inputs produces an approximately optimal codebook for the Gaussian AR source and hence a robust codebook for all sources with approximately the same second order moments. The design of the AR filters is a form of MDI clustering, and we have seen that this indeed results in an approximately optimal classified quantizer.

The ideas here suggest that similar methods will work for image coding and image segmentation, and simulations are underway to study these possibilities. Eventually it would be of interest to compare the MDI clustering approach to the traditional EM algorithm approach to the design of Gauss mixtures. An application where the two techniques could be compared in terms of overall performance is that of a vector quantization based system since Hedelin and Skoglund [17] used EM to design a vector quantizer based on Gauss mixture models. For the applications described here, however, it is the MDI approach that is the most natural: if a classifier or coder is to select a model using the minimum MDI distortion rule, then it is natural to design the codebook of models so as to minimize the average MDI distortion, i.e., to use a clustering algorithm.

On the theoretical side, the development of this chapter rests on Gersho's conjecture, which is widely believed but yet unproved. The basic high rate quantization results have been rigorously proved by completely different techniques which do not involve the concept of a quantizer point density function [13]. Similar techniques have been applied to recently provide a rigorous proof of the mismatch result, but this result has not yet been validated by by scrutiny of colleagues and reviewers. A preprint of the proof of the basic mismatch result and its applications to developing the heuristic results developed here may be found in [14].

We close with a few thoughts on further possible extensions. If in addition to estimating w_l of mixture components from relative frequencies in order to construct Gauss mixture models, one could also estimate conditional probabilities for the index of a Gauss component given the previous index, providing a hidden Markov model (HMM) without resort to EM or forward-backward algorithms. The basic results described here all require an assumption of high rate, so a natural question is whether the approximations remain valid for moderate or low rates. There is anecdotal evidence to the effect that as with many high rate results, these results often hold approximately in low rate situations.

The use of clustered models suggests an approach to some simple image segmentation problems that mimics a simple and powerful early approach to simple speech recognition, the recognition of isolated utterances such as the alphabet and numbers. Given a labeled training set, design a separate quantizer for each class using a classified VQ, i.e., each class has a Gauss mixture model associated with it, not just a single Gaussian component. Classify new data by encoding it using all of the quantizers in parallel and select the one with the smallest overall distortion. This might provide reasonable performance, for example, in texture recognition applications. This technique is currently being tested for content-addressable browsing and for image segmentation applications.

References

[1] *Robust Image Compression using Gauss Mixture Models*, PhD Dissertation, Department of Electrical Engineering, Stanford University, June 2001.

[2] I. Csiszár, "Information-type measures of difference of probability distributions and indirect observations," *Studia Scientiarum Mathematicarum Hungarica*, Vol. 2, pp. 299–318,, 1967.

[3] I. Csiszár, "On an extremum problem of information theory," *Studia Scientiarum Mathematicarum Hungarica*, pp. 57–70, 1974.

[4] P.A. Chou, T. Lookabaugh, and R.M. Gray, "Entropy-constrained vector quantization," *IEEE Trans. Acoust. Speech and Signal Proc.*, Vol. 37, pp. 31–42, January 1989.

[5] Philip A. Chou and B. Betts, "When optimal entropy-constrained quantizers have only a finite number of codewords," IEEE International Symposium on Information Theory. Full paper available at *http://research.microsoft.com/ pachou/publications.htm*.

[6] T.M. Cover and J.A. Thomas, *Elements of Information Theory*, Wiley, 1991.

[7] A. Gersho, "Asymptotically optimal block quantization," *IEEE Trans. Inform. Theory.*, vol. 25, pp. 373–380, July 1979.

[8] A. Gersho and B. Ramamurthi, "Image coding using vector quantization," *Proc. Intl. Conf. on Acoust. Speech, and Signal Processing*, vol. 1, pp. 428–431, Paris, April 1982.

[9] R. M. Gray, *Entropy and Information Theory*, Springer–Verlag, 1990.

[10] R.M. Gray, "Gauss Mixture Vector Quantization," *Proceedings ICASSP 2001.*

[11] R.M. Gray and D.L. Neuhoff, "Quantization," *IEEE Transactions on Information Theory*, Vol. 44, pp. 2325–2384, October 1998.

[12] R.M. Gray and J. Li, "On Zador's Entropy-Constrained Quantization Theorem," *Proceedings Date Compression Conference 2001*, IEEE Computer Science Press, Los Alimitos, pp. 3–12, March 2001.

[13] R.M. Gray, T. Linder, and J. Li, "A Lagrangian Formulation of Zador's Entropy-Constrained Quantization Theorem," *IEEE Transactions on Information Theory*, pp. 695–707, Vol. 48, Number 3, March 2002.

[14] R.M. Gray and T. Linder, "Mismatch in high rate entropy constrained vector quantization," preprint available at http://ee.stanford.edu/~gray/mismatch.pdf.

[15] R.M. Gray, J.C. Young, and A. K. Aiyer, "Minimum discrimination information clustering: modeling and quantization with Gauss mixtures," *Proceedings 2001 IEEE International Conference on Image Processing*, Thessaloniki, Greece, October 2001.

[16] R.M. Gray, A.H. Gray, Jr., G. Rebolledo, and J.E. Shore, "Rate distortion speech coding with a minimum discrimination information distortion measure," *IEEE Transactions on Information Theory*, vol. IT–27, no. 6, pp. 708–721, Nov. 1981.

[17] P. Hedelin and J. Skoglund, "Vector Quantization based on Gaussian mixture models," *IEEE Transactions on Speech and Audio Processing*, Vol.8, no.4, p.385–401, July 2000.

[18] F. Itakura and S. Saito, "Analysis synthesis telephony based on the maximum likelihood method," *Proc. 6th Int'l Congress of Acoustics*, pp. C-17–C-20, Tokyo, Japan, August 1968.

[19] S. Kullback. *Information Theory and Statistics*, Dover, New York, 1968. (Reprint of 1959 edition published by Wiley.)

[20] A. Lapidoth, "On the role of mismatch in rate distortion theory," *IEEE Trans. Inform. Theory*, vol. 43, pp. 38–47, Jan. 1997.

[21] S. P. Lloyd. Least squares quantization in PCM. Unpublished Bell Laboratories Technical Note. Portions presented at the Institute of Mathematical Statistics Meeting Atlantic City New Jersey September 1957. Published in special issue on quantization, *IEEE Trans. Inform. Theory*, vol. 28, pp. 129–137, Mar. 1982.

[22] S. Na and D. L. Neuhoff, " Bennett's integral for vector quantizers," *IEEE Trans. Inform. Theory*, vol. 41, pp. 886–900, July 1995.

[23] M. S. Pinsker, *Information and information stability of random variables and processes*, Holden Day, San Francisco, 1964. (Translated by A. Feinstein from the Russian edition published in 1960 by Izd. Akad. Nauk. SSSR.)

[24] D. J. Sakrison, "Worst sources and robust codes for difference distortion measures," *IEEE Trans. Inform. Theory*, vol. 21, pp. 301–309, May 1975.

[25] L. Vandenberghe, S. Boyd, and S.-P. Wu, "Determinant maximization with linear matrix inequality constraints," *SIAM Journal on Matrix Analysis and Applications*, Vol. 19, 499-533, 1998.

[26] Yuhong Yang and Andrew Barron "An Asymptotic Property of Model Selection Criteria," *IEEE Transaction on Information Theory*, vol. 44, pp. 95-116, 1998.

[27] Y. Yoo, A. Ortega, and B. Yu, "Image compression via joint statistical characterization in the wavelet domain," *IEEE Transactions on Image Processing*, Vol. 8, pp. 1702–1715, December 1999.

[28] P. L. Zador, "Topics in the asymptotic quantization of continuous random variables," Bell Laboratories Technical Memorandum, 1966.

11

Extended Linear Modeling with Splines

Jianhua Z. Huang and Charles J. Stone[1]

Summary

Extended linear models form a very general framework for statistical modeling. Many practically important contexts fit into this framework, including regression, logistic and Poisson regression, density estimation, spectral density estimation, and conditional density estimation. Moreover, hazard regression, proportional hazard regression, marked point process regression, and diffusion processes, all perhaps with time-dependent covariates, also fit into this framework. Polynomial splines and their tensor products provide a universal tool for constructing maximum likelihood estimates for extended linear models. The theory of rates of convergence for such estimates as it applies both to fixed knot splines and to free knot splines will be surveyed, and the implications of this theory for the development of corresponding methodology will be discussed briefly.

11.1 Introduction

Polynomial splines are useful in statistical modeling and data analysis. The theoretical framework of extended linear modeling has evolved through a long-term investigation, starting in the mid-eighties, of the properties of spline-based estimates in various contexts. Some early results in this effort are Stone [24, 25, 26, 27, 28] and Kooperberg, Stone and Truong [15, 17]. Hansen [5] expanded and unified the then existing theory. The resulting synthesis played a key role in the Stone, Hansen, Kooperberg and Truong [30], which reviewed the theory and corresponding methodology. Shortly after that, Huang [6, 7] substantially simplified and extended

---

[1] Jianhua Z. Huang is Assistant Professor, Department of Statistics, The Wharton School, University of Pennsylvania, Philadelphia, PA 19104-6302 (Email: jianhua@wharton.upenn.edu). Charles J. Stone is Professor, Department of Statistics, University of California, Berkeley, CA 94720-3860 (E-mail: stone@stat.berkeley.edu); he was supported in part by National Science Foundation grant DMS-9802071.

the theoretical approach. These improvements helped lead to Huang and Stone [9] and Huang, Kooperberg, Stone and Truong [10]. More recently, Huang [8] provided a fresh theoretical synthesis of extended linear modeling. To the extent that this work pertained to spline-based methods, it was restricted to fixed knot splines. Still more recently, Stone and Huang [31, 32] have used the framework of Huang [8] to investigate the theoretical properties of extended linear modeling with free knot splines.

Section 11.2 gives a detailed description of the theoretical framework of extended linear models. Some examples of such models are presented in Section 11.3. Section 11.4 describes asymptotic results for maximum likelihood estimates, such as consistency and rates of convergence; the focus is on fixed knot spline estimates. Section 11.5 discusses using functional analysis of variance to construct structural models in order to tame the curse of dimensionality. Free knot splines are studied in Section 11.6. Some implications of the theory for the development of corresponding methodology will be mentioned in Section 11.7.

11.2   Extended Linear Models: Theoretical Framework

Consider a $\mathcal{W}$-valued random variable $\boldsymbol{W}$, where $\mathcal{W}$ is an arbitrary set. The probability density $p(\eta, \boldsymbol{w})$ of $\boldsymbol{W}$ depends on an unknown function η. The function η is defined on a domain $\mathcal{U}$, which may or may not be the same as $\mathcal{W}$. We assume that $\mathcal{U}$ is a compact subset of some Euclidean space and that it has positive volume $\mathrm{vol}(\mathcal{U})$. The problem of interest is estimation of η based on a random sample from the distribution of $\boldsymbol{W}$.

Corresponding to a candidate function h for η, the log-likelihood is given by $l(h, \boldsymbol{w}) = \log p(h, \boldsymbol{w})$. The expected log-likelihood is defined by $\Lambda(h) = E[l(h, \boldsymbol{W})]$, where the expectation is taken with respect to the probability measure corresponding to the true function η. There may be some mild restrictions on h for $l(h, \boldsymbol{w})$, $\boldsymbol{w} \in \mathcal{W}$, and $\Lambda(h)$ to be well-defined. It follows from the information inequality that η is the essentially unique function on $\mathcal{U}$ that maximizes the expected log-likelihood. (Here two functions on $\mathcal{U}$ are regarded as essentially equal if their difference equals zero except on a subset of $\mathcal{U}$ having Lebesgue measure zero.)

In many applications, we are interested in a function η that is related to but need not totally specify the probability distribution of $\boldsymbol{W}$. In such applications, we can modify the above setup by taking $l(h, \boldsymbol{w})$ to be the logarithm of a conditional likelihood, a pseudo-likelihood, or a partial likelihood, depending on the problem under consideration.

Consider, for example, the estimation of a regression function $\eta(\boldsymbol{x}) = E(Y|\boldsymbol{X} = \boldsymbol{x})$. In terms of the above notation, $\boldsymbol{W}$ consists of a pair of random variables $\boldsymbol{X}$ and Y, and $\mathcal{U}$ is the range of $\boldsymbol{X}$. We can take $l(h, \boldsymbol{w})$

to be the negative of the residual sum of squares; that is, $l(h, \boldsymbol{W}) = -[Y - h(\boldsymbol{X})]^2$ with $\boldsymbol{W} = (\boldsymbol{X}, Y)$. If the conditional distribution of Y given $\boldsymbol{X}$ is assumed to be normal with constant variance, then l is (up to additive and multiplicative constants) the conditional log-likelihood. Even if this conditional distribution is not assumed to be normal, we can still think of l as the logarithm of a pseudo-likelihood. In either case, the true regression function η maximizes $\Lambda(h) = E[l(h, \boldsymbol{W})] = -E[\eta(\boldsymbol{X}) - h(\boldsymbol{X})]^2$.

From now on, we will adopt this broad view of $l(h, \boldsymbol{w})$. For simplicity, we will still call $l(h, \boldsymbol{w})$ the log-likelihood and $\Lambda(h)$ the expected log-likelihood. To relate the function of interest to the log-likelihood, we assume that, subject to mild conditions on $l(h, \boldsymbol{w})$, the function η is the essentially unique function that maximizes the expected log-likelihood.

Let $\mathbb{H}$ be a linear space of square-integrable functions on $\mathcal{U}$ such that if two functions on $\mathcal{U}$ are essentially equal and one of them is in $\mathbb{H}$, then so is the other one. We refer to $\mathbb{H}$ as the *model space* and to $l(h, \boldsymbol{W})$, $h \in \mathbb{H}$, as forming an *extended linear model*. If $\mathbb{H}$ is the space of all square-integrable functions on $\mathcal{U}$ or differs from this space only by the imposition of some identifiability restrictions as in the context of density estimation (see Section 11.3), we refer to $\mathbb{H}$ as being *saturated*. Otherwise, we refer to this space as being *unsaturated*.

The use of unsaturated spaces allows us to impose structural assumptions on the extended linear model. Suppose $\mathcal{U}$ is the Cartesian product of compact intervals $\mathcal{U}_1, \ldots, \mathcal{U}_L$, each having positive length. We can impose an additive structure by letting $\mathbb{H}$ be the space of functions of the form $h_1(u_1) + \cdots + h_L(u_L)$, where h_l is a square-integrable function on $\mathcal{U}_l$ for $1 \leq l \leq L$. This and more general ANOVA structures will be considered in Section 11.5. Alternatively, we can impose an additive, semilinear structure by letting $\mathbb{H}$ be the space of functions of the form $h_1(u_1) + b_2 u_2 + \cdots + b_L u_L$, where h_1 is a square-integrable function on $\mathcal{U}_1$ and $b_2, \ldots, b_L$ are real numbers.

The extended linear model is said to be *concave* if the following two properties are satisfied: (i) The log-likelihood function is concave; that is, given any two functions $h_1, h_2 \in \mathbb{H}$ whose log-likelihoods are well-defined, $l(\alpha h_1 + (1 - \alpha)h_2, \boldsymbol{w}) \geq \alpha l(h_1, \boldsymbol{w}) + (1 - \alpha)l(h_2, \boldsymbol{w})$ for $0 < \alpha < 1$ and $\boldsymbol{w} \in \mathcal{W}$. (ii) The expected log-likelihood function is strictly concave; that is, given any two essentially different functions $h_1, h_2 \in \mathbb{H}$ whose expected log-likelihoods are well-defined, $\Lambda(\alpha h_1 + (1 - \alpha)h_2) > \alpha \Lambda(h_1) + (1 - \alpha)\Lambda(h_2)$ for $0 < \alpha < 1$. Here, we implicitly assume that the set of functions such that $l(h, \boldsymbol{w})$ and $\Lambda(h)$ are well-defined is a convex set.

As mentioned above, the model space $\mathbb{H}$ incorporates structural assumptions (e.g., additivity) on the true function of interest. Such structural assumptions are not necessarily true and are considered rather as approximations. Thus, it is natural to think that any estimation procedure will estimate the best approximation to the true function with the imposed structure. This "best approximation" can be defined formally using the expected log-likelihood. Observe that if $\eta \in \mathbb{H}$, then $\eta = \text{argmax}_{h \in \mathbb{H}} \Lambda(h)$ since η maximizes the expected log-likelihood by assumption. More generally, we think of $\eta^* = \text{argmax}_{h \in \mathbb{H}} \Lambda(h)$ as the "best approximation" in $\mathbb{H}$ to η. Typically, when the expected log-likelihood function is strictly concave, such a best approximation exists and is essentially unique. If $\eta \in \mathbb{H}$, then η^* is essentially equal to η.

In the regression context η^* is the orthogonal projection of η onto $\mathbb{H}$ with respect to the L_2 norm on $\mathbb{H}$ given by $\|h\|^2 = E[h^2(\boldsymbol{X})]$; that is, $\eta^* = \text{argmin}_{h \in \mathbb{H}} \|h - \eta\|^2$. Here, to guarantee the existence of η^*, we need to assume that $\mathbb{H}$ is a Hilbert space; that is, it is closed in the metric corresponding to the indicated norm.

We now turn to estimation. Let $\boldsymbol{W}_1, \ldots, \boldsymbol{W}_n$ be a random sample of size n from the distribution of $\boldsymbol{W}$. Let $\mathbb{G} \subset \mathbb{H}$ be a finite-dimensional linear space of bounded functions, whose dimension may depend on the sample size. We estimate η by using maximum likelihood over $\mathbb{G}$, that is, we take $\hat{\eta} = \text{argmax}_{g \in \mathbb{G}} \ell(g)$, where $\ell(g) = (1/n) \sum_{i=1}^{n} l(g, \boldsymbol{W}_i)$ is the normalized log-likelihood. Here the space $\mathbb{G}$ should be chosen such that the function of interest η can be approximated well by some function in $\mathbb{G}$. Thus $\mathbb{G}$ will be called the approximation space. Since $\mathbb{G}$ is where the maximum likelihood estimation is carried out, it will also be called the estimation space. In this setup we do not specify the form of $\mathbb{G}$; any linear function space with good approximation properties can be used. When $\mathbb{H}$ has a specific structure, $\mathbb{G}$ should be chosen to have the same structure. For example, if $\mathbb{H}$ consists of all square-integrable additive functions, then $\mathbb{G}$ should not contain any non-additive functions. A detailed discussion of constructing the model and estimation spaces using functional ANOVA decompositions to incorporate structural assumptions will be given in Section 11.5. In our application, $\mathbb{G}$ will be chosen as a space built by polynomial splines and their tensor products. That polynomial splines and their tensor products enjoy good approximation power has been extensively studied and documented; see de Boor [1], Schumaker [22], and DeVore and Lorentz [2].

11.3   Examples of Concave Extended Linear Models

The log-likelihood function in an extended linear model takes into account the probability structure of the estimation problem. It is advantagous that the log-likelihood of an extended linear model be concave (and, in fact, suitably strictly concave with probability close to one). The maximum likelihood estimate in a finite-dimensional estimation space is then unique if it exists. Moreover, the Newton–Raphson algorithm when suitably adjusted (for example, by step-halving) is guaranteed to converge to the global maximum. Although the concavity restriction on the log-likelihood may look restrictive, it turns out that the collection of intrinsically suitably concave extended linear models is very rich, including of course ordinary regression. We present in this section a number of other contexts in which the concavity assumption on the log-likelihood is automatically satisfied.

11.3.1 Generalized Regression.

The ordinary (least squares) regression model, which was introduced in the last section, does not impose any structure on the conditional variance of the response given the covariates. On the other hand, in generalized regression the conditional variance depends on the conditional mean in some specified way as in an exponential family.

Consider a random pair $W = (X, Y)$, where the random vector X of covariates is $\mathcal{X}$-valued with $\mathcal{X} = \mathcal{U}$ and the response Y is real-valued. Suppose the conditional distribution of Y given that $X = x \in \mathcal{X}$ has the form of an exponential family

$$P(Y \in dy | X = x) = \exp[B(\eta(x))y - C(\eta(x))]\Psi(dy), \qquad (11.1)$$

where $B(\cdot)$ is a known, twice continuously differentiable function on $\mathbb{R}$ whose first derivative is strictly positive on $\mathbb{R}$, Ψ is a nonzero measure on $\mathbb{R}$ that is not concentrated at a single point, and $C(\eta) = \log \int_{\mathbb{R}} \exp[(B(\eta)y]\Psi(dy) < \infty$ for $\eta \in \mathbb{R}$. Observe that $B(\cdot)$ is strictly increasing and $C(\cdot)$ is twice continuously differentiable on $\mathbb{R}$.

Here the function of interest is the response function $\eta(\cdot)$, which specifies the dependence on x of the conditional distribution of the response Y given that the value of the vector X of covariates equals x. The mean of this conditional distribution is given by

$$\mu(x) = E(Y | X = x) = A(\eta(x)) = \frac{C'(\eta(x))}{B'(\eta(x))} \qquad x \in \mathcal{X}. \qquad (11.2)$$

The (conditional) log-likelihood is given by

$$l(h, X, Y) = B(h(X))Y - C(h(X)),$$

and its expected value is given by

$$\Lambda(h) = E[B(h(X))\mu(X) - C(h(X))],$$

which is essentially uniquely maximized at $h = \eta$. The log-likelihood is automatically concave if $B(\eta)$ is a linear function of η and in certain other specific cases as well (e.g., in probit models).

The family (11.1) includes as special cases many useful distributions such as Bernoulli, Possion, Gaussian, gamma, and inverse-Gaussian.

When the underlying exponential family is the Bernoulli distribution with parameter $\pi(x)$ and the function of interest is $\eta(x) = \mathrm{logit}(\pi(x)) = \log(\pi(x)/(1 - \pi(x)))$, we get logistic regression, which is closely connected to classification. Here $P(Y = 1 | X = x) = \pi(x)$, $P(Y = 0 | X = x) =$

$1 - \pi(\boldsymbol{x})$, Ψ is concentrated on $\{0, 1\}$ with $\Psi(\{0\}) = \Psi(\{1\}) = 1$, $B(\eta) = \eta$, $C(\eta) = \log(1 + \exp(\eta))$, and $\mu(\boldsymbol{x}) = \pi(\boldsymbol{x}) = \exp \eta(\boldsymbol{x})/(1 + \exp \eta(\boldsymbol{x}))$.

When the underlying exponential family is the Poisson distribution with parameter $\lambda(\boldsymbol{x})$ and the function of interest is $\eta(\boldsymbol{x}) = \log \lambda(\boldsymbol{x})$, we get Poisson regression. Here $P(Y = y | \boldsymbol{X} = \boldsymbol{x}) = \lambda^y \exp(-\lambda(\boldsymbol{x})/y!$ for $y \in \mathcal{Y} = \{0, 1, 2, \ldots\}$, Ψ is concentrated on $\mathcal{Y}$ with $\Psi(\{y\}) = 1/y!$ for $y \in \mathcal{Y}$, $B(\eta) = \eta$, $C(\eta) = \exp \eta$, and $\mu(\boldsymbol{x}) = \lambda(\boldsymbol{x}) = \exp \eta(\boldsymbol{x})$.

When the underlying exponential family is the normal distribution with mean $\mu(\boldsymbol{x})$ and known variance σ^2 and the function of interest is the regression function $\mu(\boldsymbol{x})$, we get ordinary regression as discussed in Section 11.2. Here $P(Y \in (y, y + dy) | \boldsymbol{X} = \boldsymbol{x}) = (1/\sqrt{2\pi\sigma^2}) \exp\{-(y - \mu(\boldsymbol{x}))^2/\sigma^2\} \, dy$ for $y \in \mathbb{R}$, $B(\eta) = \eta/\sigma^2$, $C(\eta) = -\eta^2/\sigma^2$, and $\eta(\boldsymbol{x}) = \mu(\boldsymbol{x})$.

If the conditional distribution of Y is not fully specified as in (11.1), $l(h, \boldsymbol{X}, Y)$ can be thought of as a quasi log-likelihood. In connecting the unknown function to the log-likelihood, we assume that (11.2) holds. This assumption guarantees that the expected log-likelihood is essentially maximized at the true function η of interest and thereby validates maximizing the sample log-likelihood.

11.3.2   Polychotomous Regression

Polychotomous regression, which is closely connected to multiple classification, is an extension of logistic regression. Let Y be a qualitative random variable having $K + 1$ possible values. Without loss of generality, we can think of this random variable as ranging over $\mathcal{Y} = \{1, \ldots, K + 1\}$. Suppose that $P(Y = k | \boldsymbol{X} = \boldsymbol{x}) > 0$ for $\boldsymbol{x} \in \mathcal{X}$ and $k \in \mathcal{Y}$. Set

$$\eta_k(\boldsymbol{x}) = \log \frac{P(Y = k | \boldsymbol{X} = \boldsymbol{x})}{P(Y = K + 1 | \boldsymbol{X} = \boldsymbol{x})}, \qquad 1 \leq k \leq K.$$

The log-likelihood is given by

$$\begin{aligned} l(h, \boldsymbol{X}, Y) \quad = \quad & h_1(\boldsymbol{X}) I_1(Y) + \cdots + h_K(\boldsymbol{x}) I_K(Y) \\ & - \log(1 + \exp h_1(\boldsymbol{X}) + \cdots + \exp h_K(\boldsymbol{X})), \end{aligned}$$

where $I_k(Y)$ equals one or zero according as $Y = k$ or $Y \neq k$ and $h = (h_1, \ldots, h_K)$ is a candidate for $\eta = (\eta_1, \ldots, \eta_K)$. Here $\boldsymbol{W} = (\boldsymbol{X}, Y)$ and $\mathcal{U} = \mathcal{X}$.

11.3.3   Hazard Estimation and Regression

Estimation of a hazard function and its dependence on covariates is important in survival analysis. Consider a positive survival time T, a positive censoring time C, the observed time $\min(T, C)$, and an $\mathcal{X}$-valued random vector $\boldsymbol{X}$ of covariates. Let $\delta = \text{ind}(T \leq C)$ be the indicator random variable that equals one or zero according as $T \leq C$ (T is uncensored) or $T > C$

(T is censored), and set $Y = \min(T, C)$ and $\boldsymbol{W} = (\boldsymbol{X}, Y, \delta)$. Suppose T and C are conditionally independent given $\boldsymbol{X}$. Suppose also that $P(C \leq \tau) = 1$ for a known positive constant τ. Let

$$\eta(\boldsymbol{x}, t) = \log \frac{f(t|\boldsymbol{x})}{1 - F(t|\boldsymbol{x})}, \qquad t > 0,$$

denote the logarithm of the conditional hazard function, where $f(t|\boldsymbol{x})$ and $F(t|\boldsymbol{x})$ are the conditional density function and conditional distribution function, respectively, of T given that $\boldsymbol{X} = \boldsymbol{x}$. Then

$$1 - F(t|\boldsymbol{x}) = \exp\left(-\int_0^t \exp \eta(\boldsymbol{x}, u)\, du\right), \qquad t > 0,$$

and hence

$$f(t|\boldsymbol{x}) = \exp\left(\eta(\boldsymbol{x}, t) - \int_0^t \exp \eta(\boldsymbol{x}, u)\, du\right), \qquad t > 0.$$

Hazard regression concerns estimation of the conditional hazard function $\eta(\boldsymbol{x}, t)$. Since the likelihood equals $f(T \wedge C | \boldsymbol{X})$ for an uncensored case and $1 - F(T \wedge C | \boldsymbol{X})$ for a censored case, it can be written as

$$[f(T \wedge C | \boldsymbol{X})]^\delta [1 - F(T \wedge C | \boldsymbol{X})]^{1-\delta}$$
$$= \left(\frac{f(T \wedge C | \boldsymbol{X})}{1 - F(T \wedge C | \boldsymbol{X})}\right)^\delta [1 - F(T \wedge C | \boldsymbol{X})]$$
$$= [\exp \eta(\boldsymbol{X}, T \wedge C)]^\delta \exp\left(-\int_0^{T \wedge C} \exp \eta(\boldsymbol{X}, t)\, dt\right).$$

Thus the log-likelihood for a candidate h for η is given by

$$l(h, \boldsymbol{W}) = \delta h(\boldsymbol{X}, Y) - \int_0^Y \exp h(\boldsymbol{X}, t)\, dt.$$

Here, $\mathcal{U} = \mathcal{X} \times [0, \tau]$. Hazard estimation corresponds to the above setup with the random vector $\boldsymbol{X}$ of covariates ignored.

11.3.4 Density Estimation

Let $\boldsymbol{Y}$ have an unknown positive density function on $\mathcal{Y}$. Suppose we want to estimate the log-density ϕ. Since ϕ is subject to the intrinsic non-linear constraint $\int_\mathcal{Y} \exp \phi(\boldsymbol{y})\, d\boldsymbol{y} = 1$, it is convenient to write $\phi = \eta - C(\eta)$ and model η as a member of some linear space; here $C(h) = \log \int_\mathcal{Y} \exp h(\boldsymbol{y})\, d\boldsymbol{y}$. Note that η is determined up to an arbitrary constant as the log-density function ϕ. By imposing a linear constraint such as $\int_\mathcal{Y} \eta(\boldsymbol{y})\, d\boldsymbol{y} = 0$, we can determine η uniquely and thus make the map $\sigma : \eta \mapsto \phi$ one-to-one. The log-likelihood is given by $l(h, \boldsymbol{Y}) = h(\boldsymbol{Y}) - C(h)$. Here $\boldsymbol{W} = \boldsymbol{Y}$ and $\mathcal{U} = \mathcal{Y}$.

11.3.5   Conditional Density Estimation

Consider a random pair (X, Y), where X is $\mathcal{X}$-valued, Y is $\mathcal{Y}$-valued, and the conditional distribution of Y given that $X = x$ has a positive density. Since the corresponding log-density ϕ satisfies the nonlinear constraint $\int_{\mathcal{Y}} \exp \phi(y|x)\, dy = 1$ for $x \in \mathcal{X}$, it is not natural to model ϕ as a member of a linear space. To overcome this difficulty, we write $\phi(y|x) = \eta(y|x) - C(x;\eta)$ and model η as a member of some linear space; here $C(x;\eta) = \log \int_{\mathcal{Y}} \exp \eta(y|x)\, dy$. By imposing a suitable linear constraint on η (such as $\int_{\mathcal{Y}} \eta(y|x)h(x)\, dx = 0$ for all square-integrable functions h of x) we can make the map $\sigma : \eta \mapsto \phi$ one-to-one. Then the problem of estimating ϕ is reduced to that of estimating η and can thereby be cast into the framework of extended linear modeling. The (conditional) log-likelihood is given by $l(h, X, Y) = h(Y|X) - C(X;h)$. Here $W = (X, Y)$ and $\mathcal{U} = \mathcal{X} \times \mathcal{Y}$.

11.3.6   Diffusion Process Regression

Diffusion type processes form a large class of continuous time processes that are widely used for stochastic modeling with application to physical, biological, medical, economic, and social sciences; see Prakasa Rao [20, 21]. Consider a one-dimensional diffusion type process $Y(t)$ that satisfies the stochastic differential equation

$$dY(t) = \eta(t, X(t)) + \sigma(t)\, dW(t), \qquad 0 \le t \le \tau,$$

where $0 < \tau < \infty$ and $W(t)$ is a Wiener process. It is assumed that the diffusion coefficient $\sigma^2(t)$ at time t is a known, predictable, random function of time. It is also assumed that the value at time t of the drift coefficient is an unknown function $\eta(t, X(t))$ of t and the value at time t of a predictable covariate process $X(t) = (X_1(t), \ldots, X_L(t))$, $0 \le t \le \tau$. We refer to η as the *regression function*. Let $Z(t)$, $0 \le t \le \tau$, be a predictable $\{0, 1\}$-valued process. The process $Z(t)$ can be thought of as a censoring indicator: the processes $X(t)$ and $Y(t)$ are only observed when $Z(t) = 1$. The problem of interest is to estimate the function η based on a random sample of n realizations of $W = \{(X(t), Y(t)) : 0 \le t \le \tau \text{ and } Z(t) = 1\}$. The (partial) log-likelihood corresponding to a candidate h for η based on a single observation is given by

$$l(h) = \int Z(t) \frac{h(t, X(t))}{\sigma^2(t)}\, dY(t) - \frac{1}{2} \int Z(t) \frac{h^2(t, X(t))}{\sigma^2(t)}\, dt.$$

This can be seen either by passing to the limit from a discrete-time approximation or by modeling $(X(t), Y(t))$, $0 \le t \le \tau$, as a multidimensional diffusion process and determining the appropriate partial log-likelihood.

11.3.7   Other Contexts

Counting process regression (Huang, [8]), event history analysis (Huang and Stone [9]), marked point process regression (Li [18]), proportional hazards regression (Huang, Kooperberg, Stone and Truong [10]), robust regression (Stone [29]), and spectral density estimation (Kooperberg, Stone and Truong [17]), can also be cast into the framework of concave extended linear models.

11.4   Consistency and Rate of Convergence

In this section we present results on the asymptotic properties of the maximum likelihood estimate $\widehat{\eta}$ in concave extended linear models. As discussed in Section 11.2, the best approximation η^* in $\mathbb{H}$ to the function η of interest can be thought as a general target of estimation whether or not $\eta \in \mathbb{H}$. The existence of η^* has been established in various contexts in papers cited in Section 11.1. We say that $\widehat{\eta}$ is consistent in estimating η^* if $\|\widehat{\eta} - \eta^*\| \to 0$ in probability for some norm $\|\cdot\|$. We will state conditions that ensure consistency and also determine the rates of convergence of $\widehat{\eta}$ to η^*. In the asymptotic analysis, it is natural to let the dimension N_n of the estimation space $\mathbb{G}$ grow with the sample size. The growing dimensionality of the estimation space introduces improved approximation power of this space for increasing sample size.

We assume that the log-likelihood $l(h, \boldsymbol{w})$ and expected log-likelihood $\Lambda(h)$ are well-defined and finite for every bounded function h on $\mathcal{U}$. Since the estimation space $\mathbb{G} \subset \mathbb{H}$ is a finite-dimensional linear space of bounded functions, $\ell(h, \boldsymbol{w})$ and $\Lambda(h)$ are well-defined on $\mathbb{G}$.

Since $\widehat{\eta}$ maximizes the normalized log-likelihood $\ell(g)$, which should be close to the expected log-likelihood $\Lambda(g)$ for $g \in \mathbb{G}$ when the sample size is large, it is natural to think that $\widehat{\eta}$ is directly estimating the best approximation $\bar{\eta} = \mathrm{argmax}_{g \in \mathbb{G}} \Lambda(g)$ in $\mathbb{G}$ to η. If $\mathbb{G}$ is chosen such that $\bar{\eta}$ is close to η^*, then $\widehat{\eta}$ should provide a reasonable estimate of η^*. This motivates the decomposition

$$\widehat{\eta} - \eta^* = (\bar{\eta} - \eta^*) + (\widehat{\eta} - \bar{\eta}),$$

where $\bar{\eta} - \eta^*$ and $\widehat{\eta} - \bar{\eta}$ are referred to, respectively, as the *approximation error* and the *estimation error*.

Given a function h on $\mathcal{U}$, let $\|h\|_\infty = \sup_{\boldsymbol{u} \in \mathcal{U}} |h(\boldsymbol{u})|$ denote its L_∞ norm. Let $\|\cdot\|$ be the normalized L_2 norm relative to Lebesgue measure on $\mathcal{U}$; that is, $\|h\| = \{\int_{\mathcal{U}} h^2(\boldsymbol{u})\, d\boldsymbol{u}/\mathrm{vol}(\mathcal{U})\}^{1/2}$. Note that $\|h\| \le \|h\|_\infty$. In our asymptotic theory we will use $\|\widehat{\eta} - \eta^*\|$ to measure the discrepancy between $\widehat{\eta}$ and η^*.

Sometimes it is more natural to use other norms to measure the discrepancy. In the regression context, for example, one would use $\|h\|_0^2 =$

$E[h^2(\boldsymbol{X})]$ where the expectation is with respect to the distribution of the covariates $\boldsymbol{X}$. Such a norm is closely related to the mean prediction error. Precisely, the mean prediction error of a candidate h for the regression function η is defined by $\mathrm{PE}(h) = E\{[Y^* - h(\boldsymbol{X}^*)]^2\}$, where $(\boldsymbol{X}^*, Y^*)$ is a pair of observations independent of the observed data and having the same distribution as $(\boldsymbol{X}, Y)$. It is easily seen that

$$\mathrm{PE}(h) = E[\mathrm{var}(Y|\boldsymbol{X})] + \|h - \eta\|_0^2.$$

Under mild conditions (for example, if the density of $\boldsymbol{X}$ is bounded away from zero and infinity), the norm $\|\cdot\|_0$ is equivalent to the normalized L_2-norm $\|\cdot\|$, that is, there are positive constants c_1 and c_2 such that $c_1\|h\| \leq \|h\|_0 \leq c_2\|h\|$ for any square-integrable function h on $\mathcal{U}$. Thus our asymptotic results, presented for the norm $\|\cdot\|$, can also be stated in terms of the more natural norm $\|\cdot\|_0$. (This is generally true for other concave extended linear models, though we illustrated this point only in the regression case.)

Given random variables V_n for $n \geq 1$, let $V_n = O_P(b_n)$ mean that $\lim_{c\to\infty} \limsup_n P(|V_n| \geq cb_n) = 0$ and let $V_n = o_P(b_n)$ mean that $\lim_n P(|V_n| \geq cb_n) = 0$ for $c > 0$. Set

$$\rho_n = \inf_{g\in\mathbb{G}} \|g - \eta^*\|_\infty.$$

Under various mild assumptions on η^* and $\mathbb{G}$, $\rho_n \to 0$ as $n \to \infty$.

Proposition 1. (Huang [8]) *Under appropriate conditions, $\bar{\eta}$ exists uniquely for n sufficiently large and*

$$\|\bar{\eta} - \eta^*\|^2 = O(\rho_n^2).$$

Moreover, $\widehat{\eta}$ exists uniquely except on an event whose probability tends to zero as $n \to \infty$ and

$$\|\widehat{\eta} - \bar{\eta}\|^2 = O_P\left(\frac{N_n}{n}\right).$$

Consequently,

$$\|\widehat{\eta} - \eta^*\|^2 = O_P\left(\frac{N_n}{n} + \rho_n^2\right).$$

In particular, $\widehat{\eta}$ is consistent in estimating η^*; that is, $\|\widehat{\eta} - \eta^*\| = o_P(1)$.

According to this result, the squared norm of the estimation error is bounded in probability by the inverse of the number of observations per parameter, while the squared norm of the approximation error is bounded above by a multiple of the best obtainable approximation rate in the estimation space to the target function.

The main technical requirement for Proposition 1 is that the log-likelihood be suitably concave. (See the cited paper for details.) Specifically, it is required that the following two conditions hold:

C1. For any positive constant K, there are positive numbers M_1 and M_2 such that

$$-M_1\|h_2 - h_1\|^2 \leq \frac{d^2}{d\alpha^2}\Lambda(h_1 + \alpha(h_2 - h_1)) \leq -M_2\|h_2 - h_1\|^2$$

for $h_1, h_2 \in \mathbb{H}$ with $\|h_1\|_\infty \leq K$ and $\|h_2\|_\infty \leq K$ and $0 \leq \alpha \leq 1$.

C2. (i)

$$\sup_{g \in \mathbb{G}} \frac{\left|\frac{d}{d\alpha}\ell(\bar{\eta} + \alpha g)\Big|_{\alpha=0}\right|}{\|g\|} = O_P\left(\left(\frac{N_n}{n}\right)^{1/2}\right);$$

(ii) for any positive constant K, there is a positive number M such that

$$\frac{d^2}{d\alpha^2}\ell(g_1 + \alpha(g_2 - g_1)) \leq -M\|g_2 - g_1\|^2, \qquad 0 \leq \alpha \leq 1,$$

for $g_1, g_2 \in \mathbb{G}$ with $\|g_1\|_\infty \leq K$ and $\|g_2\|_\infty \leq K$, except on an event whose probability tends to zero as $n \to \infty$.

Proposition 1 treats a general estimation space $\mathbb{G}$. It is readily applicable to fixed knot spline estimates when the knot positions are prespecified but the number of knots is allowed to increase with the sample size. Suppose $\mathcal{U}$ is the Cartesian product of compact intervals $\mathcal{U}_1, \ldots, \mathcal{U}_L$. Consider the saturated model, in which η is a bounded function and no structural assumptions are imposed on η (that is, $\mathbb{H}$ is the space of all square-integrable functions on $\mathcal{U}$). Correspondingly, $\eta^* = \eta$. Suppose η has bounded p-th derivative (for an integer p), or more generally, suppose that η is p-smooth for a specified positive number p; that is, η is k times continuously differentiable on $\mathcal{U}$, where k is the greatest integer less than p, and all the kth-order mixed partial derivatives of η satisfy a Hölder condition with exponent $p - k$.

Let $\mathbb{G}_l$ be a linear space of splines having degree $q \geq p - 1$ for $1 \leq l \leq L$ and let $\mathbb{G}$ be the tensor product of $\mathbb{G}_1, \ldots, \mathbb{G}_L$, which is the space spanned by functions of the form $g_1(u_1) \cdots g_L(u_L)$, where g_l runs over $\mathbb{G}_l$ for $1 \leq l \leq L$. Suppose the knots have bounded mesh ratio (that is, the ratios of the differences between consecutive knots are bounded away from zero and infinity uniformly in n). Let a_n denote the smallest distance between two consecutive knots. For two sequences of positive numbers b_{1n} and b_{2n}, let $b_{1n} \asymp b_{2n}$ mean that the ratio b_{1n}/b_{2n} is bounded away from 0 and infinity. Then $N_n \asymp a_n^{-L}$ and $\rho_n \asymp a_n^p \asymp N_n^{-p/L}$. According to Proposition 1,

$$\|\widehat{\eta} - \eta\|^2 = O_P\left(\frac{1}{na_n^L} + a_n^{2p}\right).$$

In particular, for $a_n \asymp n^{-1/(2p+L)}$, we have that

$$\|\widehat{\eta} - \eta\|^2 = O_P(n^{-2p/(2p+L)}).$$

The choice of $a_n \asymp n^{-1/(2p+L)}$ balances the contributions to the error bound from the estimation error and the approximation error, that is, $1/(na_n^L) \asymp a_n^{2p}$. The resulting rate of convergence $n^{-2p/(2p+L)}$ actually is optimal: no estimate has a faster rate of convergence uniformly over the class of p-smooth functions (Stone [23]). The rate of convergence depends on two quantities: the specified smoothness p of the target function and the dimension L of the domain on which the target function is defined. Note the dependence of the rate of convergence on the dimension L: given the smoothness p, the larger the dimension, the slower the rate of convergence; moreover, the rate of convergence tends to zero as the dimension tends to infinity. This provides a mathematical description of a phenomenon commonly known as the "curse of dimensionality."

The following question arises naturally. What if we restrict attention to additive estimates

$$\widehat{\eta}(\boldsymbol{x}) = \widehat{\eta}_1(x_1) + \cdots + \widehat{\eta}_L(x_L)$$

of the best additive approximation

$$\eta^*(\boldsymbol{x}) = \eta_1^*(x_1) + \cdots + \eta_L^*(x_L)$$

to η? Can we now achieve the rate of convergence $n^{-p/(2p+1)}$?

11.5   Functional ANOVA

Imposing structures such as additivity on an unknown multivariate function indeed can imply faster rates of convergence of the corresponding estimate and thus tame the curse of dimensionality. In this section we will discuss how to impose general structural assumptions using functional ANOVA decompositions. We will also study rates of convergence of the maximum likelihood estimates when the model and estimation spaces are constructed in some structured way.

To introduce the notion of functional ANOVA, it is helpful to look at a simple example. Suppose that $\mathcal{U} = \mathcal{U}_1 \times \mathcal{U}_2 \times \mathcal{U}_3$, where $\mathcal{U}_1$, $\mathcal{U}_2$ and $\mathcal{U}_3$ are compact intervals, each having positive length. Any square-integrable function on $\mathcal{U}$ can be decomposed as

$$
\begin{aligned}
\eta(\boldsymbol{u}) \ =\ & \eta_\emptyset + \eta_{\{1\}}(u_1) + \eta_{\{2\}}(u_2) + \eta_{\{3\}}(u_3) + \eta_{\{1,2\}}(u_1, u_2) \quad (11.3) \\
& + \eta_{\{1,3\}}(u_1, u_3) + \eta_{\{2,3\}}(u_2, u_3) + \eta_{\{1,2,3\}}(u_1, u_2, u_3).
\end{aligned}
$$

For identifiability, we require that each nonconstant component be orthogonal to all possible values of the corresponding lower-order components relative to an appropriate inner product. The expression (11.3) can then be viewed as a functional analysis of variance (ANOVA) decomposition. Correspondingly, we call $\eta_\emptyset$ the constant component; $\eta_{\{1\}}(u_1), \eta_{\{2\}}(u_2)$ and $\eta_{\{3\}}(u_3)$ the main effect components; $\eta_{\{1,2\}}(u_1, u_2), \eta_{\{1,3\}}(u_1, u_3)$ and

$\eta_{\{2,3\}}(u_2, u_3)$ the two-factor interaction components; and $\eta_{\{1,2,3\}}(u_1, u_2, u_3)$ the three-factor interaction component. The right side of (11.3) is referred to as the ANOVA decomposition of η.

If no structural assumption is imposed on η, we need to consider all the components in the above ANOVA decomposition. The resulting model is saturated. However, the desire to tame the curse of dimensionality leads us to employ unsaturated models, which discard some terms in the ANOVA decomposition. For example, removing all the interaction components in the above ANOVA decomposition of η, we get the additive model

$$\eta(\boldsymbol{u}) = \eta_\emptyset + \eta_{\{1\}}(u_1) + \eta_{\{2\}}(u_2) + \eta_{\{3\}}(u_3). \tag{11.4}$$

We can also include some selected interactions in the model and still keep the model manageable. For example, the following model includes just the interaction between u_1 and u_2:

$$\eta(\boldsymbol{u}) = \eta_\emptyset + \eta_{\{1\}}(u_1) + \eta_{\{2\}}(u_2) + \eta_{\{3\}}(u_3) + \eta_{\{1,2\}}(u_1, u_2). \tag{11.5}$$

To fit these models using maximum likelihood, it is necessary to choose the estimation space $\mathbb{G}$ to respect the imposed structure on η. As a result the estimate will have the same structure. For example, by choosing $\mathbb{G}$ appropriately, the maximum likelihood estimate will have the forms

$$\widehat{\eta}(\boldsymbol{u}) = \widehat{\eta}_\emptyset + \widehat{\eta}_{\{1\}}(u_1) + \widehat{\eta}_{\{2\}}(u_2) + \widehat{\eta}_{\{3\}}(u_3)$$

and

$$\widehat{\eta}(\boldsymbol{u}) = \widehat{\eta}_\emptyset + \widehat{\eta}_{\{1\}}(u_1) + \widehat{\eta}_{\{2\}}(u_2) + \widehat{\eta}_{\{3\}}(u_3) + \widehat{\eta}_{\{1,2\}}(u_1, u_2).$$

for models (11.4) and (11.5), respectively.

In general, suppose that $\mathcal{U} = \mathcal{U}_1 \times \cdots \times \mathcal{U}_L$ for some positive integer L, where each $\mathcal{U}_l$ is a compact subset of some Euclidean space and it has positive volume in that space. If η is square-integrable, we can define its ANOVA decomposition in a similar manner as above. Selecting certain terms in its ANOVA decomposition in the modeling process corresponds to imposing a particular structural assumption on η. Specifically, let $\mathcal{S}$ be a hierarchical collection of subsets of $\{1, \ldots, L\}$; by hierarchical we mean that if $s \in \mathcal{S}$, then $r \in \mathcal{S}$ for every subset r of s. Consider the model space

$$\mathbb{H} = \left\{ \sum_{s \in \mathcal{S}} h_s : h_s \in \mathbb{H}_s \text{ for } s \in \mathcal{S} \right\}, \tag{11.6}$$

where $\mathbb{H}_s$ is the space of square-integrable functions that depends only on u_l, $l \in s$. Note that the set $\mathcal{S}$ describes precisely which interaction terms are included in the model. For example, the additive model (11.4) and the model (11.5) with a single interaction component correspond to $\mathcal{S} = \{\emptyset, \{1\}, \{2\}, \{3\}\}$ and $\mathcal{S} = \{\emptyset, \{1\}, \{2\}, \{3\}, \{1, 2\}\}$, respectively. Note that the best approximation η^* in $\mathbb{H}$ to η, which is the general target for estimation in an extended linear model, has the form

$$\eta^* = \sum_{s \in \mathcal{S}} \eta_s^*, \tag{11.7}$$

where η_s^* is a member of $\mathbb{H}_s$ for $s \in \mathcal{S}$ and $\eta_\emptyset^*$ is a constant.

Again, the maximum likelihood method can be used to do the estimation and the estimation space $\mathbb{G}$ is chosen to take the form

$$\mathbb{G} = \left\{ \sum_{s \in \mathcal{S}} g_s : g_s \in \mathbb{G}_s \text{ for } s \in \mathcal{S} \right\}, \tag{11.8}$$

where $\mathbb{G}_s$ is the tensor product space of $\mathbb{G}_l, l \in s$, and for each $1 \le l \le L$, $\mathbb{G}_l$ is an appropriate finite-dimensional space of functions of u_l that contains all constant functions. The resulting estimate should have the form

$$\widehat{\eta} = \sum_{s \in \mathcal{S}} \widehat{\eta}_s, \tag{11.9}$$

where $\widehat{\eta}_s$ is a member of $\mathbb{G}_s$ for $s \in \mathcal{S}$ and $\widehat{\eta}_\emptyset$ is a constant.

The asymptotic theory concerns the consistency of $\widehat{\eta}$ in estimating η^* and the consistency of the components $\widehat{\eta}_s$ of $\widehat{\eta}$ in estimating the corresponding components of η_s^* of η^*. (Suitable identifiability constraints must be imposed on the components of $\widehat{\eta}$ and η^* as indicated in the discussion following (11.3) and in the discussion at the end of this section.) The consistency of the estimated components is desirable since one would like examination of the components of $\widehat{\eta}$ to shed light on the shape of η^* and its components.

We should allow the dimensions of $\mathbb{G}_l$ and thus those of $\mathbb{G}_s$ to depend on the sample size. Set $N_s = \dim(\mathbb{G}_s)$ and $\rho_s = \inf_{g \in \mathbb{G}_s} \|g - \eta_s^*\|_\infty$ for $s \in \mathcal{S}$.

Proposition 2. (Huang [8]) *Under appropriate conditions,*

$$\|\widehat{\eta} - \eta^*\|^2 = O_P\left(\sum_{s \in \mathcal{S}} \left(\frac{N_s}{n} + \rho_s^2 \right) \right)$$

and

$$\|\widehat{\eta}_s - \eta_s^*\|^2 = O_P\left(\sum_{s \in \mathcal{S}} \left(\frac{N_s}{n} + \rho_s^2 \right) \right), \qquad s \in \mathcal{S}.$$

Suppose each η_s^*, $s \in \mathcal{S}$, in (11.7) is p-smooth. Let the estimation space be given by (11.8) with each $\mathbb{G}_l$ being a linear space of degree q splines on $\mathcal{U}_l$ as in the previous section with $q \ge p-1$. Let $\#(B)$ denote the cardinality (number of members) of a set B. Then $N_s \asymp a_n^{-\#(s)}$ and $\rho_s \asymp a_n^p$, $s \in \mathcal{S}$. Set $d = \max_{s \in \mathcal{S}} \#(s)$. According to Proposition 2,

$$\|\widehat{\eta} - \eta^*\|^2 = O_P\left(\frac{1}{n a_n^d} + a_n^{2p} \right)$$

and

$$\|\widehat{\eta}_s - \eta_s^*\|^2 = O_P\Big(\frac{1}{na_n^d} + a_n^{2p}\Big), \qquad s \in \mathcal{S}.$$

In particular, for $a_n \asymp n^{-1/(2p+d)}$, we have that

$$\|\widehat{\eta} - \eta^*\|^2 = O_P(n^{-2p/(2p+d)})$$

and

$$\|\widehat{\eta}_s - \eta_s^*\|^2 = O_P(n^{-2p/(2p+d)}), \qquad s \in \mathcal{S}.$$

The rate of convergence $n^{-2p/(2p+d)}$ for an unsaturated model should be compared with the rate $n^{-2p/(2p+L)}$ for the saturated model. Note here that d is the maximum order of interaction among the components of η^* and $\widehat{\eta}$. For the additive model, we have $d = 1$, so the rate is $n^{-2p/(2p+1)}$, which is the same as that for estimating a one-dimensional target function. For models with interaction components of order two and no higher-order interaction components, we have $d = 2$ and corresponding rate of convergence $n^{-2p/(2p+2)}$, the same rate for estimating a two-dimensional target function. Hence, for large L, we can achieve much faster rates of convergence by considering structural models involving only low-order interactions in the ANOVA decomposition of the target function and thereby tame the curse of dimensionality. (Of course, if η does not itself possess the imposed structure, then the faster rates of convergence are obtained at the expense of estimating η^* rather than η.)

As discussed at the beginning of this section, we define the ANOVA decomposition of a function by forcing each nonconstant component to be orthogonal to all possible values of the corresponding lower-order components relative to an appropriate inner product. Usually, one uses a theoretical inner product to decompose η^* and an empirical inner product to decompose $\widehat{\eta}$. For example, in the regression case, it is natural to define the theoretical and empirical inner products by $\langle h_1, h_2 \rangle = E[h_1(\boldsymbol{X})h_2(\boldsymbol{X})]$ and $\langle h_1, h_2 \rangle_n = (1/n)\sum_i [h_1(\boldsymbol{X}_i)h_2(\boldsymbol{X}_i)]$. The reason for using different inner products is that the theoretical inner product is often defined in terms of the data-generating distribution and hence depends on unknown quantities, while the empirical inner product must be totally determined by the data since it will be used to decompose the estimate. An important necessary condition for Proposition 2 is that the two inner products or the corresponding norms are close; precisely, $\sup_{g \in \mathbb{G}} \big| \|g\|_n / \|g\| - 1 \big| = o_P(1)$.

11.6   Free Knot Splines in Extended Linear Models

In this section we extend the results in Section 11.4 to handle free knot splines; that is, the knot positions are treated as free parameters to be determined from the data.

Consider a concave extended linear model specified by the log-likelihood $l(h, \boldsymbol{W})$ and model space $\mathbb{H}$. Let $\mathbb{G}_\gamma$, $\gamma \in \Gamma$, be a collection of finite-dimensional linear subspaces of $\mathbb{H}$. We assume that the functions in each such space $\mathbb{G}_\gamma$ are bounded and call $\mathbb{G}_\gamma$ an estimation space. Here γ can be thought of as the knot positions when the estimation space consists of spline functions, and our interest lies in choosing the knot positions using the data. It is assumed that the spaces $\mathbb{G}_\gamma$, $\gamma \in \Gamma$, have a common dimension and that the index set Γ is a compact subset of $\mathbb{R}^J$ for some positive integer J. We allow $\dim(\mathbb{G}_\gamma)$, Γ, and J to vary with the sample size n.

For each fixed $\gamma \in \Gamma$, the maximum likelihood estimate is given by $\widehat{\eta}_\gamma = \max_{g \in \mathbb{G}_\gamma} \ell(g)$. In order to let the data select which estimation space to use, we choose $\widehat{\gamma} \in \Gamma$ such that $\ell(\widehat{\eta}_{\widehat{\gamma}}) = \max_{\gamma \in \Gamma} \ell(\widehat{\eta}_\gamma)$. (Such a $\widehat{\gamma}$ exists under mild conditions.) We will study the benefit of allowing the flexibility to select the estimation space from among a big collection of such spaces. Specifically we will study the rate of convergence to zero of $\widehat{\eta}_{\widehat{\gamma}} - \eta^*$, where η^* is the best approximation to η in $\mathbb{H}$. For $\gamma \in \Gamma$, set $N_n = \dim(\mathbb{G}_\gamma)$ and $\rho_{n\gamma} = \inf_{g \in \mathbb{G}_\gamma} \|g - \eta^*\|_\infty$.

It follows from Proposition 1 that, $\|\widehat{\eta}_\gamma - \eta^*\|^2 = O_P(\rho_{n\gamma}^2 + N_n/n)$ for each fixed $\gamma \in \Gamma$. Let γ^* be such that $\rho_{n\gamma^*} = \inf_{\gamma \in \Gamma} \rho_{n\gamma}$. (Such a γ^* exists under mild conditions.) Then

$$\|\widehat{\eta}_{\gamma^*} - \eta^*\|^2 = O_P\left(\rho_{n\gamma^*}^2 + \frac{N_n}{n}\right) = O_P\left(\inf_{\gamma \in \Gamma} \rho_{n\gamma}^2 + \frac{N_n}{n}\right).$$

Thus,

$$\inf_{\gamma \in \Gamma} \|\widehat{\eta}_\gamma - \eta^*\|^2 \leq \|\widehat{\eta}_{\gamma^*} - \eta^*\|^2 = O_P\left(\inf_{\gamma \in \Gamma} \rho_{n\gamma}^2 + \frac{N_n}{n}\right).$$

It is natural to expect that, with γ estimated by $\widehat{\gamma}$, the squared L_2 norm $\|\widehat{\eta}_{\widehat{\gamma}} - \eta^*\|^2$ of the difference between the estimate and the target will be not much larger than the ideal quantity $\inf_{\gamma \in \Gamma} \|\widehat{\eta}_\gamma - \eta^*\|^2$. Hence we hope that $\|\widehat{\eta}_{\widehat{\gamma}} - \eta^*\|^2$ will be not much larger than $\inf_{\gamma \in \Gamma} \rho_{n\gamma}^2 + N_n/n$ in probability. This is confirmed by the next result.

Let $V_{n\gamma} = O_P(b_{n\gamma})$ uniformly over $\gamma \in \Gamma$ mean that

$$\lim_{c \to \infty} \limsup_n P(|V_{n\gamma}| \geq c b_{n\gamma} \text{ for some } \gamma \in \Gamma) = 0,$$

where $b_{n\gamma} > 0$ for $n \geq 1$ and $\gamma \in \Gamma$. As in Section 11.4, it is enlightening to decompose the error into a stochastic part and a systematic part for each fixed $\gamma \in \Gamma$:

$$\widehat{\eta}_\gamma - \eta^* = (\widehat{\eta}_\gamma - \bar{\eta}_\gamma) + (\bar{\eta}_\gamma - \eta^*),$$

where $\widehat{\eta}_\gamma - \bar{\eta}_\gamma$ is referred to as the *estimation error* and $\bar{\eta}_\gamma - \eta$ as the approximation error.

Proposition 3. (Stone and Huang [31]) *Under appropriate conditions, for* n *sufficiently large,* $\bar{\eta}_\gamma$ *exists uniquely for* $\gamma \in \Gamma$ *and*

$$\|\bar{\eta}_\gamma - \eta^*\|^2 = O(\rho_{n\gamma}^2)$$

uniformly over $\gamma \in \Gamma$. *Moreover, except on an event whose probability tends to zero as* $n \to \infty$, $\hat{\eta}_\gamma$ *exists uniquely for* $\gamma \in \Gamma$ *and*

$$\sup_{\gamma \in \Gamma} \|\hat{\eta}_\gamma - \bar{\eta}_\gamma\|^2 = O_P\left(\frac{N_n}{n}\right).$$

Consequently,

$$\|\hat{\eta}_\gamma - \eta^*\|^2 = O_P\left(\rho_{n\gamma}^2 + \frac{N_n}{n}\right)$$

uniformly over $\gamma \in \Gamma$. *In addition,*

$$\|\hat{\eta}_{\hat{\gamma}} - \eta^*\|^2 = O_P\left(\inf_{\gamma \in \Gamma} \rho_{n\gamma}^2 + (\log n)\frac{N_n}{n}\right).$$

In the previous theoretical results for fixed knot splines, the squared norms of the approximation error and the estimation error were shown to be bounded above by multiples of $\rho_{n\gamma}^2$ and N_n/n, respectively. Here these results are shown to hold uniformly over the free knot sequences $\gamma \in \Gamma$. Finally, combining the results for the approximation error and the estimation error and incorporating a corresponding result for the maximum likelihood estimation of the knot positions, we get the rate of convergence for the free-knot spline estimate..

The benefit of using free-knot splines is that a smaller approximation error can be achieved; presumably, for certain functions the best approximation rate obtainable for free-knot splines for a collection of knot positions (i.e., $\inf_{\gamma \in \Gamma} \rho_{n\gamma}$) would be much smaller than the best approximation rate obtainable for fixed-knot splines (i.e., $\rho_{n\gamma}$ for a fixed γ). The cost is a small inflation of the variance; there is an extra $\log n$ term in the variance bound for the free-knot spline estimate. We are currently unable, in any context, to verify either that this extra $\log n$ factor is necessary or that it is unnecessary.

As a simple illustration of the improved rate of convergence of free-knot spline estimate over fixed-knot spline estimate, consider estimating the regression function $\eta(x)$ in the regression model $Y = \eta(X) + \epsilon$, where X has a uniform distribution on $[-1, 1]$ and ϵ has a normal distribution with mean 0 and variance σ^2. Suppose $\eta(x) = x^2 + |x|$. Clearly, the first derivative of η at 0 does not exist, which implies a slow fixed-knot spline approximation rate if there is no knot very close to 0. Specifically, let the estimation space $\mathbb{G}$ be the space of linear splines on $[-1, 1]$ with $2J_n$ equally spaced knots located at $\pm(2k-1)/(2J_n-1)$, $k = 1, \ldots, J_n$, and let $\hat{\eta}$ be the least squares estimate on $\mathbb{G}$. It can be shown that for some positive constant c and large n, $P(\|\hat{\eta} - \eta\|^2 > cJ_n/n + cJ_n^{-3}) > 1/3$ and thus for any choice of J_n,

$P(\|\widehat{\eta} - \eta\|^2 > cn^{-3/4}) > 1/3$. On the other hand, let $\mathbb{G}_\gamma$ be the space of linear splines on $[-1, 1]$ with $2J_n - 1$ knots located at γ and $\pm(2k-1)/(2J_n-1)$, $k = 2, \ldots, J_n$, where $-1/(2J_n - 1) \leq \gamma \leq 1/(2J_n - 1)$. Here, we simply replace two fixed knots $\pm 1/(2J_n - 1)$ in the previous setup by one free-knot at γ. Using Proposition 3, we can show that $\|\widehat{\eta}_{\widehat{\gamma}} - \eta\|^2 = O_P(J_n^{-4} + J_n \log n/n)$. Hence, for $J_n \asymp (n/\log n)^{1/5}$, the convergence rate of the free-knot spline estimate satisfies $\|\widehat{\eta}_{\widehat{\gamma}} - \eta\|^2 = O_P((n^{-1} \log n)^{4/5})$, which is faster than that of the fixed-knot spline estimate. Technical details of this example can be found in Stone and Huang [32].

The main technical requirement for Proposition 3 is that Conditions C1 and C2 in Section 11.4 be strenthened to hold uniformly over $\gamma \in \Gamma$. For data driven choices of γ, it is also required that

$$|\ell(\bar{\eta}_\gamma) - \ell(\eta^*) - [\Lambda(\bar{\eta}_\gamma) - \Lambda(\eta^*)]|$$
$$= O_P\left(\log^{1/2} n \left[\|\bar{\eta}_\gamma - \eta^*\| \left(\frac{N_n}{n}\right)^{1/2} + \frac{N_n}{n}\right]\right)$$

uniformly over $\gamma \in \Gamma$.

These conditions have been verified separately in each of various contexts—regression, logistic regression, density estimation in Stone and Huang [31], diffusion process regression in Stone and Huang [32], and marked point process regression in Li [18].

11.7   Methodological Implications

The successful development of theory, such as that surveyed in this chapter, suggests that closely related adaptive methodology based on stepwise selection of basis functions should be worth pursuing in practice. Similarly, the successful development of a theoretical synthesis that applies to a number of seemingly different contexts suggests that the corresponding adaptive methodologies for these contexts should have similar performance.

Since concavity is crucial in the theory that has been presented here, it is presumably also helpful in developing corresponding methodologies, for example, to avoid getting stuck in local optima that are far from being globally optimal. Since additive and more general models containing only low-order interactions have faster rates of convergence than models containing higher-order interactions, the corresponding methodologies are also presumably worthy of pursuit.

These implications of theory for methodology have largely been borne out in practice. For various methodological developments that have been influenced, to one extent or another, by the theory surveyed in this chapter, see Friedman and Silverman [4]; Friedman [3]; Kooperberg and Stone [11]; Kooperberg, Stone and Truong [14, 16]; Kooperberg, Bose and Stone [12]; and Kooperberg and Stone [13].

So far, there has been no statistical theory that applies rigorously to stepwise selection of basis functions. Now that there is a statistical theory for modeling with free knot splines, it is reasonable to view stepwise knot selection, which can be thought of as a special case of the stepwise selection of basis functions, as a computationally efficient shortcut to modeling with free knot splines as in Lindstrom [19].

References

[1] de Boor, C. (1978). *A Practical Guide to Splines*. Springer-Verlag, New York.

[2] DeVore, R. A. and Lorentz, G. G. (1993). *Constructive Approximation*. Springer-Verlag, Berlin.

[3] Friedman, J. H. (1991). Multivariate adaptive regression splines (with discussion). *The Annals of Statistics* **19**, 1–141.

[4] Friedman, J. H. and Silverman, B. W. (1989). Flexible parsimonious smoothing and additive modeling (with discussion). *Technometrics* **31**, 3–39.

[5] Hansen, M. H. (1994). *Extended Linear Models, Multivariate Splines, and ANOVA*. Ph.D. Dissertation, Department of Statistics, University of California at Berkeley.

[6] Huang, J. Z. (1998a). Projection estimation in multiple regression with application to functional ANOVA models. *The Annals of Statistics* **26**, 242–272.

[7] Huang, J. Z. (1998b). Functional ANOVA models for generalized regression. *Journal of Multivariate Analysis* **67**, 49–71.

[8] Huang, J. Z. (2001). Concave extended linear modeling: a theoretical synthesis. *Statistica Sinica* **11**, 173–197.

[9] Huang, J. Z. and Stone, C. J. (1998). The L_2 rate of convergence for event history regression with time-dependent covariates. *Scandinavian Journal of Statistics* **25**, 603–620.

[10] Huang, J. Z., Kooperberg, C., Stone, C. J. and Truong, Y. K. (2000). Functional ANOVA modeling for proportional hazards regression. *The Annals of Statistics* **28**, 960–999.

[11] Kooperberg, C. and Stone, C. J. (1992). Logspline density estimation for censored data. *Journal of Computational and Graphical Statistics* **1**, 301–328.

[12] Kooperberg, C., Bose, S. and Stone, C. J. (1997). Polychotomous regression. *Journal of the American Statistical Association* **92**, 117–127.

[13] Kooperberg, C. and Stone, C. J. (1999). Stochastic optimization methods for fitting polyclass and feed-forward neural network models. *Journal of Computational and Graphical Statistics* **8**, 169–189.

[14] Kooperberg, C., Stone, C. J. and Truong, Y. K. (1995a). Hazard regression. *Journal of the American Statistical Association* **90**, 78–94.

[15] Kooperberg, C., Stone, C. J. and Truong, Y. K. (1995b). The L_2 rate of convergence for hazard regression. *Scandinavian Journal of Statistics* **22**, 143–157.

[16] Kooperberg, C., Stone, C. J. and Truong, Y. K. (1995c). Logspline estimation of a possibly mixed spectral distribution. *Journal of Time Series Analysis* **16**, 359–388.

[17] Kooperberg, C., Stone, C. J. and Truong, Y. K. (1995d). Rate of convergence for logspline spectral density estimation. *Journal of Time Series Analysis* **16**, 389–401.

[18] Li, W. (2000). *Modeling Marked Point Processes with an Application to Currency Exchange Rates*. Ph. D. Dissertation, Department of Statistics, University of California at Berkeley.

[19] Lindstrom, M. J. (1999). Penalized estimation of free-knot splines. *Journal of Computational and Graphical Statistics* **8**, 333–352.

[20] Prakasa Rao, B. L. S. (1999a). *Semimartingales and their Statistical Inference*. Chapman & Hall/CRC, Boca Raton, Florida.

[21] Prakasa Rao, B. L. S. (1999b). *Statistical Inference for Diffusion Type Processes*. Oxford University Press, New York.

[22] Schumaker, L. L. (1981). *Spline Functions: Basic Theory*. Wiley, New York.

[23] Stone, C. J. (1982). Optimal global rates of convergence for nonparametric regression. *The Annals of Statistics* **10**, 1348–1360.

[24] Stone, C. J. (1985). Additive regression and other nonparametric models. *The Annals of Statistics* **13**, 689–705.

[25] Stone, C. J. (1986). The dimensionality reduction principle for generalized additive models. *The Annals of Statistics* **14**, 590–606.

[26] Stone, C. J. (1990). Large-sample inference for log-spline models. *The Annals of Statistics* **18**, 717–741.

[27] Stone, C. J. (1991). Asymptotics for doubly flexible logspline response models. *The Annals of Statistics* **19**, 1832–1854.

[28] Stone, C. J. (1994). The use of polynomial splines and their tensor products in multivariate function estimation (with discussion). *The Annals of Statistics* **22**, 118–184.

[29] Stone, C. J. (2001). Rate of convergence for robust nonparametric regression. Manuscript.

[30] Stone, C. J., Hansen, M., Kooperberg, C. and Truong, Y. K. (1997). Polynomial splines and their tensor products in extended linear modeling (with discussion). *The Annals of Statistics* **25**, 1371–1470.

[31] Stone, C. J. and Huang, J. Z. (2002a). Free knot splines in concave extended linear modeling. *Journal of Statistical Planning and Inference C. R. Rao Volume, Part II*. To appear.

[32] Stone, C. J. and Huang, J. Z. (2002b). Statistical modeling of diffusion processes with free knot splines. *Journal of Statistical Planning and Inference*. To appear.

Part II

Shorter Papers

12

Adaptive Sparse Regression

Mário A. T. Figueiredo[1]

Summary

In sparse regression, the goal is to obtain an estimate of the regression coefficients in which several of them are set exactly to zero. Sparseness is a desirable feature in regression problems, for several reasons. For example, in linear regression, sparse models are interpretable, that is, we find which variables are relevant; in kernel-based methods, like in *support vector* regression, sparseness leads to regression equations involving only a subset of the learning data. In all approaches to sparse regression, it is necessary to estimate parameters which will ultimately control the degree of sparseness of the obtained solution. This commonly involves cross-validation methods which waste learning data and are time consuming. In this chapter we present a sparseness inducing prior which does not involve any (hyper)parameters that need to be adjusted or estimated. Experiments with several publicly available benchmark data sets show that the proposed approach yields state-of-the-art performance. In particular, our method outperforms support vector regression and performs competitively with the best alternative techniques, both in terms of error rates and sparseness, although it involves no tuning or adjusting of sparseness-controlling hyper-parameters.

12.1 Introduction

The goal of supervised learning is to infer a functional relationship $y = f(\mathbf{x})$, based on a set of (possibly *noisy*) training examples $\mathcal{D} = \{(\mathbf{x}_1, y_1), \ldots, (\mathbf{x}_n, y_n)\}$. Usually, the inputs are vectors, $\mathbf{x}_i =$

---

[1]Mário A. T. Figueiredo is with the Institute of Telecommunications, and Department of Electrical and Computer Engineering Instituto Superior Técnico, 1049-001 Lisboa, Portugal (Email: mtf@lx.it.pt). This research was partially supported by the Portuguese Foundation for Science and Technology (FCT), under project POSI/33143/SRI/2000.

$[x_{i,1}, ..., x_{i,d}]^T \in I\!\!R^d$. When y is continuous (typically $y \in I\!\!R$), we are in the context of *regression*, whereas in *classification*, y is of categorical nature (*e.g.*, $y \in \{-1, 1\}$). Usually, the structure of $f(\cdot)$ is assumed fixed and the objective is to estimate a vector of parameters $\boldsymbol{\beta}$ defining it; accordingly we write $y = f(\mathbf{x}, \boldsymbol{\beta})$.

To achieve good *generalization* (*i.e.* to perform well on yet unseen data) it is necessary to control the *complexity* of the learned function (see [5], [6], [23] and [27], and the many references therein). In Bayesian approaches, complexity is controlled by placing a prior on the function to be learned, *i.e.*, on $\boldsymbol{\beta}$. This should not be confused with a *generative* (*informative*) Bayesian approach, since it involves no explicit modelling of the joint probability $p(\mathbf{x}, y)$. A common choice is a zero-mean Gaussian prior, which appears under different names, like *ridge regression* [11], or *weight decay*, in the neural learning literature [2]. Gaussian priors are also used in non-parametric contexts, like the Gaussian processes (GP) approach [6], [18], [28], [29], which has roots in earlier spline models [13] and regularized radial basis functions [22]. Very good performance has been reported for methods based on Gaussian priors [28], [29]. Their main disadvantage is that they do not control the structural complexity of the resulting functions. That is, if one of the components of $\boldsymbol{\beta}$ (say, a weight in a neural network) happens to be irrelevant, a Gaussian prior will not set it exactly to zero, thus pruning that parameter, but to some small value.

Sparse estimates (*i.e.*, in which irrelevant parameters are set exactly to zero) are desirable because (in addition to other learning-theoretic reasons [27]) they correspond to a structural simplification of the estimated function. Using Laplacian priors (equivalently, l_1-penalized regularization) is known to promote sparseness [4], [10], [25], [31]. *Support vector machines* (SVM) also lead to sparse regressors without explicitly adopting a sparseness inducing prior [6], [27]. Interestingly, however, it can be shown that the SVM and l_1-penalized regression are closely related [10].

Both in approaches based on Laplacian priors and in SVMs, there are hyper-parameters which control the degree of sparseness of the obtained estimates. These are commonly adjusted using cross-validation methods which do not optimally utilize the available data, and are time consuming.

In this chapter, we propose an alternative approach which involves no hyper-parameters. The key steps of our proposal are: (i) a hierarchical Bayes interpretation of the Laplacian prior as a *normal/independent* distribution (as used in robust regression [14]); (ii) a Jeffreys' non-informative second-level hyper-prior (in the same spirit as [9]) which expresses scale-invariance and, more importantly, is parameter-free [1]; (iii) a simple *expectation-maximization* (EM) algorithm which yields a *maximum a posteriori* (MAP) estimate of $\boldsymbol{\beta}$, and of the observation noise variance.

Our method is related to the *automatic relevance determination* (ARD) concept [18], [16], which underlies the recently proposed *relevance vector machine* (RVM) [3], [26]. The RVM exhibits state-of-the-art performance, and it seems to outperform SVMs both in terms of accuracy and sparseness [3], [26]. However, we do not resort to a *type-II maximum likelihood* approximation [1] (as in ARD and RVM); rather, our modelling assumptions lead to a marginal *a posteriori* probability function on $\boldsymbol{\beta}$ whose mode is located by a very simple EM algorithm. Related hierarchical-Bayes models were proposed in [12] and [24]; in those papers inference is carried out by *Markov chain Monte Carlo* (MCMC) sampling.

Experimental evaluation of the proposed method, both with synthetic and real data, shows that it performs competitively with (often better than) RVM and SVM.

12.2   Bayesian Linear Regression

12.2.1   *Gaussian prior and ridge regression*

We consider functional representations which are linear with respect to $\boldsymbol{\beta}$, that is,

$$f(\mathbf{x}, \boldsymbol{\beta}) = \boldsymbol{\beta}^T \mathbf{h}(\mathbf{x});$$

we will denote the dimensionality of $\boldsymbol{\beta}$ as k. This form includes:

(i) classical linear regression, where $\mathbf{h}(\mathbf{x}) = [1, x_1, ..., x_d]^T$;

(ii) nonlinear regression via a set of k basis functions, in which case $\mathbf{h}(\mathbf{x}) = [\phi_1(\mathbf{x}), ..., \phi_k(\mathbf{x})]^T$; this is the case, for example, of radial basis functions (with fixed basis functions), spline functions (with fixed knots), or even free knot splines (see [21]);

(iii) kernel regression, with $\mathbf{h}(\mathbf{x}) = [1, K(\mathbf{x}, \mathbf{x}_1), ..., K(\mathbf{x}, \mathbf{x}_n)]^T$, where $K(\mathbf{x}, \mathbf{y})$ is some (symmetric) kernel function [6] (as in SVM and RVM regression), though not necessarily verifying Mercer's condition.

We follow the standard assumption that

$$y_i = f(\mathbf{x}_i, \boldsymbol{\beta}) + w_i,$$

for $i = 1, ..., n$, where $[w_1, ..., w_n]$ is a set of independent zero-mean Gaussian variables with variance σ^2. With $\mathbf{y} \equiv [y_1, ..., y_n]^T$, the likelihood function is then

$$p(\mathbf{y}|\boldsymbol{\beta}) = \mathcal{N}(\mathbf{y}|\mathbf{H}\boldsymbol{\beta}, \sigma^2 \mathbf{I}),$$

where $\mathbf{H}$ is the $(n \times k)$ *design matrix* which depends on the $\mathbf{x}_i$s and on the adopted function representation, and $\mathcal{N}(\mathbf{v}|\boldsymbol{\mu}, \mathbf{C})$ denotes a Gaussian density of mean $\boldsymbol{\mu}$ and covariance $\mathbf{C}$, evaluated at $\mathbf{v}$.

With a zero-mean Gaussian prior with covariance $\mathbf{A}$, that is $p(\boldsymbol{\beta}|\mathbf{A}) = \mathcal{N}(\boldsymbol{\beta}|0, \mathbf{A})$, it is well known that the posterior is still Gaussian; more specifically,

$$p(\boldsymbol{\beta}|\mathbf{y}) = \mathcal{N}(\boldsymbol{\beta}|\widehat{\boldsymbol{\beta}}, \mathbf{D})$$

with mean and mode at

$$\widehat{\boldsymbol{\beta}} = (\sigma^2 \mathbf{A}^{-1} + \mathbf{H}^T \mathbf{H})^{-1} \mathbf{H}^T \mathbf{y}.$$

When $\mathbf{A}$ is proportional to identity, say $\mathbf{A} = \mu^2 \mathbf{I}$, this is equivalent to *ridge regression* (RR) [11], although RR was proposed in a non-Bayesian context.

12.2.2   Laplacian prior, sparse Regression, and the LASSO

Let us now consider a Laplacian prior for $\boldsymbol{\beta}$,

$$p(\boldsymbol{\beta}|\alpha) = \prod_{i=1}^{k} \frac{\alpha}{2} \exp\{-\alpha\,|\beta_i|\} = \left(\frac{\alpha}{2}\right)^k \exp\{-\alpha\,\|\boldsymbol{\beta}\|_1\},$$

where $\|\mathbf{v}\|_1 = \sum_i |v_i|$ denotes the l_1 norm. The posterior $p(\boldsymbol{\beta}|\mathbf{y})$ is no longer Gaussian. The *maximum a posteriori* (MAP) estimate is now given by

$$\widehat{\boldsymbol{\beta}} = \arg\min\{\|\mathbf{H}\boldsymbol{\beta} - \mathbf{y}\|_2^2 + 2\,\sigma^2\alpha\,\|\boldsymbol{\beta}\|_1\}, \tag{12.1}$$

where $\|\mathbf{v}\|_2$ is the Euclidean (l_2) norm. In linear regression, this is called the LASSO (*least absolute shrinkage and selection operator*) [25]. The main effect of the l_1 penalty is that some of the components of $\widehat{\boldsymbol{\beta}}$ may be exactly zero. If $\mathbf{H}$ is an orthogonal matrix, (12.1) can be solved separately for each β_i, leading to the *soft threshold* estimation rule, widely used in wavelet-based signal/image denoising [7]. The sparseness inducing nature of the Laplacian prior (or equivalently, of the l_1 penalty) has been exploited in several other contexts [4], [31], [21], [15], [19].

12.3   Hierarchical Interpretation of the Laplacian

Let us consider an alternative model: let each β_i have a zero-mean Gaussian prior $p(\beta_i|\tau_i) = \mathcal{N}(\beta_i|0, \tau_i)$, with its own variance τ_i (like in ARD and RVM). Now, rather than adopting a *type-II maximum likelihood* criterion (as in ARD and RVM [26]), let us consider hyper-priors for the τ_is and integrate them out. Assuming exponential hyper-priors $p(\tau_i|\gamma) = (\gamma/2)\exp\{-\gamma\,\tau_i/2\}$ (for $\tau_i \geq 0$, because these are variances) we obtain

$$p(\beta_i|\gamma) = \int_0^\infty p(\beta_i|\tau_i)p(\tau_i|\gamma)\,d\tau_i = \frac{\sqrt{\gamma}}{2} \exp\{-\sqrt{\gamma}\,|\beta_i|\}.$$

This shows that the Laplacian prior is equivalent to a 2-level hierachical model: zero-mean Gaussian priors with independent exponentially distributed variances. This equivalence has been previously exploited in robust *least absolute deviation* (LAD) regression [14].

The hierarchical decomposition of the Laplacian prior allows using the EM algorithm to implement the LASSO criterion in (12.1) by simply regarding $\boldsymbol{\tau} = [\tau_1, ..., \tau_k]$ as *hidden/missing data*. Let us define the following diagonal matrix: $\boldsymbol{\Upsilon}(\boldsymbol{\tau}) = \text{diag}(\tau_1^{-1}, ..., \tau_m^{-1})$. In the presence of $\boldsymbol{\Upsilon}(\boldsymbol{\tau})$, the complete log-posterior (with a flat prior for σ^2),

$$\log p(\boldsymbol{\beta}, \sigma^2 | \mathbf{y}, \boldsymbol{\tau}) \propto -n\sigma^2 \log \sigma^2 - \|\mathbf{y} - \mathbf{H}\boldsymbol{\beta}\|_2^2 - \sigma^2 \boldsymbol{\beta}^T \boldsymbol{\Upsilon}(\boldsymbol{\tau})\boldsymbol{\beta}, \quad (12.2)$$

would be easy to maximize with respect to $\boldsymbol{\beta}$ and σ^2. Since this complete log-posterior is liner with respect to $\boldsymbol{\Upsilon}(\boldsymbol{\tau})$, the E-step reduces to the computation of the conditional expectation of $\boldsymbol{\Upsilon}(\boldsymbol{\tau})$, given current (at iteration t) estimates $\widehat{\sigma^2}_{(t)}$ and $\widehat{\boldsymbol{\beta}}_{(t)}$. It can easily be shown that this leads to

$$\begin{aligned} \mathbf{V}_{(t)} &\equiv E[\boldsymbol{\Upsilon}(\boldsymbol{\tau}) | \mathbf{y}, \widehat{\sigma^2}_{(t)}, \widehat{\boldsymbol{\beta}}_{(t)}] \\ &= \gamma \, \text{diag}(|\widehat{\beta}_{1,(t)}|^{-1}, ..., |\widehat{\beta}_{k,(t)}|^{-1}). \end{aligned} \quad (12.3)$$

Finally, the M-step consists in updating the estimates of σ^2 and $\boldsymbol{\beta}$ by maximizing the complete log-posterior, with $\mathbf{V}_{(t)}$ replacing the missing $\boldsymbol{\Upsilon}(\boldsymbol{\tau})$. This leads to

$$\widehat{\sigma^2}_{(t+1)} = \frac{1}{n} \|\mathbf{y} - \mathbf{H}\widehat{\boldsymbol{\beta}}_{(t)}\|_2^2 \quad (12.4)$$

and

$$\widehat{\boldsymbol{\beta}}_{(t+1)} = (\widehat{\sigma^2}_{(t+1)} \mathbf{V}_{(t)} + \mathbf{H}^T \mathbf{H})^{-1} \mathbf{H}^T \mathbf{y}. \quad (12.5)$$

This EM algorithm is not the most efficient way to solve (12.1); faster special-purpose methods have been proposed in [20], [25]. Our main goal is to open the way to the adoption of different hyper-priors that do not corresponf to LASSO estimates.

12.4   The Jeffreys Hyper-Prior

One question remains: how to adjust γ, which is the main parameter controlling the degree of sparseness of the estimates? Our proposal is to remove γ from the model, by replacing the exponential hyper-prior by a non-informative Jeffreys hyper-prior

$$p(\tau_i) \propto \tau_i^{-1}. \quad (12.6)$$

This prior expresses ignorance with respect to scale (see [9], [1]) and, most importantly, it is parameter-free. Of course this is no longer equivalent to a Laplacian prior on $\boldsymbol{\beta}$, but to some other prior [9]. As will be shown experimentally, this prior strongly induces sparseness and yields state-of-the-art performance in regression applications. Computationally, this choice leads to a minor modification of the EM algorithm described above: matrix

$\mathbf{V}_{(t)}$ is now given by

$$\mathbf{V}_{(t)} = \mathrm{diag}(|\widehat{\beta}_{1,(t)}|^{-2}, ..., |\widehat{\beta}_{k,(t)}|^{-2}). \qquad (12.7)$$

(instead of Eq. (12.3)). Notice the absence of parameter γ.

Since several of the components of $\widehat{\beta}$ are expected to go to zero, it is not convenient to deal with $\mathbf{V}_{(t)}$ as defined in Eq. (12.7). However, defining a new diagonal matrix

$$\mathbf{U}_{(t)} = \mathrm{diag}(|\widehat{\beta}_{1,(t)}|, ..., |\widehat{\beta}_{k,(t)}|),$$

we can re-write Eq. (12.5) in the M-step as

$$\widehat{\beta}_{(t+1)} = \mathbf{U}_{(t)}(\widehat{\sigma^2}_{(t+1)}\mathbf{I} + \mathbf{U}_{(t)}\mathbf{H}^T\mathbf{H}\mathbf{U}_{(t)})^{-1}\mathbf{U}_{(t)}\mathbf{H}^T\mathbf{y}. \qquad (12.8)$$

This form of the algorithm avoids the inversion of the elements of $\widehat{\beta}_{(t)}$. Moreover, it is not necessary to invert the matrix, but simply to solve the corresponding linear system, whose dimension is only the number of non-zero elements in $\mathbf{U}_{(t)}$.

12.5   Experiments

Our first example illustrates the use of the proposed method for variable selection in standard linear regression. Consider a sequence of 20 true βs, having from 1 to 20 non-zero components (out of 20): from $[3, 0, 0, ..., 0]$ to $[3, 3, ..., 3]$. For each β, we obtain 100 random (50×20) design matrices, following the procedure in [25], and for each of these, we obtain data points with unit noise variance. Fig. 12.1 shows the mean number of estimated non-zero components, as a function of the true number. Our method exhibits a very good ability to find the correct number of nonzero components in β, in an adaptive manner.

Table 12.1. Relative (%) improvement in modelling error of several mehods.

Method	β_a	β_b
Proposed method	28%	74%
LASSO (CV)	13%	69%
LASSO (GCV)	30%	65%
Subset selection	13%	77%

We now consider two of the experimental setups used in [25]: $\beta_a = [3, 1.5, 0, 0, 2, 0, 0, 0]$, with $\sigma = 3$, and $\beta_b = [5, 0, 0, 0, 0, 0, 0, 0]$, with $\sigma = 2$. In both cases, $n = 20$, and the design matrices are generated as in [25]. In Table 12.1, we compare the relative modelling error ($ME = E[\|\mathbf{H}\widehat{\beta} - \mathbf{H}\beta\|^2]$) improvement (with respect to the least squares solution) of our method and of several methods studied in [25]. Our method

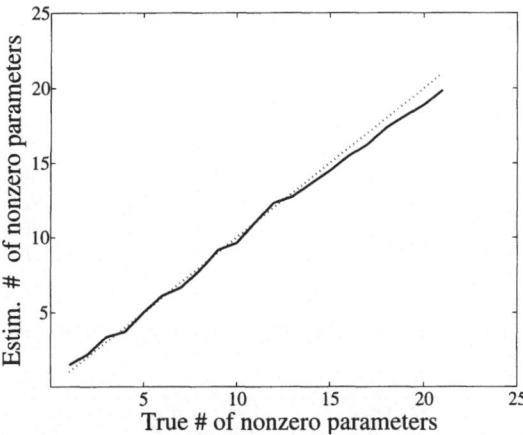

Figure 12.1. Mean number of nonzero components in $\widehat{\beta}$ versus the number of nonzero components in β (the dotted line is the identity).

performs comparably with the best method for each case (LASSO tuned by generalized cross-validation, for β_a, and subset selection, for β_b), although it involves no tuning or adjustment of parameters, and is computationally faster.

We now study the performance of our method in kernel regression, using Gaussian kernels, *i.e.*, $K(\mathbf{x}, \mathbf{x}_i) = \exp\{-\|\mathbf{x} - \mathbf{x}_i\|^2/(2h^2)\}$. We begin by considering the synthetic example studied in [3] and [26], where the true function is $y = \sin(x)/x$ (see Fig. 12.2). To compare our results to the RVM and the variational RVM (VRVM), we ran the algorithm on 25 generations of the noisy data. The results are summarized in Table 12.2 (which also includes the SVM results from [3]). Of course the results depend on the

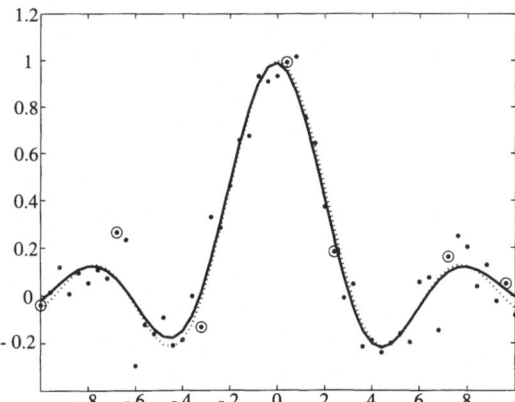

Figure 12.2. Kernel regression. Dotted line: true function $y = \sin(x)/x$. Dots: 50 noisy observations ($\sigma = 0.1$). Solid line: the estimated function. Circles: data points corresponding to the non-zero parameters.

Table 12.2. Mean root squared errors and mean number of kernels for the "sin(x)/x" function example.

Method	MSE	No. kernels
New method	0.0455	7.0
SVM	0.0519	28.0
RVM	0.0494	6.9
VRVM	0.0494	7.4

Table 12.3. Mean root squared errors and mean number of kernels for the "Boston housing" example.

Method	MSE	No. kernels
New method	9.98	45.2
SVM	10.29	235.2
RVM	10.17	41.1
VRVM	10.36	40.9

choice of kernel width h (as do the RVM and SVM results), which was adjusted by cross validation.

Finally, we have also applied our method to the well-known *Boston housing* data-set (20 random partitions of the full data-set into 481 training samples and 25 test samples); Table 12.3 shows the results, again versus SVM, RVM, and VRVM regression (as reported in [3]). In these tests, our method performs better than RVM, VRVM, and SVM regression, although it doesn't require any tuning.

12.6   Concluding remarks

We have introduced a new sparseness inducing prior for regression problems which is related to the Laplacian prior. Its main feature is the absence of any hyper-parameters to be adjusted or estimated. Experiments have shown state-of-the-art performance, although the method involves no tuning or adjusting of sparseness-controlling hyper-parameters.

It is possible .to apply the approach herein described to classification problems (*i.e.*, when the response variable is of categorical nature) using a generalized linear model [17]. In [8] we show how a very simple EM algorithm can be used to address the classification case, leading also to state-of-the-art performance.

One of the weak points of our approach (which is only problematic in kernel-based methods) is the need to solve a linear system in the M-step, whose computational requirements make it impractical to use with very large data sets. This issue is of current interest to researchers in kernel-based methods (*e.g.*, [30]), and we also intend to focus on it.

References

[1] J. Berger, *Statistical Decision Theory and Bayesian Analysis.* Springer-Verlag, 1980.

[2] C. Bishop, *Neural Networks for Pattern Recognition.* Oxford University Press, 1995.

[3] C. Bishop and M. Tipping, "Variational relevance vector machines," in *Proceedings of the 16th Conference in Uncertainty in Artificial Intelligence*, pp. 46–53, Morgan Kaufmann, 2000.

[4] S. Chen, D. Donoho, and M. Saunders, "Atomic decomposition by basis pursuit," *SIAM Journal of Scientific Computation*, vol. 20, no. 1, pp. 33–61, 1998.

[5] V. Cherkassky and F. Mulier, *Learning from Data: Concepts, Theory, and Methods.* John Wiley & Sons, New York, 1998.

[6] N. Cristianini and J. Shawe-Taylor, *Support Vector Machines and Other Kernel-Based Learning Methods.* Cambridge University Press, 2000.

[7] D. L. Donoho and I. M. Johnstone, "Ideal adaptation via wavelet shrinkage," *Biometrika*, vol. 81, pp. 425–455, 1994.

[8] M. Figueiredo, "Adaptive sparseness using Jeffreys prior", in T. Dietterich, S. Becker, and Z. Ghahramani (editors), *Advances in Neural Information Processing Systems 14*, MIT Press, Cambridge, MA, 2002.

[9] M. Figueiredo and R. Nowak, "Wavelet-based image estimation: an empirical Bayes approach using Jeffreys' noninformative prior," in *IEEE Transactions on Image Processing*, vol. 10, pp. 1322-1331, 2001.

[10] F. Girosi, "An equivalence between sparse approximation and support vector machines," *Neural Computation*, vol. 10, pp. 1445–1480, 1998.

[11] A. Hoerl and R. Kennard, "Ridge regression: Biased estimation for nonorthogonal problems," *Technometrics*, vol. 12, pp. 55–67, 1970.

[12] C. Holmes and D. Denison, "Bayesian wavelet analysis with a model complexity prior," in *Bayesian Statistics 6*, J. Bernardo, J. Berger, A. Dawid, and A. Smith (editors), Oxford University Press, 1999.

[13] G. Kimeldorf and G. Wahba, "A correspondence between Bayesian estimation of stochastic processes and smoothing by splines," *Annals of Math. Statistics*, vol. 41, pp. 495–502, 1990.

[14] K. Lange and J. Sinsheimer, "Normal/independent distributions and their applications in robust regression," *Journal of Computational and Graphical Statistics*, vol. 2, pp. 175–198, 1993.

[15] M. Lewicki and T. Sejnowski, "Learning overcomplete representations," *Neural Computation*, vol. 12, pp. 337–365, 2000.

[16] D. MacKay, "Bayesian non-linear modelling for the 1993 energy prediction competition," in *Maximum Entropy and Bayesian Methods*, G. Heidbreder (editor), pp. 221–234, Kluwer, 1996.

[17] P. McCullagh and J. Nelder, *Generalized Linear Models*. Chapman and Hall, London, U.K., 1989.

[18] R. Neal, *Bayesian Learning for Neural Networks*. Springer Verlag, New York, 1996.

[19] B. Olshausen and D. Field, "Emergence of simple-cell receptive field properties by learning a sparse code for natural images," *Nature*, vol. 381, pp. 607–609, 1996.

[20] M. Osborne, B. Presnell, and B. Turlach, "A new approach to variable selection in least squares problems," *IMA Journal of Numerical Analysis*, vol. 20, pp. 389–404, 2000.

[21] M. Osborne, B. Presnell, and B. Turlach, "Knot selection for regression splines via the LASSO", in *Dimension Reduction, Computational Complexity, and Information*, S. Weisberg (editor), pp. 44-49, Interface Foundation of North America, Fairfax Station, VA, 1998.

[22] T. Poggio and F. Girosi, "Networks for approximation and learning," *Proceedings of the IEEE*, vol. 78, pp. 1481–1497, 1990.

[23] B. Ripley, *Pattern Recognition and Neural Networks*. Cambridge University Press, 1996.

[24] M. Smith and R. Kohn, "Nonparametric regression via Bayesian variable selection," in *Journal of Econometrics*, vol. 75, pp. 317-344, 1996.

[25] R. Tibshirani, "Regression shrinkage and selection via the lasso," *Journal of the Royal Statistical Society (B)*, vol. 58, 1996.

[26] M. Tipping, "The relevance vector machine," in *Advances in Neural Information Processing Systems 12*, S. Solla, T. Leen, and K.-R. Müller (editors), pp. 652-658, MIT Press, 2000.

[27] V. Vapnik, *Statistical Learning Theory*. John Wiley & Sons, New York, 1998.

[28] C. Williams, "Prediction with Gaussian processes: from linear regression to linear prediction and beyond," in *Learning in Graphical Models*, MIT Press, Cambridge, MA, 1998.

[29] C. Williams and D. Barber, "Bayesian classification with Gaussian priors," *IEEE Trans. on Pattern Analysis and Machine Intelligence*, vol. 20, no. 12, pp. 1342–1351, 1998.

[30] C. Williams and M. Seeger, "Using the Nyström method to speedup kernel machines," in *Advances in Neural Information Processing Systems 13*, T. Leen, T. Dietterich, and V. Tresp (editors), MIT Press, Cambridge, MA, 2001.

[31] P. Williams, "Bayesian regularization and pruning using a Laplace prior," *Neural Computation*, vol. 7, pp. 117–143, 1995.

13

Multiscale Statistical Models

Eric D. Kolaczyk and Robert D. Nowak[1]

Summary

We present an overview of recent efforts developing a framework for a new class of statistical models based on the concept of multiscale likelihood factorizations. This framework blends elements of wavelets, recursive partitioning, and graphical models to derive a probabilistic analogue of an orthogonal wavelet decomposition. The casting of these results within a likelihood-based context allows for the extension of certain key properties of classical wavelet based estimators to a setting that includes, within a single unified perspective, models for continuous, count, and categorical datatypes.

13.1   Introduction

Much of the recent work in adaptive, nonlinear statistical methods employs a paradigm that begins with the traditional "signal plus noise" model and then seeks an alternative representation with respect to an orthonormal basis or redundant expansion, appropriate for the assumed underlying structure – Fourier and wavelet-based methods, of course, being canonical examples. However, in many nonparametric statistical problems the "signal plus noise" model can be less applicable or even inappropriate, yet a more general likelihood-based model may suggest itself in a natural manner. In cases like these, the concept of an orthogonal basis decomposition may be replaced by analogy with a factorization of the data likelihood. We have in mind, in particular, models for count and categorical data, such as Poisson or multinomial. Count data of this sort arises in a variety of contexts, such as high-energy astrophysics or medical imaging, while good examples of such categorical data might be binned observations in density estimation or the images found in landcover classification from remote sensing data.

[1]Eric D. Kolaczyk is Assistant Professor, Department of Mathematics and Statistics, Boston University. Robert D. Nowak is Assistant Professor, Electrical and Computer Engineering, George R. Brown School of Engineering Rice University.

13.1.1   Modeling Scale

Our focus here will be on settings in which a phenomena being studied has structure at multiple scales or granularities (e.g., of time, space, space-time, etc.). Examples are numerous and touch upon fields across many disciplines, including engineering, physics, geography, astronomy, and more. Hence *multiscale models* i.e., models that seek explicitly to represent complex phenomena as a compilation of simpler components indexed across various scales, by now make up a rather large literature.

Wavelet-based methods, of course, have had a decided impact on this area. From the perspective of the field of statistics and the area of nonparametric function estimation, much of this impact can be attributed to important fundamental characteristics such as the near-optimality of their risk properties (in a minimax sense), their adaptivity to various ranges of unknown degrees of smoothness, and the availability of simple and efficient algorithms for practical implementation. See [8], for example, and the discussions therein.

The observation model upon which much of this work is based is the standard Gaussian 'signal-plus-noise' model i.e., $Y_i = \theta_i + Z_i$, where the Z_i are independent Gaussian random variables with zero mean and constant variance and the θ_i typically are equi-spaced samples of a function f. If an orthonormal wavelet transform of the data is computed i.e., $\boldsymbol{W} = \mathcal{W}\boldsymbol{Y}$, the result is a vector $\boldsymbol{W}$ whose components too are i.i.d. Gaussian. Hence, from a likelihood based perspective we can write

$$\prod_{i=0}^{N-1} \Pr(Y_i|\theta_i) = \prod_{j,k} \Pr(W_{j,k}|\omega_{j,k}) \ , \tag{13.1}$$

where (j,k) refers to the standard scale-position indexing that accompanies the definition of orthonormal wavelet functions $\psi_{j,k}(t) = 2^{j/2}\psi(2^j t - k)$ with respect to a single function ψ, and $\boldsymbol{\omega} = \mathcal{W}\boldsymbol{\theta}$.

The key point to note here is that the joint (i.e., N dimensional) likelihood is *factorized* in both the time (left hand side) and time-scale (right hand side) domains into a product of N component likelihoods. And furthermore, each component relies on a single pairing of observation and parameter — Y_i with θ_i in the time domain and $W_{j,k}$ with $\omega_{j,k}$ in the multiscale domain. This deceptively simple fact has both analytical and computational implications. For example, it can be seen to motivate the standard idea of thresholding individual empirical wavelet coefficients $W_{j,k}$ in order to denoise the signal $\boldsymbol{Y}$ as a whole, which is essentially an $O(N)$ algorithm (e.g., [7], but see also [9] for extensions to certain types of correlated data). And much of the corresponding analysis of the statistical risk of such estimators boils down to understanding the aggregate behavior of the individual risks associated with such thresholding. Additionally, most Bayesian methods in this context consist, for similar reasons, of making a posterior inference on $\boldsymbol{\theta}$ implicitly through component-wise posterior inferences on the $\omega_{j,k}$ (e.g., [4, 5, 1]).

Now consider the case in which the same wavelet transform is applied to a vector $\boldsymbol{X}$ of independent Poisson observations i.e., $\boldsymbol{W} = \mathcal{W}\boldsymbol{X}$. With the change from Gaussian to Poisson observations, the orthogonality of $\mathcal{W}$ is no longer sufficient to ensure the statistical independence of the components of $\boldsymbol{W}$. Hence, a factorization of the form given in (13.1) does not hold. And while the effect of this point on thresholding methods might be simply to adjust the level of the thresholds used, its impact on the development of Bayesian methodologies is more substantial, as the full likelihood must be used in an explicit manner. On the other hand, there does exist a factorization of the Poisson data likelihood, also with respect to a multiscale analysis, that involves not the wavelet coefficients $\boldsymbol{W}$ but rather certain conditional probabilities localized in position and scale [10, 17].

13.1.2 Summary of Results

In this chapter we present an overview of recent work by the authors and colleagues on a likelihood-based framework for multiscale statistical models, whose structure revolves around the type of factorizations just mentioned. The nature of the results we present is three-fold:

1. MULTIRESOLUTION ANALYSIS (MRA):   A set of sufficient conditions for a certain multiscale factorization of the data likelihood to be possible. These conditions are analogous to, yet distinct from, those conditions underlying the classical MRA for wavelets – hence we label our own conditions an MRA for likelihoods.

2. EFFICIENT ALGORITHMS FOR MPLE: Computationally efficient, polynomial-time algorithms for calculating solutions to a variety of maximum penalized likelihood estimation (MPLE) problems, for likelihoods assumed to satisfy the conditions of our MRA.

3. RISK ANALYSIS: Minimax risk results demonstrating properties of near-optimality and adaptivity for these estimators, in recovering objects in a range of smoothness classes.

These three categories of results are described in more detail in Sections 13.2, 13.3, and 13.4, respectively, while Section 13.5 contains some additional remarks and discussion.

13.2   Multiresolution Analysis for Likelihoods

The notion of a *multiresolution analysis (MRA)* is fundamental to much of the work in the wavelet literature, and can be viewed as a set of sufficient

Table 13.1. Comparison of Wavelet and Likelihood MRA.

Wavelet MRA	Likelihood MRA
(A) Hierarchy of Nested Subspaces	(A*) Hierarchy of Recursive Partitions
(B) Orthonormal Basis within V_0	(B*) Independence within $\mathcal{P}_N$
(C) Scalability Between Subspaces	(C*) Reproducibility Between Subpartitions
(D) Translation within Subspaces	(D*) "Decoupling" of Parameters with Partitions (i.e. Cuts)

conditions for the existence of a classical orthonormal wavelet basis for $L^2(\mathbb{R})$. The left-hand column of Table 13.1 summarizes these conditions under four simple headings convenient for our purposes. A hierarchy of nested subspaces i.e., $\{\ldots \subset V_{j-1} \subset V_j \subset V_{j+1} \ldots\}$, establishes the notion of multiple resolutions (Condition A), which are then linked by a concept of scaling of their functional elements (Condition C), and equipped with orthonormal bases (Condition B). A natural requirement of closure within subspaces under translation (Condition D) rounds out the group. In this context the fundamental object is a function $f \in L^2(\mathbb{R})$ for which one desires an alternative representation with respect to an orthonormal basis of wavelets $\{\psi_{j,k}\}$ i.e.,

$$f(t) = \sum_{(j,k) \in \mathbf{Z}^2} \omega_{j,k}\, \psi_{j,k}(t) \ . \tag{13.2}$$

The conditions of the wavelet MRA guarantee the existence of this basis and provide defining formulas.

Now let the fundamental object be a multivariate probability density or mass function $p(\boldsymbol{X}|\boldsymbol{\theta})$ for a random vector $\boldsymbol{X} \equiv (X_0, \ldots, X_{N-1})$, parameterized by some vector $\boldsymbol{\theta} \in \Theta^N \subseteq \mathbb{R}^N$. The right-hand side of Table 13.1 gives four conditions sufficient for the factorization of $p(\cdot|\boldsymbol{\theta})$ in the form

$$p(\boldsymbol{X}\,|\,\boldsymbol{\theta}) = p(X_{I_0}\,|\,\theta_{I_0}) \prod_{I \in NT(\mathcal{P}^*)} p(X_{ch(I),l}|X_I, \omega_I) \ , \tag{13.3}$$

where $\mathcal{P}^*$ is a complete recursive partition [6] of the data-space (which we assume to be $I_0 \equiv [0, 1]$ for convenience), $NT(\mathcal{P}^*)$ are all non-terminal intervals in $\mathcal{P}^*$, $X_I \equiv \sum_{i:I_i \subseteq I} X_i$, for $I \in NT(\mathcal{P}^*)$, with $(ch(I), l)$ indicating the left-child sub-interval of I, and $\boldsymbol{\theta} \longmapsto (\theta_{I_0}, \boldsymbol{\omega})$ defines a multiscale re-parameterization.

The expression in (13.3) serves as a probabilistic analogue of that in (13.2), with the wavelet function/coefficient pairs $(\psi_{j,k}, \omega_{j,k})$ replaced by the random variable/parameter pairs $(X_{ch(I),l}|X_I , \omega_I)$. Condition A* is analogous to condition A in replacing the hierarchy of subspaces by a complete recursive partitioning of the data-space. In condition B*, the concept of statistical independence replaces that of orthogonality (i.e., which is "independence" in the context of a basis for a vector space). Similarly, a scale relationship between partitions in $\mathcal{P}^*$ is induced by assuming reproducibility in condition C* – that is, a common univariate marginal family form for the elements of X that is preserved under simple summation of the X_i's. Finally, condition D* is special to the context of likelihoods, being equivalent to a certain definition of sufficiency, and is what underlies the re-parameterization from θ to ω.

These conditions are developed in detail in [11], where they are used to characterize members of certain classes of natural exponential families allowing factorizations like that in (13.3). Three distributions of particular importance are the Gaussian, Poisson, and multinomial families, as canonical examples of models for continuous, count, and categorical measurements. (The multinomial arises when it is found that the independence condition B* is sufficient but not necessary.) We summarize these models here as follows, where Θ is a function space (e.g., Besov, BV, etc.) to be defined later.

(G) GAUSSIAN MODEL   Let $\theta \in \Theta$, and define $\theta_i = N \int_{I_i} \theta(t)dt$ to be the average of θ over $I_i \equiv [(i-1)/N, i/N)$. Sample the X_i independently as $X_i|\theta_i \sim \text{Gaussian}(\theta_i, \sigma^2)$, where σ^2 is assumed fixed and known. The multiscale components in (13.3) then take the form

$$X_{ch(I),l}|X_I, \omega_I \sim \text{Gaussian}\left(\frac{N_{ch(I),l}}{N_I}x_I - \omega_I, c_I\sigma^2\right) ,$$

with $N_I \equiv \#\{i : I_i \subseteq I\}$,

$$\omega_I = c_I\left(\frac{\theta_{ch(I),r}}{N_{ch(I),r}} - \frac{\theta_{ch(I),l}}{N_{ch(I),l}}\right) ,$$

for $c_I = N_{ch(I),l} N_{ch(I),r}/N_I$. The coarse scale component takes the form
$X_{I_0}|\theta_{I_0} \sim \text{Gaussian}\left(\theta_{I_0}, N_{I_0}\sigma^2\right)$.

(P) POISSON MODEL   Let $\theta \in \Theta$, where $\theta(t) \in [c, C], \forall t \in [0, 1]$, for $0 < c < C$. Define $\theta_i = N \int_{I_i} \theta(t)dt$ to be the average of θ over I_i. Sample the X_i independently as $X_i \sim \text{Poisson}(\theta_i)$. The multiscale components in (13.3) then take the form

$$X_{ch(I),l}|X_I, \omega_I \sim \text{Binomial}(X_I; \omega_I) ,$$

with $\omega_I = \theta_{ch(I),l}/\theta_I$, while the coarse scale component takes the form $X_{I_0}|\theta_{I_0} \sim \text{Poisson}(\theta_{I_0})$.

(M) MULTINOMIAL MODEL   Let $\theta \in \Theta$, where $\theta(t) \in [c, C], \forall t \in [0, 1]$, for $0 < c < C$, and $\int_0^1 \theta(t)dt = 1$. Define $\theta_i = \int_{I_i} \theta(t)dt$. Let the components X_i arise through (singular) multinomial sampling i.e., $\boldsymbol{X} \sim \text{Multinomial}(n; \boldsymbol{\theta})$, for some $n \sim N$.

The multiscale components in (13.3) then take the form

$$X_{ch(I),l} | X_I, \omega_I \sim \text{Binomial}(X_I; \omega_I) ,$$

with $\omega_I = \theta_{ch(I),l}/\theta_I$, as in the Poisson model, but the coarse scale component is now a point mass at n.

Model (G) is just the Gaussian "signal plus noise" model with average-sampling of the underlying function $\theta(\cdot)$, while model (P) is the Poisson analogue. Model (M) can be viewed as a discretized density estimation model, where a sample of size n from the density θ is implicit. The fact that the Poisson and multinomial models share the same multiscale components follows from their shared membership in the class of sum-symmetric power series distributions, all members of which allow factorizations like that in (13.3). Readers are referred to [11] for details.

13.3   Maximum Penalized Likelihood Estimation

The analogues between our multiscale likelihood factorizations and classical orthonormal wavelet basis decompositions are more than just structural, in that they also allow for the creation of a variety of estimators of $\boldsymbol{\theta}$ that are appealing from both practical and theoretical viewpoints. For example, consider estimators of the form

$$\hat{\boldsymbol{\theta}}(\boldsymbol{X}) \equiv \arg \min_{\boldsymbol{\theta}' \in \Gamma} \left\{ -\log \Pr(\boldsymbol{X}|\boldsymbol{\theta}') + 2\,\text{pen}(\boldsymbol{\theta}') \right\} , \qquad (13.4)$$

$$\text{pen}(\boldsymbol{\theta}') = \lambda \,\#(\boldsymbol{\theta}') = \lambda \,\#\{\omega_I(\boldsymbol{\theta}') \text{ non-trivial } \} , \qquad (13.5)$$

where, in particular, "non-trivial" means non-zero in the Gaussian case and different from $\rho_I \equiv N_{ch(I,l)}/N_I$ in the Poisson and multinomial cases.

In other words, in all three models there is a value of the multiscale parameter ω_I that is analogous to that of a wavelet coefficient being zero, and we penalize having too many of these parameters different from this value. In the context of wavelet-based estimation and the Gaussian noise model (i.e., recall equation (13.1)), the type of optimization described by (13.4) and (13.5) corresponds to the well-known "hard thresholding" when the space Γ is the collection of all possible values for the wavelet coefficients. Similarly, Donoho [6] shows how the use of Haar-like wavelets with so-called "hereditary" constraints on their inclusion/exclusion amounts to recursive partitioning equivalent to a simplified version of CART^{TM} [3]. Analogues of both of these estimators may be defined in our own setting here, through appropriate definition of the space Γ. In particular, it is possible to define hard thresholding estimators $\hat{\theta}_T$, recursive dyadic partitioning estimators $\hat{\theta}_{RDP}$, and general (non-dyadic) recursive partitioning estimators $\hat{\theta}_{RP}$. See [12] for details.

Of course, estimators of this type will be particularly practical if accompanied by efficient computational algorithms. Such is the case, as summarized below.

Proposition 4. *For models (G), (P), and (M),*

1. *$\hat{\theta}_T$ may be computed using an $O(N)$ thresholding procedure,*

2. *$\hat{\theta}_{RDP}$ may be computed using an $O(N)$ optimal pruning algorithm,*

3. *$\hat{\theta}_{RP}$ may be computed using an $O(N^3)$ message passing algorithm.*

In fact, these results hold more generally than for just these three estimators. A variety of alternative estimators may be defined using penalities other than the simple counting function in (13.5), such as those deriving from Bayesian or minimum description length (MDL) formalisms. As long as the data likelihood admits a factorization like that in (13.3) and the penalty function is additive in the indexing of the multiscale parameters ω_I, the corresponding optimization problem will involve an objective function of the form

$$\sum_{I \in NT(\mathcal{P}^*)} h\left(d_I(\boldsymbol{x}) \,;\, \mathcal{M}_I(\mathcal{P})\right) \;, \tag{13.6}$$

for some function $h(\cdot)$ deriving from the data likelihood and penalty and parameters $\mathcal{M}_I(\mathcal{P})$ acting essentially as indicator functions associated with I. When only a single $\mathcal{P}^*$ is considered, the thresholding and partitioning algorithms are $O(N)$. On the other hand, if the search is expanded to include the library of all possible $\mathcal{P}^*$, meaning an optimization over all subsets of all $(N-1)!$ complete recursive partitionings of the data space, it is still possible to accomplish this task in polynomial time i.e., $O(N^3)$. Examples and additional details may be found in [11]. Figure 13.1 shows an illustration of a variation on the estimator $\hat{\theta}_{RP}$, in which a Poisson time series is segmented into piecewise constant regions and the underlying intensity function is estimated using this segmentation and the resulting sub-interval means.

13.4   Risk Analysis

It is possible to characterize the performance of the three estimators defined in Section 13.3 and to demonstrate, in particular, properties of

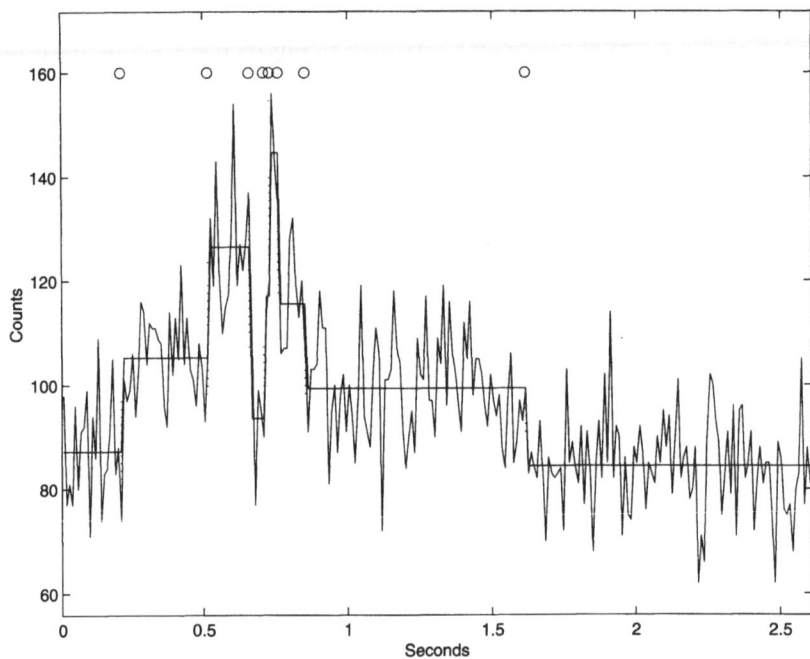

Figure 13.1. Segmentation (dotted vertical lines) and piecewise constant fit (solid horizontal lines) to a gamma-ray burst signal from high-energy astrophysics, using a variation on the estimator $\hat{\theta}_{RP}$ that can be derived from either Bayesian or MDL principles. Data are binned photon arrival times, gather by the BATSE instruments on board NASA's former CGRO satellite, modeled as a vector of Poisson counts. [Software available at http://math.bu.edu/people/kolaczyk/astro.html .]

near-optimality and adaptivity that hold uniformly for models (G), (P), and (M).

Let the loss associated with estimating θ by $\hat{\theta}$ be defined as the squared Hellinger distance between the corresponding densities i.e.,

$$L(\hat{\theta}, \theta) \equiv H^2(p_{\hat{\theta}}, p_{\theta}) = \int \left[\sqrt{p(x|\hat{\theta})} - \sqrt{p(x|\theta)} \right]^2 \nu(x) , \qquad (13.7)$$

where ν is the dominating measure and x denotes an arbitrary element from the range of the observation variable X. Taking expectations, let the corresponding risk be defined as $R(\hat{\theta}, \theta) \equiv (1/N)E[L(\hat{\theta}, \theta)]$. We then have the following.

Theorem 3. *Let* $\Theta = BV(C)$ *be a ball of functions of bounded variation, and assume the conditions of either model (G), (P), or (M). Let the constant* λ *in (13.5) be of the form* $\gamma \log_e(N)$, *for* $\gamma \geq 3/2$ *and* $N \geq 3$. *Then*

1. the risks of the estimators $\hat{\theta}_T$ and $\hat{\theta}_{RP}$ are bounded above by $O((\log N/N)^{2/3})$, while the risk of the estimator $\hat{\theta}_{RDP}$ is bounded above by $O((\log^2 N/N)^{2/3})$,

2. the minimax risk obeys a lower bound of $O(N^{-2/3})$, and hence

3. the estimators are within logarithmic factors of the optimal minimax risk.

Proof of these results may be found in [12], in which recent arguments of Li [14] and Li and Barron [15] for estimation by mixtures of densities are adapted to the present setting, to produce bounds in the spirit of say Birgé and Massart (e.g., see [2]). The rate of decay in these results is driven by our use of partitioning (i.e., through the complete partition $\mathcal{P}^*$ which underlies our likelihood MRA), which can be matched in one-to-one correspondence with unbalanced Haar wavelet bases. The space BV therefore is a natural one to consider, as are Besov spaces of certain smoothness.

Corollary 1. *Suppose* $\Theta = B_q^\xi(L_2[0,1])$ *is a Besov space with smoothness* $0 < \xi < 1$ *and* $1/q = \xi + 1/2$. *Then the conclusions of Theorem 3 hold with the exponent* $2/3$ *replaced by* $2\xi/(2\xi + 1)$.

Hence we demonstrate adaptivity, as the estimators herein are able to achieve near-optimal performance across a range of smoothness classes, without knowledge as to which particular class the function θ derives. These results are completely analgous to those obtained for wavelet-based estimators in the standard signal-plus-noise model (e.g., see [8]). In particular, they provide an illustration of how properties of classical wavelet-based estimators can be obtained in a single, unified framework that includes models for continuous, count, and categorical data types.

13.5   Discussion

The results described above are primarily an overview of [11] and [12]. These two papers, in turn, serve to provide some degree of explanation of and justification for the performance of other earlier work by the authors and colleagues with multiscale factorizations in specific methodological contexts, such as the analysis of Poisson time series [10] and images [17], Poisson linear inverse problems [16], and the spatial analysis of continuous and count data in geography [13].

At the core of all of these works is the concept of a multiscale likelihood factorization, which can be viewed as blending elements from the areas of wavelets, recursive partitioning, and graphical models. This particular framework in itself is quite general, which allows for a variety of extensions beyond the perspective laid out here. For example, extensions to irregular partitioning and spatio-temporal analysis have been studied in [13]; extensions to the use of piecewise linear surfaces are developed in [18]; and extensions to multivariate data models are, at least in principle,

References

[1] F. Abramovich, T. Sapatinas, and B. W. Silverman. Wavelet thresholding via a Bayesian approach. *Journal of the Royal Statistical Society, Series B*, 60:725–749, 1998.

[2] L. Birgé and P. Massart. From model selection to adaptive estimation. In D. Pollard, E. Torgersen, and G. Yang, editors, *Festschrift for Lucien Le Cam: Research Papers in Probability and Statistics*. Springer-Verlag, 1997.

[3] L. Breiman, J. Friedman, R. Olshen, and C.J. Stone. *Classification and Regression Trees*. Wadsworth, Belmont, CA, 1983.

[4] H. A. Chipman, E. D. Kolaczyk, and R. E. McCulloch. Adaptive Bayesian wavelet shrinkage. *Journal of the American Statistical Association*, 92:1413 – 1421, 1997.

[5] M. Clyde, G. Parmigiani, and B. Vidakovic. Multiple shrinkage and subset selection in wavelets. *Biometrika*, 85:391–402, 1998.

[6] D. L. Donoho. Cart and best-ortho-basis: A connection. *Annals of Statistics*, 25:1870 – 1911, 1997.

[7] D. L. Donoho and I. M. Johnstone. Ideal spatial adaptation via wavelet shrinkage. *Biometrika*, 81:425–455, 1994.

[8] D.L. Donoho, I.M. Johnstone, G. Kerkyachrian, and D. Picard. Wavelet shrinkage: Asymptopia? (with discussion). *Journal of the Royal Statistical Society, Series B*, 57:301 – 370, 1995.

[9] I.M. Johnstone and B.W. Silverman. Wavelet threshold estimators for data with correlated noise. *Journal of the Royal Statistical Society, Series B, Methodological*, 59:319–351, 1997.

[10] E. D. Kolaczyk. Bayesian multi-scale models for poisson processes. *Journal of the American Statistical Association*, 94:920–933, 1999.

[11] E. D. Kolaczyk and R. D. Nowak. A multiresolution analysis for likelihoods: Theory and methods. *Technical Report*, 2000.

[12] E. D. Kolaczyk and R. D. Nowak. Risk analysis for multiscale penalized maximum likelihood estimators. *Submitted*, 2001.

[13] E.D. Kolaczyk and H. Huang. Multiscale statistical models for hierarchical spatial aggregation. *Geographical Analysis*, 33:2:95 – 118, 2001.

[14] Q. J. Li. *Estimation of Mixture Models*. PhD thesis, Yale University, Department of Statistics, 1999.

[15] Q. J. Li and A. R. Barron. Mixture density estimation. In S.A. Solla, T.K. Leen, and K.R. Muller, editors, *Advances in Neural Information Processing Systems 12*. MIT Press, 2000.

[16] R. D. Nowak and E. D. Kolaczyk. A statistical multiscale framework for poisson inverse problems. *IEEE Transactions on Information Theory*, 46:5:1811–1825, 2000.

[17] K. E. Timmerman and R. D. Nowak. Multiscale modeling and estimation of Poisson processes with applications to photon-limited imaging. *IEEE Transactions on Information Theory*, 45:846–62, 1999.

[18] R. Willett and R. D. Nowak. Platelets: A multiscale approach for recovering edges and surfaces in photon limited medical imaging. *IEEE Transactions on Medical Imaging*, 2001. (submitted).

14

Wavelet Thresholding on Non-Equispaced Data

Maarten Jansen[1]

Summary

This chapter tackles the problem of non-equispaced data smoothing using the non-linear, non-parametric estimation technique of wavelet thresholding. Wavelets on irregular grids can be constructed, using the so-called lifting scheme. The lifting scheme is *grid-adaptive* in that it provides smooth basis functions on the given data set. We demonstrate however that this scheme pays little attention to the numerical condition of the transform. This bad condition originates from the fact that a straightforward application of a lifting scheme on irregular point sets mixes up scales within one level of transformation. Experiments illustrate that stability is a crucial issue in threshold or other shrinking algorithms. Unstable transforms blow up the bias in the smoothing curve. We propose a *scale-adaptive* scheme that avoids scale-mixing, leading to a smooth and close fit.

14.1   Introduction

A classical ('first generation') wavelet transform assumes the input to be a regularly sampled signal. In many applications however, data are not available on a regular grid, but rather as non-equidistant samples. Used as an input to one of the standard wavelet shrinkers, irregular point set leads to a wiggly output, since the wavelet algorithm seeks for outputs which are smooth on equidistant grids. Therefore, if the input at hand is not equidistant, the irregularity of the grid is reflected in the output curve.

---

[1]Maarten Jansen is postdoctoral research fellow of the Fund for Scientific Research Flanders (Belgium) (FWO). He is currently with Department of Mathematics and Computer Science, T.U. Eindhoven (The Netherlands).

Nearly all existing wavelet based regression of non-equispaced data combines a traditional equispaced algorithm for fitting with a "translation" of the input into an equispaced problem [9, 11, 7, 4, 1, 13, 12, 3, 2]. In general, preprocessing the data changes their statistics. White noise, for instance, becomes coloured, or stationarity may be lost. Also, working in two phases (pre- or post-processing, before or after the actual smoothing) may reduce the overall potential: a good result may become suboptimal after post-processing.

Therefore, we opt for an integrated approach, in which the basis functions are adapted to the irregular point set. One approach to doing this involves "squeezing" techniques, starting from (pieces of) standard wavelets [8]. We take a different way, based on the so called lifting scheme [15]. This scheme constructs a wavelet transform on a given data grid in consecutive steps, starting from a trivial transform. Every step adds new, smoothness properties. This gradual construction motivates the name *lifting*. Apart from a few publications that we know of, the use of lifting [6, 10, 16] in smoothing is quite new. The next section reviews the basic concepts of lifting and thresholding. Next we explore and analyse the stability problems following from a straightforward implementation of the lifting scheme on an arbitrary grid. We propose a relaxed scheme that overcomes this instability, leading to a smooth and close fit. Simulations and real data examples illustrate that the modified scheme even adapts automatically to the case of simultaneous observations.

14.2   Lifting and thresholding

A classical wavelet decomposition proceeds in consecutive steps — or levels — all corresponding to a particular scale. The action within one level can always be implemented using lifting [5]. The first stage in a lifting scheme is splitting the data into an even and odd sequence. This corresponds to the decimation in the classical implementation. Next in the lifting scheme is an alternating sequence of so called dual and primal lifting steps, as in Figure 14.1. This chapter concentrates on decompositions with one dual and one primal lifting step. The first, dual, step can be interpreted as a *prediction* of the odd points by neighbouring evens. The difference between prediction and true value is detail information, it is the wavelet coefficient. In the Haar transform, this prediction is simply the value of the left even neighbour. Other popular predictions are linear or cubic interpolation in the evens.

Whereas the dual step computes the detail *coefficients*, the primal lifting or *update* changes the values of the low-pass coefficients. It thereby actually creates the wavelet *basis functions*. This mechanism is visualised for the Haar case in [10, Section 2.6.1]. It can also be understood by running

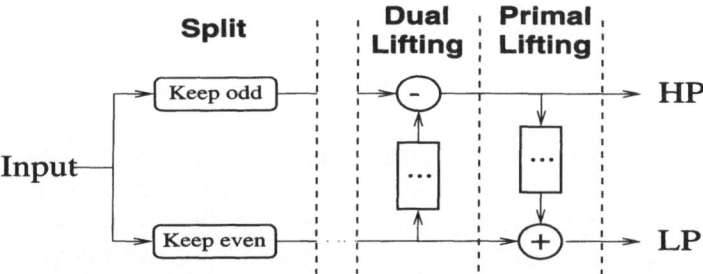

Figure 14.1. A lifting implementation of one level in a wavelet transform. Lifting operations come in three types: splitting, dual lifting and primal lifting. Just as in a classical wavelet transform, the output consists of detail, high pass (HP), or wavelet coefficients and low pass coefficients (LP), which are further processed in the following steps.

the inverse transform. The inverse transform starting from a zero input, except for a single coefficient, converges towards the wavelet basis function associated with that coefficient. This is generally true (not only for lifted transforms) and is known as the subdivision scheme [16]. It is the classical trick to make plots of wavelet functions. Applied to lifting, we see that the first step of the inverse transform is undoing the last update (primal lifting). This update filter shows a response to the only non-zero detail (wavelet) coefficient in the HP-branch. Next, this response is being subtracted from the zero sequence in the LP-branch. Suppose this filter response has two non-zeros $A_{j,k}$ and $B_{j,k}$. We then obtain the same basis function if we omit the last update step, and run the inverse transform with a single non-zero in the HP-branch *plus* a sequence with two non-zeros, $-A_{j,k}$ and $-B_{j,k}$, in the LP-branch. As a consequence, the wavelet function $\psi_{j,k}$ at scale j and location k equals

$$\psi_{j,k} = \psi_{j,k}^{[0]} - B_{j,k}\varphi_{j,k} - A_{j,k}\varphi_{j,k+1},$$

where $\varphi_{j,k}$ is the scaling function at the same scale and location, and $\psi_{j,k}^{[0]}$ is the wavelet basis function without update step. The same trick can be used to see that this wavelet function is nothing but the odd scaling function $\varphi_{j+1,2k+1}$ at the previous, fine scale $j+1$. Indeed, a single non-zero in the HP-branch would proceed unchanged and with no effect to the LP-branch in the next subdivision step. All together, we have

$$\psi_{j,k} = \varphi_{j+1,2k+1} - B_{j,k}\varphi_{j,k} - A_{j,k}\varphi_{j,k+1}. \tag{14.1}$$

The two degrees of freedom $A_{j,k}$ and $B_{j,k}$ can be used to make the wavelet have a zero integral and one more vanishing moment. Both dual and primal lifting steps are by no means limited to equidistant point sets.

A classical wavelet transform has strong decorrelating properties. It uses the correlation between neighbouring samples to obtain a sparse representation of the noise free signal. The main part of the coefficients is close to zero and the essential information is captured by a limited number of large, important coefficients. Replacing the small coefficients with an absolute value below a certain threshold with zero reduces the noise without affecting the noise free signal too much. A central issue in this kind of smoothing procedures is how to find a suitable value for the smoothing parameter, in this case the threshold λ. This article opts for a minimum mean square error (MSE) approach. The expected MSE (also known as *risk*) is a sum of bias and variance. The variance stands for the noise: it decreases when the threshold grows. The bias on the other hand increases when the threshold grows. The minimum MSE threshold is the best trade-off between variance and bias in ℓ_2-norm sense.

Figure 14.2 has an experiment with the well known test function 'heavisine':

$$f(x) = 4\sin(4\pi x) - \text{sign}(x - 0.3) - \text{sign}(0.72 - x)$$

sampled on 2048 points. Those points were drawn from a uniform density on $[0, 1]$. Such a uniform selection does not have huge gaps in absolute terms, but observations may be arbitrarily close to each other. This implies that extra small scale phenomena are observed within a relatively wide "gap". To this signal, white and stationary noise was added with a standard deviation of $\sigma = 0.3$. Figure 14.2(middle) shows the output from thresholding the 5 finest resolution level of a classical wavelet transform. The wavelet transform here was constructed by lifting with cubic prediction followed by 2-taps-update, but no grid structure was taken into account. A much smoother result follows if the transform is based on the grid. Figure 14.2(bottom) however shows that the reconstruction from these second generation wavelets contains strange artifacts, or — in statistical terminology — a strange bias.

14.3   The problem: instability

A classical wavelet transform guarantees a norm semi-equivalence between the input and the wavelet coefficients: if $\boldsymbol{w}$ is the wavelet transform of $\boldsymbol{y}$, then the ℓ_2 norms of these vectors satisfy:

$$c \cdot \|\boldsymbol{w}\| \leq \|\boldsymbol{y}\| \leq C \cdot \|\boldsymbol{w}\|,$$

with $0 < c, C < \infty$ independent of the vector length. This relates to the concept of Riesz bases. Loosely speaking, a Riesz basis, also known as stable basis, is a basis in which the basis vectors or functions cannot be arbitrarily close to each other. This notion becomes important in vector spaces with infinite dimension, or, as in our case, when dealing with situations where the dimension is finite but arbitrarily large. The constants c and C are closely related to the condition number of the wavelet transform matrix.

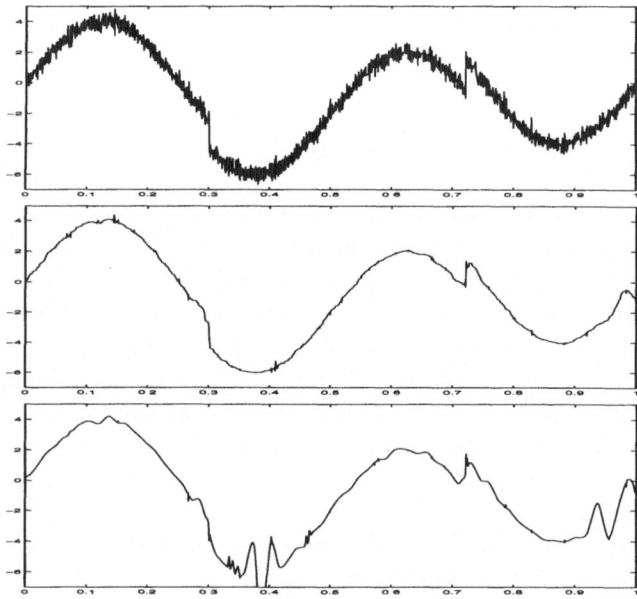

Figure 14.2. (top) A noisy version of the 'heavisine' test function. This is a piece-wise continuous signal, a typical member of the class of signals for which wavelets are the optimal approach. (middle) Classical wavelets cannot take the grid structure into account. Remapping the estimation to this grid makes its irregularity show up in the result. (bottom) Second generation wavelets, based on lifting, are smooth on the actual grid but lead to a tremendously biased reconstruction.

The extension to non-equispaced data through lifting gives no guarantee for the preservation of this comfortable Riesz basis background. High condition numbers mean that a small modification of wavelet coefficient values may result in an unpredictable effect on the output. In the case of thresholding, this means that a small coefficient may carry substantial signal information. Not only do some individual coefficients have a wide impact, the interaction between the coefficients may be unpredictable, due to the fact that the transform is far from orthogonal. This bad condition can only be detected with a global analysis: basis functions my be pairwise far from each other and yet all together, they constitute an arbitrarily oblique basis.

14.4 Stable transforms with interpolating prediction

Since the instability is so hard to detect and localise, we examine which steps in the transform initiate the problems. The lifting scheme as such does not pay any particular attention to stability. If the update coefficients in (14.1) are large, say if,

$$B_{j,k} \gg \frac{\|\psi_{j,k}^{[0]}\|}{\|\varphi_{j,k}\|},$$

the lifted wavelet $\psi_{j,k}$ nearly falls within the vector space spanned by its neighbouring scaling functions at the same scale. This creates a detail space which is far from orthogonal to the coarse scaling space. When the scaling functions are further decomposed into a wavelet basis at coarser scales, the immediate correlation between basis functions becomes hidden.

The Haar transform remains orthogonal, so in that case, the update step (with a single non-zero element in its filter) has definitely a stabilising effect: without update, the transform would not be orthogonal. (The Haar transform on irregular data is sometimes called — rather confusingly — *Unbalanced Haar*)

An update step with two non-zeros aiming at two vanishing moments (zero integral and first moment) satisfies:

$$A_{j,k} = \frac{M_{j+1,2k+1}}{M_{j,k}} \frac{\overline{x}_{j,k+1} - \overline{x}_{j+1,2k+1}}{\overline{x}_{j,k+1} - \overline{x}_{j,k}}$$

$$B_{j,k} = \frac{M_{j+1,2k+1}}{M_{j,k+1}} \frac{\overline{x}_{j+1,2k+1} - \overline{x}_{j,k}}{\overline{x}_{j,k+1} - \overline{x}_{j,k}}.$$

In these expressions, $M_{j,k}$ stands for the scaling function integrals:

$$M_{j,k} = \int_{-\infty}^{\infty} \varphi_{j,k}(x)dx,$$

and $\overline{x}_{j,k}$ stands for the first normalised moment:

$$\overline{x}_{j,k} = \frac{\int_{-\infty}^{\infty} x\varphi_{j,k}(x)dx}{\int_{-\infty}^{\infty} \varphi_{j,k}(x)dx}.$$

In the case of linear interpolating prediction, the scaling basis functions are piecewise linear (first order splines) as depicted in Figure 14.3. The classical update step aims at vanishing moments only. Unlike the Haar update, this update step is not a full orthogonalisation. Nevertheless, it does have a stabilising effect:

Theorem 4. *In a lifting scheme with linear prediction and update with two non-zero elements, both used for vanishing moments, the unlifted detail basis function (i.e. the odd scaling function at fine scale) is closer to the space spanned by the coarse scale scaling functions than the lifted detail basis function.*

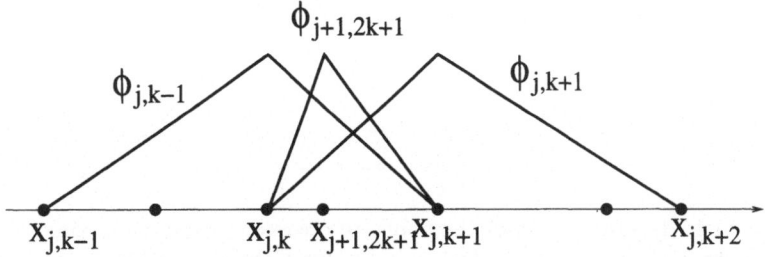

Figure 14.3. Scaling basis functions for linear prediction.

Proof. The projection of $\varphi_{j+1,2k+1}$ onto $\text{Span}\{\varphi_{j,k}, \varphi_{j,k+1}\}$ can be written as:

$$\varphi_{j+1,2k+1} = \varphi^{\perp}_{j+1,2k+1} + \varphi^{//}_{j+1,2k+1}$$
$$= \varphi^{\perp}_{j+1,2k+1} + b_{j,k}\varphi_{j,k} + a_{j,k}\varphi_{j+1,k}$$

The wavelet function can then be decomposed into:

$$\psi_{j,k} = \varphi_{j+1,2k+1} - B_{j,k}\varphi_{j,k} - A_{j,k}\varphi_{j,k+1}$$
$$= \varphi^{\perp}_{j+1,2k+1} + (b_{j,k} - B_{j,k})\varphi_{j,k} + (a_{j,k} - A_{j,k})\varphi_{j+1,k}$$
$$=: \varphi^{\perp}_{j+1,2k+1} + \psi^{//}_{j,k}.$$

So, both have the same orthogonal component. If $\|\psi^{//}_{j,k}\| \leq \|\varphi^{//}_{j+1,2k+1}\|$, then it follows that the angle between $\psi_{j,k}$ and $\text{Span}\{\varphi_{j,k}, \varphi_{j,k+1}\}$ is greater than that between $\varphi_{j+1,2k+1}$ and $\text{Span}\{\varphi_{j,k}, \varphi_{j,k+1}\}$. For normalised scaling functions, we have:

$$\|\varphi^{//}_{j+1,2k+1}\| = a_{j,k}^2 + b_{j,k}^2 + 2a_{j,k}b_{j,k}\cos\alpha_{j,k},$$

where α is the angle between $\varphi_{j,k}$ and $\varphi_{j,k+1}$. We now verify that all four $a_{j,k}, b_{j,k}, A_{j,k}, B_{j,k}$ are positive and that the latter two are bounded by:

$$A_{j,k} \leq 2a_{j,k} \text{ and } B_{j,k} \leq 2b_{j,k}. \tag{14.2}$$

To check this, and without loss of generality, we put: $x_{j,k} = 0$, $x_{j,k+1} = 1$, $x_{j+1,2k+1} = z$, $x_{j,2k} = 1+\Delta$ and $x_{j,k-1} = -\Gamma$. Thanks to the closed form of the scaling functions, all four quantities $a_{j,k}, b_{j,k}, A_{j,k}, B_{j,k}$ can then be expressed as functions of z, Δ, Γ and the inequalities (14.2) follow for all real values of these independent variables. These inequalities suffice to prove the requested norm inequality. $\square$

This result is a relative one in the sense that it compares the stability of the basis after primal lifting with that before primal lifting. This basis may be unstable in the first place, if two data points are arbitrarily close to each other. In that case however, even the orthogonal Haar has difficulties: the non-updated Haar basis can be arbitrarily unstable and in the limit, the

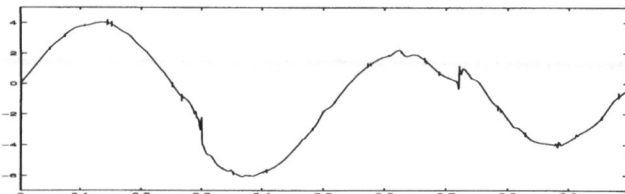

Figure 14.4. Reorganisation of the splitting operation prevents mixing of scale. This leads to more stable transforms and reduces the output bias.

updated basis function would be zero, since that is the only function with zero integral on an interval with three data points of which two coincide.

The proof heavily depends on the closed form of the basis functions. Higher order prediction schemes do not lead to closed forms (the basis functions corresponding to cubic interpolation are definitely not cubic splines), and hence proving results for those schemes will be hard.

For the classical higher order lifting, there is little to prove anyway, since these schemes turn out to be arbitrarily bad with respect to numerical condition. Bad condition is the combined effect of splitting, prediction and update. Polynomial prediction is nothing but the Deslauriers-Dubuc subdivision scheme. This scheme as such is not unstable, but it may lead to large update coefficients. Small update coefficients are necessary for good condition. The key to bounded update coefficients is the values of the normalised moments of the scaling functions. Linear prediction leads to positive basis functions, and the normalised moments are then the 'mean values' of these functions. From Figure 14.3, it follows that those 'mean values' appear in a good order, which makes the update coefficients to be bounded.

Higher order prediction schemes however, lead to negative side lobes, which can be heavy if a large gap without data is followed or preceded by a densely sampled region. The even-odd-splitting disregards this mixture of scales. To avoid this scale mixture, an alternative splitting scheme takes out all odd points that are too far away from immediate neighbours [16]. Prediction of these points is deferred to coarser scales. The effect of this reorganisation is tremendous, as Figure 14.4 illustrates.

14.5   Average interpolating prediction

So far, this chapter has concentrated on lifting schemes with one, interpolating, prediction step. A slightly more complicated scheme interprets the input not as point samples, but rather as short time averages. The odd indexed averages are predicted by an *average interpolating polynomial*. This is constructed by first taking the mean of every even-odd input pair. These mean values are obviously also the averages on the joint even-odd-intervals. Next, we look for polynomials that average out at those means on a fixed

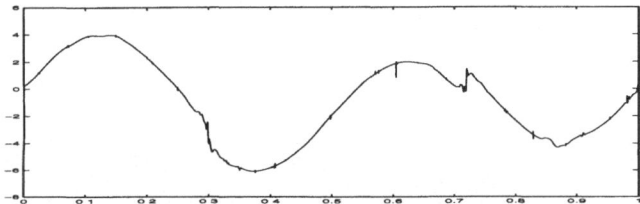

Figure 14.5. Average interpolating prediction is far less sensitive to instabilities due to scale mixing.

number of consequent intervals. The average of such a polynomial on its middle, odd, interval is then used as prediction for the input value on that interval. This operation can be decomposed into a Haar transform followed by a dual lifting step. One more update step is then necessary to create more primal vanishing moments (Haar giving us the first).

Experiments as in Figure 14.5 show that this construction is more robust to irregular grids than interpolating prediction. On "very" irregular grids (i.e. when samples are far from uniformly distributed on the observation interval), this approach however equally fails. Reorganising this transform to obtain more stable smoothing is subject of current research. One possible method is to use some of the degrees of freedom in the last update step for a local pseudo-orthogonalisation [6]. This obviously applies in the context of polynomial prediction as well. We could also consider alternative splitting schemes, where no prediction is performed on a joint even-odd interval if this interval is considerably wider than its neighbouring joint intervals (e.g. twice as wide).

14.6   Real data example

We end this discussion with a real data set, in which some measurements occurred simultaneously [14]. The classical second generation transform crashes when three or more abscis points coincide (the return values were ∞). Thanks to the separate handling of phenomena at different scales, no problem arises in the improved splitting scheme, as shown in the output in Figure 14.6.

14.7   Conclusions

This chapter has analysed the stability of multiscale data smoothing on irregular point sets. The lifting scheme itself provides a *grid-adaptive* smoothing scheme, but it tends to mix up phenomena at different scales, which leads to unstable — and hence biased — curve fitting. This

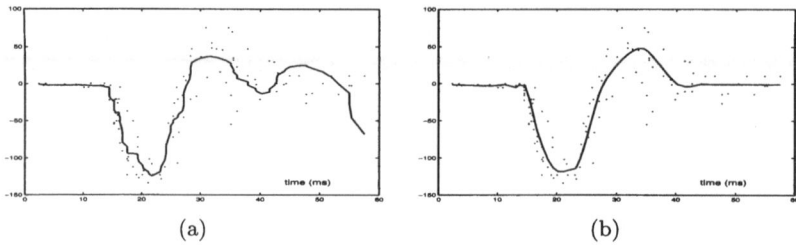

(a)   (b)

Figure 14.6. Fitting the motorcyclists' acceleration data. (a) Using a classical first generation wavelet transform (cubic interpolating prediction, 2-taps update, no grid information) leads to non-smooth fitting. (b) The result from the proposed procedure is smooth and relatively unbiased.

chapter proposes a modified, *scale*-adaptive approach, which even deals automatically with simultaneous observations.

References

[1] A. Antoniadis, G. Grégoire, and W McKeague. Wavelet methods for curve estimation. *J. Amer. Statist. Assoc.*, 89:1340–1353, 1994.

[2] A. Antoniadis, G. Grégoire, and P. Vial. Random design wavelet curve smoothing. *Statistics and Probability Letters*, 35:225–232, 1997.

[3] A. Antoniadis and D.-T. Pham. Wavelet regression for random or irregular design. *Computational Statistics and data analysis*, 28(4):333–369, 1998.

[4] T. Cai and L.D. Brown. Wavelet shrinkage for nonequispaced samples. *Annals of Statistics*, 26(5):1783–1799, 1998.

[5] I. Daubechies and W. Sweldens. Factoring wavelet transforms into lifting steps. *J. Fourier Anal. Appl.*, 4(3):245–267, 1998.

[6] V. Delouille, J. Simoens, and R. von Sachs. Smooth design-adapted wavelets for nonparametric stochastic regression. Technical report, Université Catholique de Louvain, 2001.

[7] B. Delyon and A. Juditsky. On the computation of wavelet coefficients. *J. of Approx. Theory*, 88:47–79, 1997.

[8] G. Donovan, J. Geronimo, and D. Hardin. Squeezable orthogonal bases: Accuracy and smoothness. Technical report, Department of Mathematics, Vanderbilt University, 2001. preprint.

[9] P. Hall and B. A. Turlach. Interpolation methods for nonlinear wavelet regression with irregularly spaced design. *Annals of Statistics*, 25(5):1912 – 1925, 1997.

[10] M. Jansen. *Noise reduction by wavelet thresholding*, volume 161 of *Lecture Notes in Statistics*. Springer, 2001.

[11] A. Kovac and B. W. Silverman. Extending the scope of wavelet regression methods by coefficient-dependent thresholding. *J. Amer. Statist. Assoc.*, 95:172–183, 2000.

[12] M. Pensky and B. Vidakovic. On non-equally spaced wavelet regression. Preprint, Duke University, Durham, NC, 1998.

[13] S. Sardy, D.B. Percival, A.G. Bruce, H-Y. Gao, and W. Stuetzle. Wavelet de-noising for unequally spaced data. *Statistics and Computing*, 9:65–75, 1999.

[14] B. Silverman. Some aspects of the spline smoothing approach to nonparametric regression curve fitting. *Journal of the Royal Statistical Society, Series B*, 47:1–52, 1985.

[15] W. Sweldens and P. Schröder. Building your own wavelets at home. In *Wavelets in Computer Graphics*, ACM SIGGRAPH Course Notes, pages 15–87. ACM, 1996.

[16] E. Vanraes, M. Jansen, and A. Bultheel. Stabilized lifting steps in noise reduction for non-equispaced samples. In M. A. Unser, A. Aldroubi, and Laine A. F., editors, *Wavelet Applications in Signal and Image Processing IX*, volume 4478 of *SPIE Proceedings*, pages 105–116, July 2001.

15

Multi-Resolution Properties of Semi-Parametric Volatility Models

Enrico Capobianco[1]

Summary

Wavelet approximation and decomposition techniques are applied to non-stationary high frequency financial time series by means of global and local optimization algorithms implemented through function dictionaries. One of the objectives is to verify what possible role wavelets might have in the domain of finance, where signal processing methods may be built so to detect the latent structures characterizing volatility processes. Here the observed data have features which result difficult to handle and whose extraction by standard volatility models may not be possible. We measure the performance of computational algorithms, test their approximation power and control their effectiveness with respect to the multi-resolution pursuit.

15.1   Introduction

This work is about latent variable systems endowed with complex dynamics, non-gaussian and non-stationary behavior. One of the most recent directions of research in various disciplines is selecting relevant information from sparsely represented signals [11, 18, 21]. Sparse signals can be represented by a small number of expansion coefficients.

Given a stochastic process, the decomposition of its observed structures in statistically independent or least dependent components may be of relevance in some applications; wavelets can play an important role, since they yield [17, 1] de-correlating and stationarizing effects on the expansion coefficient sequences.

---

[1]Enrico Capobianco is with CWI, Kruislaan 413, 1098 SJ, Amsterdam (The Netherlands).

With regard to financial time series analysis, in previous work [5, 6] some wavelet-based methodologies have been proposed; in particular, algorithms like the *Matching Pursuit* (MP) [19] have been seen to effectively detect features in high frequency financial time series.

Here we show that *Independent Component Analysis* (ICA) [7, 9] might be adopted in combination with the MP algorithm so to artificially learn the structure of a class of volatility processes by using a bank of sources from the signals decomposed according to *ad hoc* transforms and through dictionaries of *wavelet* and *cosine packets* (WP and CP, respectively). The approach here followed allows for a separation of sources to occur in the expansion coefficients domain so that the signal is reconstructed from resolution levels selected as least dependent ones.

In our applications we refer to the Nikkei stock return index trading activity in 1990 (among many other years available and with similar behavior), with observations collected every minute (1m). The sample has 35,463 observations, where intra- daily trading prices cover the working week, and exclude holidays and weekends. With the 1m stock returns, computed as $r_t = ln(p_t/p_{t-1}) \times 100$, we also form a temporally aggregated time series of correspondent five-minute (5m) data, which consists of 7092 observations.

We show in Figure 15.1 the plots for the autocorrelation function (ACF) of the 1m/5m return values, together with their absolute and squared transforms[2]. The values of the ACFs are persistently significant, i.e. even when observed at distant lags. They also suggest the possible presence of long memory for the only initially fast exponential decay of the functions, in agreement with an hyperbolic behavior. In the squared returns there is evidence of possible periodic behavior, due to the presence of recurrent spikes, which on the other hand might be due to non-stationarity, thus resulting in spurious components.

15.2   Review of Wavelets

Given a so-called *scaling function* (or *father wavelet*) ϕ, such that its dilates and translates constitute orthonormal bases for all the V_j sub-spaces obtained as scaled versions of the sub-space V_0 to which ϕ belongs, and given a so-called *mother wavelet* ψ and its derived terms indicated with ψ_{jk}, where j is the dilation or level index and k is the translation or shift index $\psi_{jk}(x) = 2^{\frac{j}{2}}\psi(2^j x - k)$ obtained as differences among approximations computed at successively coarser resolution levels, we can form a *Multiresolution Analysis* (MRA) in many functional spaces which are important for applications. Details and technicalities can be found mainly in [10, 20, 12, 13, 14, 16].

---

[2]The experiments were conducted with the *wavelets module* [4], written in *S-Plus*.

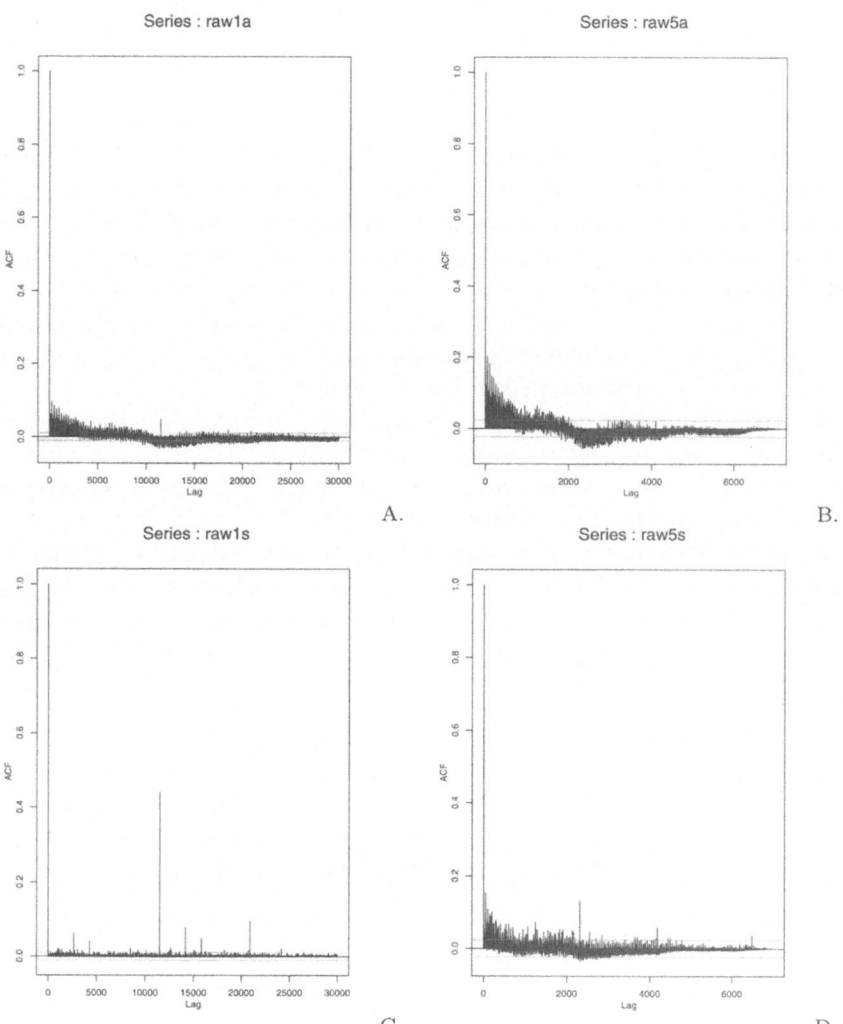

Figure 15.1. Absolute (indexed by *a*) and squared (indexed by *s*) raw 1m and 5m returns.

With a *Discrete Wavelet Transform* (DWT), we map $f \to w$ from the signal domain to the wavelet coefficient domain, or in other words we apply the transformation $w = Wf$.

A sequence of smoothed signals and of details, giving information at finer and finer resolution levels, may thus be used in a signal expansion:

$$f(x) = \sum_k c_{j0,k}\phi_{j0,k}(x) + \sum_{j>j0}\sum_k d_{j,k}\psi_{j,k}(x) \qquad (15.1)$$

where $\phi_{j0,k}$ is a scaling function with the corresponding coarse scale coefficients $c_{j0,k}$ and $d_{j,k}$ are the detail (fine scale) coefficients, i.e. $c_{j,k} = \int f(x)\phi_{jk}(x)dx$ and $d_{jk} = \int f(x)\psi_{jk}(x)dx$.

With wavelet packets, because of the presence of an oscillation index b related to a periodic behaviour in the series and because of a richer combination of wavelet functions, we get a better domain of wavelets from which to select a basis to represent the signal. With the *Cosine Packet Transform* we use instead cosine functions localized in time, which form smooth basis functions. An advantage over the classic *Discrete Cosine Transform* is that the latter defines an orthogonal transformation and thus maps a signal from the time to the frequency domain, but it is not localized in time and thus is not able to adapt well to non-stationary signals.

In particular, when we look at long financial time series of stock returns, we usually see patterns related to market activity which reflect the well-known *clusters of volatility*, similar in their shape to chirp-like signals, and spikes or jumps, similar to Diracs.

In order to design optimal algorithms, adaptive signal approximation techniques and sparse representations are usually required. The MP algorithm has been widely and successfully implemented in studies and requires that a signal is decomposed as a sum of atomic waveforms. For WP and CP tables, the signal are respectively represented as:

$$f(t) = \sum_{jok} w_{j,o,k} W_{j,o,k}(t) + \mathrm{res}_n(t)$$

and

$$f(t) = \sum_{jok} c_{j,o,k} C_{j,o,k}(t) + \mathrm{res}_n(t).$$

They offer advantages in terms of the flexibility in the type of approximating kernel adopted and with regard to the type of function used, i.e. localized cosine functions and variably oscillating wavelets. In summary, the MP algorithm approximates a function with a sum of n elements, called atoms or atomic waveforms, which are indicated with H_{γ_i} and depend from a dictionary Γ; these functions should ideally adapt to the characteristics of the signal at hand. The MP decomposition exists in orthogonal or redundant version and refers to a greedy algorithm which at successive steps decomposes the residual term left from a projection of the signal onto the elements of a selected dictionary, in the direction of that one allowing for the best fit.

At each time step the decomposition in (2) is computed, together with the coefficients h_i which represent the projections, and the residual component, which will be then re-examined and iteratively re-decomposed according to:

$$f(t) = \sum_{i=1}^{n} h_i H_{\gamma_i}(t) + \mathrm{res}_n(t) \tag{15.2}$$

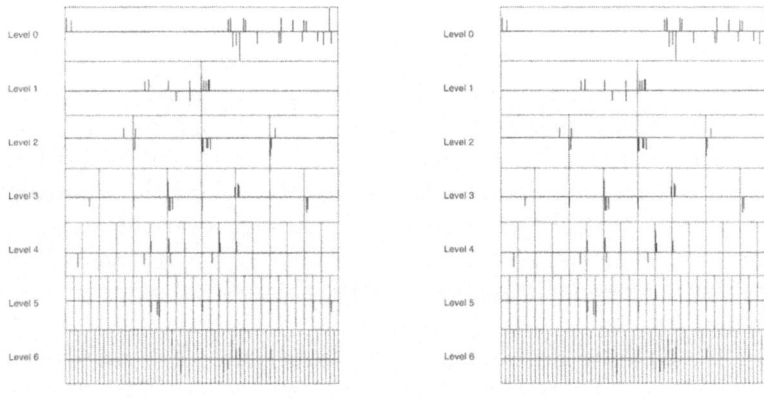

A.   B.

Figure 15.2. Signal approximation with the 100 largest coefficients for for MP run on CP (A) and WP (B).

15.3 Exploratory Analysis

In Figure 15.2 we report the top-100 largest coefficients approximation obtained with the MP algorithm applied on CP and WP dictionaries, respectively in (A) and in (B). Since in WP tables blocks are ordered by frequency, and within blocks wavelet coefficients are ordered by time, the low frequency information in the signal is expected to be concentrated on the left side and the high frequency information on the right side of the table[3]. For CP tables, the blocks are ordered by time and cosine coefficients within blocks are ordered by frequency; thus, the high frequency part of the signal is now expected on the left side, while the low frequency behavior should appear on the right side.

Model design for financial time series involves the representation of features such as short and long range dependence, hidden periodicities, external shocks, surprise variable effects and other factors with impact on prices and returns. These components are usually not easy to interpret, given the presence of non-stationarity by the evidence of spurious features in the data.

We adopt a strategy which aims to pre-process the return series with ad-hoc filtering, i.e. targeted to deal with the hidden periodic components. The goal is that of getting residual returns where the only dynamics left are those strictly related to the volatility process. In practical terms, we have in mind a two-stage process where the battery of wavelet-based techniques

---

[3]The oscillation index b goes from 0 to $2^J - 1$ from the left to the right.

and the classes of functions available through the selected dictionaries may enable a de-seasonalization step followed by a de-volatilization step.

We thus apply a methodology that first computes the WP and CP transforms for the 1m and 5m series; then, runs the MP algorithm so to find the best possible signal representation; last, compares the residual return series obtained by subtracting from the original 1m and 5m return the results obtained from reconstructing the signals by means of the MP algorithm.

We have run some experiments with the aim of checking the approximation power of the MP algorithm when it works through a variable number of approximating elements, up to 500 atoms selected by the WP and CP dictionaries. This choice should suggest whether the residual autocorrelation may be better detected.

We have computed values obtained with a procedure that first segments the WP and CP tables in two parts[4], and then computes the MP at various degrees of approximation power. In general, and particularly for the WP case, the high resolution levels are those capturing most of the energy.

From the plots in Figure 15.3, we notice that a residual autocorrelation and a certain persistence still remain, regardless the different approximation power which is considered and used. The low and high frequency informative content in the form of dependencies and periodicities is not totally explained, with a substantial part of autocorrelation left in the de-seasonalized residual returns.

15.4   Fine Tuning of Resolution Pursuit

An algorithm known as *High Resolution Pursuit* (HRP) [15] has been proposed as an alternative to the MP procedure so to improve the local fit power by using information from the highest scales, i.e. the atoms belonging to low scales can be represented as averages of higher resolution level atoms. This strategy delivers advantages of HRP over MP, but suggests also limitations and assumes constraints[5].

We focus on how to make the best use of the information coming from the various resolution levels, in a way that the most independent coordinates might be of help for understanding and interpreting the structure of volatility.

There are still relatively few ICA applications in the domain of finance. The goal is obtaining a set of statistically independent components (IC)

---

[4]Apart from the choice of working through a simple case of segmentation, there is a computational convenience in keeping only few segments, with a segmentation applied according to the sample splitting rule which restricts to sample sizes divisible by 2^J.

[5]In particular, there is not adaptivity with regard to scale selection rules and for fixing stopping rules which avoid overfitting, where instead case-dependent solutions are adopted.

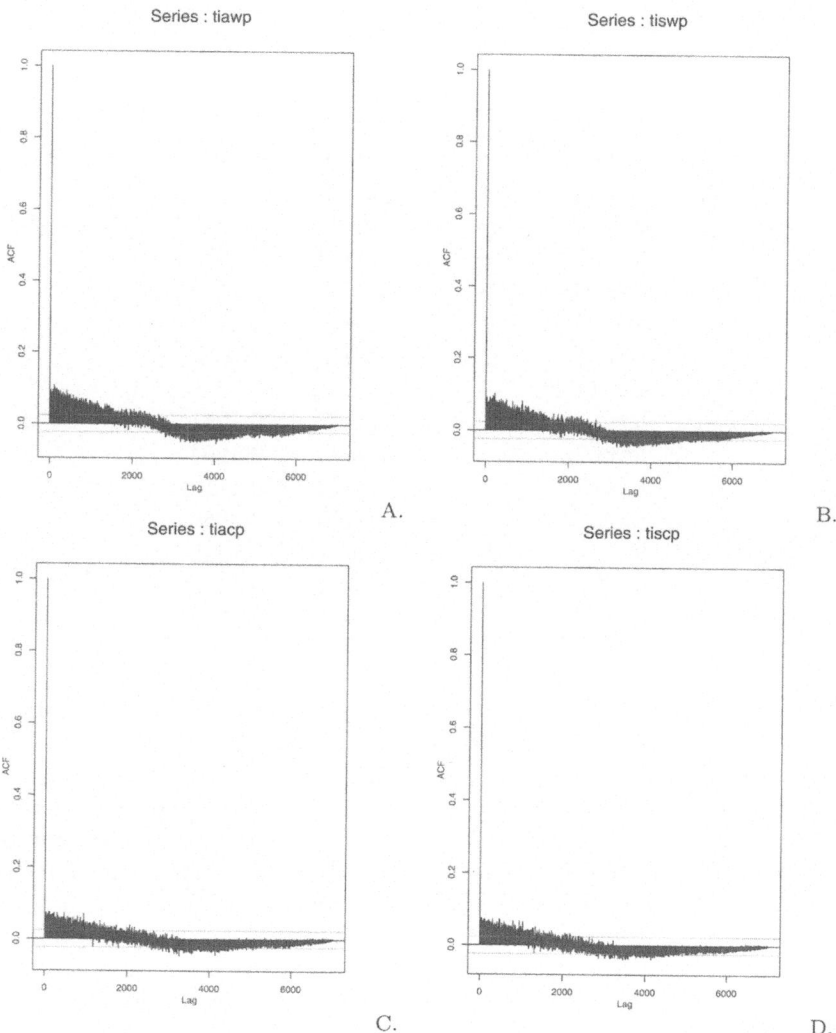

Figure 15.3. ACF of absolute (A,C) and squared 5m residuals from MP with 500 atoms on WP and CP tables (respectively indicated at the end of the series titles.

with which to interpret the main driving forces behind the volatility process, for instance. ICA is related to linear and nonlinear mixture models, including the case of convolutive mixing, and refers to noise-free or noisy data applications. It generalizes the well-known Principal Component Analysis, but unlike the latter which uses statistical information coming from the first two moments of the involved probability distributions, it decorrelates the data by using statistical information of higher order and thus becomes suitable for non-Gaussian contexts. Thus, ICA is a latent variable

Table 15.1. Estimates from mixing matrix A (absolute values) and IC's factor influences for signal resolution levels.

WP table	↔	→	level	abs. value
level 0	↔		level0	0.2218
level 1	↔		level1	0.1951
level 2	↔		level2	0.167
level 3	↔		level3	0.1438
level 4	↔		level4	0.1318
level 5		→	level6	0.1147
level 6		→	level5	0.121
CP table				
level 0		→	level5	0.1261
level 1		→	level6	0.1204
level 2		→	level3	0.1712
level 3		→	level0	0.1868
level 4		→	level1	0.1832
level 5		→	level4	0.1482
level 6		→	level2	0.1748

statistical model where linear or non-linear transforms of non-Gaussian and independent variables deliver the observed data.

By assuming that the sensor outputs are indicated by $x_i, i = 1, \ldots, n$ and represent a combination of independent, non-Gaussian and unknown sources $s_i, i = 1, \ldots, m$, a non-linear system $Y = f(X)$ could be approximated by a linear one AS, where $X = AS$. Instead of computing $f(X)$ one may now work for estimating the sources S together with the $m \times m$ mixing matrix A, where usually $m \ll n$, with n the number of sensor signals, but with $m = n$ holding in many cases too.

The *Joint Approximate Diagonalization of Eigenmatrices for Real signals* or *JadeR* algorithm [8] is one of the algorithms which implements ICA, and is the one that we have applied. It delivers an estimate for the separating or de-mixing matrix B, obtained from $Y = BX$, such that when $B = A^{-1}$ a perfect separation would be obtained. This in general cannot happen, being just an ideal setting, and thus solutions hold approximately up to permutation and scaling. De-correlation and rotation steps are implemented by *JadeR* so to deal with these aspects, and a set of m ICs is obtained.

The approach we have adopted here works in combination with wavelet signal decomposition techniques. We start from considering the detail signals obtained through WP and CP transforms: the series of scaled signals bring a different degree of resolution and refer to specific information obtained by the transforms by switching between resolution levels.

We estimate the mixing matrix A for WP and CP dictionaries. In our applications these scaled signals have been de-seasonalized and are given

Table 15.2. Energy percentage distribution among the 4 finest resolution levels for residual 5m series obtained in WP/CP tables and computed via the MP algorithm at the approximation power of 50, 100, 200 and 500 atoms.

T = number of Atoms	50	100	200	500
WP table.				
level 0	0.228	0.268	0.339	0.472
level 1	0.139	0.088	0.135	0.120
level 2	0.1	0.146	0.125	0.126
level 3	0.533	0.497	0.401	0.282
CP table.				
level 0	0.819	0.637	0.704	0.722
level 1	0.021	0.135	0.145	0.150
level 2	0.081	0.127	0.084	0.068
level 3	0.079	0.101	0.067	0.060

as inputs to ICA for the extraction of "m" possible sources, which we set equal to the number of sensors.

At each available resolution levels corresponds a row of the estimated mixing matrix A which represents a series of weights for the IC. We extract from each level an approximate value used to represent the contribution of the resolution dependent component to the signal features, in the most independent way compared to the other levels. We thus focus on the highest "energetic" values computed; these elements, as said, suggest what are the dominant ICs on a scale-dependent basis, without identifying their specific nature, i.e. the underlying economic factors, system dynamics or pure shocks.

From the WP estimated mixing matrix A we note a strong within-level factor always dominating apart from levels 5 and 6, where a mutual cross-influence appears. From the CP extimated mixing matrix A things change substantially, since each level depends mainly from out-of-level factors, i.e. components inherent to other resolution levels, and is only negligibly influenced by within-level factors. We summarise these results in Table 15.1, with the help of the symbols → for indicating the out-of-level most dominant factor influence and ↔ for indicating the within-level one.

Considering the results obtained with the ICA application we now investigate the performance of the MP algorithm when we adapt its domain of atoms according to some selected resolution levels. We choose to restrict the MP range of application to the four finest resolution levels of the WP and CP tables.

We adopt the same flexible degree of approximation power of MP as before, up to 500 atoms and compare the energy percentage distribution obtained after the MP runs. In Table 15.2 these results are reported.

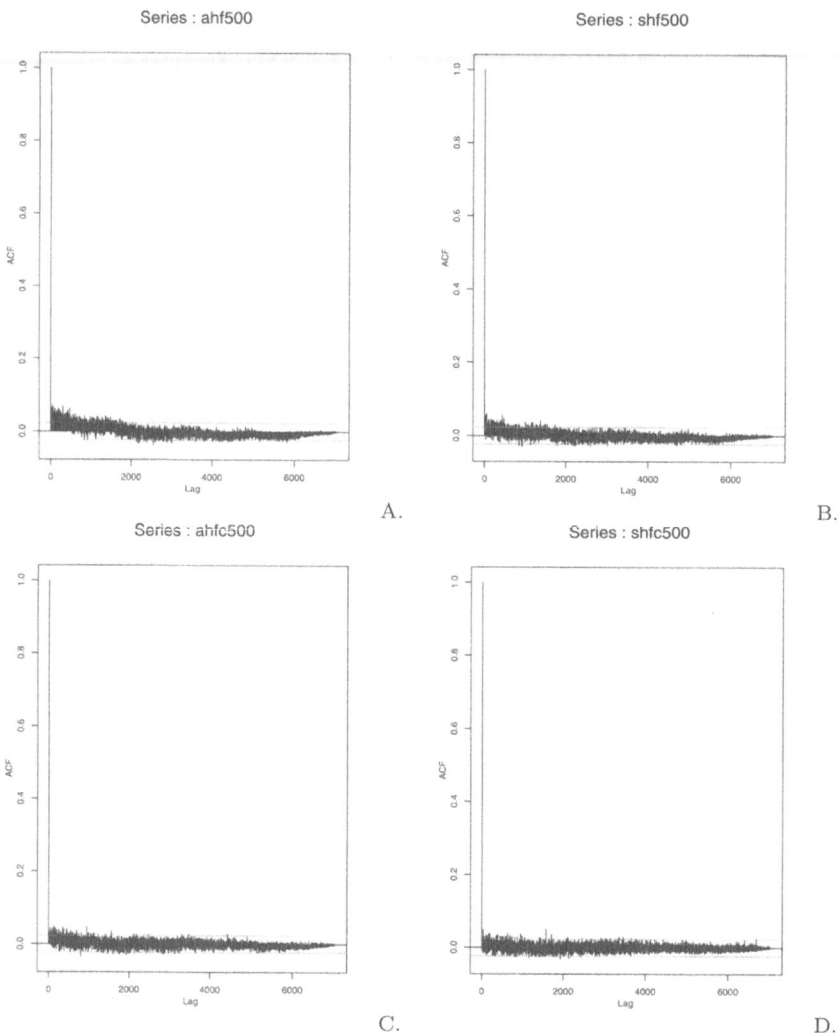

Figure 15.4. ACF of absolute (left side) and squared 5m residuals from MP with 500 atoms on WP (A,B) and CP tables through the finest four resolution levels.

In general, things change more, as expected, with the WP table. This because as explained by [15], certain families of functions, like wavelets packets, allow for low scale atoms in the dictionary to be represented as averages of higher scale atoms; in Figure 15.4 we report the ACF of the *absolute* and *squared* residuals obtained with MP running with 500 atoms through the finest resolution levels of the the WP and CP tables.

Thus, the features we wanted to detect are well captured, and with computational savings, by simply concentrating the MP activity only on the finest resolution levels.

Our simple ICA-based procedure of pre-selecting the resolution levels over which MP runs, it also suggests a simple strategy aimed to bypass the use of modified algorithms bringing limitations and contraints into the analysis. The MP algorithm may be very effective by just limiting its range of activity, in this case the domain of resolution levels obtained from a signal decomposition. Exploiting the independent information content of MR signals, as indicated by the ICA stage, may represent an efficient procedure and a near-optimal way of tuning the resolution pursuit.

15.5   Conclusions

Our experiments with high frequency signals such as financial time series suggest that interesting results are obtained through the Matching Pursuit procedure based on wavelet and cosine packets as a tool for function approximation. Return data may be analysed in two steps, where the first one is a filtering procedure removing all the hidden periodicities, thus de-seasonalizing the volatility process, and the second step is played by Independent Component Analysis which finds what scales appear to have informative content, based on the independent contribution coming from each detail or MRA level to the signal structure. Wavelet and Cosine Packets offer excellent covariance diagonalizing bases that form a useful ground for the Matching Pursuit algorithm runs.

References

[1] P. Abry, P. Flandrin, M.S. Taqqu and D. Veitch, Wavelets for the analysis, estimation and synthesis of scaling data. In: Park, C. & Willinger, W. (eds), *Self-similar network traffic and performance evaluation*. New York: Wiley 2000, pp 39-88.

[2] T. Andersen and T. Bollerslev, Intraday periodicity and volatility persistence in financial markets, *Journal of Empirical Finance*, 4 (1997) 115-158.

[3] T. Andersen and T. Bollerslev, Heterogeneous Information Arrivals and Return Volatility Dynamics: Uncovering the Long-Run in High Frequency Returns, *The Journal of Finance*, LII(3) (1997) 975-1005.

[4] A. Bruce and H.V. Gao, *S+Wavelets*, (StaSci Division, MathSoft Inc., Seattle, 1994).

[5] E. Capobianco, Wavelets for High Frequency Financial Time Series, *Proceedings Interface '99*, 31 (1999) 373-378.

[6] E. Capobianco, Statistical Analysis of Financial Volatility by Wavelet Shrinkage, *Methodology and Computing in Applied Probability*, I(4) (1999) 423-443.

[7] J. Cardoso, Source separation using higher order moments, Proceedings Conference on Acoustic, Speech and Signal Processing, (1989) 2109-2112.

[8] J. Cardoso and A. Souloumiac, Blind beamforming for non-Gaussian signals, IEE Proceedings F., 140(6) (1993) 771-774.

[9] P. Comon, Independent Component Analysis - a new concept?, *Signal Processing*, 36(3) (1994) 287-314.

[10] I. Daubechies, *Ten Lectures on wavelets*, (SIAM, Philadelphia, 1992).

[11] D. Donoho, Unconditional Bases and Bit-Level Compression. *Applied Computational Harmonic Analysis*, 3 (1996) 388-392.

[12] D.L. Donoho and I. M. Johnstone, Ideal Spatial Adaptation via Wavelet Shrinkage, Biometrika 81 (1994) 425-455.

[13] D.L. Donoho and I.M. Johnstone, Adapting to unknown smoothness via wavelet shrinkage, *Journal of American Statistical Association*, 90 (1995) 1200-1224.

[14] D.L. Donoho and I.M. Johnstone, Minimax Estimation via Wavelet Shrinkage, *The Annals of Statistics*, 26 (1998) 879-921.

[15] S. Jaggi, W.C. Karl, S. Mallat and A.S. Willsky, High Resolution Pursuit for Feature Extraction, *Applied and Computational Harmonic Analysis*, 5(4) (1998) 428-449.

[16] I.M. Johnstone, Wavelet shrinkage for correlated data and inverse problems: adaptivity results, *Statistica Sinica*, 9 (1999) 51-83.

[17] I.M. Johnstone and B.W. Silverman, Wavelet threshold estimators for data with correlated noise, *Journal Royal Statistical Society B.*, 59 (1997) 319-351.

[18] M.S. Lewicki and T.J. Sejnowski, Learning Overcomplete Representations, *Neural Computation*, 12(2) (2000) 337-365.

[19] S. Mallat and Z. Zhang, Matching Pursuit with time frequency dictionaries, *IEEE Transactions Signal Processing*, 41 (1993) 3397-3415.

[20] I. Meyer, *Wavelets: algorithms and applications*, (SIAM, Philadelphia, 1993).

[21] M. Zibulewsky and B.A. Pearlmutter, Blind Source Separation by Sparse Decomposition in a Signal Dictionary, *Neural Computation*, 13(4) (2001) 863-882.

16

Confidence Intervals for Logspline Density Estimation

Charles Kooperberg and Charles J. Stone[1]

Summary

Several ways to obtain pointwise confidence intervals corresponding to logspline density estimation are studied. These methods
include a variety of approaches based on estimation using free knot
splines, a couple of approaches based on the bootstrap, and a
Bayesian approach. It is concluded that a variation of the bootstrap,
in which only a limited number of bootstrap simulations are used to
estimate standard errors that are combined with standard normal
quantiles, seems to perform the best, especially when coverages and
computing time are both taken into account.

16.1 Introduction

Getting confidence intervals corresponding to function estimates that are
obtained using an adaptive polynomial spline method is a notoriously hard
problem. After model selection has been carried out, the estimated function
has a simple parametric form [12]. However, treating the final model as a
fixed parametric model, ignoring the large amount of model selection that
may have occurred, yields confidence intervals with too low coverage.

Recently, Kooperberg and Stone [9] described an algorithm for logspline
density estimation with free knots. This is a modification to previous

---

[1]Charles Kooperberg is Member, Division of Public Health Sciences, Fred
Hutchinson Cancer Research Center, Seattle, WA 98109-1024 (E-mail: clk@fhcrc.org).
Charles J. Stone is Professor, Department of Statistics, University of California, Berkeley,
CA 94720-3860 (E-mail: stone@stat.berkeley.edu). Charles Kooperberg was supported
in part by National Institutes of Health grant CA74841. Charles J. Stone was supported
in part by National Science Foundation grant DMS-9802071.

logspline density algorithms [7, 8, 12], in which the knots are not selected by a greedy stepwise algorithm, but are viewed as additional parameters. Two reasons for studying logspline density estimation with free knot splines are that (i) stepwise selection algorithms can be seen as crude approximations to the free knot algorithm and (ii) coverages of (pointwise) confidence intervals based on the free knot algorithm may be more accurate since they reflect uncertainty in the knot placement. It was concluded that the coverages of nominal 95% confidence intervals using the free knot algorithm, while closer to 95% than the coverages ignoring knot selection, are still well below 95%.

In the current chapter we investigate alternative methods for obtaining confidence intervals corresponding to logspline density estimation. In particular, we investigate whether an expansion of the free knot intervals improves the coverage, and we also discuss bootstrap and Bayesian methods for obtaining confidence (credible) intervals.

In Section 16.2 we briefly review logspline density estimation in general and the procedure with free knots in particular. In Section 16.3 we discuss the various approaches to obtaining confidence intervals. The approaches based on expansion of the standard errors for free knot splines and on the bootstrap are compared in a simulation study in Section 16.4. In Section 16.5, the various approaches are applied to a real example. We end with a brief discussion.

16.2   Logspline density estimation with free knots

Given the free knots $-\infty < L < \gamma_1 < \cdots < \gamma_J < U < \infty$, set $\gamma = (\gamma_1, \ldots, \gamma_J)$ and let $\mathbb{G}_\gamma$ denote the space of cubic splines on $[L, U]$ corresponding to the knot sequence γ and satisfying the usual tail linear constraints. Thus a function g on $[L, U]$ is a member of $\mathbb{G}$ if and only if it is twice continuously differentiable on $[L, U]$, its restriction to each of the intervals $[L, \gamma_1], [\gamma_1, \gamma_2], \ldots, [\gamma_{J-1}, \gamma_J], [\gamma_J, U]$ is a cubic polynomial, $g''(L) = 0$, and $g''(U) = 0$. Observe that $\mathbb{G}_\gamma$ is a $(J + 2)$-dimensional linear space. Set $p = J + 1$, and let $1, B_{\gamma 1}, \ldots, B_{\gamma p}$ be a basis of $\mathbb{G}_\gamma$. Given $\theta = (\theta_1, \ldots, \theta_p) \in \Theta = \mathbb{R}^p$, set

$$\eta_\gamma(y; \theta) = \theta_1 B_{\gamma 1}(y) + \cdots + \theta_p B_{\gamma p}(y) - C_\gamma(\theta), \qquad L \leq y \leq U,$$

where

$$C_\gamma(\theta) = \log \left(\int_L^U \exp(\theta_1 B_{\gamma 1}(y) + \cdots + \theta_p B_{\gamma p}(y)) \, dy \right).$$

Note that $\exp \eta_\gamma(y; \theta)$ is a positive density function on $[L, U]$ for every γ and θ.

Let $Y_1, \ldots, Y_n$ be a random sample of size n from a distribution having density f and log-density $\eta = \log f$. Consider the log-likelihood

$$\ell_{\gamma}(\boldsymbol{\theta}) = \sum_{i=1}^{n} \eta_{\gamma}(Y_i; \boldsymbol{\theta}) = \sum_{j=1}^{p} \theta_p \sum_{i=1}^{n} B_{\gamma j}(Y_i) - nC_{\gamma}(\boldsymbol{\theta}).$$

Let $\widehat{\gamma}$ and $\widehat{\boldsymbol{\theta}}$ denote the maximum likelihood estimates of γ and $\boldsymbol{\theta}$, so that

$$\widehat{\ell} = \ell_{\widehat{\gamma}}(\widehat{\boldsymbol{\theta}}) = \operatorname*{argmax}_{\gamma, \boldsymbol{\theta}} \ell_{\gamma}(\boldsymbol{\theta}).$$

Observe that for the free knot procedure [9] the positive integer parameter J must also be chosen. Let $\widehat{\gamma}_J$, $\widehat{\boldsymbol{\theta}}_J$, and $\widehat{\ell}_J$ now indicate the dependence of $\widehat{\gamma}$, $\widehat{\boldsymbol{\theta}}$, and $\widehat{\ell}$, respectively, on J. For choosing J, we will employ the Akaike Information Criterion $\mathrm{AIC}_{J,a} = -2\widehat{\ell}_J + (2J+1)a$ [1], which depends on the complexity parameter a. (Note that $\widehat{\gamma}_J$ has J free parameters and $\widehat{\boldsymbol{\theta}}_J$ has $p = J+1$ free parameters.) We select the value $\widehat{J}$ of J that minimizes $\mathrm{AIC}_{J,2}$. Set $\widehat{\gamma} = \widehat{\gamma}_{\widehat{J}}$ and $\widehat{\boldsymbol{\theta}} = \widehat{\boldsymbol{\theta}}_{\widehat{J}}$. We refer to $\widehat{\eta}(y) = \eta_{\widehat{\gamma}}(y; \widehat{\boldsymbol{\theta}})$ as the maximum (penalized) likelihood estimate of the log-density η at y and to $\widehat{f}(y) = \exp \widehat{\eta}(y)$ as the logspline estimate with free knots of the density f at y.

Computing the maximum likelihood estimates with free knots is a highly nontrivial numerical problem, as the likelihood function $\ell_{\widehat{\gamma}}(\widehat{\boldsymbol{\theta}})$ is severely multimodal, and degenerate solutions exist when too many of the knots γ_j get close together.

For a procedure with fixed knots we would have to specify a set of knots. Rather than specifying a complete set in advance, we select knots by a stepwise procedure. Such a procedure can be either a stepwise deletion procedure or a stepwise addition and deletion procedure. For the former, we initially position a large number of knots and remove the "least significant" knot one at a time [8]. For the later we add knots one at a time, to increase the log-likelihood as much as possible, until a maximum number of knots is reached, after which we carry out a stepwise deletion procedure [12]. For either of the two stepwise procedures we use the AIC criterion with $a = \log n$, as in the Bayesian Information Criterion [10] to select the number J of knots.

16.3   Confidence intervals

16.3.1   Free knot splines

In [9] we proposed obtaining confidence intervals for the density using a "standard" maximum likelihood approach. In particular, let $\widehat{\nabla}_J \eta(y)$ denote the $(2J + 1)$-dimensional gradient of $\eta_{\gamma}(y; \boldsymbol{\theta})$ at the maximum likelihood estimate, and let $\widehat{H}_J$ denote the $(2J+1) \times (2J+1)$ Hessian matrix of the log-likelihood at the maximum likelihood estimate when there are J free knots. Set $\widehat{\nabla}\eta(y) = \widehat{\nabla}_{\widehat{J}}\eta(y)$ and $\widehat{H} = \widehat{H}_{\widehat{J}}$. The standard error in the estimate $\widehat{\eta}(y)$ is given by

$$\mathrm{SE}(\widehat{\eta}(y)) = \sqrt{[\widehat{\nabla}\eta(y)]^T(-\widehat{H})^{-1}\widehat{\nabla}\eta(y)}. \qquad (16.1)$$

This leads to the nominal 95% confidence interval

$$\Big(\exp\big(\widehat{\eta}(y) - 1.96\mathrm{SE}(\widehat{\eta}(y))\big),\ \exp\big(\widehat{\eta}(y) + 1.96\mathrm{SE}(\widehat{\eta}(y))\big)\Big) \qquad (16.2)$$

for $f(y)$.

The distribution function corresponding to f is given by $F(y) = \int_L^y \exp\eta(z)\,dz$ for $L \leq y \leq U$, which can be estimated by $\widehat{F}(y) = \int_L^y \exp\widehat{\eta}(z)\,dz$. The corresponding standard error is given by

$$\mathrm{SE}(\widehat{F}(y)) = \sqrt{[\widehat{\nabla}F(y)^T](-\widehat{H})^{-1}\widehat{\nabla}F(y)},$$

where $\widehat{\nabla}F(y) = \int_L^y \widehat{\nabla}\eta(z)\exp\widehat{\eta}(z)\,dz$.

Simulation studies were carried out, which suggested that the actual coverage of nominal 95% confidence intervals using these standard errors is about 87% for the density and 93% for the distribution function. This coverage is, however, much better than when the uncertainty in the knots is ignored. Let $\mathrm{SEFX}(\widehat{\eta}_i(y))$ and $\mathrm{SEFX}(\widehat{F}_i(y))$ be the standard errors assuming that the knots are fixed (so that they make use only of the $(\widehat{J}_i + 1) \times (\widehat{J}_i + 1)$ Hessian matrix for the coefficients). The coverage of the nominal 95% confidence intervals using these standard errors was only about 81% for the density and 92% for the distribution function. To improve the coverage we will also investigate confidence intervals

$$\Big(\exp\big(\widehat{\eta}(y) - 1.96\alpha\mathrm{SE}(\widehat{\eta}(y))\big),\ \exp\big(\widehat{\eta}(y) + 1.96\alpha\mathrm{SE}(\widehat{\eta}(y))\big)\Big), \qquad (16.3)$$

in which the standard errors are expanded by the factor α for some $\alpha > 1$ and using similarly expanded confidence intervals for the distribution function in this chapter.

16.3.2   The bootstrap

Alternatively, we can employ the bootstrap in combination with either the stepwise knot deletion algorithm of [8] or the stepwise addition and deletion algorithm of [12] to obtain confidence intervals corresponding to logspline density estimates. In this chapter we use the former algorithm and examine the coverage of bootstrap percentile intervals [3] for both the log-density and the distribution function. That is, we take B (we used $B = 1000$) samples $\mathbf{Y^i}$ with replacement of size n from the data $Y_1, \ldots, Y_n$, and for each sample $\mathbf{Y^i}$ we obtain the logspline density estimate. The 95% pointwise confidence interval for $\widehat{\eta}(y)$ $(F(y))$ is then from the 2.5th to the 97.5th percentile of the B bootstrap estimates for the log-density (distribution function).

Clearly, the bootstrap is a computationally time consuming procedure for getting confidence intervals, as we need to fit B logspline densities. However, it is still slightly faster than using the algorithm developed in [9] for fitting logspline densities with free knots.

A considerably cheaper approach is to hope that the logspline estimates of the log-density and distribution function have approximately a normal distribution, but that the estimates of the standard errors that are obtained using standard techniques are too small. If so, we can get by with a much smaller number B of bootstrap estimates (say $B = 25$) by using these estimates to obtain pointwise bootstrap estimates of $\text{SE}^{B}(\widehat{\eta}(y))$ and $\text{SE}^{B}(\widehat{F}(y))$ and then using equation (16.2) or the equivalent to obtain confidence intervals for η and F.

16.3.3 A Bayesian approach

Hansen and Kooperberg [6] describe a Bayesian approach to logspline density estimation, which involves a prior $p(J)$ on the dimension of the model, a prior $p(\gamma \mid J)$ on the location of the knots, and a prior $P(\boldsymbol{\theta} \mid \gamma, J)$ on the coefficients. Given the data $Y_1, \ldots, Y_n$, the posterior distribution of $(J, \gamma, \boldsymbol{\theta})$ is explored using a reversible jump Markov chain Monte Carlo [5] algorithm. At each step of the algorithm a new density is proposed by either adding a knot, deleting a knot, moving a knot, or updating the coefficients. This new proposed density is always accepted if the posterior probability is higher than the previous density; otherwise it still has a positive probability of being accepted. The acceptance probability is governed by the reversible jump algorithm. The algorithm of [6] for logspline density estimation is similar to algorithms for univariate regression using polynomial splines proposed by [2] and [11].

To make (pointwise) 95% credible intervals about the logspline density estimate obtained from this Bayesian procedure, we use as endpoints the 2.5th and 97.5th percentiles of all Markov chain Monte Carlo simulations. Credible intervals have a different interpretation from (frequentist) confidence intervals. For confidence intervals we are 95% confident that the confidence interval will cover the true value of the density; for a 95% credible interval, there is 95% (posterior) probability that the density falls within the interval.

Hansen and Kooperberg [6] point out that, depending on how priors are selected, a Bayesian procedure can be similar in performance to a greedy stepwise procedure using AIC to select the number of knots when a geometric prior on the number of knots is used, or it can be similar to a smoothing spline approach when a uniform prior on the number of knots and a particular multivariate normal prior on the coefficients are used.

16.4   A simulation study

In this section we augment the results of the simulation study in [9], in which we generated 250 samples of size 250 and 250 samples of size 1000 from each of four distributions:

Normal 2 A mixture of two normal distributions, so that the true density of Y is given by

$$f(y) = c\left(\frac{1}{3}f_{Z_1}(y) + \frac{2}{3}f_{Z_2}(y)\right)\text{ind}(-4, 8),$$

where Z_1 has a normal distribution with mean 0 and standard deviation 0.5, Z_2 has a normal distribution with mean 2 and standard deviation 2, ind($\cdot$) is the usual indicator function, and c is the multiplier to correct for the truncation to $(-4, 8)$.

Normal 4 As in example 1, but the mean of Z_2 is 4 and Y is truncated to $(-2, 10)$.

Normal 6 As in example 1, but the mean of Z_2 is 6 and Y is truncated to $(-1.5, 12)$.

Gamma 2 A gamma distribution with shape parameter 2 and mean 1, with Y truncated to the interval $(0, 9)$.

The Normal 2 density has one mode, but a clear second hump; Normal 4 has two, not very well separated, modes; Normal 6 has two well separated modes; and the Gamma 2 density is unimodal.

In Table 16.1 we compare the coverages of four approaches for getting confidence intervals using the free knot spline methodology. The first two columns are taken from [9]. These columns are the coverages obtained by using the regular SE (see equation (16.1)) or SEFX, for which it is assumed that the knots are fixed. As can be seen from this table, the coverages are well below the nominal 95% level. For the third and fourth columns, these standard errors are expanded by $\alpha = 1.34$ for SE and $\alpha = 1.55$ for SEFX, respectively (see equation (16.3)). These expansion factors were chosen so that the average coverage over these eight simulations is exactly 95%; thus, it could be argued that these columns do not provide a completely fair comparison. The last two columns are using bootstrap samples for the logspline density estimation procedure of [8]. The fifth column is based on 1000 bootstrap samples, and the confidence intervals are from the 2.5th through the 97.5th (pointwise) percentiles. For the sixth column we generated only 25 bootstrap samples, computed the pointwise standard errors for the log-density, and then used (16.2) to obtain the confidence intervals.

Table 16.1. Coverages for six different approaches to obtaining confidence intervals for a log-density, estimated using logspline.

| Density | Free Knot Standard Error | | | | Bootstrap | |
| | Nominal | | Expanded | | | |
	SE	SEFX	1.34SE	1.55SEFX	Percentiles	SE
$n = 250$						
Normal 2	84.0	77.4	91.9	91.9	97.4	95.2
Normal 4	88.8	82.5	95.5	94.9	97.4	96.4
Normal 6	89.0	84.0	96.9	97.2	96.5	94.6
Gamma 2	86.2	81.2	94.8	96.0	97.8	97.3
$n = 1000$						
Normal 2	89.2	79.6	95.8	93.9	96.8	94.4
Normal 4	89.3	82.7	97.4	97.0	98.0	94.7
Normal 6	86.2	81.4	95.2	96.2	96.3	92.9
Gamma 2	84.0	77.3	92.6	92.8	97.4	95.4
Average	87.1	80.7	95.0	95.0	97.2	95.1

It is clear from this table that the confidence intervals based on the free knot spline standard error or the fixed knot spline standard error have too low coverage. With an appropriate expansion factor it is possible to get the coverage to be about 95%. With this approach the problem is, naturally, to find the right expansion factor α. If we had chosen separate expansion factors for each density and each sample size, we would have had factors that for SE varied between $\alpha = 1.2$ for Normal 4 with $n = 1000$ to $\alpha = 1.63$ for Normal 2 with $n = 250$. On the other hand, the expansion factor that gives an overall coverage of about 95% for the four distributions being studied essentially does not depend on n for the two sample sizes being studied. Actually, there seems to be little advantage of the expanded SE over the expanded SEFX in this case, except that the expansion factors are larger for SEFX.

Surprisingly, the coverages for the bootstrap percentile intervals are consistently too high. It is our impression that this is due to some instability in the stepwise logspline algorithm when there are many repeat observations, causing the intervals to be occasionally too large. That is in line with what we will see for the income data in the next section. Interestingly, the coverages in the sixth column of Table 16.1, corresponding to the bootstrap SE approach, not only are very close to 95% on average, but have considerably less variation than those in columns 3 and 4 based on expanded SE's.

For the distribution function all approaches yielded somewhat better results (coverage closer to nominal, less variation between different distri-

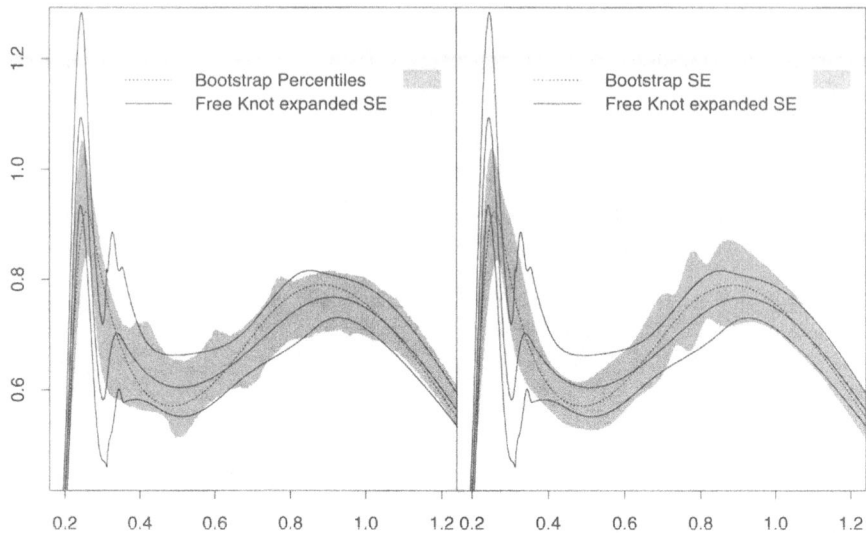

Figure 16.1. Comparison of the expanded free knot spline pointwise confidence intervals and bootstrap pointwise confidence intervals for the income data. The solid lines are estimate and confidence bounds for the free knot procedure, the dashed line is the estimate for the stepwise procedure and the grey area are the bootstrap intervals (left side percentiles, right side SE).

butions) than for the (log-)density, except for the bootstrap SE approach, for which the average coverage was down to 93.6%. This is not surprising, since the logspline estimate of the distribution function is presumably not approximately normal in the tails. A logistic transformation may improve the results here.

16.5   An example

In this section we further analyze the income data, which was also discussed in [9] and [6]. In Figure 16.1 we show the 95% free knot (pointwise) confidence intervals, expanded by $\alpha = 1.34$ as in the previous section, together with the corresponding logspline density estimate (solid lines for the estimate, the lower and upper confidence bounds). We also show the stepwise logspline density estimate with knot deletion (dashed line) along with the 95% bootstrap percentile confidence intervals (left side, grey area) and the 95% bootstrap confidence intervals using 25 samples to estimate the standard error (right side, grey area). As can be seen from these plots, the bootstrap SE approach and the bootstrap percentile approach yield intervals that are approximately the same size as the expanded free knot approach, but which are slightly less smooth. Averaged over the region shown, the average size of all three intervals are within 5% of each other.

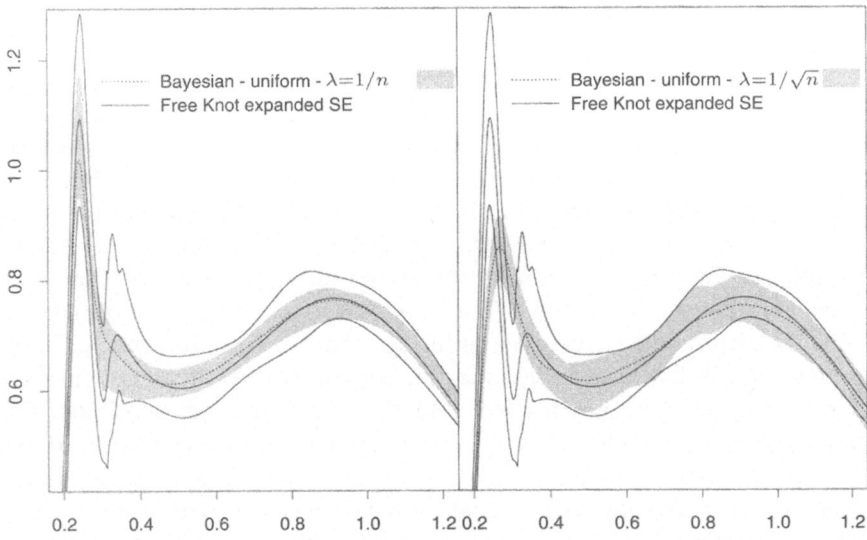

Figure 16.2. Comparison of the expanded free knot spline pointwise confidence intervals and Bayesian pointwise credible intervals for the income data. The solid lines are the same as in Fig. 16.1, the dashed lines are the estimates for the Bayesian procedures and the grey areas are the corresponding credible intervals.

Overall, these intervals agree with the conclusion from the previous section: the bootstrap SE approach yields reasonable confidence intervals at a computing price that is mush smaller than free knot splines or a full bootstrap approach.

In Figure 16.2 we show the same expanded free knot intervals as in Figure 16.1, but this time we added 95% credible intervals from the Bayesian algorithm described in [6] (dashed lines and grey area). The algorithms shown have a uniform prior for the number of knots and a multivariate normal prior on the coefficients. The variance of this later prior (proportional to the λ parameter indicated in these plots) plays a role as a smoothing parameter. The results shown in this figure are based on a run of 100,000 MCMC iterations, which takes a cpu time that is comparable to the bootstrap percentile approach, and which is considerably larger than what is needed to obtain good point estimates. The estimates with $\lambda = 1/n$ were the ones with the largest value of λ that gave a reasonable estimate for the height of the peak, as argued in [6]. The corresponding 95% credible intervals are still considerably smaller than the 95% expanded free knot intervals, suggesting that the coverages of the former intervals may be significantly under 95%. Even when $\lambda = 1/\sqrt{n}$, so large that the height of the peak gets reduced to about 0.86, rather than the "correct" height of between 1.00 and 1.10, the credible intervals still appear too small.

16.6   Discussion

Several ways for obtaining confidence or credible intervals for logspline density estimates were studied here. Free knot and fixed knot confidence intervals that are not expanded yield substantially too low coverages. These intervals can be expanded to give reasonable coverage, but it is not obvious how well the expansion factors used in the simulation study reported here would work for other choices of the underlying density or sample size. Bayesian credible intervals for density estimates that look reasonable appear too small, while those intervals that are wide enough seem to correspond to density estimates that smooth too much. Bootstrap percentile intervals appear ragged, suggesting that very large numbers of bootstrap samples are needed, and their coverages are too high. The bootstrap SE approach—estimating the standard error based on a limited number of bootstrap estimates and using "1.96" to obtain 95% confidence intervals—seems to have the best performance. The coverage is about right, the computational expense is low, and the pointwise confidence intervals are fairly smooth. This performance came as a pleasant surprise to us and suggests that the bootstrap SE approach deserves a more thorough investigation.

References

[1] Akaike, H. (1973) "Information theory and an extension of the maximum likelihood principle", in *Second International Symposium on Information Theory* (eds B. N. Petrov and F. Csáki), Budapest: Akademia Kiadò, 267–281.

[2] Denison, D. G. T., Mallick, B. K., and Smith, A. F. M. (1998), "Automatic Bayesian curve fitting," *J. Roy. Statist. Soc.*, Ser. B., **60**, 333–350.

[3] Efron, B. and Tibshirani, R. J. (1993), *An Introduction to the Bootstrap*, New York: Chapman & Hall.

[4] Family Expenditure Survey (1968–1983), *Annual Base Tapes and Reports (1968–1983)*, London: Department of Employment Statistics Division, Her Majesty's Stationary Office.

[5] Green, P. J. (1995) Reversible jump Markov chain Monte Carlo computation and Bayesian model determination. *Biometrika* **82** 711–732.

[6] Hansen, M. H. and Kooperberg, C. (2002). "Spline adaptation in extended linear models (with discussion)", *Statist. Science*, **17**, 2-51.

[7] Kooperberg, C. and Stone, C. J. (1991), "A study of logspline density estimation," *Comp. Statist. Data Anal.*, **12**, 327–347.

[8] Kooperberg, C. and Stone, C. J. (1992), "Logspline density estimation for censored data," *J. Comp. Graph. Statist.*, **1**, 301–328.

[9] Kooperberg, C. and Stone, C. J. (2002), "Comparison of Parametric, Bootstrap, and Bayesian Approaches to Obtaining Confidence Intervals for Logspline Density Estimation," manuscript.

[10] Schwarz, G. (1978), "Estimating the dimension of a model", *Ann. Statist.*, **6**, 461–464.

[11] Smith, M. and Kohn R. (1996), "Nonparametric regression using Bayesian variable selection," *J. Environ.*, **75**, 317–344.

[12] Stone, C. J., Hansen M., Kooperberg, C. and Truong, Y. K. (1997) "Polynomial splines and their tensor products in extended linear modeling" (with discussion), *Ann. Statist.* **25**, 1371–1470.

17

Mixed-Effects Multivariate Adaptive Splines Models

Heping Zhang[1]

Summary

A mixed-effects multivariate adaptive splines model is presented for analyzing longitudinal or growth curves data that may or may not have been collected through a regular measurement schedule. The MASAL (an acronym for multivariate adaptive splines for the analysis of longitudinal data) algorithm by Zhang [19, 20, 21] is used to determine the nonparametric fixed-effects in the mixed-effects multivariate adaptive splines model. The original MASAL algorithm requires the characterization and specification of the within subject autocorrelation structure, which is usually a tedious while not always rewarding process. In contrast, the idea of mixed-effects is introduced to the MASAL algorithm in this work, leading to an automated procedure for analysis of longitudinal and growth curves data. To demonstrate the great potential of this new procedure, I re-analyzed a data set on the effect of cocaine use by pregnant women on the growth of their infants after birth. The numerical results are remarkable as opposed to a previously published analysis by [21] in terms of the dissection of random effects.

17.1   Introduction

Longitudinal data arise from numerous applications. The standard method for analyzing such data is the mixed-effects linear models [9]. Diggle, Liang, and Zeger [4] offers an excellent description of the related methods with examples. An issue that has attracted a great deal of attention recently is the introduction of nonparametric spline smoothing into the mixed-effects models. Most of this effort has been concentrating on using kernel smoothing along the time axis and keeping the covariates intact with linear effects (see, e.g., [16], [1], [18], [8], [11], and [10]). Because it is a one-dimensional

[1]Heping Zhang is with the Department of Epidemiology and Public Health, Yale University School of Medicine, New Haven, CT 06520. This research is supported in part by NIH grants DA12468 and AA12044.

smoothing and the covariates remain to be a linear component, the computation is relatively straightforward and, in the meantime, the asymptotic results with respect to the covariate effects can be generalized from the mixed-effects linear models to the mixed-effects semi-parametric models. Despite these advantages, this smoothing along the time does not adequately address the need in practice where the role of the covariates and their interactions with the time are not always straightforward to specify, and at least the effects are not necessarily linear. Thus, a fully nonparametric form of fixed effects (including time) is particularly useful. For this consideration, this work is based on [19, 20, 21] in which multivariate adaptive splines for the analysis of longitudinal data (MASAL) was proposed. In contrast to the kernel smoothing, the adaptive splines are practically appealing because the resulting model is readily interpretable.

The early version of MASAL emphasized on the nonparametric estimation of mean surface (fixed-effects) and did not provide a systematic way of constructing the correlation structure for longitudinal data. The purpose of this work is to incorporate random effects into MASAL. I will demonstrate the great potential of this idea by re-analyzing a data set on infant growth from a retrospective study conducted by Dr. John Leventhal and his colleagues at Yale University School of Medicine, New Haven, Connecticut. They collected data from 298 children born at Yale-New Haven Hospital after reviewing the medical records for all women who had deliveries from September 1, 1989 through September 30, 1990. Detailed eligibility criteria have been reported previously elsewhere such as [17], and [15]. One of the primary goals of the original study is to examine the potential impact of the mother's cocaine use during the pregnancy on an infant's growth after the birth. Thus, the study subjects were carefully selected to ensure that the mother's use of cocaine during the pregnancy was correctly classified: one group is regular cocaine users and the other group never used. The present analysis characterizes the infant growth pattern of weight for the first one and half years by identifying variables that influence the pattern and defining the stages within each of which the growth velocity is relatively stable. In addition, the model presented here does not reveal a significance effect of the prenatal exposure to cocaine due to the mother's use on the infant growth in terms of weight.

17.2   The Model

Suppose that we observe the data from n subjects and that for the ith subject observations are made on T_i occasions at time of t_{ij} $(j = 1, \ldots, T_i, i = 1, \ldots, n)$. Let y_{ij} and $x_{k,ij}$ be the response variable and the kth covariate collected from the ith subject at occasion j $(k = 1, \ldots, p, j = 1, \ldots, T_i, i = 1, \ldots, n)$, respectively.

I consider a mixed-effects multivariate adaptive splines model

$$y_{ij} = f(x_{1,ij}, \ldots, x_{p,ij}, t_{ij}) + \sum_{u=1}^{v} b_{iu}\phi_u(t_{ij}) + e_{ij}, \qquad (17.1)$$

where f is an unknown smooth function, $\phi_u(t)$ is a pre-specified function (usually a lower order polynomial), b_{iu} is a Gaussian random variable with mean zero and variance σ_u^2 $(u = 1, \ldots, v)$, and e_{ij} is the random measurement Gaussian error with mean zero and variance σ^2 $(j = 1, \ldots, T_i, i = 1, \ldots, n)$. Furthermore, we assume that all b_{iu}'s and e_{ij}'s are independent. Thus, the (j, k)-element of the within ith subject autocorrelation matrix Σ_i is

$$(\Sigma_i)_{jk} = \sum_{u=1}^{v} \sigma_u^2 \phi_u(t_{ij})\phi_u(t_{ik}) + \sigma^2 \delta_{jk}, (i = 1, \ldots, n)$$

where δ_{jk} equals 1 if $j = k$ and 0 otherwise. The covariance matrix for all observations is a block diagonal matrix with the Σ_i's on the diagonal.

To estimate the mean function f in (17.1), MASAL uses a member from the following class of functions:

$$\{f : f(\mathbf{x}) = \sum_{k=0}^{M} \beta_k B_k(\mathbf{x}), M = 0, 1, \ldots\}, \qquad (17.2)$$

where B_k is a special basis function of the p covariates $\mathbf{x} = (x_1, \cdots, x_p)'$, and β_k is the regression coefficient $(k = 0, 1, \ldots, M)$. Unlike parametric regression, the number of terms M and the individual basis B_k need to be estimated from the data. Particularly, $B_k(\mathbf{x})$ is made of the following two functions

$$(x_k - t)^+ \text{ and } x_k, \ k = 1, \ldots, p, \qquad (17.3)$$

where, for any number a, $a^+ = \max(a, 0)$. $B_k(\mathbf{x})$ can either be one of the functions in (17.3) or the product of those functions with distinct covariates such as $(x_1 - t_1)^+ x_2(x_3 - t_2)^+$.

When Σ_i $(i = 1, \ldots, n)$ is given, [20] described in detail how to select a model from the class (17.2) and use it to estimate the fixed-effect function f in (17.1). Briefly, a forward step is taken first to gradually accumulate terms made of the functions in (17.3), and the terms are added to minimize the (weighted) sum of squared residuals:

$$\sum_{i=1}^{n} (\mathbf{y}_i - \hat{\mathbf{y}}_i)' \Sigma_i^{-1} (\mathbf{y}_i - \hat{\mathbf{y}}_i),$$

where $\mathbf{y}_i = (y_{i1}, \cdots, y_{iT_i})'$, $\hat{\mathbf{y}}_i$ contains the fitted values of $\mathbf{y}_i$, $i = 1, \ldots, n$. The forward step is then followed by a backward step. During the backward step, the least significant term from the model is removed from the model one at a time. The final model is chosen to yield the smallest value of the generalized cross-validation [7, 20, 21, 23].

The primary interest of this work is to estimate random coefficients b_{iu} and their variances, $u = 1, \ldots, v$, $i = 1, \ldots, n$, because the estimation of fixed effects has been addressed previously by [20, 21]. To this end, we first initialize $\sigma_u = 0$, which amounts to assuming the independence of all observations. Then, the f function can be estimated as described above. Next, the residual r_{ij} between y_{ij} and its predicted value from the estimated f can be used to estimate Σ_i by the maximum likelihood method, $j = 1, \ldots, T_i, i = 1, \ldots, n$. Note that Σ_i is determined by an array of variance parameters $(\sigma_1^2, \cdots, \sigma_v^2, \sigma^2)$, $i = 1, \ldots, n$. Either the simplex algorithm of Nelder and Mead [13] or a quasi-Newton algorithm can be used to maximize the following likelihood function:

$$-\frac{1}{2} \sum_{i=1}^{n} \{T_i \log(2\pi) + \log(|\Sigma_i|) + \mathbf{r}_i' \Sigma_i^{-1} \mathbf{r}_i\}, \qquad (17.4)$$

where $\mathbf{r}_i = (r_{i1}, \cdots, r_{iT_i})'$, $i = 1, \ldots, n$.

After the covariance matrix is estimated, we can decompose the residuals into terms involving random coefficients, i.e., $b_{iu} \phi_u(t_{ij})$, and leave the remaining to be e_{ij}, $u = 1, \ldots, v$, $j = 1, \ldots, T_i, i = 1, \ldots, n$. Specifically, the kernel of the joint log likelihood of $(\mathbf{y}_i, b_{i1}, \cdots, b_{iv})$ is

$$-T_i \log \sigma - \sum_{u=1}^{v} \log \sigma_u - \frac{1}{2} [\sum_{u=1}^{v} \frac{b_{iu}^2}{\sigma_u^2} + \sum_{j=1}^{T_i} \{r_{ij}$$

$$- \sum_{u=1}^{v} b_{iu} \phi_u(t_{ij})\}^2 / \sigma^2].$$

If we replace $(\sigma, \sigma_1, \ldots, \sigma_v)$ with their estimates $(\hat{\sigma}, \hat{\sigma}_1, \ldots, \hat{\sigma}_v)$, maximizing the likelihood above over $(b_{i1}, \cdots, b_{iv})$ leads to the estimates

$$\hat{b}_{iu} = \hat{\sigma}_u (\phi_u(t_{i1}), \cdots, \phi_u(t_{iT_i})) \hat{\Sigma}_i^{-1} \mathbf{r}_i, \quad i = 1, \ldots, n, \qquad (17.5)$$

and

$$e_{ij} = r_{ij} - \sum_{u=1}^{v} \hat{b}_{iu} \phi_u(t_{ij}), \quad j = 1, \ldots, T_i, i = 1, \ldots, n.$$

This procedure for estimating the random coefficients are sometimes referred to as a two-stage estimation in mixed-effects linear models (see [3], Chapter 6). The equation (17.5) actually gives rise to the empirical Bayes estimates of the random coefficients, namely, their conditional expectations given the covariance structure and the data, [2]. In addition, the estimates in (17.5) would be called the best linear unbiased predictor (BLUP) under a mixed linear model [14].

Recall that the residuals r_{ij} $(j = 1, \ldots, T_i, i = 1, \ldots, n)$ used above were obtained by assuming independent observations, i.e., $\sigma_u = 0, u = 1, \ldots, v$. Then, the variance parameters σ_u $(u = 1, \ldots, v)$ and σ were updated by maximizing the likelihood (17.4). MASAL can use these variance estimates to derive another estimate for the fixed-effect function f and to reproduce the residuals r_{ij} $(j = 1, \ldots, T_i, i = 1, \ldots, n)$. This leads to another iteration of estimating the covariance matrix and the random coefficients. Based on empirical evidence, [20, 21], and some theoretical studies in related subjects, [1, 11], further iterations are usually not warranted.

17.3 Application

In this section, we re-analyze a data set consisting of the growth curves from 298 infants. These growth curves were collected in a retrospective study by Dr. John Leventhal and his colleagues at Yale University School of Medicine, New Haven, Connecticut. All 298 children were born at Yale-New Haven Hospital from September 1, 1989 through September 30, 1990. They were recruited into the study because their mothers were either regular cocaine users or clearly not cocaine users. The classification for cocaine use was determined by the clinicians' review of the medical records of the mothers and toxicology screens of the infants. Detailed eligibility criteria have been reported previously elsewhere such as [17] and [15].

The response variable for the present analysis is the measurements of weight (in kilograms) from the birth to one and half years old. Race (either white or black, x_1), number of previous pregnancies of the mother (x_2), age of the mother at delivery (x_3), infant's gender (x_4), mother's cocaine use (x_5), and gestational age (x_6) are the covariates. The purpose is to characterize the infant growth pattern and examine its relationship to the covariates.

As described above, I started the estimation process by assuming independent observations. The f function is estimated as follows:

$$0.863 + 0.031t - 0.017(t - 151)^+$$
$$+(0.00024x_2 - 0.0015x_4)t + 0.225(x_6 - 27)^+$$
$$-0.0003(x_6 - 35)^+t + (0.0082x_5 - 0.022)x_3. \qquad (17.6)$$

Then, residuals were calculated from (17.6). Based on those residuals, the covariance matrix and random coefficients were estimated. Specifically, I used two ϕ functions and chose $\phi_1(t) = 1$ and $\phi_2(t) = t$. Using the estimated covariance matrix, MASAL re-estimated the f function with

$$0.465 + 0.033t - 0.008(t - 61)^+ - 0.012(t - 140)^+$$
$$+\{0.21 - 0.0003(t - 353)^+\}(x_6 - 27)^+. \qquad (17.7)$$

The final variance estimates are $\hat{\sigma}_1^2 = 0.22, \hat{\sigma}_2 = 7/10^6$, and $\hat{\sigma}^2 = 0.16$.

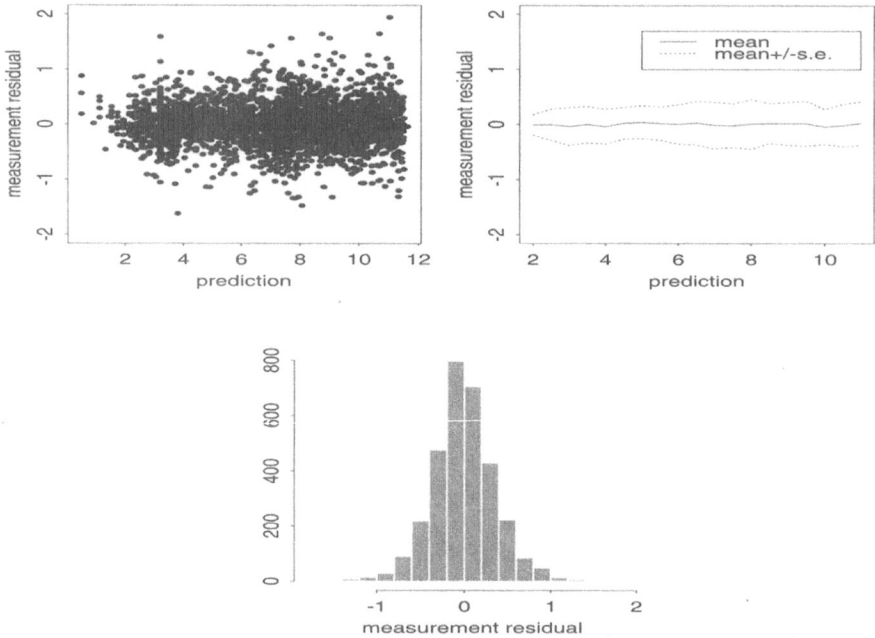

Figure 17.1. Residual Plots. The top-left panel plots the measurement residuals against the predicted values. The top-right panel displays the mean and the mean ± standard error curves for the predicted values in a fixed neighborhood. At the bottom is the histogram of the measurement residuals.

Figure 17.1 displays the measurement residuals $(\hat{e}_{ij})$ against the the prediction (the top-left panel). Because the plot is very crowded and may mislead our vision, I calculated the means and standard errors of the residuals within fixed neighborhoods of the predicted values. The result is displayed in the top-right panel of Figure 17.1. The mean is clearly near zero along the predicted value and the standard error remains relatively constant. In addition, the histogram at the bottom of Figure 17.1 shows a reasonable resemblance of the residual distribution to a normal distribution.

Figure 17.2 plots the random intercept estimates and shows an independent pattern of the U_i's among all subjects. The histogram is also consistent with the normal assumption. Similar conclusions can be drawn for the random slope estimates based on Figure 17.3. Thus, it is evident from Figures 17.1 to 17.3 that the independence and normality assumptions with regard to the random components are appropriate and validated.

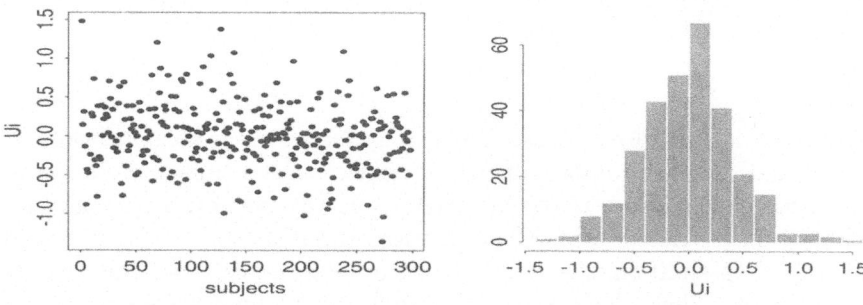

Figure 17.2. Random Intercept U_i and Its Histogram

In the previous work, [20, 21], there is no systematic mechanism to diagnose the appropriateness of the covariance structure. As demonstrated by Figures 17.1 to 17.3, the setup of model (17.1) makes this diagnosis possible.

The interpretation of model (17.7) is rather straightforward. Age (t) is an obvious factor of growth. It is interesting to note that the growth gradually slows down and can be divided into four major stages during the period of one and half years. These four stages are defined by three time points approximately at 2 months, 5 months, and 1 year. One growth modifying factor is gestational age (x_6). The term $\{0.21 - 0.0003(t - 353)^+\}(x_6 - 27)^+$ suggests that infants born in the last trimester are heavier and their weights are proportional to their gestational age. However, the older their gestational age, the slower they they grow after one year old.

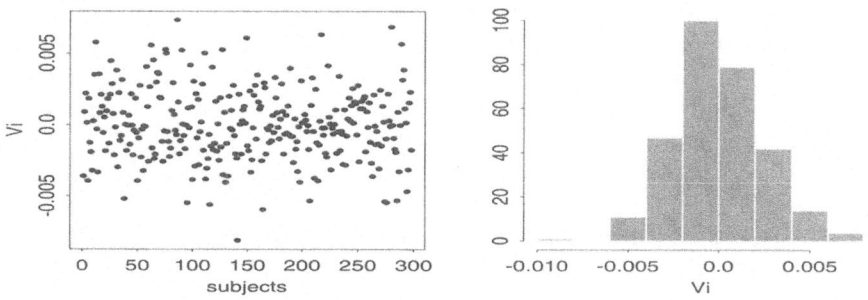

Figure 17.3. Random Slope V_i and Its Histogram

17.4 Discussion

I have introduced the mixed-effects multivariate adaptive splines model. The fixed effects in this model is a nonparametric component and is fitted by multivariate adaptive splines. To my knowledge, this is the only available model that does not impose functional restrictions on time and covariates *a priori*. Not only does it accommodate time varying covariates, but also it allows un-restricted interactions among covariates and between time and covariates. In addition, the fitted models are easy to calculate and readily interpretable. Thus, the mixed-effects multivariate adaptive splines model is handy for data-mining longitudinal data.

The mixed-effects multivariate adaptive splines model is further developed from MASAL by introducing random effects as a systematic way to fit the correlation structure underlying the longitudinal data. In this work, the random coefficients are associated with time only and incorporated in a way similar to [12]. However, it is important to note that the model requires a pre-specification of the random effects. In the future research, further steps must be taken so that the random effects can depend on time and other covariates and that they are data-driven. This effort would lead to a completely nonparametric model for longitudinal data analyses.

The purpose of this work is to estimate both the fixed and random effects for longitudinal data. An important issue that needs to be resolved in the future is how to conduct statistical inference for the estimates of fixed effects. This involves significance tests for curves or high dimensional surfaces (see, e.g., [6]). Unlike mixed-effects linear or semiparametric models for which the fixed effects are pre-specified as linear, the fixed effects under the mixed-effects multivariate adaptive splines model are not specified and are likely nonlinear. Finding an ideal solution is a theoretical challenge. However, some remedies are worth considerations before a theoretical solution is available. For a continuous covariate, we can re-sample the data using the bootstrap method [5] and examine the variability of the knot locations with respect to the bases associated with this continuous covariate. If the knots are within a close vicinity, we can perform the inference as if the knots were fixed. For a categorical variable, it enters into the model through the dummy variables that are created from it. Since the dummy variables are dichotomous, the knot locations are not an issue when they are selected. Therefore, we treat them as if they were prespecified in the inference.

Even though such remedies are not completely satisfactory, it would be a misperception to think automatically that they are less rigorous than what is done in a mixed-effects linear model. The assumption of linear effects is rarely appropriate in applications and it is a matter of the extent to which they are inappropriate. When the linear assumption is violated, the inference based on this assumption is misleading. On the other hand, the mixed-effects multivariate adaptive splines model provides closer approximations to the fixed effects so that the inference is at least built on a reasonable basis.

One concern with the post hoc inference by treating the bases as if they were fixed *a priori* is that the significance tests are potentially over optimistic, due to the extensive basis selection. Although this is a valid concern, the sensitivity and specificity analysis reported in [22] should alleviate such a concern.

References

[1] Altman, N. S. (1992). An iterated Cochrane-Orcutt procedure for nonparametric regression. *Journal of Statistical Computation and Simulation* **40**, 93–108.

[2] Carlin, B.P. and Louis, T.A. (1996) *Bayes and Empirical Bayes Methods for Data Analysis,* New York: Chapman and Hall.

[3] Crowder, M. J. and Hand, D. J. (1991). *Analysis of Repeated Measures,* New York: Chapman and Hall.

[4] Diggle, P.J., Liang, K.Y., and Zeger, S.L. (1994). *Analysis of Longitudinal Data.* New York: Oxford University Press, Inc.

[5] Efron, B. and Tibshirani, R. (1993) *An Introduction to the Bootstrap,* Chapman and Hall, New York.

[6] Fan, J. and Lin, S.K. (1998) "Test of significance when data are curves," *Journal of the American Statistical Association,* **93**, 1007-1021.

[7] Friedman, J. H. (1991). Multivariate adaptive regression splines. *The Annals of Statistics* **19**, 1-141.

[8] Hoover, D.R., Rice, J.A., Wu, C.O., and Yang, L.P. (1998). Nonparametric smoothing estimates of time-varying coefficient models with longitudinal data. *Biometrika* **85**, 809-822.

[9] Laird, N.M. and Ware, J.H. (1982). Random-effects models for longitudinal data. *Biometrics* **38**, 963-974.

[10] Lin, D.Y. and Ying, Z. (2000). Semiparametric and nonparametric regression analysis of longitudinal data. *Journal of the American Statistical Association,* in press.

[11] Lin, X.H. and Carroll, R.J. (2000). Nonparametric function estimation for clustered data when the predictor is measured without/with error. *Journal of the American Statistical Association* **95**, 520-534.

[12] Meredith, W. and Tisak, J. (1990). Latent curve analysis. *Psychometrika*, **55**, 107-122.

[13] Nelder, J.A. and Mead, R. (1965) A simplex method for function minimisation. *The Computer Journal* **7**, 303-313.

[14] Robinson, G.K. (1991) That BLUP is a good thing: The estimation of random effects. *Statistical Science*, **6**, 15-51.

[15] Stier, D.M., Leventhal, J.M., Berg, A.T., Johnson, L., and Mezger, J. (1993). Are children born to young mothers at increased risk of maltreatment? *Pediatrics* **91**, 642-648.

[16] Truong, Y. K. (1991). Nonparametric curve estimation with time series errors. *Journal of Statistical Planning and Inference* **28**, 167-183.

[17] Wasserman, D.R. and Leventhal, J.M. (1993). Maltreatment of children born to cocaine-dependent mothers. *American Journal of Diseases of Children* **147**, 1324-1328.

[18] Zeger, S.L. and Diggle, P.J. (1994). Semi-parametric models for longitudinal data with application to CD4 cell numbers in HIV seroconverters. *Biometrics* **50**, 689-699.

[19] Zhang, H. P. (1994). Maximal correlation and adaptive splines. *Technometrics* **36**, 196-201.

[20] Zhang, H. P. (1997). Multivariate adaptive splines for the analysis of longitudinal data. *Journal of Computational and Graphical Statistics* **6**, 74-91.

[21] Zhang, H. P. (1999). Analysis of infant growth curves using multivariate adaptive splines. *Biometrics* **55**, 452-459.

[22] Zhang, H. P. (2000). Mixed-effects multivariate adaptive splines models. *Amer. Statist. Assoc. 2000 Proceedings of the Biometrics Section*, 20-29.

[23] Zhang, H. P. and Singer, B. (1999). *Recursive Partitioning in the Health Sciences*, Springer, New York.

18

Statistical Inference for Simultaneous Clustering of Gene Expression Data

Katherine S. Pollard and Mark J. van der Laan [1]

Summary

Current methods for analysis of gene expression data are mostly based on clustering and classification of either genes or samples. We offer support for the idea that more complex patterns can be identified in the data if genes and samples are considered simultaneously. We formalize the approach and propose a statistical framework for simultaneous clustering. A simultaneous clustering parameter is defined as a function $\theta = \Phi(P)$ of the true data generating distribution P, and an estimate is obtained by applying this function to the empirical distribution P_n. We illustrate that a wide range of clustering procedures, including generalized hierarchical methods, can be defined as parameters which are compositions of individual mappings for clustering patients and genes. This framework allows one to assess classical properties of clustering methods, such as consistency, and to formally study statistical inference regarding the clustering parameter.

18.1 Motivation

Gene expression studies are swiftly becoming a very significant and prevalent tool in biomedical research. The microarray and gene chip technologies

---

[1]Mark J. van der Laan is Professor in Biostatistics and Statistics and Katherine S. Pollard is a doctoral candidate in Biostatistics in the School of Public Health, Division of Biostatistics, University of California, Berkeley. This research has been supported by a grant from the Life Sciences Informatics Program with industrial partner Chiron Corporation, Emeryville, CA. Author for correspondence: Mark J. van der Laan, Div. of Biostatistics, Univ. of California, School of Public Health, Earl Warren Hall #7360, Berkeley, CA 94720-7360.

allow researchers to monitor the expression of thousands of genes simultaneously. A typical experiment results in an observed data matrix X whose columns are n copies of a p-dimensional vector of gene expression measurements, where n is the number of observations and p is the number of genes. Consider, for example, a population of cancer patients from which we take a random sample of n patients, each of whom contributes p gene expression measurements. For microarrays, each measurement is a ratio, calculated from the intensities of two flourescently labeled mRNA samples cohybridized to arrays spotted with known cDNA sequences. Gene chips produce similar data, except each element is a quantitative expression level rather than a ratio. Data preprossessing may include background subtraction, combining data from replicated spots representing the same cDNA sequence, normalization, log transformation, and truncation.

Given data from such an experiment, researchers are interested in identifying groups of differentially expressed genes which are *significantly correlated with each other*, since such genes might be part of the same causal mechanism or pathway. For example, healthy and cancerous cells can be compared within subjects in order to learn which genes tend to be differentially expressed together in the diseased cells; regulation of such genes could produce effective cancer treatment and/or prophylaxis [5, 6, 14, 16]. In addition to identifying interesting clusters of genes, researchers often want to find subgroups of samples which share a common gene expression profile. Examples of such studies include classifying sixty human cancer cell lines [19], distinguishing two different human acute leukemias [10], dissecting and classifying breast cancer tumors [17], and classifying sub-types of B-cell lymphoma [1] and cutaneous malignant melanoma [2].

Most gene expression research to date has focused on either the gene problem or the sample problem, or possibly on the two problems as separate tasks. Many researchers have performed two-way clustering by applying algorithms to both genes and samples separately and possibly reordering the rows and columns of the data matrix according to the clustering results. Tibshirani *et al.* [20], for example, illustrate several methods for two-way visualization of a reordered data matrix based on separately clustering genes and samples. They also propose applications of block clustering and principal components analysis to simultaneously cluster genes and samples. Their gene shaving methodology addresses the problem that different subsets of genes might cluster samples in different ways. With the same problem in mind, Getz, Levine and Dommany [9] propose an algorithm (coupled two-way clustering) for simultaneously identifying stable and significant subsets of both genes and samples.

We want to emphasize the importance of identifying simultaneous subsets of genes and samples. Consider, for example, that different gene clusters represent different biological mechanisms or states, so that there may be clusters of genes which are very good for distinguishing different types of samples. In addition to an overall clustering label, sub-groups of samples could be more accurately characterized by their expression pattern for each of these gene clusters. In this way, we might find that in an experiment consisting of three types of cancer, several gene clusters show similar expression across all three types (generic cancer genes), one gene cluster shows two types of cancer clustering together, and another gene cluster shows a different two types of cancer clustering together. It is possible that clustering the samples using all genes might not have revealed the three types of cancer, whereas the characterization based on the patient clustering within separate gene clusters would make the distinction clear. Similarly, first clustering samples and then genes within sample clusters could offer more insight than simply clustering genes. For example, in a heterogeneous sample of patients, clustering patterns seen in individual patient groups might disappear when averaged across groups or the profile of the more numerous patient group might dominate. Concern over failing to identify these more subtle and complicated patterns in gene expression data has motivated the development of a general statistical framework for simultaneous clustering.

18.2   Background

Approaches to gene expression data analysis rely heavily on results from cluster analysis (e.g., k-means, self-organizing maps and trees), supervised learning (e.g., recursive partitioning), and classification and regression trees (CART). For unsupervised clustering, we recommend the clustering algorithm Partitioning Around Medoids (PAM) [12], because it is nonparametric and more robust to outliers than many other methods. PAM also has the advantage of taking a general dissimilarity matrix (based on any distance metric of interest) as input. The statistical framework we present can be applied with any choice of clustering algorithm.

All exploratory techniques are capable of identifying interesting patterns in data, but they do not inherently lend themselves to statistical inference. It is necessary to define important statistical notions such as parameter, parameter estimate, consistency, and confidence. The ability to assess reliability in an experiment is particularly crucial with the high dimensional data structures and relatively small samples presented by gene-expression experiments. In the biotechnology industry, for example, where thousands of potential drug targets are screened simultaneously, it is imperative to have a method to determine the significance of gene clusters found in a single microarray experiment.

Others have noted this need for statistical rigor in gene expression data analysis [11, 15, 9]. Recent papers propose methods which apply both the jackknife [23] and the bootstrap [21, 13] approaches to perform statistical inference with gene expression data. van der Laan and Bryan [21], present a general statistical framework for clustering genes using a deterministic subset rule applied to (μ, Σ), the mean and covariance of the gene expression distribution. A typical subset rule will draw on "screens" and "labellers". A screen is used to eliminate certain genes from the subset. A labeller will apply labels, such as the output of a clustering routine. Meaningful analyses can be done with various combinations of screens and labellers or even with a screen or labeller alone. For example, one might apply PAM to all genes at least two-fold differentially expressed. The target subset $S(\mu, \Sigma)$ is the subset (with cluster labels) the subset rule would select if the true data generating distribution were known, and it is estimated by the observed sample subset $S(\hat{\mu}_n, \hat{\Sigma}_n)$, where the empirical mean and covariance are substituted for the true parameters. Most currently employed clustering methods fit into this framework, since they need only be deterministic functions of the empirical mean and covariance $\hat{\mu}_n, \hat{\Sigma}_n$. The authors also provide measures and graphs of gene cluster stability based on the parametric bootstrap using a truncated multivariate normal distribution. Finally, they establish consistency under $\frac{n}{\log(p)} \to \infty$ of:

1. $\hat{\mu}_n, \hat{\Sigma}_n$ and hence smooth functions $S(\hat{\mu}_n, \hat{\Sigma}_n)$,

2. the parametric bootstrap for the limiting distribution of $\sqrt{n}(\hat{\mu}_n - \mu, \hat{\Sigma}_n - \Sigma)$ and simple convergence of the bootstrap subset to the true subset.

18.3   Simultaneous Clustering Parameter

In this chapter we discuss a generalization of the method of van der Laan and Bryan [21] such that the subset rule may include a multi-stage clustering method which involves simultaneous clustering of genes and samples. In order to be concrete, we will refer to the samples as patients, but the methodology applies to any i.i.d. experimental units. Define a simultaneous clustering parameter as a function of the true data generating distribution which is a composition of a mapping involving clustering of patients and a mapping involving clustering of genes. An estimate of the simultaneous clustering parameter is obtained by applying this function to the empirical distribution. This formal framework allows us to assess classical properties of the clustering method such as consistency (in the novel context of $n << p$) and also allows us to study statistical inference regarding the clustering parameter.

18.3.1   Clustering Patients

Given k_1 and an m-variate distribution P, let $\Phi_1(P) = (P_j(P), p_j(P) : j = 1, \ldots, k_1)$ be an algorithm that maps P into k_1 m-variate distributions $P_1, \ldots, P_{k_1}$ and corresponding mixing proportions $p_1, \ldots, p_{k_1}$. This mapping $\Phi_1(P)$ represents clustering of patients.

For example, one could fit to P a mixture of k_1 m-variate normal distributions. In other words, let $f(x \mid (\mu_j, \Sigma_j, p_j), j = 1, \ldots, k_1)$ be the density of a mixture $\sum_j p_j N(\mu_j, \Sigma_j)$ of k_1 multivariate normal distributions. Now, we could define $\Phi_1(P) = \max^{-1} \int \log(f(X \mid (\mu_j, \Sigma_j, p_j), j = 1, \ldots, k_1)) dP(x)$ as the distribution (i.e., parameters) which maximizes the log-likelihood, where the maximum is taken over all mixtures of multivariate normal distributions. In this case, the algorithm involves maximum likelihood estimation over a mixture of parametric families, which could be carried out with the EM-algorithm. This clustering methodology is carried out by Fraley and Raftery [8], and it is referred to as model based clustering.

The alternative to model based clustering is nonparametric clustering, which is much less computer intensive and allows specification of a user-supplied distance matrix. One particular method of nonparametric clustering is PAM. In this case, the easiest way to define $\Phi_1(P)$ is by simulation. In other words,

1. sample an infinitely large number (say N) observations from P,

2. compute a $N \times N$-distance matrix containing the pairwise distances for each pair of observations for some specified distance metric,

3. apply the clustering algorithm PAM to split the N observations into k_1 groups,

4. for each group j, report the proportions of observations p_j and the empirical cumulative distribution function P_j of these observations, $j = 1, \ldots, k_1$.

If $N = \infty$, then the output of this simulation algorithm is a function $\Phi_1(P) = (P_j, p_j : j = 1, \ldots, k_1)$.

18.3.2 Clustering Genes

Given k_2 and an m-variate distribution Q, we define an algorithm $\Phi_2(Q) = (m_j(Q), S_j(Q), G_j(Q) : j = 1, \ldots, k_2)$ that maps Q into k_2 $m_j(Q)$-variate subdistributions $G_1(Q), \ldots, G_{k_2}(Q)$ of Q corresponding with subsets $S_j(Q)$ of $\{1, \ldots, m\}$ of sizes m_j. This algorithm represents the clustering of m genes into k_2 clusters S_j of genes of sizes m_j, $j = 1, \ldots, k_2$. We might also extend the definition of S_j so that it includes not just the clusters, but also other output of a clustering algorithm such as probabilities that genes belong to a cluster (fuzzy membership).

As highlighted in van der Laan and Bryan [21], the finding of the actual clusters $s_j(Q)$ of genes, $j = 1, \ldots, k_2$, can typically be based on the m-variate mean vector $\mu(Q)$ and the $m \times m$-covariance matrix $\Sigma(Q)$ of Q. In that case, $s_j(Q) = s_j(\mu(Q), \Sigma(Q))$, $j = 1, \ldots, k_2$. For example, $\Phi_2(Q)$

can be defined by applying PAM to a $m \times m$-distance matrix (Euclidean, correlation, absolute correlation, etc.) calculated from the $m \times m$-covariance matrix $\Sigma(Q)$.

18.3.3 Compositions

To summarize we have:

$$\Phi_1(P) = (p_j(P), P_j(P) : j = 1, \ldots, k_1)$$
$$\Phi_2(Q) = (m_j(Q), S_j(Q), G_j(Q) : j = 1, \ldots, k_2),$$

where Φ_1 and Φ_2 are defined by what algorithm you want to use (e.g., clustering with a specified distance metric). Note that Φ_1 splits the population into k_1 subpopulations so that applied to patients it will split the sample into k_1 subsamples, while Φ_2 splits the dimension into subdimensions. These mappings $\Phi_1(P)$ and $\Phi_2(Q)$, representing clustering of patients and genes, are the building blocks for simultaneous clustering parameters $\Phi(P)$.

Define the composition $\Phi_2 \circ \Phi_1(P)$ as a $k_1 \times k_2$-matrix with (I, J)-th element being $(p_I(P), P_I(P), m_J(P_I(P)), S_J(P_I(P)), G_J(P_I(P)))$, i.e. it reports the I-subpopulation distribution of patients computed by $\Phi_1(P)$ and the corresponding J-th cluster of genes. Similarly, define the composition $\Phi_1 \circ \Phi_2(P)$ as a $k_2 \times k_1$-matrix with (I, J)-element being $(m_I(P), S_I(P), G_I(P), p_J(G_I(P)), P_J(G_I(P)))$, i.e. it reports the I-th cluster of genes and clusters patients just based on the genes in this I-th cluster. Then, $\Phi_2 \circ \Phi_1(P)$ represents clustering genes within each cluster of patients and $\Phi_1 \circ \Phi_2(P)$ represents clustering patients within each cluster of genes. Figure 18.1 illustrates how a data set is recursively partitioned in the composition $\Phi_1 \circ \Phi_2(P)$. Stopping at this stage often reveals interesting patterns in the data.

One might find it of interest to iterate these compositions in order to design more "aggressive" algorithms. The mappings can be composed alternately. For example, one might 1) cluster genes, 2) within each cluster of genes cluster patients, and 3) within each cluster of patients cluster genes again. The resulting simultaneous clustering parameter $\Phi_2 \circ \Phi_1 \circ \Phi_2(P)$ reports the I-th cluster of genes based on all patient data, the J-th cluster of patients just based on I-th cluster of genes and the K-th cluster of genes just based on this J-th cluster of patients. It is natural to think of such a multi-way-array as a tree which is many levels deep so that to get to the (I, J, K)-th element of the tree, for example, one goes to branch I at the first node, branch J at the second node and branch K at the third node. One could also iterate one mapping repeatedly. Define

1. $\Phi_1 \circ \Phi_1(P)$ as a $k_1 \times k_1$-matrix with (I, J)-th element being

$$(p_I(P), P_I(P), p_J(P_I(P)), P_J(P_I(P))),$$

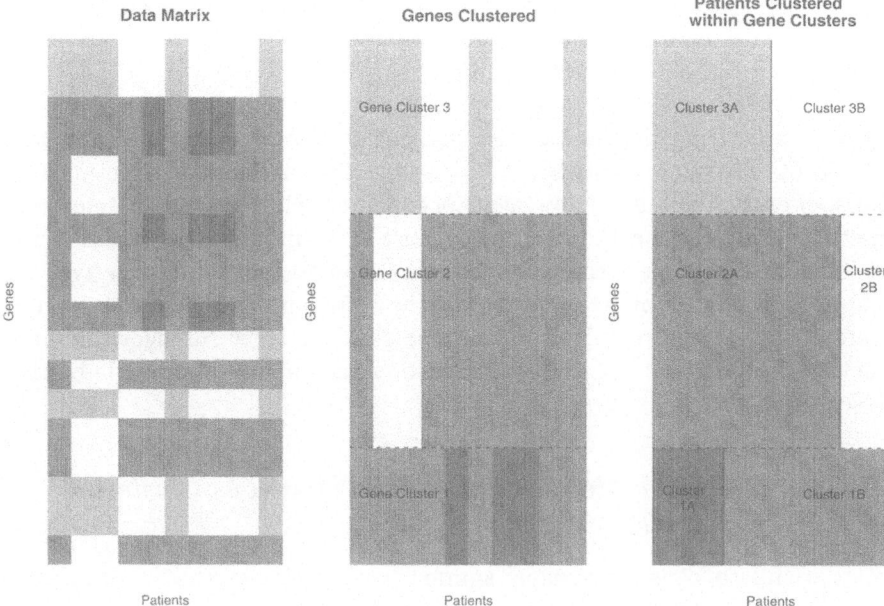

Figure 18.1. Clustering patients within clusters of genes $(\Phi_1 \circ \Phi_2(P))$ involves first splitting the rows (genes) into clusters. The rows can be rearranged so that the genes in a cluster lie together. Then, the columns (patients) are split into clusters using only the rows in each gene cluster.

 i.e. it reports the I-th cluster of patients based on all data and the J-th cluster of patients just based on this I-th cluster of patients.

2. $\Phi_2 \circ \Phi_2(P)$ as a $k_2 \times k_2$-matrix With (I, J)-th element being

$$(m_I(P), S_I(P), G_I(P), m_J(G_I(P)), S_J(G_I(P)), G_J(G_I(P))),$$

 i.e. it reports the I-th cluster of genes based on all data and the J-th cluster of genes just based on this I-th cluster of genes.

Thus, clustering of samples (or genes) is defined within an earlier obtained cluster of samples (or genes). This is top-down (or divisive) hierarchical clustering. It is easy to see that repeatedly applying Φ_1 or Φ_2 results a final hierarchical tree structure of the type produced by Eisen's application of bottom-up (agglomerative) hierarchical clustering [7]. Note that contrary to his algorithm, we are not necessarily restricted to binary splits, making this approach more general. The block clustering method for gene expression data described in Tibshirani *et al.* [20] can also be expressed as a hierarchical iteration which includes both Φ_1 and Φ_2. It is important to remember that applying such iterated algorithms to the actual data yields very aggressive search algorithms. We must keep in mind that as more compositions are taken, the mapping becomes less smooth so that there is a trade off between stability (i.e., sample size needed) and an aggressive search (i.e., finding patterns in the data).

Now, we can define a simultaneous clustering parameter $\theta = \Phi(P)$ as some (possibly iterative) composition of Φ_1, Φ_2. The parameter depends on the underlying distribution P, the specific composition, and also the choice of clustering algorithm(s) (including the method for selecting the number of clusters) used in Φ_1, Φ_2. Since different clustering algorithms and methods for selecting the number of clusters define different parameters, it is important to perform simulations in order to understand the type of parameter defined by a particular method and in what contexts such a parameter is of interest. Here we assume that the choice of algorithm and method for selecting the number of clusters has already been made by the researcher.

18.3.4 Summary Measures of the Simultaneous Clustering Parameter

It is useful to consider various summary measures of the simultaneous clustering parameter θ. Examples of such summary measures include:

- **Cluster Membership (Genes only):** probabilities (fuzzy clustering), assignments (hard clustering). Recall that patient cluster labels are not a parameter.

- **Cluster Size:** proportion in each cluster

- **Cluster Profile:** means, medoids, indicator that a gene cluster shows strong differential expression (uniform) in at least one of the patient clusters.

- **Cluster Strength:** diameter, separation, silhouettes, measures of how uniformly the genes behave across patients.

Many of these measures can be summarized together in a picture. For example, we can order the gene clusters within each cluster of patients, and then order patients and genes within clusters. Ordering can be based on distance between clusters, silhouette, or some dimension reducing projection (multidimensional scaling, principal components). For each cluster of patients, we can visualize (i) the gene-by-patient reordered data matrix and (ii) the gene-by-gene reordered correlation (or dissimilarity) matrix.

18.3.5 Estimation and Remarks Regarding Asymptotic Consistency

Let P_n be the empirical distribution of the data. We estimate $\theta = \Phi(P)$ with $\theta_n = \Phi(P_n)$. Note that this is a nonparametric estimate since it is not based on model assumptions. In particular, if $\Phi : (D_1, \| \cdot \|_1) \to (D_1, \| \cdot \|_2)$

is continuous w.r.t to a metric $\| \cdot \|$ for which $\| P_n - P \| \to 0$ in probability and p is fixed, then $\| \theta_n - \theta \|_1 \to 0$ in probability.

Since, in practice, the number of genes being studied p continues to grow and to grow much more rapidly than sample size n, it is of interest to establish consistency of summary measures of θ_n in the context that the number of genes $p = p(n)$ increases with n such that $n \to \infty$ and $n/\log(p(n)) \to \infty$. In general, simultaneous clustering parameters will be much less smooth than $S(\mu, \Sigma)$ from van der Laan and Bryan [21], and thus consistency, sample size and asymptotic validity of the bootstrap are issues we need to address. One benefit of having a composition is that we only need consistency of each mapping alone in order to show consistency of their composition. Since we already have consistency for gene clustering, it remains to look at patient clustering.

In this section, we will explain why in principal under appropriate regularity conditions we can extend the consistency and sample size proofs of van der Laan and Bryan [21] to the simultaneous clustering context. Let $\theta_j = \theta_j(P)$ be a real valued parameter of the data generating distribution P of the p-dimensional gene expression profiles X, where j indexes a large set of such parameters, say $j = 1, \ldots, r(p)$. Here $r(p)$ is monotone function with bounded derivative in the number of genes p. In particular, we can view θ_j as one of the many parameters of a simultaneous clustering $\Phi(P) = \Phi_1 \circ \Phi_2(P)$. For a fixed number of genes p and increasing sample size n one will typically have that the empirical estimate $\theta_{jn} = \theta_j(P_n)$ is asymptotically linear:

$$\theta_{jn} - \theta_j = \frac{1}{n} \sum_{i=1}^{n} IC_j(X_i) + r_{jn},$$

where $| r_{jn} | = o_P(1/\sqrt{n})$. Here $Y \to IC_j(X)$ is the influence curve of the estimator θ_{jn} and $E\{IC_j(X)\} = 0$. Let $\sigma_j^2 = \mathrm{VAR} IC_j(X)$ denote the variance of the influence curve $IC_j(X)$.

Now we can state conditions under which the parameters θ_{jn} are uniformly consistent in j, even in the realistic setting where the arrays keep getting larger and larger. Let's now assume that (1) the log gene expressions are truncated, so that they are bounded from above by a universal constant M, (2) the influence curves IC_j are uniformly bounded in all possible X's by a universal constant C, and (3) the second order term $\sqrt{n} r_{jn}$ converges to zero uniformly in j in probability. Condition 3 requires typically the same assumptions as condition 2. In other words, once one makes sure that denominators in the influence curve of the estimator θ_{jn} are uniformly bounded away from zero, one will often also be able to prove a uniform bound on the second order terms.

Let $\tilde{\theta}_{jn} \equiv \theta_j + \frac{1}{n} \sum_i IC_j(X_i)$ be the first order approximation of θ_{jn}. In van der Laan and Bryan [21] it is shown that if the number of genes

$p = p(n)$ is such that $n/\log(p(n)) \to \infty$ as $n \to \infty$, then, as $n \to \infty$,

$$\max_j |\tilde{\theta}_{jn} - \theta_j| \to 0 \text{ in probability,}$$

so that, by condition 3,

$$\max_j |\theta_{jn} - \theta_j| \to 0 \text{ in probability.}$$

The same Bernstein's Inequality argument as used in van der Laan and Bryan [21] leads to a sample size formula based on the first order approximation $\tilde{\theta}_{jn}$ of θ_{jn} in the more concrete setting of a fixed value of the number of genes p. Define n^* with the following formula:

$$n^*(p, \epsilon, \delta, C, \sigma^2) \;=\; \frac{1}{c}(\log p + \log \frac{2}{\delta}),$$

where $c = c(\epsilon, \sigma^2, C) = \frac{\epsilon^2}{2\sigma^2 + 2C\epsilon/3}$. In the above, $\sigma^2 = \max_j \sigma_j^2$ and δ is a user-specified value between 0 and 1 that can be thought of as 1 minus the "power". If $n > n^*$, then

$$P\left(\max_j |\tilde{\theta}_{jn} - \theta_j| > \epsilon\right) < \delta.$$

It is of interest to see that the effect of the number of genes on this sample size formula (and the truly needed sample size) is very minimal; in other words, if one needs a certain sample size for 10 genes, then adding 50 subjects to the sample will guarantee the same uniform precision based on 100000 genes. This teaches us that achievable sample sizes will allow complete trust in *each* of the elements of the observed estimates θ_{jn}, which will become essential if one is interested in selecting association pathways between genes.

18.4   Statistical Inference with the Bootstrap

Though $\theta = \Phi(P)$ generates a large set of methods for finding clustering patterns in the true data-generating distribution, once applied to empirical data P_n, it is likely to find patterns due to noise. To deal with this issue, one needs methods for assessing the variability of θ_n and, in particular, assessing the variability of the important summary measures of θ_n. One also needs to be able to test if certain components of θ_n are significantly different from the value of these components in a specified null-experiment.

18.4.1   The Bootstrap Method

To assess the variability of the estimator θ_n we propose to use the bootstrap. The idea of the bootstrap method is to estimate the true data generating distribution P with some estimate $\mathbf{P}_n$ and estimate the distribution of θ_n with the distribution of $\theta_n^\# = \Phi(\mathbf{P}_n^\#)$, where $\mathbf{P}_n^\#$ is the empirical distribution based on an i.i.d. bootstrap sample (i.e., a sample of n i.i.d. observations from $\mathbf{P}_n$). The distribution of $\theta_n^\#$ is obtained by applying $\theta = \Phi(\cdot)$ to $\mathbf{P}_n^\#$ from each of B bootstrap samples, keeping track of parameters of interest. The distribution of a parameter is approximated by its empirical distribution over the B samples.

There are several common methods for generating bootstrap samples.

- **Nonparametric:** Resample n columns from X with replacement.

- **Parametric:** Fit a model (e.g., multivariate normal, mixture of multivariate normals) and generate observations from the fitted distribution.

- **Convex pseudo-data:** For $d \in \{0, 0.5\}$, choose $\epsilon \in \{0, d\}$. Then use ϵ to form new samples as convex combinations of pairs of randomly sampled columns of X. This is a smoothed version of the nonparametric bootstrap proposed by Breiman [3].

The nonparametric bootstrap has the advantage of being computationally much easier than the parametric bootstrap. In addition, the nonparametric bootstrap avoids distributional assumptions about the parameter of interest, whereas the estimation of the distribution of $\sqrt{n}(\Sigma_n - \Sigma)$ using the parametric bootstrap is only consistent under the model assumption. There is reason to believe, however, that the parametric bootstrap might perform better in the gene-expression context (where the number of observations n is typically very small relative to the dimension p), because the empirical distribution $\mathbf{P}_n$ (i.e. nonparametric bootstrap) might be an inappropriate estimate of P.

18.4.2   Assymptotic Validity of the Bootstrap

The performance of the bootstrap is measured by how well the distribution of $\theta_n^\#$ approximates the distribution of θ_n. This performance is mainly dependent on how close $\mathbf{P}_n$ is to P. We have conducted simulation studies to assess the asymptotic validity of the nonparametric, convex, and parametric bootstraps for estimating the distribution of a simultaneous clustering parameter. A summary of the results is given here, and we refer the interested reader to other published work [18] for more details. Since θ can take many forms but is always a composition of the mappings Φ_1 and Φ_2, we designed the simulations to look at Φ_1 and Φ_2 separately.

The nonparametric and parametric bootstraps can be used to assess the variability of summary measures of gene clustering with thousands of genes (see also van der Laan and Bryan [21]). Estimated variability in the mean vector is quite accurate and estimated variability in the correlation is increasingly accurate for larger sample sizes. We are able to use the bootstrap to assess the variability of subset rules of the form $S(\mu, \Sigma)$ accurately (or at least conservatively) for reasonable sample sizes ($n < 50$).

We have also shown that the nonparametric and parametric bootstraps can be used to assess the variability of many summary measures of patient clustering for reasonable sample sizes ($n < 50$) and thousands of genes. The asymptotic validity of the bootstrap for estimating the variability of all summary measures of patient clustering is clear as we increase the sample size. This is the true purpose of the bootstrap. Larger sample sizes ($n > 200$) may be needed, however, to avoid bias in estimates of the less smooth summary measures of patient clustering compared to the smoother summary measures of gene clustering. We found that for estimating the distributions of certain parameters (i.e., those that can not be expected to act like a normally distributed random variable for $n < 50$), the parametric bootstrap requires many fewer samples to perform well and hence might be worth using (despite the extra computational effort) in these cases.

By varying both the sample size and the number of genes, we were able to confirm the theoretical result that consistency of estimates and the bootstrap are driven mainly by n with dependence on p only through $log(p)$. Overall, the nonparametric bootstrap performs relatively well (but was some times more conservative) compared to the parametric. The convex bootstrap had inconsistent performance and can not be recommended uniformly as an improvement to the nonparametric bootstrap.

18.5   Conclusions

We have demonstrated that a large family of simultaneous clustering methods can be regarded as compositions of mappings involving clustering of genes and/or mappings involving clustering of patients. In this way, we can define a simultaneous clustering parameter as a function of the underlying data generating distribution that produced the results of a gene expression experiment. This statistical formalism allows us to understand and perform both estimation and inference using many commonly employed clustering methods. By forming iterative compositions, we can also design more aggressive algorithms for finding patterns in data. The beauty of this framework is that statistical inference, using methods such as the bootstrap, allows us to assess the reliability of patterns found by such "greedy" algorithms. In the context of gene expression data, where the dimension of the problem (p = number of genes) far exceeds the sample size (n), the need for this statistical rigor is particularly important and all to often overlooked by researchers keen to identify biological relationships between genes clustered together.

We have illustrated how bootstrap methods can be used to estimate the variability of simultaneous clustering parameters. Our simulations identified several interesting points about the bootstrap. We were pleased to see that for sample sizes in the range of $n = 40$ to 200, the bootstrap is a valid method for estimate the variability of simultaneous clustering parameters. The nonparametric bootstrap performs relatively well compared to the parametric, although it is quite conservative for estimating the variance of some summary measures of θ_n. We were surprised to find that using convex pseudo-data was not an improvement over the nonparametric bootstrap.

Experience applying these methods to data sets [18, 22, 4] supports the idea that simultaneous clustering provides additional insight over one-way clustering results. We have seen that even when the goal is to cluster samples, first clustering genes and then samples within gene clusters can help to identify important relationships amongst samples that correspond with distinct biological states and to highlight genes which are members of a single biochemical/causal pathway. Pictures of the data matrix and particularly the dissimilarity matrix emphasize that these insights are of real interest, since simultaneous clustering labels correspond well with visual patterns seen in the data. Applying the bootstrap helps to understand the reliability of these patterns.

References

[1] Alizadeh A. A., Eisen M. B., Davis R. E., Ma C., Lossos I. S., Rosenwald A., Boldrick J. C., Sabet H., Tran T., Yu X., Powell J. I., Yang L., Marti G. E.,Moore T., Hudson T. Jr., Lu L., Lewis D. B., Tibshirani R., Sherlock G., Chan W. C., Greiner T. C., Weisenberger D. D., Armitage J. O., Warnke R., Levy R., Wilson W., Grever M. R., Byrd J. C., Botstein D., Brown P. O., Staudt L. M., Distinct types of diffuse large B-cell lymphoma identified by gene expression profiling. *Nature* **403** (2000) pp. 503–511.

[2] Bittner M., Meltzer P., Chen Y., Jiang Y., Seftor E., Hendrix M., Radmacher M., Simon R., Yakhini Z., Ben-Dor A., Sampas N., Dougherty E., Wang E., Marincola F., Gooden C., Lueders J., Glatfelter A., Pollock P., Carpten J., Gillanders E., Leja D., Dietrich K., Beaudry C., Berens M., Alberts D., Sondak V., Hayward N., Trent J., Molecular classification of cutaneous malignant melanoma by gene expression profiling. *Nature* **406** (2000) pp. 536–540.

[3] Breiman L., Using convex pseudo-data to increase prediction accuracy. Technical Report no. 513, U. C. Berkeley, Dept. of Statistics, March 1998.

[4] Bryan J., Pollard K. S., van der Laan M. J., Paired and unpaired comparison and clustering with gene expression data. *Statistica Sinica***12** (2002) pp. 87–110.

[5] Debouck C. and Goodfellow P. N., DNA microarrays in drug discovery and development. *Nature Genetics* **21**: 1, suppl. (1999) pp. 48–50.

[6] DeRisi J., Penland L., Brown P. O., Bittner M. L., Meltzer P. S., Ray M., Chen Y., Su Y. A., Trent J. M., Use of a cDNA microarray to analyse gene expression patterns in human cancer. *Nature Genetics* **14** (1996) pp. 456–460.

[7] Eisen M. B., Spellman P. T., Brown P. O., Botstein D., Cluster analysis and display of genome-wide expression patterns. *PNAS* **95** (1998) pp. 14863–14868.

[8] Fraley C. and Raftery A. E., Model-based clustering, discriminant analysis, and density estimation. Technical Report no. 380, Univ. of Washington, Dept. of Statistics, Oct. 2000.

[9] Getz G., Levine E., Domany E., Coupled two-way clustering analysis of gene microarray data. *Proc. Natl. Acad. Sci.* **97** (2000) pp. 12079-12084.

[10] Golub T. R., Slonim D. K., Tamayo P., Huard C., Gaasenbeek M., Mesirov J. P., Coller H., Loh M. L., Downing J. R., Caligiuri M. A., Bloomfield C. D., Lander E. S., Molecular Classification of Cancer: Class Discovery and Class Prediction by Gene Expression Monitoring. *Science* **286** (1999) pp. 321–531.

[11] Hughes T. R., Marton M. J., Jones A. R., Roberts C. J., Stoughton R., Armour C. D., Bennett H. A., Coffey E., Dai H., He Y. D., Kidd M. J., King A. M., Meyer M. R., Slade D., Lum P. Y., Stepaniants S. B., Shoemaker D. D., Gachotte D., Chakraburtty K., Simon J., Bard M., Friend S. H., Functional discovery via a compendium of expression profiles. *Cell* **102** (2000) pp. 109–126.

[12] Kaufman L. and Rousseeuw P. J. *Finding Groups in Data: An Introduction to Cluster Analysis* (John Wiley & Sons, New York, 1990).

[13] Kerr M. K. and Churchill G. A., Bootstrapiing cluster analysis: Assessing the reliability of conclusions from microarray experiments. *Proc. Natl. Acad. Sci. Early Edition* (24 July, 2001).

[14] Lillie J., Probing the genome for new drugs and targets with DNA arrays. *Drug Development Research* **41** (1997) pp. 160-172.

[15] Lockhart D. J. and Winzeler E. A., Genomics, gene expression and DNA arrays. *Nature* **405** (2000) pp. 827–836.

[16] Marton M. J., DeRisi J. L., Bennett H. A., Iyer V. R., Meyer M. R., Roberts C. J., Stoughton R., Burchard J., Slade D., Dai H., Bassett D. E. Jr., Hartwell L. H., Brown P. O., Friend S. H., Drug target validation and identification of secondary drug target effects using DNA microarrays. *Nature Medicine* **4** (1998) pp. 1293–1301.

[17] Perou C. M., Jeffrey S. S., Van de Rijn M., Rees C. A., Eisen M. B., Ross D. T., Pergamenschikov A., Williams C. F., Zhu S. X., Lee J. C. F., Lashkari O., Shalon D., Brown P. O., Botstein D., Distinctive gene expression patterns in human mammary epithelial cells and breast cancers. *Proc. Natl. Acad. Sci.* **96** (1999) pp. 9212–9217.

[18] Pollard K. S. and van der Laan M. J., Statistical inference for simultaneous clustering of gene expression data. *Mathematical Biosciences* **176** (2002) pp. 99–121.

[19] Ross D. T., Scherf U., Eisen M. B., Perou C. M., Rees C., Spellman P., Iyer V., Jeffrey S. S., Van de Rijn M., Waltham M., Pergamenschikov A., Lee J.C. F., Lashkari D., Shalon D., Myers T. G., Weinstein J. N., Botstein D., Brown P. O., Systematic variation in gene expression patterns in human cancer cell lines. *Nature Genetics* **24** (2000) pp. 227–235.

[20] Tibshirani R., Hastie T., Eisen M., Ross D., Botstein D., Brown P., Clustering methods for the analysis of DNA microarray data. Technical Report, Stanford University, Oct. 1999.

[21] van der Laan M. J. and Bryan J. F. Gene Expression Analysis with the Parametric Bootstrap. *Biostatistics* **2** (2001) pp. 1-17.

[22] van der Laan M. J. and Pollard K. S., Hybrid clustering of gene expression data with visualization and the bootstrap. Technical Report no. 93, U. C. Berkeley, Group in Biostatistics, May 2001. (To appear in *Journal of Statistical Planning and Inference*)

[23] Yeung K. Y., Haynor D. R., Ruzzo W. L., Validating clustering for gene expression data. *Bioinformatics* **17** (2001) pp. 309-318.

19

Statistical Inference for Clustering Microarrays

Jörg Rahnenführer[1]

19.1 Introduction

Clustering high dimensional data and complexity reduction are often guided by practical intuitive approaches, although the concept of information loss is well-known in statistical decision theory, see LeCam [5] for an early reference. A meaningful connection between data compression and statistical inference has recently been presented by Pötzelberger and Strasser [7] and by Strasser [10]. They introduce a quantization concept based on maximum support plane partitions (MSP-partitions). The goal is not to find density clusters in empirical data but rather to obtain good partitions with respect to some information measure, even when the data set is not nicely structured by density clusters. In this sense, the clustering step is seen as an explorative method for data compression and hypotheses generation. A data set is decomposed into m disjoint groups, a partition $\mathcal{B} = (B_1, \ldots, B_m)$, where the number of clusters $m \in I\!N$ is fixed. The quality of the procedure is then measured in terms of information loss.

In the next paragraph, we specify these ideas and briefly describe the integration of the notions MSP-partition and f-information within a general theory, as developed by Strasser [10]. He shows that, for compressed data, the optimization of a specific functional to discriminate between two probability measures is equivalent to minimizing information loss. We subsequently report results of a simulation study for finite sample size performed by Rahnenführer [8] that confirmed the theoretical expectations. Parameter specifications and the main conclusions are reviewed.

---

[1]Jörg Rahnenführer is with the Mathematical Institute at the Heinrich-Heine-Universität Düsseldorf. He is currently visiting the Department of Biostatistics and the Department of Statistics, University of California, Berkeley. This work was supported by the German Research Foundation (Deutsche Forschungsgemeinschaft, RA 870/2-1).

In the last paragraph, the methods are applied to microarray data, where the gene expression of some thousand genes is measured simultaneously. This genome-wide approach has already produced many promising results, including classification of cancer tissue samples on a purely molecular basis, see Eisen et al. [2], Golub et al. [3] or Perou et al. [6]. Here, we are interested in the discrimination problem and class discovery. Comparing MSP-partitions, we demonstrate that robust clustering methods such as the Kohonen algorithm are superior to the non-robust k-means algorithm, if the goal is to detect small differences in gene expression patterns.

19.2   Data compression and statistical inference – Clustering with MSP-partitions

In this paragraph, clustering with MSP-partitions is explained. First, the central notion of this concept, the f-information, is introduced and motivated.

Definition 1. *(f-Information)*
Let $f : \mathbb{R}^d \to \mathbb{R}$ be a convex function and P an arbitrary distribution on $\mathbb{R}^d$. The f-information of the partition $\mathcal{B} = (B_1, \ldots, B_m)$ (with respect to P) is given by

$$I_f(\mathcal{B}) := \sum_{j=1}^{m} P(B_j)\, f(m(B_j)), \tag{19.1}$$

where $m(B_j) := \frac{1}{P(B_j)} \int_{B_j} x\, P(dx)$ is the mean of the set B_j of the partition, if $P(B_j) > 0$.

The restriction to convex functions ensures that a division of one partition set into two new ones cannot decrease the f-Information.

Our goal is the maximization of this information, i.e. to find a partition $\mathcal{B} = (B_1, \ldots B_m)$ such that

$$I_f(\mathcal{B}) = \texttt{Max}! \quad \text{under } |\mathcal{B}| \leq m. \tag{19.2}$$

Strasser [11] gives a decision theoretic justification for the introduction of the information $I_f(\mathcal{B})$ and the formulation of (19.2). Consider the classical discriminant problem between two probability measures P and Q. Let E_P and V_P be the expectation and variance with respect to P and γ a fixed constant. A solution of the classical optimization problem

$$E_Q(t) - E_P(t) - \gamma V_P(t) = \texttt{Max}! \tag{19.3}$$

is given by the centered likelihood ratio $h = \frac{dQ}{dP} - 1$. Let $h(x) = x$, which holds for example for normal distributions. Assume now that first a data set is clustered before tackling the discriminant problem. The goal is to find an optimal partition $\mathcal{B}$, in the decision theoretic sense of (19.3). Strasser shows that for $h(x) = x$ this problem is equivalent to maximizing $I_f(\mathcal{B})$ with $f(x) = |x|^2/2$. Here $|.|$ denotes the Euclidean norm. For other cases with $h(x) \neq x$, a preceding data transformation can bring the two problems in agreement. In other words, a suitable data transformation with subsequent maximization (19.2) of the information (with quadratic f) yields a solution of the discriminant problem (19.3). Finally, also other choices for the convex function f than $f(x) = |x|^2/2$ can be justified. If a penalty term for outliers is introduced, then the optimal convex function f can be determined depending on this penalty term. Then again, (19.2) and (19.3) are equivalent.

For the special case $f(x) = |x|^2/2$ the f-information can be written as $I_f(\mathcal{B}) = \int |x|^2 \, P(dx) - \sum_{j=1}^m \int_{B_j} |x - m(B_j)|^2 \, P(dx)$. Therefore, maximizing $I_f(\mathcal{B})$ equals minimizing the inner dispersion of $\mathcal{B}$. A solution of this problem is called minimum variance partition. The k-means algorithm is known to search for such partitions. Thus it equivalently aims at maximizing the information $I_f(\mathcal{B})$. k-means improves a starting solution by alternating between two steps until convergence. It computes centers of a given partition and constructs a new partition by assigning every point to its closest of these centers in Euclidean distance. In the one-dimensional case, Bock [1] presented a generalization of the k-means algorithm for arbitrary convex functions f. Pötzelberger and Strasser [7] extended this approach to higher dimensions. They define fixpoint algorithms that coincide with k-means for $f(x) = |x|^2/2$ and with the Kohonen [4] algorithm for $f(x) = |x|$. The Kohonen algorithm is known from competitive learning theory. The resulting partition does not depend on the distance of the data points from 0. Thus outliers have no influence on the result.

For the introduction of MSP-Partitions, some notions have to be provided. Remember that for a convex function $f : \mathbb{R}^d \to \mathbb{R}$ the function $f^c(a) := \sup_x (a'x - f(x))$ for $a \in \mathbb{R}^d$ is called the conjugate convex function of f. The support of this function is given by $A := K(f^c) = \{x \in \mathbb{R}^d : f^c(x) < \infty\}$. It is known that $f^c(a)$ is maximal with $l(x) := a'x - f^c(a) \leq f(x)$. The affine linear function l is called *maximal support function* of f in x with slope a.

Definition 2. *(MSP-partition)*
A partition $\mathcal{B} = (B_1, B_2, \ldots, B_m)$ is called a MSP-partition (maximum support plane partition) if there exists $\mathbf{a} \in A^m$ such that for $j = 1, 2, \ldots, m$:

$$x \in B_j \quad \Rightarrow \quad a_j'x - f^c(a_j) = \max_{1 \leq k \leq m} \left(a_k'x - f^c(a_k) \right). \tag{19.4}$$

MSP-partitions are constructed by maximal support planes of the convex function f. For given slopes a_j, $j = 1, \ldots, m$ each point is assigned to the partition with the maximal value of its support plane at that point. Figure 19.1 demonstrates the proceeding for $f(x) = |x|^{1.5}/1.5$ in dimension 1.

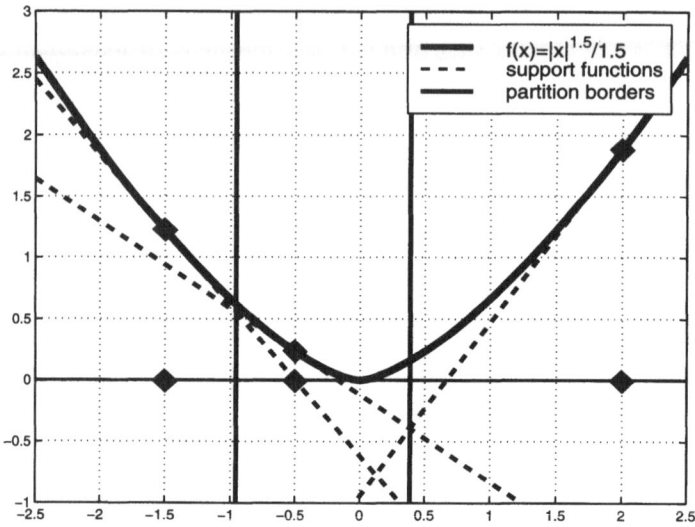

Figure 19.1. MSP-partition for $f(x) = |x|^{1.5}/1.5$

Here, centers of the partition are given by the points -1.5, -0.5 and 2. At these three data points, tangents are fitted to the convex function. Then the partition borders are calculated as the intersections of these support lines. Every data point belongs to the center with maximal according support function.

In higher dimensions $d > 1$ the tangents are substituted by d-dimensional planes. To demonstrate the effect of the convex function f, Figure 19.2 shows MSP-partitions in dimension $d = 2$ with $f(x) = \frac{1}{2}|x|^2$ and $f(x) = |x|$. The data set consists of 400 realizations of a standard normal distribution and is clustered into $m = 20$ partition sets. For given centers, the partition is obtained by assigning each point to its closest center using a specific distance function. For $f(x) = \frac{1}{2}|x|^2$ this is the Euclidean distance. For $f(x) = |x|$ it is the angle between the observations or in mathematical terms the cosine of the normalized inner product. This leads to a purely directional oriented classification. One of the strengths of the approach is its flexibility. Using other convex functions f like $f_p(x) = \frac{1}{p}|x|^p$, $p \in (1, 2)$, any weighting of the two mentioned extremes can be chosen. Figure 19.1 shows such a compromise. The center $c = 2$ claims most of the area on the right side, because it is the only center with $c > 0$.

The fixpoint algorithm introduced by Pötzelberger and Strasser [7] is specifically constructed to produce MSP-partitions.

Definition 3. *(Implementation of fixpoint algorithm)*
Select a random sample of m initial prototypes c_i, $i = 1, \ldots, m$ out of the points of the data set. Repeat the following steps until the algorithm terminates:

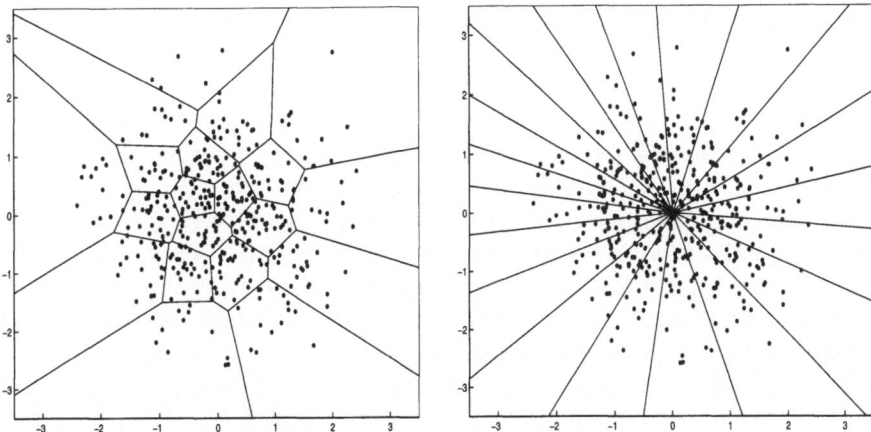

Figure 19.2. MSP-Partition for 400 normal distributed data, compressed to $m = 20$ clusters, $f(x) = \frac{1}{2}|x|^2$ (left) and $f(x) = |x|$ (right)

1. *Evaluate the maximal support functions in the points c_i, $i = 1, \ldots, m$ and determine the corresponding MSP-partition $\mathcal{B} = (B_1, \ldots, B_m)$.*

2. *If $\mathcal{B}$ contains partitions with no points, continue splitting the set B_k with the highest number of points h_k in two sets by replacing the prototype with no representing points by a random point of B_k.*

3. *Calculate the midpoints c_i of the new partition. They represent the new prototypes. Stop if the midpoints are not changed from the last step.*

It can be proven that the f-Information strictly increases in each iteration step of the fixpoint algorithm. Since for discrete data sets the number of possible segmentations is finite, the algorithm also stops after a finite number of iterations. The constant increase of the f-information also guarantees that the final solution is a local optimum of the optimization problem (19.3) $I_f(\mathcal{B}) = \texttt{Max!}$. This emphasizes the suitability of the fixpoint algorithm and of the concept of MSP-partitions within the current framework.

19.3   Simulation study

In this part we report on a detailed Monte Carlo simulation study of Rahnenführer [8] that confirmed the hypothesis formulated in the decision theoretic section above. To compare the performance of the fixpoint algorithm for specific convex functions the multivariate k-sample problem with shift alternatives was used. In this testing problem k samples are drawn from the same multivariate distribution, differing only with respect to a location parameter. Under the null hypothesis H_0, all distributions are equal. Under the alternative H_A, different constants are added to the k samples.

First, the fixpoint algorithm was applied to the pooled sample for complexity reduction of the data set. No sample membership information was used. For computing test statistics the data points were then replaced by their conditional expectations with respect to the MSP-partition, i.e. by their prototypes, respectively. To address the data driven clustering step we used multivariate permutation tests. This ensured that the tests did hold the level $\alpha = 0.05$. Power functions of the test procedures were simulated by counting in how many of 1000 Monte Carlo simulations the test rejected the hypothesis. For every single test 1000 permutations of the data were calculated.

We analyzed data sets in dimension $d = 2, 5, 10$ with $k = 2, 3, 10$ samples consisting of 100 or 200 data points that were clustered to $m = 6, 10, 20$ clusters. The convex functions were $f_p(x) = |x|^p/p$, $p = 1, 1.5, 2$ and $f(x) = \ln \cosh(|x|/2)$. The latter is a smooth function that behaves like $|x|^2$ close to 0 and like $|x|$ for $|x| \to \infty$. As underlying distributions we considered normal, uniform, double exponential, cauchy and mixtures of normal and exponential distributions.

Two test statistics were considered. The χ^2 statistic compares the frequencies of the samples within the clusters. As long as data from different samples are clustered in different groups, this test can detect it. The F-test statistic is the ratio of the mean squared errors between and within the samples, which is asymptotically optimal for normal distributions. Remember that we replace the data points by their cluster prototypes before using them in the test statistic. Let $n^{(i)}$ be the number of data points in sample i and $n_j^{(i)}$ the number of data points in sample i and in partition set B_j. Let $m^{(i)}$ denote the mean of data points in sample i, calculated after data points were replaced by their prototypes. Let m_j be the mean of data points in partition set B_j and $\bar{m}$ the pooled sample mean. Then the F-test statistic is given by

$$T = \frac{1}{k-1} \sum_{i=1}^{k} n_i \, |m^{(i)} - \bar{m}|^2 \Big/ \left(\frac{1}{n-k} \sum_{i=1}^{k} \sum_{j=1}^{m} n_j^{(i)} \, |m_j - m^{(i)}|^2 \right).$$

Apparently only information available after the clustering step is used in the test statistic. This is important for the applicability of the permutation tests, see also Strasser and Weber [11], who obtained asymptotically optimality results of such tests with the help of LAN-theory. The following conclusions could be drawn in Rahnenführer [8]:

- For normal and uniform distributions the k-means algorithm ($p = 2$) and mixtures (like $p = 1.5$) had almost equal power, whereas the Kohonen algorithm ($p = 1$) had a power loss of up to 0.1. This was true without exception for all tested parameter constellations.

- For exponential distributions it was the other way around, but far more drastic. The Kohonen algorithm ($p = 1$) was always the most efficient one, for $p = 1.5$ a power loss of up to 0.6 was observed, for k-means ($p = 2$) even up to 0.85. These values dropped to 0.25 and 0.6 using the χ^2 test instead of the F-test. For cauchy distributions the power difference was up to 0.45.

- Due to the heavy tails of these distributions, for non-robust methods the outer clusters contained only few data points. More significant differences in the inner part were covered only by a small number of clusters. Even for mixture distributions with 90% normal and only 10% exponential data, the small number of factual outliers did impair the performance of the k-means algorithm.

- Altogether the order of the quality of the cluster algorithms depended only on the tails of the distribution of the data. No other parameter could influence this order. In the presence of extreme data points, the more an algorithm was direction oriented, the more the discriminatory power was increased.

19.4   Application to gene expression microarray data

Microarray experiments measure relative expression levels of several thousand genes at the same time. This technology has become very popular, since it represents the first genome-wide view at the status of a cell. Especially promising results were obtained regarding classification of cancer tumor tissue samples, see e.g. Eisen et al. [2], Golub et al. [3], Perou et al. [6] and Ross et al. [9]. Particular goals are class prediction and class discovery. The goal of class prediction is to find fast and reliable methods to distinguish different classes of cancer, which is crucial for the following treatment, in order to maximize efficiency and minimize toxicity. Class discovery means finding new classes of cancer by molecular markers, here by using gene expression measurements.

Our approach is a little different. We present an analysis in the sense of the multivariate k-sample problem. Consider a data set with k cancer types and the clustering of all of the samples into m groups. In an ideal situation, choosing $m = k$ would lead to perfect concordance of cancer types and clusters, but this is often not the case. We address the question, which cluster algorithms preserve at best differences between cancer types, especially when these differences are rather small. Such small differences are a more reasonable assumption in the real world. In contrast to the largest part of recent microarray literature, we are not interested in clustering genes, but only in clustering microarray samples.

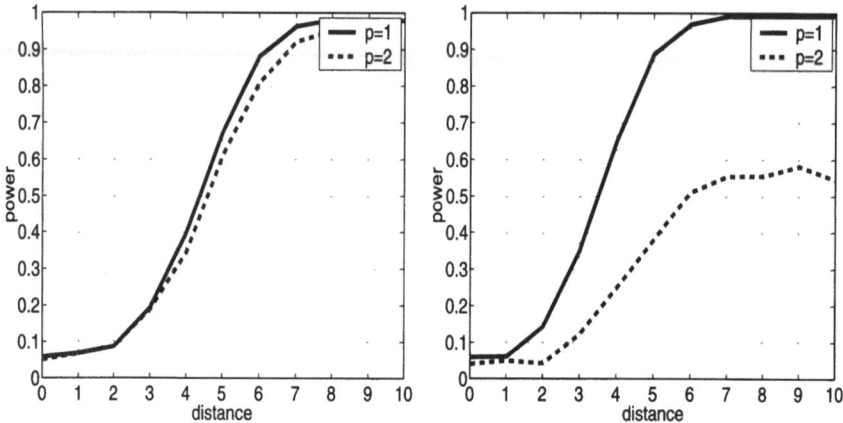

Figure 19.3. Power functions for χ^2 test and simulated microarray data, normal distribution (left) and cauchy polluted normal distribution (right)

First, pure simulations of microarray data were performed. The log expression values were drawn from a standard normal distribution. We simulated $n = 60$ samples with $d = 2000$ genes. For 30 samples, 100 genes were up-regulated and 100 genes were down-regulated. The $k = 2$ true groups were then partitioned into $m = 6$ clusters. The left side of Figure 19.3 shows power functions for the fixpoint algorithm with $p = 2$ (k-means) and $p = 1$ (Kohonen). The distance on the abscissa specifies the value added or subtracted for the up- and down-regulated genes, respectively. The ordinate gives the probability that the χ^2 test rejected the null hypothesis of no difference. The power functions on the right side of Figure 19.3 were obtained by additionally polluting some genes with cauchy noise. With probability 0.01 we added to every expression value a cauchy distributed variable, truncated at 5. The figure demonstrates the influence of outliers. For pure normal distributions no big differences exist. By adding minor cauchy noise though, the robust algorithm clearly outperforms k-means. This behavior was observed also for other parameter values, e.g. using the F-test or compressing to other numbers m of clusters.

As a second step we examined real microarray cancer data sets, explicitly the Leukemia data set of Golub et al. [3], http://www.genome.wi.mit.edu/MPR, and the NCI60 cancer data set of Ross et al. [9], http://genome-www.stanford.edu/nci60. The Leukemia data set consists of 3 cancer classes with 9, 25 and 38 cases. We used the same data preprocessing steps as in the paper. The NCI60 data set consists of 9 classes of cancer with $2 - 9$ cases, in total 63. We used all 5244 genes. Both data sets are popular, because the distinction of cancer types is easy and thus many algorithms have been applied successfully.

The real data were integrated in the simulations as more realistic models for expression differences. Because of the clear distinction between can-

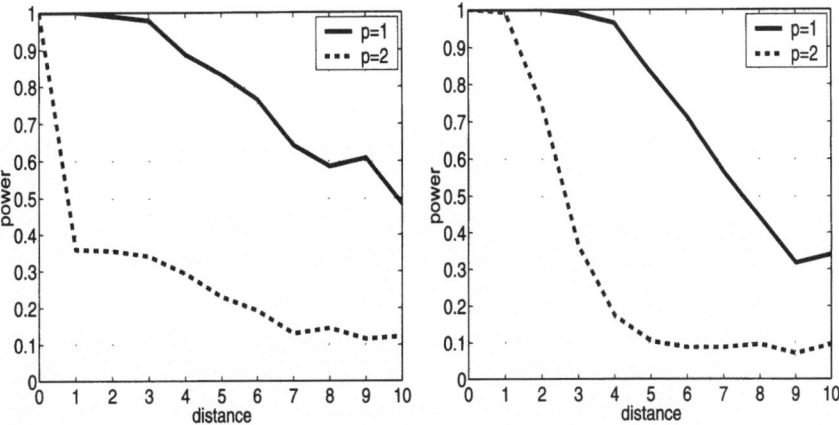

Figure 19.4. Power functions for Leukemia (left) and NCI60 (right) cancer data sets, compressed to $m = 6$ clusters

cer types, no algorithm preference can be expected for the raw data. For this reason, the original data were polluted before the comparison of the algorithms. The most natural idea is to add gaussian noise with increasing variance. The important disadvantage here is, that in this case we are just comparing shifted normal distributions, the particular characteristic structure of the microarray data plays almost no role. Not surprisingly the result looked similar as the left side of Figure 19.3. With this observation in mind, we developed a bootstrap type approach and draw gene expression values randomly within the cancer types. To introduce a measure for the amount of noise, the distance of all expression values to the mean within their cancer type was multiplied with a factor d. For $d = 0$ the data points collapse to the mean within their cancer type, for $d = 1$ they remain unchanged, and with growing d the distinction becomes more difficult. Again, we compressed both data sets to $m = 6$ groups. Figure 19.4 shows the resulting power functions. Note that the distance on the x-axis has a different meaning than in the previous figures. Smaller values for the distance d now correspond to more clearly separated clusters and lead to higher power values. We observe an extreme superiority of the fixpoint algorithm with $p = 1$ (Kohonen) compared to the fixpoint algorithm with $p = 2$ (k-means). The necessity of robustification becomes obvious.

Two general conclusions can be drawn. By simply polluting data with gaussian white noise, no differences between the quality of algorithms can be observed, due to the unsuitable statistical model. With the bootstrap type approach that uses more information contained in the data, it becomes obvious, that a robust algorithm should be preferred. The popular k-means algorithm comes along with a dramatic power loss.

References

[1] Bock, H. H. (1992): A clustering technique for maximizing ϕ-divergence, noncentrality and discriminating power. In M. Schader, editor, *Analyzing and Modeling Data and Knowledge*, p.19–36, Springer.

[2] Eisen, M. B., Spellman, P.T., Brown, P.O. and Botstein, D. (1998): Cluster analysis and display of genome-wide expression patterns, *Proc. Natl. Acad. Sci.* 95, p.14863–14868.

[3] Golub, T.R., Slonim, D.K., Tamayo, P., Huard, C., Gaasenbeek, M., Mesirov, J.P., Coller, H., Loh, M.L., Downing, J.R., Caligiuri, M.A., Bloomfield, C.D., Lander, E.S. (1999): Molecular classification of cancer: class discovery and class prediction by gene expression monitoring, *Science* 286, p.531–537.

[4] Kohonen, K. (1984): *Self organization and associative memory*, Springer.

[5] LeCam, L. (1964): Sufficiency and approximate sufficiency, *Ann. Math. Stat.* 35, p.1419–1455.

[6] Perou, C. M., Sorlie, T., Eisen, M.B., van de Rijn, M., Jeffrey, S.S., Rees, C.A., Pollack, J.R., Ross, D.T., Johnsen, H., Akslen, L.A., Fluge, O., Pergamenschikov, A., Williams, C.F., Zhu, S.X., Lonning, P.E., Borresen-Dale, A.L., Brown, P.O. and Botstein, D. (2000): Molecular portraits of human breast tumors, *Nature* 406, p.747–752.

[7] Pötzelberger, K. and Strasser, H. (2000): Clustering and Quantization by MSP-Partitions, *Statitics & Decisions* 19 (4), p.331-372.

[8] Rahnenführer, J. (2002): Multivariate permutation tests for the k-sample problem with clustered data, *Comput. Stat.* 17 (2), p.165-184.

[9] Ross, D.T., Scherf, U., Eisen, M.B., Perou, C.M., Rees, C., Spellman, P., Iyer, V., Jeffrey, S.S., van de Rijn, M., Waltham, M., Pergamenschikov, A., Lee, J.C., Lashkari, D., Shalon, D., Myers, T.G., Weinstein, J.N., Botstein, D., Brown, P.O. (2000): Systematic variation in gene expression patterns in human cancer cell lines, *Nature genetics* 24 (3), p:227–234.

[10] Strasser, H. (2000): *Reduction of Complexity*, Technical report, Department of Statistics, Vienna University of Economics and Business Administration.

[11] Strasser, H. and Weber, C. (2000): On the asymptotic theory of permutation statistics, to appear in *Math. Methods Stat.* .

20

Logic Regression - Methods and Software

Ingo Ruczinski, Charles Kooperberg, and Michael LeBlanc[1]

Summary

Logic Regression is an adaptive regression methodology that constructs predictors as Boolean combinations of binary covariates. This method, introduced by Ruczinski, Kooperberg, and LeBlanc [5] is particularly useful for problems where most covariates are binary, and the interactions between those predictors is of main interest. Here, we briefly review the methodology, describe the publicly available software, and give an example. The software is currently available from http://bear.fhcrc.org/~ingor/logic.

20.1   The Logic Regression Methodology

In most regression problems a model is developed that relates the main effects (the predictors or transformations thereof) to the response. Although interactions between predictors are sometimes considered as well, those interactions are typically kept simple (two- to three-way interactions at most). But often, especially when all predictors are binary, the interaction between many predictors is what is associated with differences in response. This issue arises, for example, in the analysis of SNP microarray data or in some data mining problems. Given a set of binary predictors X, we try to create new predictors for the response by considering combinations of those binary predictors. For example, if the response is binary as well

---

[1]Ingo Ruczinski is at the Department of Biostatistics, Bloomberg School of Public Health, Johns Hopkins University, Baltimore, MD 21205-2179 (E-mail: ingo@jhu.edu); Charles Kooperberg and Michael LeBlanc are at the Division of Public Health Sciences, Fred Hutchinson Cancer Research Center, Seattle, WA 98109-1024 (E-mail: clk@fhcrc.org, mikel@swog.fhcrc.org). This research was supported in part by National Institutes of Health grants CA74841 and CA53996.

(which is not required in general), we attempt to find decision rules such as "if X_1, X_2, X_3 and X_4 are true", or "X_5 or X_6 but not X_7 are true", then the response is more likely to be in class 0. In other words, we try to find Boolean statements involving the binary predictors that enhance the prediction for the response. Formally: let $X_1, \ldots, X_k$ be binary predictors, and let Y be a response variable. We try to fit regression models of the form $g(E[Y]) = b_0 + b_1 L_1 + \cdots + b_n L_n$, where L_j is a Boolean expression of the predictors X, such as $L_j = [(X_2 \wedge X_4^c) \vee X_7]$. The above framework includes many forms of regression, such as linear regression ($g(E[Y]) = E[Y]$) and logistic regression ($g(E[Y]) = \log(E[Y]/(1 - E[Y]))$). For every model type, we define a score function that reflects the "quality" of the model under consideration. For linear regression the score could be the residual sum of squares and for logistic regression it could be the binomial deviance. We try to find the Boolean expressions in the regression model that minimize the scoring function associated with this model type, estimating the parameters b_j simultaneously with the Boolean expressions L_j. In the Logic Regression framework any type of model can be considered, as long as a scoring function can be defined. For example, we also implemented the Cox proportional hazards model, using the partial likelihood as the score.

There are some similarities between Logic Regression and some of the so called rule induction methods developed in the field of Machine Learning. Logic Regression differs from all methods that we are aware of in one or both of two important aspects: (i) the Logic Regression methodology places no restrictions on the form of the logic expressions L_j, and (ii) the Logic Regression methodology is not specifically designed for one particular problem (in machine learning often classification) but works with any scoring function. In our experience it performs better with continuous measures, such as log-likelihoods, than with discrete measures, such as misclassification. We refer to Ruczinski et al. [5] for a comparison of Logic Regression and machine learning methods.

Any Boolean statement can be represented as a binary tree (called Logic Tree), the variables being the leaves of the tree and the logic operators $(\vee, \wedge)$ as the other knots (see [5] for details; Figure 20.3 later in this chapter displays Logic Trees). On the set of trees we define a move set by a collection of standard operations: alternating leaves, changing operators, splitting and deleting leaves, and growing and pruning the trees. The terminology used is similar to the terminology introduced by Breiman et al. [1]. Ruczinski et al. [5] also provide a comparison between CART and Logic models. Since the number of possible Logic Models for a given set of predictors can be very large, we rely on search algorithms to help us find the best scoring models. We implemented two algorithms: a greedy (stepwise) and a simulated annealing algorithm. While the greedy algorithm is very fast, it does not always find a good scoring model. Our preferred algorithm is the simulated annealing algorithm, which usually does find good scoring models, but is computationally more expensive.

As for many adaptive regression methodologies, the best scoring model often over-fits the data, and model selection is needed. We implemented several methods for model selection, using randomization tests and cross-validation. A detailed introduction to Logic Regression can be found in Ruczinski et al. [5]. See Kooperberg et al. [3] for an application of said methodology to single nucleotide polymorphism (SNP) data.

20.2   The Logic Regression Software

The Logic Regression program is a stand-alone program *xlogic* written in Fortran 90 that can be downloaded from http://bear.fhcrc.org/~ingor/logic. *xlogic* can be used to fit one logic regression model, to fit logic regression models of pre-specified sizes, to carry out cross-validation, or to do various randomization tests. Each application requires an input file, which can be edited manually or be generated from one of the online available menus. The results of *xlogic* are a number of ASCII files. These can be directly used as input to several S-Plus functions to generate graphical representations of the output.

Currently the Logic Regression methodology has scoring functions for linear regression (residual sum of squares), logistic regression (binomial deviance), classification (misclassification), and proportional hazards models (partial likelihood). A feature of the Logic Regression methodology is that it is easy to include and use ones own scoring function if that is desired. Online help is available from the website.

20.2.1   Running the Software

In the following sections we will focus on the current version of the program. A number of extensions of the methodology are planned for the near future. In Figure 20.2 is the online menu that one obtains after selecting **how to run the program** on the previous (main) menu. It displays the currently available features of the Logic Regression software.

There are currently five versions of the Logic Regression program, available on the web site. They are listed in Figure 20.2. For each of these versions a menu is available, which guides the user through the selection of the various options. We now discuss the various versions of the program.

20.2.2   Find the best scoring model of any size

To select a good scoring Logic Regression model, we use a simulated annealing (see, for example, Otten and van Ginneken [4] and van Laarhoven and Aarts [6]) search algorithm. In general, simulated annealing operates on a state space S, which is a collection of individual states, representing

Logic Regression

Logic regression is a (generalized) regression methodology that is primarily applied when most of the covariates in the data to be analyzed are binary. The goal of logic regression is to find predictors that are Boolean (logical) combinations of the original predictors. For more information follow the link **basic info about the methodology** below.

On this page you can download the software for the logic regression algorithm and find the basic info you need to run the software. Please click on the appropriate link to find out more.

basic info about the methodology
basic info about the available software
download the software
how to run the program
write your own scoring functions
description of the output format
an example to check out
sample programs

The current version of the code is 0.1.3 dated July 17, 2001 (changelog).

The logic regression methodology was developed by Ingo Ruczinski, Charles Kooperberg, and Michael LeBlanc at the Fred Hutchinson Cancer Research Center in Seattle. The copyright of the logic regression code is owned by Ingo Ruczinski, Charles Kooperberg, and Michael LeBlanc. You are free to use the software, for non-commercial purposes only, if:
(1) Copyright notices are not removed.
(2) Publications using logic regression refer to: *Ruczinski I, Kooperberg C, LeBlanc ML (2001), Logic Regression, manuscript.* or *Kooperberg C, Ruczinski I, LeBlanc ML, Hsu L (2001), Sequence Analysis using Logic Regression, Genetic Epidemiology, to appear.*

For questions please contact Ingo Ruczinski or Charles Kooperberg.

Figure 20.1. The Logic Regression menu, as of October 2001, available from http://bear.fhcrc.org/~ingor/logic/ Online, you can click any of the links, indicated by the bold face fonts, to find out more about that topic.

a configuration of the problem under investigation. The states are related by a neighborhood system, and the set of neighboring pairs in S defines a substructure M in $S \times S$. The elements in M are called moves. Two states s, s' are called adjacent, if they can be reached by a single move (i. e. $(s, s') \in M$). Similarly, $(s, s') \in M^k$ are said to be connected via a set of k moves. In our application, the state space is finite. The basic idea of the annealing algorithm is: given the current state, pick a move according to a selection scheme from the set of permissible moves, which leads to yielding a new state. Compare the scores of the old and the new state. If the score

Logic Regression - running xlogic

To tune the program you need an input file specifying all options. Such an input file is most easily generated by one of these scripts:

find the best scoring model of any size
find the best scoring models for various sizes
carry out cross-validation for model selection
carry out a randomization test to check for signal in
the data
carry out a randomization test for model selection

You can now run the code as

% xlogic < inputfile

If you want to edit input files yourself, the format of the input files is described here.

Output is (by default) written in the *Scratch* subdirectory of the directory in which **xlogic** is. Output formats are described here and there are sample programs here .

For questions please contact Ingo Ruczinski or Charles Kooperberg.

Figure 20.2. The features of the software as of October 2001, available from `http://bear.fhcrc.org/~ingor/logic/running/running.html` Online, you can click any of the links, indicated by the bold face fonts, to get the templates for the input file needed to run the program.

of the new state is better than the score of the old state, accept the move. If the score of the new state is not better than the score of the old state, accept the move with a certain probability. This acceptance probability depends on the difference of the scores of the two states under consideration and a parameter that reflects at which point in time the annealing chain is (this parameter is usually referred to as the temperature). For any pair of scores, this probability decreases during the algorithm. For infinitely long algorithms with slowly decreasing temperatures it can be established that the best state is reached. However, even when that is not the case, this algorithm generally leads to good-scoring states.

In our case, a state is a Logic Tree. Given the current tree, we randomly pick, following a pre-determined distribution, a candidate from the move set for this tree. We re-fit the parameters for the new model, and determine its score, which we then compare to the score of the previous state (Logic model), and repeat the process. There are various possibilities how to implement the annealing algorithm and fit the Logic models. This requires,

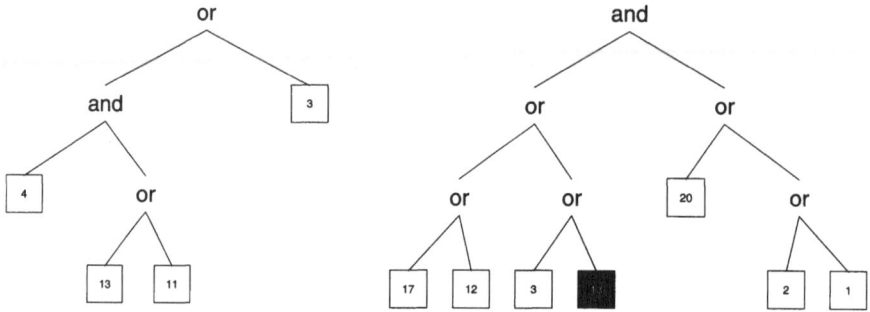

Figure 20.3. Results of letting *xlogic* find the best model of any size with two logic trees on the simulated data set

for computational reasons, that we pre-select the number t of trees. Unless we have an idea of how many trees we maximally want to fit, it may not be clear a priori what this number should be. We generally pick larger than necessary t and trim the model down if needed. Our simulated annealing algorithm has similarities with the Bayesian CART algorithm [2], in which a CART tree is optimized stochastically. Both of these algorithms are distinct from the greedy algorithm employed by CART, in that at any stage they not necessarily pick the move that improves the score the most.

Example

We simulated a data set with 500 cases and 20 binary predictors. Each predictor k is simulated from as an independent Bernoulli random variables, with success probability p_k between 0.1 and 0.9. The response variable is simulated from the model

$$Y = 3 + 1L_1 - 2L_2 + Z, \qquad (20.1)$$

where $L_1 = (X_1 \vee X_2)$ and $L_2 = (X_3 \vee X_4)$, and Z is independent standard normal noise. We use linear regression within the logic regression framework to find L_1 and L_2. The results of letting *xlogic* find the best model of any size with two logic trees is shown in Figure 20.3. The logic trees in these figures are read upside down; for example, in the left-hand side of this figure $L_1 = (X_4 \wedge (X_{13} \vee X_{11})) \vee X_3$. As can be seen, these trees are too large, and model selection needs to be carried out.

While the example in this chapter uses linear regression, all modeling options can also be applied to any other regression model with an appropriately defined score function.

20.2.3   Find the best scoring models for various sizes

In certain situations it is of interest to know what the best scoring logic regression model of a certain size is. The size of a logic regression model is defined as the total number of leaves in all Logic Trees combined, thus the

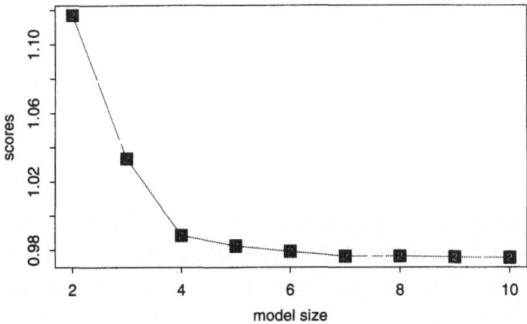

Figure 20.4. Scores of the best model of a specific size when two logic trees are fit to the simulated data set. The white 2 denotes that the logic model was fit with two trees. In general, models with various numbers of trees are fitted, but only the scores for two-tree models of sizes 2 through 10 are shown here.

model displayed in Figure 20.3 has size 11. Finding models of a fixed size is essential when using cross-validation to determine the best overall model size, as discussed below. For the simulated annealing algorithm described above, the tree or model size changes constantly, and the final model can be of any size. The straightforward solution to find the best scoring model of a fixed size would be to alter the move set, and only allow moves that keep the size of the model constant. However, this turns out to be computationally inefficient, as the resulting chains do not mix very well because the move set becomes more complicated. Thus, instead we do allow moves that increase or decrease the size of the Logic Regression model, but we prohibit moves that increase the model size when its desired size has been reached. Strictly speaking, this guarantees us only to find the best of **up to** the desired size. In reality, the maximum (desired) tree size almost always is reached, provided this size is not too large.

Example (cont.)

In Figure 20.4 we show the score of the best Logic Regression model of size for a variety of sizes two through ten, when two logic trees are fit. We note that the score improves considerably up to size four, and levels out after that. In fact,, the best logic model of size four has the correct L_1 and L_2 in model (20.1).

20.2.4 Carry out cross-validation for model selection

Searching for the globally best scoring model on the entire data, we know that the model with the best predictive capability may be smaller than the model we find via simulated annealing. We therefore want to compare the performances of the best models for different sizes. This can be done using an independent test set or by cross-validation. When sufficient data are available, we prefer the training set/test set approach. Otherwise, we can use cross-validation instead. Assume we want to assess how well the best model of size k performs in comparison to models of different sizes. We split the cases of the data set into m (approximately) equally sized groups. For each of the m groups of cases (say group i), we proceed as follows: remove the cases from group i from the data. Find the best scoring model of size k (as described in the previous section), using only the data from the remaining $m-1$ groups, and score the cases in group i under this model. This yields score ϵ_{ki}. The cross-validated (test) score for model size k is $\epsilon_k = \frac{1}{m}\sum_i \epsilon_{ki}$. We can compare the cross-validated scores for models of various sizes.

Example (cont.)

In Figure 20.5 we show both the average training and (cross-validation) test score. As can be seen, the training scores decrease as the model size increases, but the test scores are minimized for model sizes four and five.

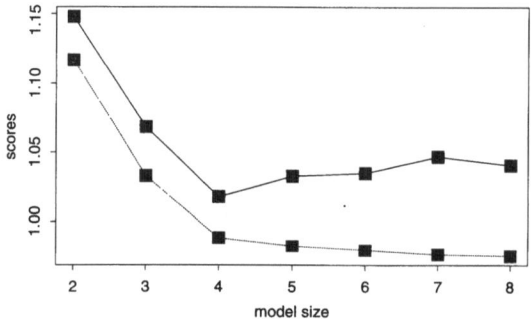

Figure 20.5. Training (dashed) and test (solid) scores of the cross-validation model selection for the simulated data set

20.2.5   Carry out a randomization test to check for signal in the data

The first step in our analysis usually is check for signal in the data. To do this, we first find the best scoring model, given the data. The null hypothesis which we want to test is: "there is no association between X the predictors and the response. If that hypothesis was true, the best model fit on the data with the response randomly permuted should yield about the same score as the best model fit on the original data. We carry out this randomization procedure as often as desired, and claim the proportion of scores better than the score of the best model on the original data as an p-value, indicating evidence against the null hypothesis.

Example (cont.)

In Figure 20.6 we show a histogram of 50 scores of the best model based on randomized data for the simulated example. As can be seen, these scores are considerably worse than the true best model, making us believe that there is signal in the data.

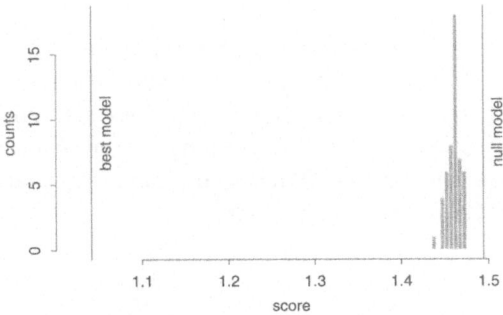

Figure 20.6. Histogram of 50 scores of the best model based on randomized data for the simulated example.

20.2.6   Carry out a randomization test for model selection

We can carry out a similar randomization test to find the model size. First, we find the best scoring model, with score ϵ^*, say. Assume that this model has size k. We also find the best scoring models of sizes 0 through k. The null hypothesis for each sequential randomization test is: "the optimal model has size j, the better score obtained by models of larger sizes is due to noise", for some $j \in \{0, \ldots, k\}$. Assume that such a null hypothesis is true, and the optimal model size is j, with score ϵ_j. We now "condition" on this model, considering the fitted values of the Logic model. For a model with p Logic Trees, there can be up to 2^p fitted classes (one for each combination of the p Logic Trees $L_1, \ldots, L_p$). We now randomly permute the response within each of those classes. The exact same model of size j considered still scores the same, say ϵ_j (other models of size j potentially could score better). If we now fit the overall best model (of any size), it will have a score ϵ_j^{**}, which is as least as good, but usually better, than ϵ_j. However, this is due to noise! If the null hypothesis was true, and the model of size j was indeed optimal, then ϵ^* would be a sample from the same distribution as ϵ_j^{**}. We can estimate this distribution as closely as desired by repeating this procedure multiple times. On the other hand, if the optimal model had a size larger than j, then the randomization would yield on average worse scores than ϵ^*.

We carry out a sequence of randomization tests, starting with the test using the null model, which is exactly the test for signal in the data as described in the previous subsection. We then condition on the best model of size one and generate randomization scores. Then we condition on the best model of size two, and so on. Comparing the distributions of the randomization scores, we can make a decision regarding which model size to pick.

Example (cont).

In Figure 20.7 we see the results of the randomization tests conditioning on models of size 3, 4 and 5. We note that the best score is considerably better than all scores based on randomized data sets conditioned on the model of size three, but that this is no longer true when we condition on the models of size four or five. This, again, suggests that the best model does indeed have the (correct) size four.

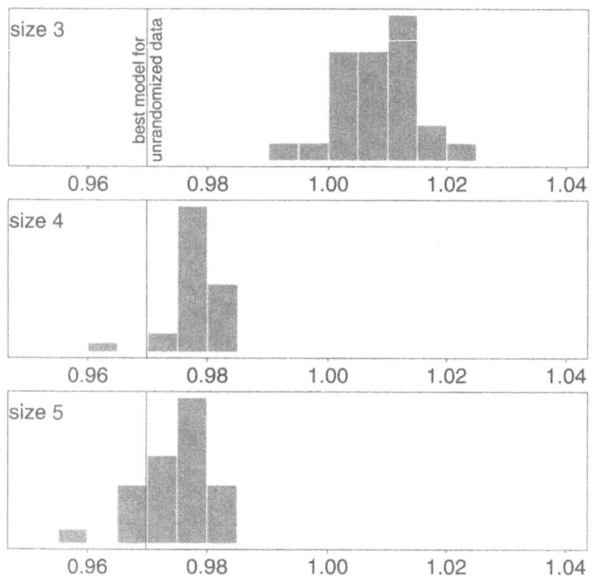

Figure 20.7. Histogram of 25 scores of the best model of fixed size for the conditional randomization tests the simulated example.

20.3   Conclusion

Logic Regression considers a novel class of models to detect interactions between binary predictors that are associated with a response variable. We developed the Logic Regression methodology with some statistical genetics problems in mind, but also found applications in other areas such as medicine and finance. Areas of current and future research include assessing model uncertainty using McMC, alternative ways for model selection, such as penalized scoring functions, and development of models that take familial dependence in genetic data into account. We also work on software improvements.

References

[1] Breiman, L., Friedman, J. H., Olshen, R. A., and Stone, C.J. (1984), *Classification and Regression Trees*, Belmont, CA: Wadsworth.

[2] Chipman, H., George, E., and McCulloch, R. (1998). Bayesian CART model search (with discussion). *Journal of the American Statistical Association*, 93, 935–960.

[3] Kooperberg, C., Ruczinski, I., LeBlanc, M. L., and Hsu, L. (2001), "Sequence Analysis using Logic Regression", *Genetic Epidemiology*, 21 (S1), 626–631.

[4] Otten, R. H., and Ginneken, L.P. (1989), *The Annealing Algorithm*, Boston: Kluwer Academic Publishers.

[5] Ruczinski, I., Kooperberg, C., LeBlanc, M. L. (2001), "Logic Regression", (under review, draft available from http://bear.fhcrc.org/~ingor/html/publications.html).

[6] van Laarhoven, P. J., and Aarts, E. H. (1987), *Simulated Annealing: Theory and Applications*, Boston: Kluwer Academic Publishers.

21

Adaptive Kernels for Support Vector Classification

Robert Burbidge[1]

Summary

Support vector machines (SVM) are a recent addition to the set of machine learning techniques available to the data miner for classification. The SVM is based on the theory of structural risk minimization [26] and as such has good generalisation properties that have been demonstrated both theoretically and empirically. The flexibility of the SVM is provided by the use of kernel functions that implicitly map the data to a higher, possibly infinite, dimensional space. A solution linear in the features in the higher dimensional space corresponds to a solution non-linear in the original features. This flexibility does not however lead to overfitting since the SVM performs automatic capacity control.

A remaining problem with SVMs is the selection of the kernel and regularizaton parameters. The commonly used Gaussian RBF kernel benefits from a good choice of the scaling parameter σ. This is usually chosen by training an SVM for a range of values of σ and using an estimate of the generalization error to do model selection. The standard approach of validation or cross-validation becomes impractical for large data sets as SVMs can perform "abysmally slow in prediction" [4]. A heuristic suggested by Jaakkola et al. [11] is used to initialize σ and this heuristic is then used to update σ during training. On benchmark data sets from the UCI Repository this algorithm achieves a cross-validated error not significantly different from the optimal value. It is also more computationally efficient than a line search for σ.

[1]Robert Burbidge is with the Statistics Section, Department of Mathematics, Imperial College, London, SW7 2AZ. This research was undertaken within the Postgraduate Training Partnership established between Sira Ltd and University College London. Postgraduate Training Partnerships are a joint initiative of the Department of Trade and Industry (DTI) and the Engineering and Physical Sciences Research Council (EPSRC). Support by DTI and EPSRC is gratefully acknowledged.

21.1   Introduction

The learning of decision functions is one of the main aims of machine learning and has many applications to data mining, pattern recognition, and design of autonomous intelligent agents [18]. The support vector machine (SVM) [26, 4] is a recent addition to the toolbox of machine learning algorithms that has shown improved performance over standard techniques in many domains both for classification [23] and regression [26, 4]. The advantages of the SVM over existing techniques are often cited as: the existence of a global optimum, improved generalization to unseen examples, sparsity of the model, and automated model selection [6]. The decision function is an expansion on a subset of the available data, known as the support vectors (SVs). The other data play no part in defining the decision boundary.

The SVM was inspired by statistical learning theory [26], which provides generalization error bounds for the learned hypotheses. Model selection can thus be automated by using expected generalization error as a criterion for searching model space. When there are many data, learning many models and choosing the best one may be time-consuming. In this context it is desirable to have some heuristics for model selection. Heuristics are also useful during the exploratory data analysis phase. For example, it is often not known whether there is anything useful to be gained from the data and heuristics can provide quick answers that have similar performance to more rigorous methods. As the principles behind the SVM are fairly simple, it is a technique that offers plenty of scope for developing such 'rules-of-thumb'.

The remainder of this paper is comprised as follows. In the next section I describe the problem of classification of labelled data, concentrating on the two-class case. The SVM for classification is briefly described, and the problem of model selection introduced. In Section 21.3 I describe a heuristic for model selection due to [11] and adapt it in order to tune σ during training. In Section 21.4 the results obtained are evaluated and discussed.

21.2   Classification

Consider learning a binary decision function on the basis of a set of training data drawn independently and identically distributed at random from some unknown distribution $p(\mathbf{x}, y)$,

$$\{(\mathbf{x}_1, y_1), \ldots, (\mathbf{x}_l, y_l)\}, \ \mathbf{x}_i \in \mathcal{R}^d, \ y_i \in \{-1, +1\} \ . \qquad (21.1)$$

One common approach [21] is to search for a pair $(\mathbf{w}, b)$ such that the decision function is given by

$$f(\mathbf{x}) = \text{sgn}(\mathbf{w}^T \mathbf{x} + b) \ . \qquad (21.2)$$

If we assume that the misclassification costs are equal, the pair $(\mathbf{w}, b)$ should be chosen so as to minimize the future probability of misclassification.

Minimizing the misclassification rate on the training set, however, does not necessarily minimize future misclassification rate. For example, the function f that takes $f(\mathbf{x}_i) = y_i$ on the training data and is random elsewhere has zero training error but does not generalize to unseen examples[2]. This is an artificial but important illustration of the problem of overfitting. One approach to avoiding overfitting is to use regularization [3], that is, one attempts to minimize training error whilst controlling the expressiveness of the hypothesis space. Roughly speaking, if the data can be classified with low error by a 'smooth' function then that function is likely to generalize well to unseen examples. This is a variation on the principle of Ockham's Razor [26].

21.2.1 Support Vector Machines

The SVM method avoids overfitting by maximizing the *margin* γ between the two classes of the training data, i.e. maximizing the distance between the separating hyperplane and the data on either side of it. This is equivalent to minimizing a regularized error function of the form:

$$\frac{1}{2}\|\mathbf{w}\|_2^2 + +\frac{1}{2}b^2 + C\sum_{i=1}^{l}\xi_i, \tag{21.3}$$

where $\|\mathbf{w}\|_2^2 = \mathbf{w}^T\mathbf{w}$ is the 2-norm of the weight vector. Separation of the two classes is achieved by enforcing the constraints $y_i f(\mathbf{x}_i) \geq 1 - \xi_i$ during training. The ξ_i measure the degree of violation of the constraints (i.e. the training error). Non-linear decision surfaces can be learned by mapping the data into a higher-dimensional feature space. This is a reproducing kernel Hilbert space [28] for some kernel K. The standard SVM optimization for the non-linear separation of two classes (once represented in the dual formulation) is a convex quadratic programming (QP) problem [6]:

$$\text{Maximize}_{\alpha} \quad \tfrac{1}{2}\alpha^T(\mathbf{y}^T K\mathbf{y} + \mathbf{y}^T\mathbf{y})\alpha - 1^T\alpha \tag{21.4}$$

$$\text{subject to} \qquad 0 \leq \alpha_i \leq C, \tag{21.5}$$

where α is a vector of Lagrange multipliers, $\mathbf{y} = (y_1, \ldots, y_l)$ and K is the $l \times l$ matrix with entries $K_{ij} = K(\mathbf{x}_i, \mathbf{x}_j)$. C is a regularization parameter that controls the trade-off between minimizing training error and minimizing $\|\mathbf{w}\|_2^2$. The decision function, resulting from the above optimization is:

$$f(\mathbf{x}_i) = \text{sgn}\left(\sum_{i=1}^{l}\alpha_i^* y_i K(\mathbf{x}, \mathbf{x}_i) + b^*\right), \tag{21.6}$$

---

[2]We assume the training set is of measure zero.

where $b^* = \alpha^T \mathbf{y}$ is the operating point. Note that this formulation corresponds to the case where the hyperplane passes through the origin in feature space. This removes the need to enforce the constraint $\alpha^T \mathbf{y} = 0$. The unbiased hyperplane has slightly worse theoretical properties [6], but has the same generalization performance as the biased hyperplane on many real-world data sets [10]. A common method of solving the optimization problem (21.4) is to decompose it into a series of smaller subproblems of size $q \ll l$. The optimization remains convex and has a global optimum. The subproblem selection method of Hsu and Lin [10], is used in this work, with $q = 10$.

Note that many of the α_i in (21.6) may be zero at the optimum of the QP (21.4). Hence, the hyperplane is an expansion on a subset of the training data. These points are known as *support vectors* (SVs). When there are few (resp. many) SVs the model is said to be *sparse* (resp. *dense*). A sparse model is desirable as it leads to less time and space requirements during prediction.

21.2.2   Kernel Functions

The free parameters for an SVM can be divided into two categories: those affecting the optimization, and those defining the model. Since the optimization problem (21.4) has a global optimum the first set of parameters do not affect the quality of the solution. This is in contrast to neural networks, where parameters such as the decay rate and momentum term affect the quality of the local optimum obtained. The parameters defining the model are the choice of kernel and its parameters, and the regularization constant C. Common choices of kernel function are the linear kernel $K(\mathbf{x}, \mathbf{z}) = \mathbf{x}^T \mathbf{z}$, the polynomial kernel $K(\mathbf{x}, \mathbf{z}) = (\mathbf{x}^T \mathbf{z}/\sigma^2 + 1)^d$, the Gaussian kernel $K(\mathbf{x}, \mathbf{z}) = \exp(-\|\mathbf{x} - \mathbf{z}\|^2/2\sigma^2)$ which learns a radial basis function (RBF) network with centres at the support vectors, and the sigmoid kernel $K(\mathbf{x}, \mathbf{z}) = (1 + \exp(v\mathbf{x}^T \mathbf{z} + c))^{-1}$ which learns a two-layer perceptron. The problem of kernel selection is not addressed here. In the following the Gaussian kernel is used. This provides a powerful learner, however it is sensitive to the choice of the scale (or width) parameter σ.

21.3   Automated Model Order Selection

When using a Gaussian kernel the scale parameter σ must be specified in advance. This parameter can be thought of as the width of the RBF. This parameter controls the trade-off between faithfulness to the training data and smoothness of the decision surface in the input space. When σ is too small the SVM overfits the training data, essentially every training point becomes an RBF centre (support vector). If σ is too large then the decision boundary is not able to model the required decision boundary. The quality of the solution is sensitive to the choice of σ.

In general, the SVM is trained for a range of values of σ and that value is chosen which minimizes some estimate of the generalization error. The generalization error may be estimated by means of bounds on its expectation provided by Vapnik-Chervonenkis theory [26], or by means of a validation set. Building many models is prohibitive in terms of training time, which for an SVM has been found empirically to scale quadratically in the sample size [19, 12].

In Section 21.3.1 below I describe some previous approaches to automatically tuning kernel parameters. In the subsequent sections I present a heuristic for updating the width parameter σ during optimization. This heuristic is shown to have good performance on some publicly available data sets.

21.3.1 Incremental Tuning

Consider the case where the kernel has one free parameter σ. The most basic approach to model order selection is to train an SVM for a range of values of σ and choose that value minimizing some estimate of the generalization error. The generalization error of an hyperplane classifier with margin γ can be upper bounded as follows [27]:

$$\varepsilon \leq O\left(\frac{R^2}{l\gamma^2}\right), \tag{21.7}$$

where R is the radius of the smallest ball containing the training set and l is the number of training points. Cristianini et al. [7] show that this bound is smooth in σ. They suggest the following procedure for Gaussian kernels.

1. Initialize σ to a very small value.

2. Maximize the margin
   - Compute the error bound
   - Increase the kernel parameter $\sigma \leftarrow \sigma + \delta\sigma$

3. Stop when a predetermined value of σ is reached, else repeat 2.

The motivation here is that for small σ convergence is rapid, and few iterations will be required to bring the solution back to maximal margin solution after each update. This approach, when implemented with the kernel-adatron algorithm [9] provides a reasonable value of σ for good generalization. The experiments reported used hard margin SVMs, for which there is no regularization parameter C. When there is more than one kernel parameter to choose the above approach becomes more computationally intensive.

A more principled approach to parameter selection is provided by Chapelle et al. [5]. Starting from small values of C and σ for a Gaussian kernel the SVM is trained and the gradient of the bound with respect to C and σ is calculated. A step is taken in parameter space in the direction of maximum decrease in the error bound and the SVM is retrained. They compared this approach to estimating error rate by five-fold cross-validation for 10 values each of C and σ, as described in [20]. The main advantage of the gradient descent approach is that the number of SVMs trained is much less than the 500 required by the cross-validation approach. A similar approach is described by Wahba in this volume [29].

21.3.2   Heuristic Approaches

The above approaches to adaptive tuning of SVMs suffer from two disadvantages. Firstly, they all require sequential retraining of the SVM for a range of values of σ. This results in an approximate doubling of the training time for the methods of Cristianini et al. and of Chapelle et al. Secondly, the gradient descent algorithms have an additional layer of machinery. This introduces extra parameters regarding the training rate and extra computation in calculating the gradients.

One aspect common to all of the above techniques is that σ is tuned according to *predicted error*. An alternative approach is to tune σ to be on the scale of the margin of separation, i.e. to let the data dictate directly the value of σ. An RBF network is usually trained by first using some clustering technique to choose the RBF centres and widths σ_i and then performing a classification (or regression) step in the feature space implicitly defined by the RBFs [1]. Bishop [3] suggests setting all σ_i equal to some multiple of the average separation of the centres. To allow for different widths in different areas of space, the widths may be set to be the average distance from each basis function to its k-nearest neighbours, for some small value k. Jaakkola et al. [11] suggest a similar heuristic for the RBF kernel of an SVM, viz. σ is set to be the median separation of each example to its nearest neighbour in the other class. That is:

$$\sigma_0 = \operatorname*{median}_{i|y_i=-1} \left(\min_{j|y_j=+1} \|\mathbf{x}_i - \mathbf{x}_j\|^2 \right), \qquad (21.8)$$

Since the SVM solution depends only on the support vectors, which are the RBF centres, it seems desirable to apply this heuristic to the support vectors only. The simplest way to do this would be to train the SVM, calculate σ from the set of support vectors obtained, and retrain with the updated σ. This process could be repeated until the solution appeared to stabilize (although there is no guarantee that this would happen). This simple procedure involves retraining the SVM for a range of values of σ

and is hence undesirable. In the following sections, I describe and analyse an online approximation to this process.

21.3.3 Approximating the Support Vector Set

Since we wish to apply the above heuristic only to the basis function centres (support vectors) an approximation to the set of support vectors is required. LAIKA (locally adaptive iterative kernel approximation) updates σ during training based on an estimate of the final support vector set. After every h_a iterations of the decompostion algorithm σ is updated according to:

$$\sigma = \underset{i \in A}{\text{median}} \left(\min_{j \in B} \| \mathbf{x}_i - \mathbf{x}_j \|^2 \right), \qquad (21.9)$$

where A and B are as follows.

$$
\begin{aligned}
A(t) &= \{ i | \alpha_i(t) > 0 \ \& \ y_i = -1 \}, \\
B(t) &= \{ i | \alpha_i(t) > 0 \ \& \ y_i = 1 \},
\end{aligned}
$$

where $\alpha_i(t)$ is the Lagrange multiplier of $\mathbf{x}_i$ at the tth iteration. Initially, the sets A and B will provide very poor approximations to the sets of SVs. Nearer convergence, the approximations will improve. Although it appears that the parameter σ has been replaced by the parameter h_a, this latter is not strictly required as the natural thing to do is update after every iteration. To speed up training h_a is set to be 10 for the experiments reported here.

21.3.4 Results and Discussion

The LAIKA heuristic was evaluated on the following four data sets.

Cancer This data set was obtained from the University of Wisconsin Hospitals, Madison from Dr. William H. Wolberg [16, 2]. There are 699 examples of tumour cells, characterized by nine physiological measurements. One of the features has 16 missing values, these are replaced by the feature mean. The task is to distinguish benign tumour cells from malignant tumour cells.

Diabetes This data set is from the National Institute of Diabetes and Digestive Kidney Diseases [25]. There are 768 examples of female Pima Indians, characterized by eight physiological descriptors. The task is to predict presence or absence of diabetes.

Heart This data set was created by Robert Detrano at the V.A. Medical Center, Cleveland. (The version used here is that used in Statlog [17]). There are 270 examples of patients characterized by 13 physiological attributes. The task is to discriminate between absence and presence of heart disease.

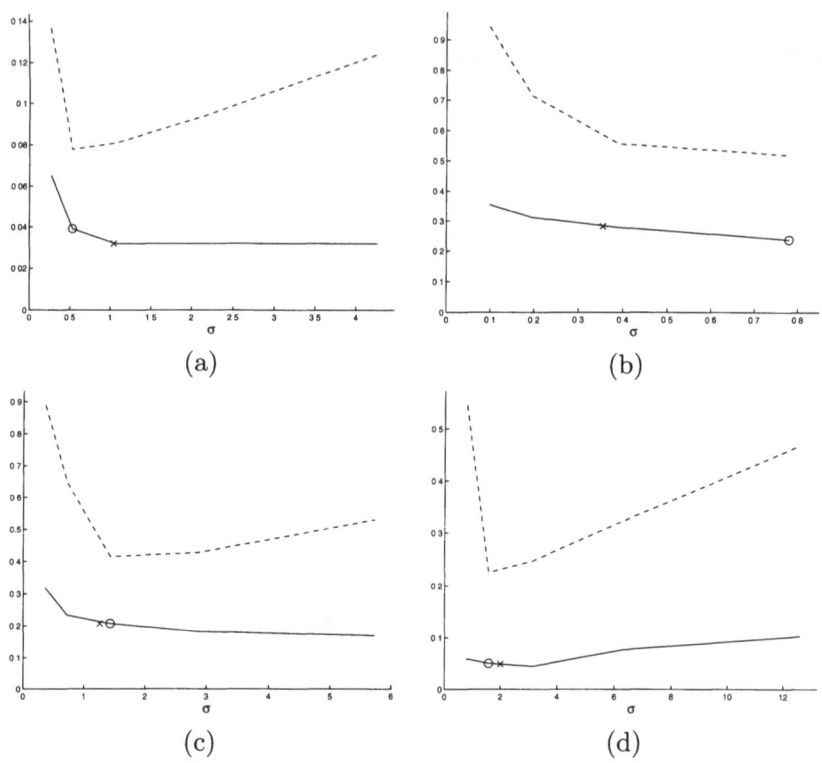

Figure 21.1. Cross-validated errors for the SVM (*circle*) and LAIKA (*cross*). For the SVM that value of σ is chosen which minimizes the $\alpha\xi$ estimator (*dotted line*). This a good predictor of the minimizer of generalization error (*solid line*). Data sets: (a) Cancer, (b) Diabetes, (c) Heart, (d) Ionosphere

Ionosphere This data set was created by Vince Sigillito at Johns Hopkins University, Maryland [24]. There are 350 examples of radar returns, characterized by 34 attributes. The problem is to discriminate between 'good' radar returns from the ionosphere and 'bad' radar returns.

An SVM with Gaussian kernel was trained for

$$\sigma \in \{1/4\sigma_0, 1/2\sigma_0, \sigma_0, 2\sigma_0, 4\sigma_0\}.$$

That value of σ was chosen which minimized the $\alpha\xi$ estimate of the generalization error [13]. (The $\alpha\xi$ estimator has been shown to be an accurate and efficient estimator for model selection [8].) This was compared to using LAIKA to automatically tune σ. Since all of the data sets have some degree of noise, the regularization parameter C was chosen as the minimizer of the $\alpha\xi$ estimator. The performance of the various classifiers was evaluated by 10 fold cross-validation (CV).

Table 21.1. Training and model selection times (in minutes) for the SVM and LAIKA

Data Set	SVM	LAIKA
Cancer	11.07	2.48
Diabetes	345.45	7.68
Heart	9.34	0.51
Ionosphere	114.91	0.54

The CV error of the standard SVM is indicated by a solid line in Figures 21.1 (a)–(d). The $\alpha\xi$ estimator is indicated by a dotted line. The value of σ minimizing the $\alpha\xi$ estimator, and the corresponding CV error of the SVM, indicated by a circle. The performance achieved with this value of σ is close to the optimal over the range tried. The value of σ chosen by LAIKA, and the corresponding CV error, is indicated by a cross. The performance of LAIKA is the same or better than the SVM except on the Diabetes data, when it is slightly worse.

The main advantage of using LAIKA over performing a grid search is the computational saving. The training and model selection times for the SVM and LAIKA are shown in Table 21.1. The computational time of LAIKA could also be reduced by updating the cached kernel computations when σ is updated [14], as opposed to recalculating them.

21.4   Conclusions

An online heuristic, termed LAIKA, has been presented for automatically tuning the width parameter σ when training a support vector classifier with a Gaussian kernel. LAIKA chooses a value of σ close to that found by a search, and with comparable performance. When learning an SVM classifier with LAIKA only one classifier is built, thus reducing training time. Work is currently in progress to assess the effect on performance and training time of the update frequency. The heuristic could also be refined to ignore noisy examples in the training data. These points can become misclassified support vectors, thus misleading the heuristic. This may be the cause of the poorer performance of LAIKA exhibited on the Diabetes data. A further possibility is to allow a different scale parameter for each dimension, in order to perform automated feature selection [15].

References

[1] M. Aizerman, E. Braverman, and L. Rozonoer. Theoretical foundations of the potential function method in pattern recognition learning. *Automations and Remote Control*, 25:821–837, 1964.

[2] K.P. Bennett and O.L. Mangasarian. Robust linear programming discrimination of two linearly inseparable sets. *Optimization Methods and Software*, 1:23–34, 1992.

[3] C. Bishop. *Neural Networks for Pattern Recognition*. Clarendon Press, 1995.

[4] C. J. C. Burges. A tutorial on support vector machines for pattern recognition. *Data Mining and Knowledge Discovery*, 2(2):1–47, 1998.

[5] O. Chapelle, V. Vapnik, O. Bousquet, and S. Mukherjee. Choosing multiple parameters for support vector machines. *Machine Learning*, 46:131–160, 2002.

[6] N. Cristianini and J. Shawe-Taylor. *Support Vector Machines*. Cambridge University Press, 2000.

[7] N. Cristianini, J. Shawe-Taylor, and C. Campbell. Dynamically adapting kernels in support vector machines. In M.S. Kearns, S.A. Solla, and D.A. Cohn, editors, *Advances in Neural Information Processing Systems, 11*, Denver, CO, 1998. The MIT Press.

[8] K. Duan, S.S. Keerthi, and A.N. Poo. Evaluation of simple performance measures for tuning SVM hyper parameters. Technical Report CD-01-11, Department of Mechanical Engineering, National University of Singapore, 10, Kent Ridge Crescent, 119260, Singapore, 2001.

[9] T.-T. Friess, N. Cristianini, and C. Campbell. The kernel-adatron: a fast and simple learning procedure for support vector machines. In J. Shavlik, editor, *Machine Learning Proceedings of the Fifteenth International Conference (ICML '98)*, pages 188–196, San Francisco, CA, 1998. Morgan Kaufmann.

[10] C.-W. Hsu and C.-J. Lin. A simple decomposition method for support vector machines. *Machine Learning*, 46:291–314, 2002.

[11] T. Jaakkola, M. Diekhans, and D. Haussler. Using the fisher kernel method to detect remote protein homologies. In T. Lengauer, R. Schneider, P. Bork, D. Brutlag, J. Glasgow, H.-W. Mewes, and R. Zimmer, editors, *Proceedings of the Seventh International Conference on Intelligent Systems for Molecular Biology*, pages 149–158, Heidelberg, Germany, 1999. AAAI Press.

[12] T. Joachims. Making large-scale SVM learning practical. In Schölkopf et al. [22], pages 169–184.

[13] T. Joachims. Estimating the generalization performance of a SVM efficiently. In P. Langley, editor, *Machine .Learning Proceedings of the Seventeenth International Conference (ICML '00)*, pages 431–438, Stanford, CA, 2000. Morgan Kaufmann.

[14] J.-H. Lee and C.-J. Lin. Automatic model selection for support vector machines. Technical report, Department of Computer Science and Information Engineering, National Taiwan University, Taipei 106, Taiwan, November 2000.

[15] D.J.C. Mackay. Bayesian methods for backpropagation networks. In J.L. van Hemmen, E. Domany, and K. Schulten, editors, *Models of Neural Networks II*. Springer, 1993.

[16] O.L. Mangasarian and W.H. Wolberg. Cancer diagnosis via linear programming. *SIAM News*, 23(5):1&18, September 1990.

[17] D. Michie, D.J. Spiegelhalter, and C.C. Taylor. *Machine Learning, Neural and Statistical Classification*. Prentice Hall, Englewood Cliffs, N.J., 1994. Data available at anonymous ftp: ftp.ncc.up.pt/pub/statlog/.

[18] T. Mitchell. *Machine Learning*. McGraw-Hill International, 1997.

[19] J. Platt. Fast training of support vector machines using sequential minimal optimization. In Schölkopf et al. [22], pages 185–208.

[20] G. Rätsch, T. Onoda, and K.-R. Müller. Soft margins for AdaBoost. *Machine Learning*, 42(3):287–320, 2001.

[21] F. Rosenblatt. The perceptron: a probabilistic model for information storage and organization in the brain. *Psychological Review*, 65:386–408, 1958.

[22] B. Schölkopf, C.J.C. Burges, and A. Smola, editors. *Advances in Kernel Methods: Support Vector Learning*. The MIT Press, 1999.

[23] B. Schölkopf, K.-K. Sung, C.J.C. Burges, F. Girosi, P. Niyogi, T. Poggio, and V.N. Vapnik. Comparing support vector machines with Gaussian kernels to radial basis function classifiers. *IEEE Transactions on Signal Processing*, 45(11), 1997.

[24] V.G. Sigillito, S.P. Wing, L.V. Hutton, and K.B. Baker. Classification of radar returns from the ionosphere using neural networks. *Johns Hopkins APL Technical Digest*, 10:262–266, 1989.

[25] J.W. Smith, J.E. Everhart, W.C. Dickson, W.C. Knowler, and R.S. Johannes. Using the ADAP learning algorithm to forecast the onset

of diabetes mellitus. In R.A. Greenes, editor, *Proceedings of the Symposium on Computer Applications in Medical Care*, pages 261–265, Washington, 1998. Los Alamitos, CA: IEEE Computer Society Press.

[26] V. Vapnik. *Statistical Learning Theory*. John Wiley & Sons, 1998.

[27] V.N. Vapnik. *The Nature of Statistical Learning Theory*. Springer-Verlag, 1995.

[28] G. Wahba. Support vector machines, reproducing kernel Hilbert spaces, and randomized GACV. In Schölkopf et al. [22], pages 69–88.

[29] G. Wahba, Y. Lin, Y. Lee, and H. Zhang. Optimal properties and adaptive tuning of standard and nonstandard support vector machines. In *Nonlinear Estimation and Classification*. Springer-Verlag, 2002.

22

Generalization Error Bounds for Aggregate Classifiers

Gilles Blanchard[1]

22.1 Introduction

The use of multiple classifiers has raised much interest in the statistical learning community in the past few years. The basic principle of multiple classifiers algorithms, also called aggregation or ensemble or voting methods, is to construct, according to some algorithm, several (generally a few dozens) different classifiers belonging to a certain family (e.g. support vector machines, classification trees, neural nets...). The "aggregate" classifier is then obtained by majority vote among the outputs of the single constructed classifiers once they are presented a new instance. For some algorithms the majority vote is replaced by a weighted vote, with weights prescribed by the aggregation algorithm. Classical references about this kind of methods include [2, 3, 9, 8, 12, 14, 19].

In particular, various versions of the so-called "Boosting" algorithm which iteratively construct a sequence of classifiers have recently received much attention because of their good reported practical results. Several authors have reported that, while a perfect aggregate classifier (that is, having zero training error) can be reached very quickly by the algorithm, adding more classifiers to the aggregate through additional iterations of the algorithm usually still results in a noticeable reduction of test error. Since this property can obviously not be forecast just by using training error statistics, people have tried to introduce more accurate statistics on training to explain these good results of Boosting. A particularly interesting lead has been opened by Freund, Shapire, Bartlett and Lee [19] using the notion of empirical margin, which we recall below. The goal of this chapter is to give the reader some taste of the topic, and point out what

[1]Gilles Blanchard is with the Département de Mathématiques et Applications, École normale supérieure, 45 rue d'Ulm 75230 Paris Cedex 05 (Email: gblancha@dma.ens.fr).

results have been obtained (along with some new refinements), and what is still unknown.

22.1.1 Setting and notations

In the present chapter, for simplicity we will restrict our attention to 2-class classification problems. X will denote a random variable taking values in a space $\mathcal{X}$, typically a real vector space of high dimension. Y will denote the "class variable", taking its values in $\{-1, 1\}$. A labeled training sample $(X_1, Y_1), \ldots, (X_N, Y_N)$ is already known, and the goal of the classification is, based on this information, to find a *classifier*, that is, a function $h : \mathcal{X} \to \{-1, 1\}$, having a low *generalization error* defined by

$$\mathcal{E}(h) = E[\mathbf{1}\{X \neq h(Y)\}], \qquad (22.1)$$

where the expectation is taken with respect to a new (unknown) example (X, Y). Usually, the empirical error of classifier h on the training sample is called *training error*, and the generalization error is estimated by testing the performance h on an independent sample; this estimate is called *test error*.

We suppose we are given a set of classifiers $\mathcal{H}$: for example $\mathcal{H}$ can be a set of classification trees (CART model), of support vector machines, of neural nets... We will call $\mathcal{H}$ the set of "base classifiers"; for simplicity, throughout the chapter we will restrict our attention to a finite set $\mathcal{H}$ of base classifiers of cardinality L (all the results in the finite case can be extended to the case where $\mathcal{H}$ is infinite, but of finite VC dimension). A classifier $h \in \mathcal{H}$ will take its values in $\{-1, 1\}$ (in some cases we will allow that $h \in \mathcal{H}$ takes its values in the full range $[-1, 1]$. This will be stated accordingly in the hypotheses).

We focus our interest on aggregate classifiers: in this setting, any weighted voting scheme can be represented by a certain family α of weights summing to one; in other words, α is a probability distribution on $\mathcal{H}$, belonging to the $L-1$ dimensional simplex S_{L-1}. In the sequel, we will denote by

$$H_\alpha(x) = \sum_{i=1}^{L} \alpha_i h_i(x)$$

the aggregate function obtained through the convex combination α: the actual aggregate classifier is obtained as the sign of H_α. The *margin M_α* associated to an example $Z = (X, Y)$ and a convex combination α is defined as

$$M_\alpha(Z) = H_\alpha(X)Y.$$

Thus it is equivalent to say that an example is misclassified by H_α or that its margin is negative. However, margins give more information about the behavior of the classifier with respect to the data than just training error: intuitively, they give an information on how "confident" the classification is (or how dominant the majority is in a majority vote). It is this intuitive idea that we try to develop in the sequel.

22.1.2 Examples of aggregation algorithms

A specific aggregation algorithm is nothing more than a protocol to build a convex combination $\widehat{\alpha}$ based on the training sample (usually obtained sequentially, in the sense that a sequence of base classifiers is built iteratively). We give here a brief review of existing algorithms.

1. **Bootstrap and AGgregate, BAGGING.** To build each classifier a bootstrap sample from the training data is produced. This weighted sample is then used to build a classifier ([6]).

2. **Bayesian sampling.** A prior distribution on classifier space is determined. The posterior distribution on classifier is determined by multiplying the prior by a likelihood which is given by the probability assigned to the data by a classifier. A procedure like Markov Chain Monte Carlo (MCMC) is often used to draw sample classifiers from this posterior and the aggregate classifier is given by the average over sampled classifiers. See [4, 10] for explicit construction in the case of classification trees.

3. **Boosting.** An initial uniform weight vector W_1 for the training data is set, $W_1(l) = 1/n, l = 1, \ldots, n$. A classifier h_1 is built to fit the training data. The training error ε_1 of the classifier is obtained. The misclassified points in the training set have their weight increased by a factor of $(1-\varepsilon_1)/\varepsilon_1$. The weights of the training data are renormalized to produce the new weight vector W_2. A new classifier h_2 is built to fit the *weighted* training sample. The weighted training error ε_2 of h_2 is evaluated in terms of the weighted training data, the weights of the examples misclassified by h_2 are increased by a factor $(1 - \varepsilon_2)/\varepsilon_2$, and so on. Each classifier h_n is finally affected a voting weight equal to $\frac{1}{2} \log((1 - \varepsilon_n)/\varepsilon_n)$. See for example [8] for an overview of this kind of methods.

4. **Randomized trees.** This algorithm is specific to classification trees. At each node a random subset of fixed size of all candidate splits is entertained. The best split from this subset is chosen. This method seems to appear first in [2, 3], a very similar procedure has recently been proposed in [9], and some other variants are available in the literature.

22.1.3 Overview

The goal of the chapter is to propose several possible methods to infer bounds on the generalization error from statistics coming from the *empirical* margin distribution. These bounds should be *independent* on the specific algorithm used. We will propose two different point of views.

In section 22.2, we give bounds on generalization error through a Chebychev inequality, following ideas initially proposed in [2]. We derive uniform controls of mean and variance of the margin function for any convex combination of base classifiers, based on their empirical counterparts. These bounds mainly involve very standard tools for control of empirical processes (Hoeffding's inequality).

In section 22.3, we present a bound on generalization error based on the empirical cumulative distribution function of the margins. It can be seen as a interpolation result between the bounds given in the original work of Shapire et al. [19] and a bound proved by Breiman [7] based on the minimum empirical margin. The result we present uses the same kind of proof technique and allows us to retrieve, up to a multiplicative factor, both of these previous bounds as special cases.

22.2 Bounds based on the empirical mean and variance of the margin

22.2.1 Motivation

In this section, we investigate how one can derive generalization error bounds based on the estimation of the margin mean and variance.

By applying Chebychev's inequality, we easily obtain the following inequality: for any $\alpha \in \mathcal{M}_1(\mathcal{H})$,

$$\mathcal{E}(H_\alpha) = P[M_\alpha(Z) \le 0] \le \frac{Var[M_\alpha(Z)]}{E[M_\alpha(Z)]^2} \tag{22.2}$$

The initial motivation behind this kind of inequality is to give a simple device able to draw a link between the generalization error and the first two moments of the margin. The expected margin can be interpreted as the average performance of a single classifier (drawn according to distribution α) on a new example; on the other hand, it is possible to give an interpretation of the variance of the margin as the average correlation given the class, of two independent classifiers (also drawn according to distribution α), see [1] for more detail. Various heuristics have indeed pointed out that aggregating classifiers that have low correlation given the class yield good generalization performance: see [2, 1, 9].

The goal of this section is to give exact inequalities for the control of mean and variance of margin distributions based on their empirical counterparts:

in other words, confidence intervals. This will allow to get from (22.2) a bound on generalization error based purely on empirical quantities. Since this control is based on a weak Chebychev's inequality, it is clear that from an asymptotic point of view, when the size of the training set tends to infinity, Chebychev-based bounds won't be accurate enough to be tight; in particular they can't (asymptotically speaking) compete with the type of bounds shown in the next section, which make use of information coming from the full empirical distribution on the margin, because the Chebychev-based bound only makes use of first and second moment statistics. However, our hope is that for small training sets, it may be more favorable in practice to only take into account these simpler statistics because their estimation is more reliable.

Since we want to obtain bounds that are valid independently of the aggregation algorithm used, we need to control estimation accuracy uniformly over all convex combinations of base classifiers. This way, the control obtained from (22.2) will be valid for any α chosen based on the actual data used for estimation.

The results presented in this section will be derived in the case of finite $\mathcal{H}$. We only make use of very classical tools and the inequalities obtained should be considered as fairly standard.

22.2.2   Uniform control of absolute deviations

We give here a control of absolute deviations of empirical mean and variance uniform over convex combinations of a finite family of bounded random variables.

Proposition 1. *Let* $(X^1, \ldots, X^L)$ *be* L *real random variables of joint probability* P *and such that* $|X^i| \leq 1$, $i = 1, \ldots, L$. *For* $\alpha \in S_{L-1}$ *belonging to the* $(L-1)$-*dimensional simplex, denote* $X^\alpha = \sum_i \alpha_i X^i$. *Let* $M_\alpha = E[X^\alpha]$, *and* $V_\alpha = Var[X^\alpha]$.

Consider an i.i.d. sample of size N *of* $(X^1, \ldots, X^L)$ *and for* $\alpha \in S_{L-1}$, *let* $\widehat{M}_\alpha$ *be the associated empirical mean of variable* X_α, $\widehat{M}_{\alpha\alpha}$ *the empirical mean of variable* X_α^2, *and* $\widehat{V}_\alpha = \widehat{M}_{\alpha\alpha} - \widehat{M}_\alpha^2$.

Then, except on a set of samples of probability less than $\delta > 0$, *the following inequalities hold simultaneously for all* $\alpha \in S_{L-1}$:

$$|M_\alpha - \widehat{M}_\alpha| \leq \sqrt{\frac{2\log(2(L+1)^2/\delta)}{N}}$$

and

$$|V_\alpha - \widehat{V}_\alpha| \leq \sqrt{\frac{32\log(2(L+1)^2/\delta)}{N}}$$

We can now apply this result to the variables $(h_1(X)Y, \ldots, h_L(X)Y)$, when $\mathcal{H}$ is the finite set $\{h_1, \ldots, h_L\}$. For a given α, M_α and V_α above then correspond to the average and variance of M_α appearing in (22.2). Since these inequalities are valid uniformly in α, they are in particular valid for the combination $\widehat{\alpha}$ built by the algorithm depending on the sample. Then the above inequalities, once used into (22.2), allow us to build a confidence interval on the generalization error based on empirical quantities only.

22.3   Bounds using the empirical margin distribution

22.3.1   Notations and previous work

In this section, we will make use of the richer statistic given by the full margin cumulative distribution function (cdf). We will use the following notations: let F_α denote the cdf of $M_\alpha(Z)$ and $\widehat{F}_\alpha^N$ denote the empirical cdf of the same variable, based on sample Z_1^N.

In [19], Shapire et al. proved the following theorem:

Theorem 5 (Freund, Bartlett, Shapire, Lee). *Let Z_1^N be an i.i.d. training sample drawn according to some distribution P on $\mathcal{X} \times \mathcal{Y}$. Assume $|\mathcal{H}| = L$, and that a classifier $h \in \mathcal{H}$ takes its values in $\{-1, 1\}$. Let $\delta > 0$, then except on a set of samples of P_N-probability less than δ, for all $\theta > 0$ and all $\alpha \in \mathcal{M}_1(\mathcal{H})$ the following inequality holds:*

$$F_\alpha(0) \leq \widehat{F}_\alpha^N(\theta) + \frac{C}{\sqrt{N}} \left(\frac{\log L \log N}{\theta^2} + \log(1/\delta) \right)^{1/2}, \qquad (22.3)$$

where C is a universal constant.

This theorem permits us to bound the generalization error by a function depending on the empirical margin distribution on the sample. Remember that $F_\alpha(0)$ is exactly the generalization error of the aggregate classifier; the upper bound is a sum of two terms, the empirical cdf at "margin" θ, and a complexity term decreasing (as a function of θ) as $1/\theta$, so that the lowest value of the bound is obtained for some tradeoff between these two terms (it is possible however to pick the best possible value for θ – which is a random quantity – since the inequality is valid simultaneously for all θ). Moreover, the same authors give empirical and theoretical evidence that the Boosting algorithm tends to make the empirical margins as high as possible. Intuitively, one can say that at each iteration, the Boosting algorithm will mainly concentrate on those examples that have the lower margins with respect to the current aggregate classifier.

Based on this results, one can think of building an algorithm specifically trying to maximize over convex combinations of classifiers the lowest observed margin on the training set, what we can call the "minimax margin" principle. Let us mention that this idea is strongly related to game theory (see [7, 13]), but our main point of interest here is the following bound established by Breiman [7] based on the minimum empirical margin:

Theorem 6 (Breiman). *Let Z_1^N be an i.i.d. training sample drawn according to some distribution P on $\mathcal{X} \times \mathcal{Y}$. Assume $|\mathcal{H}| = L$, and that a classifier $h \in \mathcal{H}$ takes its values in $[-1, 1]$. Define $\underline{M_\alpha}$ as*

$$\underline{M_\alpha} \doteq \inf_{1 \leq i \leq N} M_\alpha(Z_i)$$

Let $\delta > 0$, then except on a set of samples of P_N-probability less than δ, for all $\Delta > 0$, $\alpha \in S_{L-1}$ the following inequality holds:

$$F_\alpha(\underline{M_\alpha} - \Delta) \leq \frac{C}{N} \left(\frac{\log L \log N}{\Delta^2} + \log(1/\delta) \right), \qquad (22.4)$$

where C is an universal constant.

We thus get a bound on generalization error by taking $\Delta = \underline{M_\alpha}$ in the inequality above. Note that we are allowed to do so (although $\underline{M_\alpha}$ is a random variable) because the inequality is valid uniformly for all $\Delta > 0$. The interesting point is to compare with bound (22.3) when $\theta = \underline{M_\alpha}$: the bound obtained by Breiman is of the order of magnitude of the square of the Shapire et al. bound, and therefore in principle much more accurate. On the other hand, inequality (22.3) is more general since we can choose any value for the margin level θ and not only the minimum margin. In fact it is likely that the minimum of the right hand side of (22.3) is obtained for some $\theta > \underline{M_\alpha}$ and it is not clear in this case whether (22.4) gives a better bound or not.

22.3.2 An interpolation bound

The main result of this section is an "interpolation" bound between inequalities (22.3) and (22.4) which renders the regime transition explicit. The proof of this inequality is mainly based on the same ideas used by Shapire et al. and Breiman to prove theorems 5 and 6, and on the use of VC bounds for relative deviations.

Theorem 7. *Let Z_1^N be an i.i.d. training sample drawn according to some distribution P on $\mathcal{X} \times \mathcal{Y}$. Assume $|\mathcal{H}| = L$, and that a classifier $h \in \mathcal{H}$ takes its values in $[-1, 1]$. Then, except on a set Ω_δ of training samples of probability less than δ, for any $\theta \in [-1, 1]$ and $\Delta \in [0, 2]$, for all $\alpha \in S_{L-1}$ the following inequality holds:*

$$F_\alpha(\theta - \Delta) \leq \widehat{F}_\alpha^N(\theta) + C_1 \sqrt{\widehat{F}_\alpha^N(\theta)} \left(\frac{\log L \log N}{\Delta^2 N} + \log(1/\delta) \right)^{1/2}$$

$$+ C_2 \left(\frac{\log L \log N}{\Delta^2 N} + \log(1/\delta) \right), \qquad (22.5)$$

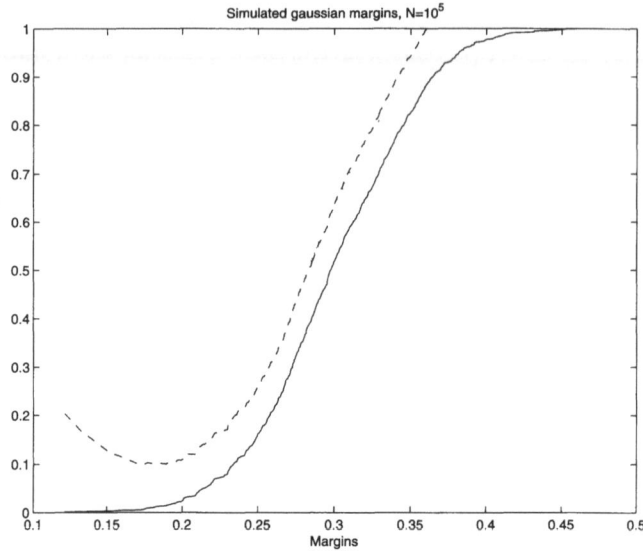

Figure 22.1. Simulated Gaussian margins: suppose the observed empirical margin cdf is given by the solid curve and $n = 10^5$. Then the controls over generalization error obtained through inequalities (22.3) and (22.5) are shown by the dotted and dashed curves, respectively. The respective minima of these curves show the best possible control obtained by the one and the other inequality.

where C_1, C_2 are universal constants.

To obtain an upper bound on the generalization error, take $\theta = \Delta$ in this bound; we can then optimize the upper bound by picking the best value for θ. It is easy to see that from inequality (22.5) we can retrieve both inequalities (22.3) and (22.4), up to a constant factor. But what kind of additional information can we extract from this interpolation inequality theorem as compared to Freund et al's and Breiman's original bounds? The interpolation bound can beat the two original inequalities when θ is picked in the "transition region" between the two regimes, that is, in the region where the empirical margin cdf is nonzero but still quite low: in that case, the more important of the two complexity terms in (22.5) is still the second one, similar to the one in (22.4), but smaller for $\theta > \underline{M_\alpha}$. If the minimum of the upper bound is attained in that region, then inequality (22.5) gives a better control on generalization error than the previous ones. Note that this situation, where the complexity term can vary in order from $\mathcal{O}(1/N)$ to $\mathcal{O}(1/\sqrt{N})$ depending on some empirical quantity (here the empirical cdf) is comparable to the "classical" analysis of generalization error on spaces of classifiers where a similar regime transition can be observed, depending on the empirical error (for a complete account, see e.g. [17]).

The interpolation bound can thus be interesting in the case where the training set is separable (zero training error) or almost separable (only a few misclassified examples), and when the empirical cdf is at first only slowly increasing (this corresponds to the case where there is a tiny proportion of training examples that have margin close to zero). This kind of behavior is illustrated in figure 22.1. In this sense, this theorem can also qualitatively justify the principle of throwing out a small number of training examples if it allows to increase the empirical margin of the others, a philosophy that is akin to the "soft margins" algorithm, see e.g. [18].

On figure 22.1, we have drawn a toy illustrative example where we suppose that the empirical margin distribution is a Gaussian (restricted to [-1,1]), centered at 0.3. We have drawn the two curves obtained by applying (22.3) and (22.5) to this case. One can see the gain obtained with the latter. However for this figure we fixed $n = 10^5$: it is a recognized fact that this kind of bound has so far remained extremely loose and informative only for extremely big sizes of training sets.

22.4   Conclusion

22.4.1   Extension to infinite base classifier spaces

All the analyses and inequalities in the present chapter can be extended, without too much effort, to the case where the base classifier space $\mathcal{H}$ is infinite, but has finite Vapnik-Chervonenkis (VC) dimension.

The VC dimension of a set of indicator functions (classifiers) is defined as a purely combinatorial quantity, and measures a form of complexity of the set of functions in terms of the cardinality of its trace on finite samples (see [17] for a complete and concise account of classical VC-theory). One crucial point about all the bounds presented in this chapter is that the complexity terms only have a mild (logarithmic) dependence in the cardinality of $\mathcal{H}$ (which represents the "complexity" of $\mathcal{H}$ in the finite case). When $\mathcal{H}$ is infinite, all the bounds are still basically valid with $\log(L)$ replaced by $d\log(N)$, where d is the VC-dimension of $\mathcal{H}$. The exact statements and proofs of these results can be found in [5] (as well as the proofs of the results stated in the present chapter).

22.4.2   Discussion and perspectives

The inequalities we have presented obviously have but a limited power. In particular, since they are very loose unless the number of training examples is really huge, their prediction power in practical applications is mostly nonexistent. This severe drawback is however usually a common feature to most rigorous bounds on generalization error found in machine learning literature, in particular the "margins bounds" for support vector machines (see [11]), which have initially inspired theorem 5. Therefore, the interest of these inequalities is mainly qualitative: they point out a link between low conditional correlation of classifiers and good generalization ability (section 22.2), which was observed empirically; and a link between empirical margin distribution and generalization error (section 22.3).

The field of research is still wide open in this area, in particular the search for tighter bounds. Recently, there have been advances in the fine study of rates of convergence that can be expected with this kind of inequalities, see [15]. An interesting link has been drawn with the so-called "PAC-Bayesian" principle, see [16]. Also, so-called "sparsity" bounds have been recently put forward. In my sense, one of the challenges of future research on related topics is to define in a sensible way the notion of complexity of a convex combination. In the "sparsity" philosophy, the size of the support of the convex combination should be a measure for its complexity. In the "smoothing" philosophy, related to the PAC-Bayesian bounds referred above, the complexity should be measured by the distance to some reference measure (e.g. uniform). These two approaches sound at odds and it an interesting problem to try finding a common framework in which they could fit.

References

[1] Y. Amit and G. Blanchard. Multiple randomized classifiers: MRCL. Technical report, University of Chicago, 2000.

[2] Y. Amit and D. Geman. Shape quantization and recognition with randomized trees. *Neural Computation*, 9:1545–1588, 1997.

[3] Y. Amit, D. Geman, and K. Wilder. Joint induction of shape features and tree classifiers. *IEEE Trans. PAMI*, 19(11):1300–1306, 1997.

[4] G. Blanchard. The "progressive mixture" estimator for regression trees.
Annales de l'I.H.P., 35(6):793–820, 1999.

[5] G. Blanchard. Mixture and aggregation of estimators for pattern recognition. Application to decision trees. PhD disseration, Université Paris-Nord, 2001. (In English, with an introductory part in French). Available at http://www.math.u-psud.fr/~blanchard/publi/these.ps.gz

[6] L. Breiman. Bagging predictors. *Machine Learning*, 26:123–140, 1996.

[7] L. Breiman. Prediction games and arcing algorithms. Technical report, Statistics department, University of California at Berkeley, December 1997.

[8] L. Breiman. Arcing classifiers. *The annals of statistics*, 26(3):801–849, 1998.

[9] L. Breiman. Random forests - random features. Technical report, University of California, Berkeley, 1999.

[10] H. Chipman, E. I. George, and E. McCulloch. Bayesian CART model search. *JASA*, 93:935–947, September 1998.

[11] N. Christianini and J. Shawe-Taylor. *An introduction to Support Vector Machines*. Cambridge University Press, 2000.

[12] T. G. Dietterich and G. Bakiri. Solving multiclass learning problems via error-correcting output codes. *J. Artificial Intell. Res.*, 2:263–286, 1995.

[13] Y. Freund and R. E. Schapire. Game theory, on-line prediciton and Boosting. In *Proceedings of the 9th annual conference on computational learning theory*, 1996.

[14] J. Friedman, T. Hastie, and R. Tibshirani. Additive logistic regression: a statistical view of boosting. *The Annals of Statistics*, 28:337–374, 2000.

[15] V. Kolchinkskii and D. Panchenko. Empirical margin distributions and bounding the generalization error of combined classifiers. To appear in Ann. Statist.

[16] J. Langford, M. Seeger, and N. Megiddo. An improved predictive accuracy bound for averaging classifiers. NIPS 2001.

[17] G. Lugosi. Lectures on statistical learning theory. Presented at the Garchy Seminar on Mathematical Statistics and Applications, available at
http://www.econ.upf.es/~lugosi, 2000.

[18] G. Rätsch, T. Onoda, and K.-R. Müller. Soft margins for AdaBoost. *Machine Learning*, 2000.

[19] R. E. Schapire, Y. Freund, P. Bartlett, and W. S. Lee. Boosting the margins: a new explanation for the effectiveness of voting methods. *Annals of Statistics*, 26(5):1651–1686, 1998.

23

Risk Bounds for CART Regression Trees

Servane Gey and Elodie Nedelec[1]

23.1 Introduction

The aim of the Classification And Regression Trees (CART) proposed by
Breiman, Friedman, Olshen and Stone [1] in 1984 is to construct some efficient algorithm which gives a piecewise constant estimator of a classifier or
a regression function from a training sample of observations. This algorithm
is based on binary tree-structured partitions and on a penalized criterion
which permits to select some "good" tree-structured estimators among a
huge collection of trees. From a practical point of view, it gives some easy
to interpret and easy to compute estimators, which are then widely used
in many applications such that Medicine, Meteorology, Biology, Pollution
or Image Coding (see [2], [3] for example).
More precisely, given a training sample of observations, the CART algorithm consists in constructing a large tree on the observations by
minimizing at each step some impurity function, and then on pruning the
so constructed tree to obtain a finite sequence of trees thanks to a penalized
criterion, which penalty term is proportional to the number of leaves.
So the question is: "Why is this penalty convenient ?". The purpose of this
paper is to give some answers to this question in the fixed design Gaussian
regression framework and in the random design bounded regression framework, where we shall see that this choice of penalty is indeed convenient.
In the classification case, it is not that clear that a good penalty should
be proportional to the number of leaves and the interested reader will find
some discussion and results about this topic in the paper by Nobel [4].

Let $\mathcal{L} = \{(X_1, Y_1); \ldots; (X_N, Y_N)\}$ be a set of independent random vari-

---

[1]Servane Gey and Elodie Nedelec are with the Laboratoire de mathématiques -
U.M.R. 8628 Université Paris XI, 91405 Orsay, France (Email: Servane.Gey@math.u-
psud.fr, Elodie.Nedelec@math.u-psud.fr).

ables, where each $(X_i, Y_i) \in \mathcal{X} \times \mathbb{R}$ follows a regression model with common regression function s. Let $\tilde{s}$ be the piecewise constant estimator of s provided by CART. We measure the performance of $\tilde{s}$ by the risk defined as follows :

$$R(\tilde{s}, s) = \frac{1}{N} \sum_{i=1}^{N} \mathbb{E}_s \left[(\tilde{s}(X_i) - s(X_i))^2 \right].$$ (23.1)

We put aside the analysis of the growing procedure to focus on the pruning procedure and see that this method used to reduce the complexity of the problem is well-chosen in the sense that it warrants a good performance of the selected estimator $\tilde{s}$ in terms of its risk $R(\tilde{s}, s)$. All our upper bounds for the risk are regarded conditionally to the growing procedure and we refer the reader who wants some results about the growing procedure to the papers by Nobel and Olshen [5] and Nobel [6] about Recursive Partitioning. We shall focus on two methods giving about the same results : let us split $\mathcal{L}$ in three independent subsamples $\mathcal{L}_1$, $\mathcal{L}_2$ and $\mathcal{L}_3$ containing respectively n_1, n_2 and n_3 observations, with $n_1 + n_2 + n_3 = N$. Suppose that either a large tree is constructed using $\mathcal{L}_1$ and then pruned using $\mathcal{L}_2$ (as done in Gelfand et al. [7]), or a large tree is constructed and pruned using the same subsample $\mathcal{L}_1$ (as done in Breiman et al. [1]). Then the final step used in both cases is to choose a subtree among the sequence obtained after the pruning procedure. One method we will study in the sequel is to make $\mathcal{L}_3$ go down each tree of the sequence and to select the tree having the minimum empirical quadratic contrast, i.e., given for any $n \leqslant N$ and any $u \in \mathbb{L}^2(\mathcal{X})$ the empirical quadratic contrast

$$\gamma_n(u) = \frac{1}{n} \sum_{i=1}^{n} (Y_i - u(X_i)),$$ (23.2)

to take the final estimator of s as follows :

$$\tilde{s} = \operatorname*{argmin}_{\{\hat{s}_{T_i}; 1 \leqslant i \leqslant K\}} \gamma_{n_3}(\hat{s}_{T_i}),$$ (23.3)

where $\hat{s}_{T_i}$ is the piecewise constant estimator of s defined on the leaves of the tree T_i and K is the number of trees appearing in the sequence.

So the purpose of this paper is to analyze the risk of $\tilde{s}$ and to prove that the penalty used in the pruning algorithm for the two above mentioned cases is convenient, using Model Selection technic, and particularly some results of Birgé, Massart [8] and Massart [9]. Our results differ from those obtained by Engel [10] and Donoho [11] in the sense that the partitions of the histograms we consider are not based on a fixed regular grid and that the bounds we obtain are nonasymptotic. For more details about asymptotic results, see also Nobel [12].

The paper is organized as follows. In the first section we recall some facts about the CART algorithm and give the results we obtain. Then we study more precisely in the second section the random design bounded regression framework, where we validate the pruning algorithm taking either $\mathcal{L}_1$ independent of $\mathcal{L}_2$ or $\mathcal{L}_1 = \mathcal{L}_2$ and finally give an upper bound concerning the final selection using $\mathcal{L}_3$ as test-sample. The third section is then devoted to some open questions.

23.2   Preliminaries and Main Result

23.2.1   The CART Algorithm

CART needs a training sample $\widetilde{\mathcal{L}}$ of the random variable $(X, Y) \in \mathcal{X} \times \mathbb{R}$ (we shall take as $\widetilde{\mathcal{L}} = \mathcal{L}_1$ or $\widetilde{\mathcal{L}} = \mathcal{L}_1 \cup \mathcal{L}_2$), and a class $\mathcal{S}$ of subsets of $\mathcal{X}$ which informs us how to split at each step of the recursive partitioning. The algorithm is computed in two steps, we shall call *Growing procedure* and *Pruning procedure*. It is computed as follows :

a) <u>Growing Procedure</u> : constructs a maximal tree T_{max} based on the data composing $\mathcal{L}_1$. Starting with the whole sample $\mathcal{L}_1$ (assimilated with the root $\{t_1\}$ of T_{max}), one makes an iterative procedure minimizing at each stage the total squared error

$$\gamma_{n_1}(\hat{s}_{|t_L}) + \gamma_{n_1}(\hat{s}_{|t_R})$$

on $sp = t_L \bigcup t_R \in \mathcal{S}$, where

$$\hat{s}_{|t} = \underset{\{a \mathbb{1}_t \; ; \; a \in \mathbb{R}\}}{\operatorname{argmin}} \gamma_{n_1}(a \mathbb{1}_t).$$

In case of ties, the choice remains arbitrary. Then one keeps on splitting stage after stage until one wants to stop. This provides the maximal tree T_{max} and one calls *terminal nodes* or *leaves* the final nodes of T_{max}.

b) <u>Pruning Procedure</u> : prunes T_{max} to provide a sequence of pruned subtrees (i.e. binary subtrees of T_{max} having the same root $\{t_1\}$ as T_{max}) containing the whole statistical information. This relationship is denoted by : if T_1 is a pruned subtree of T_2, write $T_1 \preceq T_2$. Furthermore, for a tree T, $\widetilde{T}$ denotes the set of its leaves and $|T|$ the cardinality of $\widetilde{T}$.

Next, to prune T_{max}, one uses a penalized criterion as follows. First we simply denote by n the number of data used. Also, given a tree T and S_T a set of piecewise functions in $\mathbb{L}^2(\mathcal{X})$ defined on the partition given by the leaves of T, one defines

$$\hat{s}_T = \underset{z \in S_T}{\operatorname{argmin}} \gamma_n(z).$$

Then, given a temperature α, one defines T_α the subtree of T_{max} satisfying :

(i) $T_\alpha = \mathrm{argmin}_{T \preceq T_{max}} [\gamma_n(\hat{s}_T) + \alpha|T|/n]$,

(ii) if $\gamma_n(\hat{s}_T) + \alpha|T|/n = \gamma_n(\hat{s}_{T_\alpha}) + \alpha|T_\alpha|/n$, then $T_\alpha \preceq T$.

The existence and the unicity of T_α are given in [1, pp 284-290].

Then a sequence $(T_i, \alpha_i)_{1 \leqslant i \leqslant K}$ of subtrees pruned from each other associated with some positive temperatures is obtained by making the temperature α increase so that the number of leaves of T_α decreases. This sequence verifies :

Theorem 8 (Breiman, Friedman, Olshen, Stone). $\alpha_1 = 0$, $T_K = \{t_1\}$ and $(\alpha_i)_{1 \leqslant i \leqslant K}$ is nonincreasing. Furthermore, given $k \in \{1, \ldots, K\}$, if $\beta \in [\alpha_k, \alpha_{k+1}[$, then $T_\beta = T_{\alpha_k} = T_k$.

Actually, thanks to this theorem, it is easy to check that :

Corollary 2. Let $\alpha > 0$. Then there exists some $i_0 \in \{1, \ldots, K\}$ such that $T_{i_0} = T_\alpha$.

So we easily see that this algorithm reduces efficiently the complexity of the choice of a subtree pruned from T_{max}, since by Corollary 2 the sequence of pruned subtrees contains the whole statistical information according to the choice of the penalty function used to obtain each subtree T_α. So, to validate completely this algorithm, it remains to show that this choice of penalty is convenient.

23.2.2   Main Result

Assume we observe a set of independent random variables $\mathcal{L} = \{(X_1, Y_1), \cdots \ldots, (X_N, Y_N)\}$ such that :

$$Y_i = s(X_i) + \varepsilon_i,$$

where (X_i, Y_i) lies in $\mathcal{X} \times \mathbb{R}$, ε_i is a centered noise independent of X_i and s is the regression function to be estimated. For $i \in \{1; \ldots; N\}$ let P_{X_i} denote the distribution of the observation X_i and let us define

$$\mu = \frac{1}{N} \sum_{i=1}^{N} P_{X_i}.$$

Let $\|.\|$ be the $L^2(\mathcal{X}, \mu)$-norm. Then the risk (23.1) of the final estimator $\tilde{s}$ becomes

$$R(\tilde{s}, s) = \mathbb{E}_s \left[\|\tilde{s} - s\|^2 \right].$$

Next, for a given tree T, S_T will denote the set of some piecewise constant functions defined on the partition given by the leaves of T. Thus $\hat{s}_T$ will be the minimum quadratic contrast estimator of s on S_T and $\bar{s}_T$ the $\mathbb{L}^2(\mathcal{X}, \mu)$-projection of s on S_T. Then, for a given tree T, the loss of $\hat{s}_T$ with respect to s is

$$R(\hat{s}_T, s) = \|s - \bar{s}_T\|^2 + \sigma^2 \frac{|T|}{n_1} \tag{23.4}$$

with $\sigma^2 = \text{Var}(\varepsilon)$. Moreover, $\mathbb{E}^{(j)}$ will denote the expectation conditionally to the j-th subsample, $j = 1, 2, 3$.

Then a tree-structured estimator $\hat{s}$ of s will be said to achieve the oracle if, for a nonnegative constant C,

$$\mathbb{E}^{(1)}\left[\|\hat{s} - s\|^2\right] \leqslant C \inf_{T \preceq T_{max}} R(\hat{s}_T, s).$$

Our goal is then to estimate s using the CART algorithm and to compare the performance of $\tilde{s}$ with those ones of each $\hat{s}_T$ in order to see if $\tilde{s}$ achieves the oracle or not. To proceed, two different methods can be applied :

M1: $\mathcal{L}$ is split in three independent parts $\mathcal{L}_1$, $\mathcal{L}_2$ and $\mathcal{L}_3$ containing respectively n_1, n_2 and n_3 observations. Hence T_{max} is constructed from $\mathcal{L}_1$, then pruned using $\mathcal{L}_2$ and finally a best subtree $\widehat{T}$ is selected among the sequence of pruned subtrees thanks to $\mathcal{L}_3$, and we define $\tilde{s} = \hat{s}_{\widehat{T}}$.

M2: $\mathcal{L}$ is split in two independent parts $\mathcal{L}_1$ and $\mathcal{L}_3$ containing respectively n_1 and n_3 observations. Hence T_{max} is constructed from $\mathcal{L}_1$, then pruned using $\mathcal{L}_1$ again and finally a best subtree $\widehat{T}$ is selected among the sequence of pruned subtrees thanks to $\mathcal{L}_3$, and we define $\tilde{s} = \hat{s}_{\widehat{T}}$.

Recall that the aim of this section is to prove on one hand that the complexity penalty used by Breiman *et al.* [1] in the pruning algorithm is well-chosen, and, on the other hand that the final selection among the pruned subtrees is in terms of risk not far from an optimal one. We will see that, conditionally to the construction of the sequence of subtrees pruned from T_{max}, the final estimator $\tilde{s}$ has the same behavior when using any method M1 or M2. Moreover the penalty form is the same with the two different methods, although a factor $\log n_1$ can occur in the temperature when $\mathcal{L}_1 = \mathcal{L}_2$.

The following theorem gives an upper bound for the conditional risk of $\tilde{s}$:

Theorem 9. *Given for $i = 1, \ldots, N$*

$$Y_i = s(X_i) + \varepsilon_i$$

with $(X_i, Y_i) \in \mathcal{X} \times \mathbb{R}$ and ε_i centered conditionally to X_i, we consider two different models :

1) X_i is deterministic and ε_i is $\mathcal{N}(0, \sigma^2)$-distributed, σ^2 known.

2) X_i is μ-distributed, μ unknown, and Y_i is bounded by 1.

Then we have :

(i) *if $\tilde{s}$ is constructed via M1*

There exist some nonnegative constants C_1, C_2 and C_3 such that :

$$\mathbb{E}^{(1)}\left[\|s - \tilde{s}\|^2\right] \leqslant C_1 \inf_{T \preceq T_{max}} R(\hat{s}_T, s) + \frac{C_2}{n_2} + C_3 \frac{\log n_1}{n_3}.$$

(ii) *if $\tilde{s}$ is constructed via M2*

Let $\mathcal{S}$ denote the set of all splits used in the growing algorithm and suppose that the Vapnik-Chervonenkis dimension V of $\mathcal{S}$ is finite. Moreover, in the bounded regression cases, suppose that

$$\lim_{n_1 \to +\infty}\left[\frac{n_1}{\log n_1} \inf_{t \in \tilde{T}_{max}} \mu(t)\right] = +\infty. \tag{23.5}$$

Let ξ and q be two nonnegative constants. Then there exist some nonnegative constants Σ and $C(q)$, and some absolute nonnegative constants C_1, C_2 and C_3 such that :

$$\mathbb{E}^{(1)}\left[\|s - \tilde{s}\|^2\right] \leqslant C_1 \inf_{T \preceq T_{max}}\left\{\|s - \bar{s}_T\|^2 + \sigma^2 V\left(\log \frac{n_1}{V} + 1\right)\frac{|T|}{n_1}\right\}$$
$$+C_2 \frac{\sigma^2}{n_1}(1 + \xi) + C_3 \frac{\log n_1}{n_3}$$

on a set Ω_ξ such that $P(\Omega_\xi) \geqslant 1 - \left(2\Sigma e^{-\xi} + C(q)/n_1^q\right)$, where P denotes the product distribution on $\mathcal{L}_1$ and

1) *$C(q) = 0$ in the Gaussian regression case,*

2) *$C(q) > 0$ in the bounded regression case.*

So we can notice that

- the penalty form is the one given by Breiman *et al.* [1],

- $\tilde{s}$ achieves the oracle up to some multiplicative and additive constants,

- the discrete selection adds a term of order $\log(n_1)/n_3$, which is at worst of the same order as the penalty, so which does not alter too much the accuracy of the estimation.

Furthermore the main difference between the Gaussian regression case and the bounded regression case is due to the random design. Actually, assumption (23.5) leads to the factor $1/n_1^q$ occurring in the definition of the large probability set Ω_ξ. But this assumption is expected, since we have no high hope that CART makes a good job if the observations are not sufficiently well-distributed in the space.

The two following sections give some more precise results in the random design bounded regression context mentioned above, which in particular lead to Theorem 9. We just notice that we obtain about the same results in the fixed design Gaussian regression case.

23.3 Validation of the pruning algorithm

In the following sections we consider the random design bounded regression framework, where for a given $i \in \{1; \ldots; N\}$ X_i is μ-distributed, μ unknown, Y_i is bounded by 1 and ε_i is an unknown bounded noise centered conditionally to X_i.

We will focus on the pruning algorithm and see that, for a convenient constant α, $\hat{s}_{T_\alpha}$ is not far from s in terms of the risk conditionally to $\mathcal{L}_1$.

23.3.1 $\tilde{s}$ constructed via M1

In this subsection, we consider the second subsample $\mathcal{L}_2$ of n_2 observations, which we shall call for simpler notation $\mathcal{L}_2 = \{(X_1, Y_1), \ldots, (X_{n_2}, Y_{n_2})\}$. We assume that T_{max} is constructed on the first set of observations $\mathcal{L}_1$ and then pruned with the second set $\mathcal{L}_2$ independent of $\mathcal{L}_1$. Since the set T of pruned subtrees is deterministic according to $\mathcal{L}_2$, we make a selection among a deterministic collection of models.

In the sequel, given a subtree T of T_{max}, we write S_T the subspace of $\mathbb{L}^2(\mathcal{X}, \mu)$ composed by all the piecewise constant functions bounded by 1 defined on the partition associated with the leaves of T and $\bar{s}_T$ will then be the $\mathbb{L}^2(\mathcal{X}, \mu)$ projection of s on S_T. Then we choose the estimators as follows : for any $T \preceq T_{max}$,

$$\hat{s}_T = \operatorname*{argmin}_{z \in S_T} \gamma_{n_2}(z).$$

Now we have to look after the behavior of $\hat{s}_{T_\alpha}$, where T_α is chosen as in subsection 23.2.1.

Taking (23.1) into account, the following upper bound is actually an upper bound for the risk of $\hat{s}_{T_\alpha}$ conditionally to $\mathcal{L}_1$ and $\mathcal{L}_2$:

Proposition 5. *Let $P_{\mathcal{L}_2}$ be the product distribution on $\mathcal{L}_2$. Let $\xi > 0$. Then there exist a nonnegative constant Σ and a nonexplicit positive constant α_0 such that there exist some nonnegative constants C_1' and C_2' such that, for any given $\alpha > \alpha_0$:*

$$\|s - \hat{s}_{T_\alpha}\|^2 \leqslant C_1' \inf_{T \preceq T_{max}} \left[\|s - \bar{s}_T\|^2 + \alpha \frac{|T|}{n_2} \right] + C_2' \frac{1 + \xi}{n_2}$$

on a set Ω_ξ such that $P_{\mathcal{L}_2}(\Omega_\xi) \geqslant 1 - 2\Sigma e^{-\xi}$

Remark 1. The fact that we do not know anything about the noise (except it is bounded) leads to a minimal temperature we cannot catch.

To conclude, we can notice that :

- the penalty form is the same as the one proposed by Breiman *et al.* [1] in their pruning algorithm,

- the mimimal conditional risk (23.4) over all the subtrees pruned from T_{max} is already achieved up to some multiplicative and additive constants.

23.3.2 $\tilde{s}$ constructed via M2

In this subsection we shall write for simpler notations $\mathcal{L}_1 = \{(X_1, Y_1); \ldots \ldots; (X_{n_1}, Y_{n_1})\}$ and we define the different estimators and projections exactly in the same way as subsection 23.3.1. In this case, we obtain nearly the same performance for $\hat{s}_{T_\alpha}$ despite the fact that the constants can depend on n_1.

Suppose that $V \leqslant n_1$ is the Vapnik-Chervonenkis dimension of the set of splits used to construct T_{max} and that (23.5) given by Theorem 9 holds. Then one gets :

Proposition 6. *Let P denote the product distribution on $\mathcal{L}_1$.*
Let $\xi > 0$ and for some nonexplicit $\alpha_0 > 0$ let

$$\alpha(n_1, V) = \alpha_0 \left[1 + V \left(1 + \log \frac{n_1}{V} \right) \right].$$

Then there exist a nonnegative constant Σ and some nonnegative constants C_1' and C_2' such that, for any given $\alpha > \alpha(n_1, V)$ and any given $q > 0$:

$$\|s - \hat{s}_{T_\alpha}\|^2 \leqslant C_1' \inf_{T \preceq T_{max}} \left\{ \|s - \bar{s}_T\|^2 + \alpha \frac{|T|}{n_1} \right\} + C_2' \frac{1 + \xi}{n_1}$$

on a set Ω_ξ such that $P(\Omega_\xi) \geqslant 1 - \left(2\Sigma e^{-\xi} + C(q)/n_1^q \right)$.

Finally we can conclude that, in both cases, Breiman *et al.* [1] choose a convenient penalty and then that the pruning algorithm is valid. Furthermore Corollary 2 gives another important information : the sequence of pruned subtrees contains the whole information, so it is useless to look at $\mathcal{T}$ on its whole. Then, if we wish to select a subtree in $\mathcal{T}$, we just have to consider those ones appearing in the sequence. One way to select a subtree is to proceed by test-sample, and we focus on this particular method in the next subsection.

23.4   Discrete selection

Given the sequence $(T_i, \alpha_i)_{1 \leqslant i \leqslant K}$ pruned from T_{max} as defined in subsection 23.2.1, we want to ensure that using a test-sample to select a subtree among this sequence does not alter too much the accuracy of the estimation. As above, we simply denote $\mathcal{L}_3 = \{(X_1, Y_1); \ldots; (X_{n_3}, Y_{n_3})\}$. Moreover, since we are considering only $\mathcal{L}_3$ and working conditionally to $\mathcal{L}_1$ and $\mathcal{L}_2$, the result is the same for both methods M1 and M2.

Let us recall that the final estimator $\tilde{s}$ provided by CART is defined by

$$\tilde{s} = \operatorname*{argmin}_{\{\hat{s}_{T_i}; 1 \leqslant i \leqslant K\}} \gamma_{n_3}(\hat{s}_{T_i})$$

The performance of this estimator can be compared to the performance of the subtrees $(T_i)_{1 \leqslant i \leqslant K}$ by the following :

Lemma 1. $\mathbb{E}^{(1,2)}\left[\|s - \tilde{s}\|^2\right] \leqslant \inf_{1 \leqslant i \leqslant K} \|s - \hat{s}_{T_i}\|^2 + C' \dfrac{\log K}{n_3}.$

This leads easily to Theorem 9.
So the proof of Theorem 9 is complete for the random design bounded regression case.

Remark 2. Having $\alpha(n_1, V)$ and $\alpha > \alpha(n_1, V)$ could permit, via Theorem 8, to choose a model without $\mathcal{L}_3$. In that case, the last term in the loss for the conditional risk could be removed.

23.5   Open questions

So we can conclude that pruning a maximal tree is a convenient algorithm in terms of model selection for the two regression contexts mentioned above. But there is still two questions to ask : first, "how to choose a convenient tree in the pruned sequence ?". The method we studied in this paper gives some positive results, but could it be possible to remove the third (or second) subsample in order to obtain a better upper bound for the risk of $\tilde{s}$? Actually, considering the different results we obtain, if we had the true constant α occurring in the penalty, we would only have to take in the sequence the subtree T_k such that $\alpha_k \leqslant \alpha < \alpha_{k+1}$. Then the last term in the upper bound for the risk could be removed. But in theory this α is unreachable since it depends on too many unknown parameters, such that the noise variance σ^2. We only have a minimal constant, which can be interpreted as follows : when the temperature increases, the number of leaves decreases. But, according to Propositions 5 and 6, a "good" subtree is associated with a large enough temperature. So what could happen is that there is a jump in the number of leaves when the temperature comes over the minimal constant. At this stage, we hope that the "good" subtree is over this temperature. In practice, this is a question to study. An answer should be to extract from the data the right temperature for the penalized criterion. Until now there exists no general method to do this, but there are some heuristic ones based on simulations and experiments (see Gey and Lebarbier [13] for example) which give some hope about this problem.

Second, "how to analyse the approximation quality of CART to obtain an upper bound for the complete risk ?". In fact, Nobel, Olshen [5] and Nobel [6] give some asymptotic results on recursive partitioning and more particularly Engel [10] and Donoho [11] obtain some upper bounds for the risk of the penalized estimator constructed via a recursive partitioning on a fixed dyadic grid. But some approximation results are missing concerning CART as introduced by Breiman *et al.* [1]. So this aspect of the problem remains to be analyzed.

References

[1] L. Breiman, J. H. Friedman, R. A. Olshen, and C. J. Stone, *Classification And Regression Trees.* Chapman & Hall, 1984.

[2] P. A. Chou, T. Lookabaugh, and R. M. Gray, "Optimal pruning with applications to tree-stuctured source coding and modeling," *IEEE Transactions on Information Theory*, vol. 35, no. 2, pp. 299–315, 1989.

[3] Wernecke, Possinger, Kalb, and Stein, "Validating classification trees," *Biometrical Journal*, vol. 40, no. 8, pp. 993–1005, 1998.

[4] A. B. Nobel, "Analysis of a complexity based pruning scheme for classification trees." To appear in IEEE Transactions on Information Theory, 2001.

[5] A. Nobel and R. Olshen, "Termination and continuity of greedy growing for tree-structured vector quantizers," *IEEE Trans. on Inform. Theory*, vol. 42, no. 1, pp. 191–205, 1996.

[6] A. B. Nobel, "Recursive partitioning to reduce distortion," *IEEE Trans. on Inform. Theory*, vol. 43, no. 4, pp. 1122–1133, 1997.

[7] S. B. Gelfand, C. Ravishankar, and E. J. Delp, "An iterative growing and pruning algorithm for classification tree design," *IEEE Transactions on PAMI*, vol. 13, no. 2, pp. 163–174, 1991.

[8] L. Birgé and P. Massart, "A generalized C_p criterion for Gaussian model selection," Tech. Rep. 647, Université Paris 6, 2001.

[9] P. Massart, "Some applications of concentration inequalities to statistics," *Annales de la Faculté des Sciences de Toulouse*, 2000.

[10] J. Engel, "A simple wavelet approach to nonparametric regression from recursive partitioning schemes," *Journal of Multivariate Analysis*, vol. 49, pp. 242–254, 1994.

[11] D. L. Donoho, "CART and best-ortho-basis : A connection," *The Annals of Statistics*, vol. 25, no. 5, pp. 1870–1911, 1997.

[12] A. B. Nobel, "Histogram regression estimation using data-dependent partitions," *The Annals of Statistics*, vol. 24, no. 3, pp. 1084–1105, 1996.

[13] S. Gey and E. Lebarbier, "A CART based algorithm for detection of mutiple change points in the mean." Unpublished Manuscript.

24

On Adaptive Estimation by Neural Net Type Estimators

Sebastian Döhler and Ludger Rüschendorf[1]

Summary

We investigate the quality of neural net based estimators in the model example of estimating the conditional log-hazard function in censoring models. The quality of estimators is measured in terms of convergence rates. Our bounds for the convergence rate show that for some activation functions like the threshold function and for piecewiese polynomial functions optimal convergence rates are attained by the corresponding neural net estimators. For the standard sigmoid function we however obtain only bounds which indicate suboptimal behavior. A complexity regularized version of the neural net type estimators is also considered and is shown to be approximatively adaptive in smoothness classes.

24.1   Introduction

Kooperberg, Stone, and Truong (1995) proved that sieved maximum likelihood estimators based on tensor product splines attain the minimax convergence rate in probability for estimation of the conditional log-hazard function in smoothness classes with known degree of smoothness. The same convergence rate (up to a log-factor) was also obtained for the mean integrated squared error (MISE) from a general approach to sieved estimation in Döhler (1999/2000) and in Döhler and Rüschendorf (2000) (denoted in the following by DR (2000)). In that paper also a complexity regularized version of the estimator was introduced and proved to be approximatively adaptive. For theory and application of the method of functional estimation based on sieves we refer to Devroye, Györfi, and Lugosi (1996) and to Lugosi and Zeger (1995).

---

[1]Sebastian Döhler and Ludger Rüschendorf are with the Institute for Mathematical Stochastics, University of Freiburg, Eckerstr. 1, D-79104 Freiburg, Germany (Email: ruschen@stochastik.uni-freiburg.de).

In this chapter based on the approach in DR (2000) we investigate neural net based ML-estimators. We consider in detail the standard sigmoid neural net and obtain for the corresponding sieved ML-estimator a convergence rate w.r.t. MISE of the order $O\left(\left(\frac{\log n}{n}\right)^{\frac{2p}{2p+4(k+1)}}\right)$ where p is the degree of smoothness and k is the dimension of the covariates. We also establish an adaptive complexity regularized version of the estimator. In comparison to the approximative minimax optimal rate $O\left(\left(\frac{\log n}{n}\right)^{\frac{2p}{2p+k+1}}\right)$ of the tensor product spline estimator this rate is only suboptimal. For some other activation functions like threshold functions and for piecewise polynomial functions however it turns out that the approximatively optimal rate is attained by the corresponding neural net estimators.

In the final section we give an upper bound of the stochastic error for general net sieve estimators in terms of covering numbers only, which do not need finite VC-dimension as typically assumed in the first sections. We also indicate that our estimation method applies in a similar way to minimum contrast estimation and to other estimation problems like density estimation or regression estimation. As example for regression estimation the approximatively optimal rate $E\|m_n - m\|_2^2 = O\left(\left(\frac{\log n}{n}\right)^{\frac{2p}{2p+4d}}\right)$, where p is the degree of smoothness and d the dimension of the regressor, is attained by certain neural net estimators minimizing the empirical risk.

24.2   Censoring model and neural net ML-estimator

We concentrate in this chapter on the particular example of estimation of the conditional log-hazard function in a right censoring model as introduced in Kooperberg, Stone, and Truong (1995). Let T denote a survival (failure) time, $C : \Omega \to \mathcal{T} = [0, 1]$ a censoring time on the underlying probability space Ω, $X : \Omega \to \mathcal{X} = [0, 1]^k$ a vector of covariates, $Y = T \wedge C$ the observable time and $\delta = 1_{(T \leq C)}$ the censoring indicator. Assume the conditional density $f_0(t|x)$ of the distribution of T given $X = x$ to exist and let $\alpha_0(t|x) = \log \frac{f_0(t|x)}{\overline{F}_0(t|x)}$ denote the conditional log-hazard function, where $\overline{F}_0(t|x) = 1 - F_0(t|x)$ is the conditional survival function. We assume throughout that α_0 is bounded on $\mathcal{T} \times \mathcal{X}$ and that T and C are conditionally independent given X.

Let $(t_i, c_i, x_i), 1 \leq i \leq n$, be an *iid* sample of (T, C, X). The observed sample is $(y_i, \delta_i, x_i), 1 \leq i \leq n$, and the conditional log-likelihood functional is

$$L_n(\alpha) = \sum_{i=1}^{n} \left(\delta_i \alpha(y_i|x_i) - \int_0^{y_i} \exp \alpha(u|x_i) du \right) \qquad (24.1)$$

Denote the expected conditional log-likelihood by $\Lambda(\alpha) = EL_1(\alpha)$. Then $|\Lambda(\alpha) - \Lambda(\alpha_0)|$ defines a statistical distance of α to the true underlying α_0. It has the following representation

$$|\Lambda(\alpha) - \Lambda(\alpha_0)| = \Lambda(\alpha_0) - \Lambda(\alpha) = \int_{\mathcal{T} \times \mathcal{X}} \overline{F}_C G(\alpha - \alpha_0) dP^{(T,X)}, \quad (24.2)$$

where $\overline{F}_C$ is the conditional survival function of C given X and $G(y) = \exp(y) - (1 + y)$.

Based on a sieve $(\mathcal{F}_n)$ depending on the number of observations the sieved maximum likelihood estimator $\widehat{\alpha}_n$ is defined as

$$\widehat{\alpha}_n = \underset{\alpha \in \mathcal{F}_n}{\operatorname{argmax}} L_n(\alpha). \quad (24.3)$$

For some monotonically nondecreasing activation function $\sigma : \mathbb{R} \to [0,1]$ with $\lim_{x \to \infty} \sigma(x) = 1$, $\lim_{x \to -\infty} \sigma(x) = 0$, and let $\mathcal{F}(K) = \{f : [0,1]^{k+1} \to \mathbb{R}$, $f(z) = \sum_{i=1}^{K} c_i \sigma\left(a_i^T z + b_i\right)$, $a_i \in \mathbb{R}^{k+1}$, $b_i, c_i \in \mathbb{R}$, $1 \leq i \leq K\}$ for $K \in \mathbb{N}$ denote the neural net generated by σ. In particular $\sigma_s(x) = \frac{1}{1+\exp(-x)}$ denotes the standard sigmoid activation function and $\sigma_t = 1_{[0,\infty]}$ the threshold activation function. We consider neural net sieves of the form $\mathcal{F}_n = T_L \circ \mathcal{F}(K_n) = \{T_L \circ f; f \in \mathcal{F}(K_n)\}$, where T_L denotes the truncation at level L, and $T_L(x) = x 1_{[-L,L]}(x) + L 1_{[L,\infty)}(x) - L 1_{(-\infty,L)}(x)$. L is an upper bound for the L^∞-norm of α_0 and $K_n \uparrow \infty$ the number of internal nodes to be chosen independent of the distribution.

24.3   Convergence rates for the neural net ML-estimator

For sieved ML-estimators a decomposition result for the estimation error was given in DR (2000). Let $M \geq \|\alpha_0\|_\infty$, $B_0 = \exp(\|\alpha_0\|_\infty) \cdot \exp(\exp \|\alpha_0\|_\infty)$, and consider a model $\mathcal{F} \subset \{\alpha : \mathcal{T} \times \mathcal{X} \to [-M, M]\}$. Then for the MISE of the ML-estimator (given by 24.3 with $\mathcal{F}_n = \mathcal{F}$) holds:

$$E \|\widehat{\alpha}_n - \alpha_0\|_2^2 \leq 2 \exp(2M) \inf_{\alpha \in \mathcal{F}} \|\alpha - \alpha_0\|_2^2 \quad (24.4)$$

$$+ 32 \kappa_0 B_0^2 M^2 \exp(2M) \frac{\log C_n(\mathcal{F}) + 1}{n},$$

where $\kappa_0 = \frac{2608}{3}$ and where $C_n(\mathcal{F})$ is a measure for the complexity of $\mathcal{F}$ given by

$$C_n(\mathcal{F}) = 6 \sup_{\tilde{z} \in Z^{2n}} N\left(\frac{1}{n}, \mathcal{F}, d_{L^1(\nu_{\tilde{z}})}\right) N\left(\frac{1}{n}, \mathcal{F}, d_{L^1(\tilde{\nu}_z \otimes U[0,1])}\right). \quad (24.5)$$

Here $Z = \mathcal{X} \times \mathcal{T} \times \{0,1\}$, $\nu_{\tilde{z}} = \frac{1}{n} \sum_{i=1}^{n} \delta_{z_i}$, for $z = (z_1, \ldots, z_n)$, $z_i = (x_i, y_i, \delta_i)$, $\tilde{\nu}_z = \frac{1}{n} \sum_{i=1}^{n} \delta_{x_i}$, $N(\varepsilon, \mathcal{F}, d)$ is the ε-covering number of $\mathcal{F}$ w.r.t.

d and $U[0,1]$ is the uniform distribution on $[0,1]$. Similarly, for the Λ-error holds:

$$E\,|\Lambda(\widehat{\alpha}_n) - \Lambda(\alpha_0)| \;\leq\; 2\inf_{\alpha\in\mathcal{F}} |\Lambda(\alpha) - \Lambda(\alpha_0)| \tag{24.6}$$

$$+ 8\kappa_0 B_0^2 M \exp(2M)\frac{\log\mathcal{C}_n(\mathcal{F}) + 1}{n}.$$

For VC-classes $\mathcal{F}$ the estimate (24.4) simplifies to

$$E\|\widehat{\alpha}_n - \alpha_0\|_2^2 \;\leq\; C_1(M)\inf_{\alpha\in\mathcal{F}} \|\alpha - \alpha_0\|_2^2 \tag{24.7}$$

$$+ C_2(M, B_0)\dim_{VC}(\mathcal{F})\frac{\log n}{n};$$

where $\dim_{VC}(\mathcal{F})$ is the Vapnik-Cervonenkis dimension of $\mathcal{F}$; a corresponding result holding true for the Λ-error. A related general upper bound was given in Birge and Massart (1998) where however a more complicated term including L^2- and L^∞-covering numbers is used.

Further related recent results are in Wong and Shen (1995) for density estimation, in Kohler (1997) for regression estimation based on minimal L^2-empirical risk in Yang and Barron (1998) for density estimation and in others. The estimates in (24.4), (24.6), (24.7) are extended to minimum contrast estimation for a general class of loss functions and also to a general class of estimation problems with a modified definition of the complexity $\mathcal{C}_n(\mathcal{F})$ (see DR (2000)).

We next consider in detail the case of standard sigmoid neural net estimators. Assume that $\alpha_0 \in W_{p,k+1}^q, \|\alpha_0\|_\infty \leq L$, i.e. $\alpha \in W_{p,k+1}^q(L)$ the Sobolev-class of functions on $[0,1]^{k+1}$, whose p-th partial derivatives exist a.s. and are in $L^q([0,1]^{k+1})$ and α has L^∞-bound L.

Theorem 10 (sigmoid neural net estimator).
Consider the standard sigmoid neural net ML-estimator $\widehat{\alpha}_n$ of α_0, $\widehat{\alpha}_n =$ argmax$_{\alpha\in\mathcal{F}_n} L_n(\alpha), \mathcal{F}_n = T_L \circ \mathcal{F}(K_n)$ with $K_n = \left\lceil \left(\frac{n}{\log n}\right)^{\frac{1}{4+\frac{2p}{k+1}}} \right\rceil$. Then for $2 \leq q \leq \infty$ and $\alpha_0 \in W_{p,k+1}^q(L)$ holds

$$E\|\widehat{\alpha}_n - \alpha_0\|_2^2 = O\left(\left(\frac{\log n}{n}\right)^{\frac{2p}{2p+4(k+1)}}\right). \tag{24.8}$$

If $\alpha_0 \in W_{p,k+1}^\infty$ then

$$E\,|\Lambda(\widehat{\alpha}_n) - \Lambda(\alpha_0)| = O\left(\left(\frac{\log n}{n}\right)^{\frac{2p}{2p+4(k+1)}}\right). \tag{24.9}$$

Proof. For the proof we at first estimate the approximation error. Observe that $\|\alpha_0 - T_L\alpha\|_\infty = \|T_L\alpha_0 - T_L\alpha\|_\infty \leq \|\alpha_0 - \alpha\|_\infty$. Therefore, from the approximation result for neural nets in Mhaskar (Theorem 2.1, 1996) we obtain

$$\inf_{\alpha \in \mathcal{F}_n} \|\alpha_0 - \alpha\|_q^2 \leq C \left(\frac{1}{K_n} \right)^{\frac{2p}{k+1}} \|\alpha_0\|_{W_{p,k+1}^q}. \tag{24.10}$$

For the stochastic error we use the estimate of the VC-dimension of $\mathcal{F}(K)$, in Karpinski and Macintyre (1997) for the sigmoid net with $\sigma = \sigma_s$, $\dim_{VC} \mathcal{F}(K) = O(K^4)$. Therefore, we obtain

$$\dim_{VC} \mathcal{F}_n = \dim_{VC} T_L \circ \mathcal{F}(K_n) \leq \dim_{VC} \mathcal{F}(K_n) \leq C K_n^4. \tag{24.11}$$

The estimate in (24.4) resp. (24.7) then implies that $E\|\widehat{\alpha}_n - \alpha_0\|_2^2 \leq C_1 \left(\frac{1}{K_n} \right)^{\frac{2p}{k+1}} + C_2 K_n^4 \frac{\log n}{n} = O\left(\left(\frac{\log n}{n} \right)^{\frac{2p}{2p+4(k+1)}} \right)$. For the proof of (24.9) we use (24.6). From the representation of the approximation error in (24.2) and using some properties of the function G we obtain as in DR (2000) using (24.10)

$$\inf_{\alpha \in \mathcal{F}_n} |\Lambda(\alpha) - \Lambda(\alpha_0)| \leq \inf_{\alpha \in \mathcal{F}_n} G\left(\|\alpha - \alpha_0\|_\infty \right) \leq G\left(\inf_{\alpha \in \mathcal{F}_n} \|\alpha - \alpha_0\|_\infty \right)$$

$$= O\left(\left(\inf_{\alpha \in \mathcal{F}_n} \|\alpha - \alpha_0\|_\infty \right)^2 \right) = O\left(\left(\inf_{\alpha \in \mathcal{F}(K_n)} \|\alpha - \alpha_0\|_\infty \right)^2 \right)$$

$$= O\left(\left(\frac{1}{K_n} \right)^{\frac{2p}{k+1}} \right).$$

The stochastic error part then is dealt with as in the first part of the proof. □

24.4 Complexity regularized neural net estimator

For the construction of the neural net ML-estimator in Section 24.3 knowledge of the degree of smoothness p and of L was supposed, i.e. the assumption that $\alpha \in W_{p,k+1}^q(L)$ resp. $W_{p,k+1}^q$ where $q \geq 2$. The aim of this section is to construct an adaptive estimator without knowledge of the smoothness class i.e. in the model $W = \bigcup \left\{ W_{p,k+1}^q(L); 1 \leq p < \infty, 0 < L, 2 \leq q \leq \infty \right\}$. A general construction of this type was given in DR (2000) based on the method of complexity regularization.

Consider for any $n \in \mathbb{N}$ a class of models $\{\mathcal{F}_{n,p}; p \in \mathcal{P}_n\} \subset \{\alpha = \mathcal{T} \times \mathcal{X} \to [-M, M]\}$ where $\mathcal{P}_n$ is a finite set, $M = M_n$, and assume that $\mathcal{C}_n(\mathcal{F}_{n,p}) < \infty$. Then define the complexity regularized estimator α_n^* in two steps:

1. For $n \in \mathbb{N}$, let $p_n^* \in \mathcal{P}_n$ be defined by

$$p_n^* = \operatorname*{argmin}_{p \in \mathcal{P}_n} \left(-\frac{1}{n} \sup_{\alpha \in \mathcal{F}_{n,p}} L_n(\alpha) + \operatorname{pen}_n(p) \right), \tag{24.12}$$

where $\text{pen}_n(p)$ is a penalization term with $\text{pen}_n(p) \geq 4\kappa_0 B_0^2 M \exp(2M)$ $\frac{\log C_n(\mathcal{F}_{n,p})}{n}$ and C_n is given by (24.5).

2. $$\alpha_n^* = \underset{\alpha \in \mathcal{F}_{n,p_n^*}}{\text{argmax}} \, L_n(\alpha) \in \mathcal{F}_{n,p_n^*}. \tag{24.13}$$

In DR (2000) the following error estimates were proved for α_n^* in general sieved models

$$E\|\alpha_n^* - \alpha_0\|_2^2 \leq 2 \inf_{p \in \mathcal{P}_n} \left(4M \, \text{pen}_n(p) + \exp(2M) + \inf_{\alpha \in \mathcal{F}_{n,p}} \|\alpha - \alpha_0\|_2^2 \right)$$
$$+ \frac{16\kappa_0 B_0^2 M^2 \exp(2M)}{n}(1 + \log |\mathcal{P}_n|) \tag{24.14}$$

and similarly for the Λ-error

$$E|\Lambda(\alpha_n^*) - \Lambda(\alpha_0)| \leq 2 \inf_{p \in \mathcal{P}_n} \left(\text{pen}_n(p) + \inf_{\alpha \in \mathcal{F}_{n,p}} |\Lambda(\alpha) - \Lambda(\alpha_0)| \right)$$
$$+ \frac{4\kappa_0 B_0^2 M \exp(2M)}{n}(1 + \log |\mathcal{P}_n|). \tag{24.15}$$

Some general theory of complexity regularized estimation theory has been developed in Barron, Birge, and Massart (1999). The upper bounds in their paper use some more involved L^2-L^∞-covering numbers as well as the L^1-metric with bracketing. The estimates (24.14), (24.15) hold true more generally for minimum contrast estimation in more general estimation problems; see the remarks in DR (2000). There are many related results on adaptive estimation for various types of estimation problems in the literature; see in particular Lee, Bartlett, and Williamson (1996), Wong and Shen (1995), Krzyzak and Linder (1998), Vapnik (1982), Barron and Cover (1991), and Lugosi and Zeger (1995). Typically empirical L^2-risk minimizing estimators are considered in these papers. For estimation of the conditional log hazard function let for $n \in \mathbb{N}$, $r_n = n$, $\beta_n = \frac{1}{5} \log \log n$, $\mathcal{P}_n = \{r \in \mathbb{N}; r \leq r_n\}$, and $\mathcal{F}_{n,r} = T_{\beta_n} \circ \mathcal{F}(K_{n,r})$, with $K_{n,r} = \left\lceil \left(\frac{n}{\log n} \right)^{\frac{1}{4 + \frac{2r}{k+1}}} \right\rceil$. Let α_n^* denote the complexity regularized neural net ML-estimator as defined in (24.12), (24.13) with penalization term $\text{pen}_n(r) = \frac{(\log n)^{\frac{8}{5}}}{n} K_{n,r}^4$. Then we obtain

Theorem 11. *If $\alpha_0 \in W_{p,k+1}^q(L)$, $p \geq 1$, $q \geq 2$, $L > 0$ then the complexity regularized sigmoid neural net ML-estimator α_n^* satisfies*

$$E\|\alpha_n^* - \alpha_0\|_2^2 = O\left(\log n \left(\frac{\log n}{n} \right)^{\frac{2p}{2p+4(k+1)}} \right). \tag{24.16}$$

In the case $q = \infty$ we also obtain

$$E|\Lambda(\alpha_n^*) - \Lambda(\alpha_0)| = O\left(\log n \left(\frac{\log n}{n} \right)^{\frac{2p}{2p+4(k+1)}} \right). \tag{24.17}$$

Proof. To apply the general result in (24.14), (24.15) we first have to prove that $\mathrm{pen}_n(r) \geq 4\kappa_0 B_0^2 \beta_n \exp(2\beta_n) \frac{\log \mathcal{C}_n(\mathcal{F}_{n,r})}{n}$. From the bound $\dim_{VC} \mathcal{F}_{n,r} \leq CK_{n,r}^4$ (cp. 24.11) we obtain as in the proof of Theorem 5.3 in DR (2000) the estimate $\log \mathcal{C}_n(\mathcal{F}_{n,r}) \leq 2CK_{n,r}^4 \log(n \log \beta_n)$ which implies the bound for $\mathrm{pen}_n(r)$ and with some calculations we obtain (24.16) and (24.17). $\qquad\square$

24.5   Further neural net estimators, optimal rates and regression estimation

The convergence results in Theorems 10 and 11 yield a suboptimal estimation rate $O\left(\left(\frac{\log n}{n}\right)^{\frac{2p}{2p+4(k+1)}}\right)$ for the sigmoid neural net estimator in comparison to the minimax-optimal rate $O\left(\left(\frac{1}{n}\right)^{\frac{2p}{2p+k+1}}\right)$ attained by the product spline estimator in the related but smaller Hölder classes (see Kooperberg, Stone, and Truong (1995) resp. DR (2000) for the adaptive estimator).

From the proof of our result we find that for both sieve-types the approximation error as function of the number of basis functions is of the same order. The stochastic error however is for tensor product splines of linear order while it is for the class of sigmoid neural nets of the order K^4 which leads to the term $4(k+1)$ in the exponent in comparison to $k+1$ for splines. As stated in the paper of Sontag (1998) the upper bound for the VC-dimension $\dim_{VC}(\mathcal{F}(K)) = O(K^4)$ for sigmoid neural nets possibly could be improved to some order $O(K^s)$ with $2 < s \leq 4$ but $O(K^2)$ is a strict lower bound and so the approximation order could be improved at most to some order of the form $O\left(\left(\frac{\log n}{n}\right)^{\frac{2p}{2p+s(k+1)}}\right)$ with $2 < s \leq 4$. So the optimal order is not attainable for sigmoidal neural nets by our method. The exact convergence rate of sigmoid net estimators in this problem is still unknown and an interesting open problem.

For some other activation functions with similar approximation rate and complexity growth rate as the tensor product splines the corresponding neural net estimator however can attain the approximative optimal minimax rates. For the threshold activation function $\sigma_t = 1_{[0,\infty]}$ it holds that the VC-dimension is approximatively linear, $\dim_{VC} \mathcal{F}(K) = O(K \log K)$ (see Sontag (1998)). Furthermore, it is proved in Petrushev (1998) that the threshold neural nets have the approximation order $O((\frac{1}{K})^{\frac{2p}{k+1}})$. Combining these results we obtain from our upper estimates that the threshold neural net ML-estimator as well as the complexity regularized version attain the approximative optimal minimax rates.

Corollary 12 (Threshold neural net estimator). *Let $\widehat{\alpha}_n$ (resp. α_n^*) denote the neural net based (adaptive) ML-estimator with threshold function σ_t based on the sieve $\mathcal{F}_n$ as in Sections 24.3, 24.4 with $K_n = \left\lceil \left(\frac{n}{\log n}\right)^{\frac{1}{4+\frac{2p}{k+1}}} \right\rceil$ (resp. $K_{n,r} = \left\lceil \left(\frac{n}{\log n}\right)^{\frac{1}{4+\frac{2r}{k+1}}} \right\rceil$). Then $\widehat{\alpha}_n$ (resp. α_n^*) attain the approximatively optimal minimax rates*

$$E\|\widehat{\alpha}_n - \alpha_0\|^2 = O\left(\left(\frac{\log n}{n}\right)^{\frac{2p}{2p+k+1}}\right) \quad and \qquad (24.18)$$

$$E\|\alpha_n^* - \alpha_0\|^2 = O\left(\log n \left(\frac{\log n}{n}\right)^{\frac{2p}{2p+k+1}}\right). \qquad (24.19)$$

Some general polynomial bounds on the VC-dimension of neural nets are given in Karpinski and Macintyre (1997). Approximatively linear bounds of the order $O(K \log K)$ are obtained in Bartlett, Maiorov, and Meir (1998) and Sakurai (1999) for piecewise polynomial networks with a fixed number L of layers. Also several optimal approximation rate results for neural net type functions have been established by Mhaskar (1996) (smooth nets), Petrushev (1998) (sigmoid nets and $p \geq \frac{d-1}{2}$), Delyon, Juditsky, and Benveniste (1995) (wavelet nets), Maiorov and Meir(2000) and Meir and Maiorov (2000) (neural nets in smoothness classes). For a recent survey see Pinkus (1999). As a result also further neural net type estimators, e.g. those with piecewise polynomial activation functions attain the approximative optimal minimax bounds.

Improved resp. approximatively optimal convergence rates for neural net type estimators can in some problems be obtained even without having nearly linear or even finite VC-dimension as in the previous examples. Consider general net sieves of the form

$$\mathcal{F}(\beta, K) = \left\{\alpha = \sum_{i=1}^{K} c_i f_i(t, x); \ f_i \in \mathcal{F}_0, \sum_{i=1}^{K} |c_i| \leq \beta\right\} \qquad (24.20)$$

where $\mathcal{F}_0$ is a class of basis functions, $\mathcal{F}_0 = \{\Psi(a_\vartheta(x)); \vartheta \in \Theta\}$ of $\mathcal{F}(\beta, K)$, parametrized by Θ and transformed by Ψ. Then the $L^1(\nu)$-covering number for any probability measure ν can be estimated from above

$$N(\delta, \mathcal{F}(\beta, K), d_{L^1(\nu)}) \leq C^K (\beta K)^{K(2D-1)} \left(\frac{1}{\delta}\right)^{K(2D-1)} \qquad (24.21)$$

$$\text{assuming that } N(\delta, \mathcal{F}_0, d_{L^2(\nu)}) = O\left(\left(\frac{1}{\delta}\right)^{2(D-1)}\right). \qquad (24.22)$$

The proof of (24.21) is based on some well-known rules for covering numbers. The bound in (24.22) holds true by Pollard's estimate in particular if

$\dim_{VC} \mathcal{F}_0 = D$ but may be true even in cases with infinite VC-dimension. As consequence, one obtains an estimate for the complexity of $\mathcal{F}(\beta, K)$.

Proposition 13. *Assume that*

$$N(\delta, \mathcal{F}_0, d_{L^2(\nu)}) = O\left(\left(\frac{1}{\delta}\right)^{2(D-1)}\right) \tag{24.23}$$

then for some constant $C = C(D)$

$$\mathcal{C}_n(\mathcal{F}(\beta, K)) \leq C^K (\beta K)^{2K(2D-1)} n^{2K(2D-1)}. \tag{24.24}$$

Combining this bound for the complexity with an "optimal" approximation rate of α_0 by a sieve net $\mathcal{F}(\beta_n, K_n)$ one obtains as consequence of the upper bounds in (24.4), (24.6), (24.14), (24.15) convergence rates for the corresponding sieve type estimator.

As application of this method some improved convergence rates have been derived in DR (2000) for various net sieve estimators like neural net, radial basis function, and wavelet net estimators under the assumption that α_0 has an integral representation of the form

$$\alpha_0(t, x) = \int_\Theta \Psi(a_\vartheta(t, x)) d\nu(\vartheta) \tag{24.25}$$

for some measure ν of finite variation. In this context an interesting problem is whether the approximation results in Maiorov and Meir (2000) and Meir and Maiorov (2000) which are based on an integral representation of the form (24.25) for smoothness classes are still valid under the additional assumption that approximation is considered in the classes $\mathcal{F}(\beta_n, K_n)$ with slowly increasing bounds β_n for the coefficients (like $\beta_n = \log \log n$). Maiorov and Meir (2000) and Meir and Maiorov (2000) derive nearly optimal approximation results in the class $\mathcal{F}(\infty, K_n)$ without restrictions on the coefficients. If true this result would imply improved consistency results of neural net type estimators in general smoothness classes.

The convergence rates obtained in this chapter for neural net estimators use only weak growth and approximation properties of the nets and are applicable also to further estimation problems like density estimation or regression problems. For the estimation in the regression problem e.g. it was proved in Kohler (1997) that for sieved minimum squared error regression estimate m_n w.r.t. a general VC-class $\mathcal{F}_n$ one obtains the bounds

$$E\|m_n - m\|_2^2 = O\left(\inf_{f \in \mathcal{F}_n} \|f - m\|_2^2 + \dim_{VC}(\mathcal{F}_n)\frac{\log n}{n}\right) \tag{24.26}$$

where m is the underlying regression function $m(x) = E(Y|X = x)$. As consequence we obtain as in Theorems 10 and 11 or in Corollary 4.2 in Kohler (1997) for certain neural net sieves $\mathcal{F}_n$ as for threshold nets or piecewise polynomial nets the approximatively optimal estimation rates

$$E\|m_n - m\|_2^2 = O\left(\left(\frac{\log n}{n}\right)^{\frac{2p}{2p+4d}}\right), \tag{24.27}$$

where d is the dimension of the regressor X and p is the degree of smoothness. A similar rate of convergence also holds true for the complexity regularized version of the estimator.

Résumé. As a result we obtain that neural net based estimators can compete with other nonparametric estimators and attain the (approximatively) optimal convergence rate in various classes of estimation problems. There remain several interesting open questions on the architecture on the approximation and complexity properties of neural nets to be further investigated.

References

[1] Barron, A.R. (1993). Universal approximation bounds for superpositions of a sigmoidal function. *IEEE Transactions on Information Theory 39*, 930–945.

[2] Barron, A.R., L. Birgé, and P. Massart (1999). Risk bounds for model selection via penalization. *Probability Theory and Related Fields 113*, 301–413.

[3] Barron, A.R. and T. M. Cover (1991). A bound on the financial value of information. *IEEE Transactions on Information Theory 34*, 1097–1100.

[4] Bartlett, P.L., V. Maiorov, and R. Meir (1998). Almost linear VC dimension bounds for piecewise polynomial networks. *Neural computation 10*, 2159–2173.

[5] Birgé, L. and P. Massart (1998). Minimum contrast estimators on sieves: Exponential bounds and rates of convergence. *Bernoulli 4*, 329–375.

[6] Delyon, B., A. Juditsky, and A. Benveniste (1998). Accuracy analysis for wavelet approximations. *IEEE Transactions on Information Theory 6*, 332–348.

[7] Devroye, L., L. Györfi, and G. Lugosi (1996). *A Probabilistic Theory of Pattern Recognition*. Springer, New York.

[8] Döhler, S. (1999). *Consistent hazard regression estimation by sieved maximum likelihood estimators*. Preprint, Universität Freiburg.

[9] Döhler, S. (2000). *Empirische Risiko-Minimierung bei zensierten Daten.* Dissertation, Universität Freiburg.

[10] Döhler, S. and L. Rüschendorf (2000). *Adaptive estimation of hazard functions.* Preprint, Universität Freiburg.

[11] Karpinski, M. and A. Macintyre (1997). Polynomial bounds for VC dimension of sigmoidal and general Pfaffian neural networks. *Journal of Computer and System Sciences 54*, 169–176.

[12] Kohler, M. (1997). *Nichtparametrische Regressionsschätzung mit Splines.* Dissertation, Universität Stuttgart.

[13] Kohler, M. (1999a). Nonparametric estimation of piecewise smooth regression functions. *Statistics & Probability Letters 43*, 49–55.

[14] Kohler, M. (1999b). Universally consistent regression function estimation using hierarchial B-splines. *Journal of Multivariate Analysis 68*, 138–164.

[15] Kooperberg, C., C.J. Stone, and Y.K. Truong (1995). The L_2 rate of convergence for hazard regression. *Scandinavian Journal of Statistics 22*, 143–157.

[16] Krzyzak, A. and T. Linder (1998). Radial basis function networks and computational regularization in function learning. *IEEE Transactions on Information Theory 9*, 247–256.

[17] Lee, W.S., P.L. Bartlett, and R. Williamson (1996). Efficient agnostic learning of neural networks with bounded fan-in. *IEEE Transactions on Information Theory 42*, 2118–2132.

[18] Lugosi, G. and K. Zeger (1995). Nonparametric estimation via empirical risk minimization. *IEEE Transactions on Information Theory 41*, 677–687.

[19] Maiorov, V.E. and R. Meir (2000). On the near optimality of the stochastic approximation of smooth functions by neural networks. *Advances in Computational Mathematics 13*, 79–103.

[20] Makovoz, Y. (1996). Random approximants and neural networks. *Journal of Approximation Theory 85*, 98–109.

[21] Meir, R. and V. E. Maiorov (2000). On the optimality of neural networks. Preprint, Technion, Haifa.

[22] Mhaskar, H. N. (1996). Neural networks for optimal approximation of smooth and analytical function. *Neural Computation 8*, 164–177.

[23] Modha, D. S. and E. Masry (1996). Rate of convergence in density estimation using neural networks. *Neural Computation 8*, 1107–1122.

[24] Petrushev, P.P. (1998). Approximation by ridge functions and neural networks. *SIAM Journal on Mathematical Analysis 20*, 155–189.

[25] Pinkus, A. (1999). Approximation theory of the MLP model in neural networks. *Acta Numerica 8*, 143–195.

[26] Sakurai, A. (1999). Tight bonds for the VC-dimension of piecewise polynmial networks. *Advances in Neural Information Processing Systems 11*, 324–329.

[27] Shen, X. and W.H. Wong (1995). Convergence rate of sieve estimates. *Annals of Statistics 22*, 580–615.

[28] Sontag, E.D. (1998). *VC* dimension of neural networks. In C. Bishop (Ed.), *Neural Networks and Machine Learning*, pp. 69–95. Springer Berlin.

[29] Vapnik, V. (1982). *Estimation of dependencies based on empirical data.* Series in Statistics. Springer.

[30] Wong, W.H. and X. Shen (1995). Probability inequalities for likelihood ratios and convergence rates of sieve MLES. *Annals of Statistics 23*, 339–362.

[31] Yang, Y. and A. Barron (1998). An asymptotic property of model selection criteria. *IEEE Transactions on Information Theory 44*, 95–116.

25

Nonlinear Function Learning and Classification Using RBF Networks with Optimal Kernels

Adam Krzyżak[1]

Summary

We derive mean integrated squared error (MISE) optimal kernels in the standard and normalized radial basis function network applied in nonlinear function learning and classification. The corresponding optimal MISE rate of convergence is also given.

25.1 Introduction

In this article we study the problem of nonlinear regression estimation and classification by the radial basis function (RBF) networks with k nodes and a kernel $\phi : \mathcal{R}_+ \to \mathcal{R}$:

$$f_k(x) = \sum_{i=1}^{k} w_i \phi\left(\|x - c_i\|_{A_i}\right) + w_0 \qquad (25.1)$$

where

$$\|x - c_i\|_{A_i}^2 = [x - c_i]^T A_i [x - c_i],$$

the parameters of the network are output weights $w_1 \ldots, w_k$, centers $c_1, \ldots, c_k \in \mathcal{R}^d$, and positive semidefinite covariance matrices $A_1, \ldots, A_k \in \mathcal{R}^d \times \mathcal{R}^d$ and $\phi\left(\|x - c_i\|_{A_i}\right)$ is the radial basis function. The network is shown in Figure 25.1.

RBF networks have been introduced by Broomhead and Lowe [1] and Moody and Darken [14]. Their application to regression estimation problem and classification via empirical risk minimization was studied by Krzyżak et

---

[1]Adam Krzyżak is with the Department of Computer Science, Concordia University, 1455 de Maisonneuve Blvd. W., Montreal Quebec, Canada H3G 1M8 (Email: krzyzak@cs.concordia.ca).
 This research was supported by the Alexander von Humboldt Foundation and Natural Sciences and Engineering Research Council of Canada.

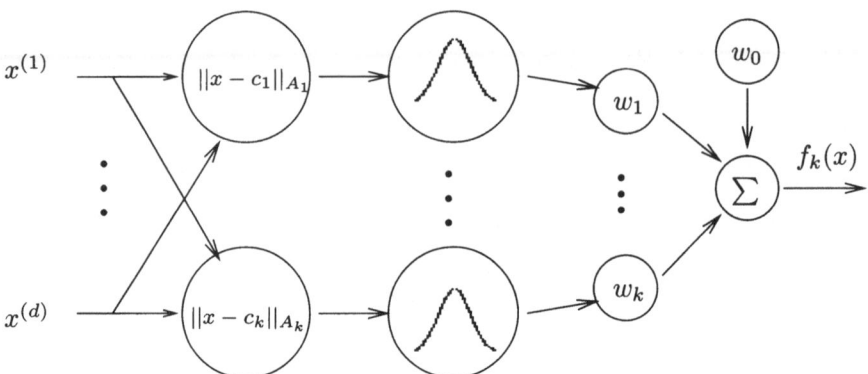

Figure 25.1. The radial basis network with one hidden layer

al [11]. The approximation rates were obtained by Park and Sandberg [15], Girosi and Anzellotti [7]. McCaffrey and Gallant [13] studied regression estimation rates in Sobolev spaces and Krzyżak and Linder [12] obatianed the rates using complexity regularization. Typical forms of radial functions encountered in estimation applications are monotonically decreasing kernels, i.e. kernels such that $K(x) \to 0$ as $x \to \infty$. Common examples of such kernels are (Figure 25.2):

- $K(x) = I_{\{x \in [0,1]\}}$ (window)

- $K(x) = \max\{(1 - x^2), 0\}$ (truncated parabolic)

- $K(x) = e^{-x^2}$ (Gaussian)

- $K(x) = e^{-x}$ (exponential kernel).

In approximation and interpolation [16] increasing kernels are common, i.e. kernels such that $K(x) \to \infty$ as $x \to \infty$. Some examples of these are (Figure 25.3):

- $K(x) = x$ (linear)

- $K(x) = x^3$ (cubic)

- $K(x) = \sqrt{x^2 + c^2}$ (multiquadric)

- $K(x) = x^{2n+1}$ (thin plate spline), $n \geq 1$

- $K(x) = x^{2n} \log x$ (thin plate spline), $n \geq 1$.

The results obtained in this paper are motivated by the study of the optimal MISE kernel in density estimation by Watson and Leadbetter [19] which was subsequently extended to a class of Parzen kernels by Davis [2].

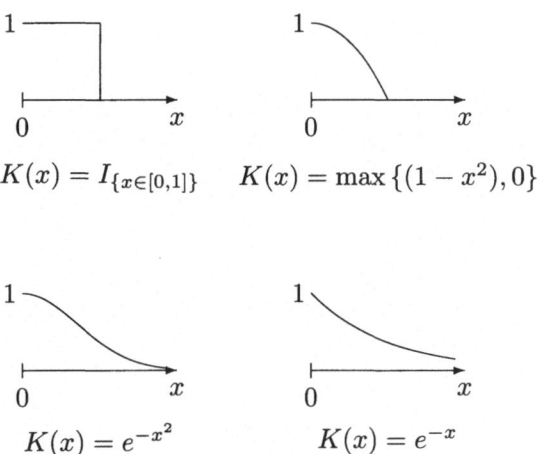

Figure 25.2. Window, truncated parabolic, gaussian, exponential kernels

25.2   MISE optimal RBF networks for known input density

In this section we will derive the optimal RBF network in the mean integrated squared error sense (MISE) and the corresponding optimal rate of convergence in the regression estimation and classification problems. Let $(X, Y) \in \mathcal{R}^d \times \mathcal{R}$ be random vector and let probability density of X be known and denoted by $f(x)$. Let $E\{Y|X = x\} = R(x)$ be the regression function of Y given X and $EY^2 < \infty$. In the sequel we will propose RBF network estimate of $G(x) = R(x)f(x)$. We will minimize MISE of the estimate of G and obtain implicit formula for the optimal kernel. We also obtain the exact expression for the MISE rate of convergence of the optimal standard RBF network estimate. We will also consider optimal kernel for the normalized RBF network regression estimate. Estimation of G is important in the following two situations:

1. Nonlinear estimation. Consider the model $Y = R(X) + Z$, where Z is zero mean noise and R is unknown nonlinear input-output mapping which we want to estimate. Clearly R is the regression function $E(Y|X = x)$. In order to estimate R we generate a sequence of i.i.d. random variables $X_1, \ldots, X_n$ from X, whose density is known (e.g. uniform on the interval on which we want to reconstruct R) and observe $Y's$. We construct estimate G_n of G. The estimate enables us to

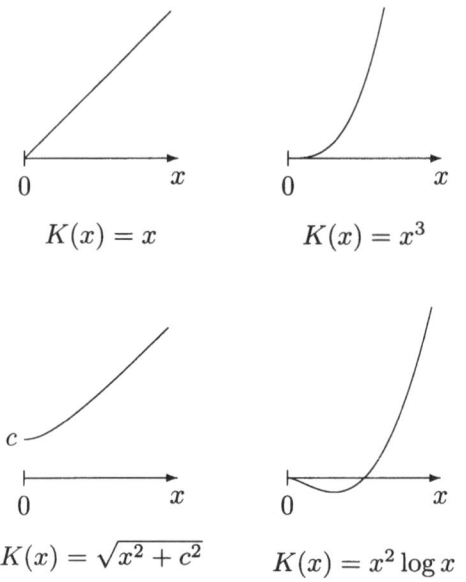

$$K(x) = x \qquad K(x) = x^3$$

$$K(x) = \sqrt{x^2 + c^2} \qquad K(x) = x^2 \log x$$

Figure 25.3. Linear, thin plate spline, multiquadric, thin plate spline kernels

recover $G(x) = R(x)f(x)$. Hence the estimate of R is trivially given by $G_n(x)/f(x)$.

2. **Classification.** In the classification (pattern recognition) problem, we try to determine a label Y corresponding to a random feature vector $X \in \mathcal{R}^d$, where Y is a random variable taking its values from $\{-1, 1\}$. The decision function is $g : \mathcal{R}^d \to \{-1, 1\}$, and its goodness is measured by the *error probability* $L(g) = \mathbf{P}\{g(X) \neq Y\}$. It is well-known that the decision function that minimizes the error probability is given by

$$g^*(x) = \begin{cases} -1 & \text{if } R(x) \leq 0 \\ 1 & \text{otherwise,} \end{cases}$$

where $R(x) = \mathbf{E}(Y|X = x)$, g^* is called the *Bayes decision*, and its error probability $L^* = \mathbf{P}\{g^*(X) \neq Y\}$ is the *Bayes risk*.

When the joint distribution of (X, Y) is unknown (as is typical in practical situations), the Bayes decision has to be learned from a training sequence

$$D_n = ((X_1, Y_1), \dots, (X_n, Y_n)),$$

which consists of n independent copies of the $\mathcal{R}^d \times \{-1, 1\}$-valued pair (X, Y). Then formally, a decision rule g_n is a function $g_n : \mathcal{R}^d \times$

$(\mathcal{R}^d \times \{-1, 1\})^n \to \{-1, 1\}$, whose error probability is given by

$$L(g_n) = \mathbf{P}\{g_n(X, D_n) \neq Y | D_n\}.$$

Note that $L(g_n)$ is a random variable, as it depends on the (random) training sequence D_n. For notational simplicity, we will write $g_n(x)$ instead of $g_n(x, D_n)$.

Pattern recognition is closely related to regression function estimation. This is seen by observing that the function R defining the optimal decision g^* is just the regression function $\mathbf{E}(Y|X = x)$. Thus, having a good estimate $R_n(x)$ of the regression function R, we expect a good performance of the decision rule

$$g_n(x) = \begin{cases} -1 & \text{if } R_n(x) \leq 0 \\ 1 & \text{otherwise.} \end{cases} \tag{25.2}$$

Indeed, we have the well-known inequality [6]

$$\mathbf{P}\{g_n(X) \neq Y | X = x, D_n\} - \mathbf{P}\{g^*(X) \neq Y | X = x\} \leqslant |R_n(x) - R(x)|$$

and in particular,

$$\begin{aligned} \mathbf{P}\{g_n(X) \neq Y | D_n\} &- \mathbf{P}\{g^*(X) \neq Y\} \\ &\leq \left(\mathbf{E}\left((R_n(X) - R(X))^2 \big| D_n\right)\right)^{1/2}. \end{aligned} \tag{25.3}$$

Therefore, any strongly consistent estimate R_n of the regression function R leads to a strongly consistent classification rule g_n via equation (25.2) and (25.3). For example, if R_n is an RBF-estimate of R based on minimizing the empirical L_2, then g_n is a strongly universally consistent classification rule. That is, for any distribution of (X, Y), it is guaranteed that the error probability of the RBF-classifier gets arbitrarily close to that of the best possible classifier if the training sequence D_n is large enough.

Bayes rule can also be obtained by assigning a given feature vector x to a class with the highest *a posteriori* probability, i.e. by assigning x to class i, if $P_i(x) = \max_j P_j(x)$, where $P_i(x) = EI_{(\theta=i, X=x)} = ER_i(X)$, θ is a class label, $R_i(x) = EI_{(\theta=i|X=x)}$ and I_A is indicator of set A [10]. Therefore it is essential to estimate G to obtain a good classification rule.

Suppose that $(X_1, Y_1), \cdots, (X_n, Y_n)$ is a sequence of i.i.d. observations of (X, Y). Consider a generalization of (25.1) by allowing each radial function $\phi(||x - c_i||_{A_i})$ to depend on k, where $|| \cdot ||$ is an arbitrary norm (not necessary Euclidean). Thus

$$f_k(x) = \sum_{i=1}^{k} w_i \phi_k(||x - c_i||_{A_i}) + w_0. \tag{25.4}$$

There are several approaches to learn parameters of the network. In empirical risk minimization approach the parameters of the network are selected so that empirical risk is minimized, i.e.

$$J_n(f_\theta) = \min_{\hat{\theta} \in \Theta_n} J_n(f_{\hat{\theta}}).$$

where

$$J_n(f_\theta) = \frac{1}{n} \sum_{j=1}^{n} |f_\theta(X_j) - Y_j|^2$$

is the empirical risk and

$$\Theta_n = \left\{ \theta = (w_0, \ldots, w_{k_n}, c_1, \ldots, c_{k_n}, A_1, \ldots, A_{k_n}) : \sum_{i=0}^{k_n} |w_i|^2 \leqslant b_n \right\}$$

is the vector of parameters. In order to avoid too close fit of the network to the data (overfitting problem) we carefully control the complexity of the network expressed in terms of the number of hidden units k as the size of the training sequence increases. This is the method of sieves of Grenander [8]. The complexity of the network can also be described by the Vapnik-Chervonenkis dimension [6]. This approach has been applied to learning of RBF networks in [11]. The learning is consistent for bounded output weights and the size of the network increasing subject to the condition $k_n^3 b_n^2 \log(k_n^3 b_n^2)/n \to 0$ as $n \to \infty$ where bound on weights b_n is allowed to grow with n. It means that the network performs data compression by using only k_n hidden units, where k_n is usually much smaller than n. k_n can be automatically learned from data in the process called complexity regularization [12] in which we control complexity (size) of the network by imposing penalty increasing with network complexity. The optimal size of RBF network with radial function of bounded variation and for smooth regressions is of order $O(\sqrt{n/\log n})$.

Empirical risk minimization is an asymptotically optimal strategy but it has high computational complexity. If gradient-based backpropagation search is used to learn parameters then we run the risk of getting stuck in a local minima. If we use global optimization algorithms such as simulated annealing, genetic or evolutionary search [17] then we pay the price in efficiency. A simpler parameter training approach consists of assigning data values to output weights and centers (plug-in approach). This approach does not offer compression but is easy to implement. Consistency of plug-in RBF networks was investigated in [20].

In this paper we focus our attention on a plug-in approach. Let parameters of (25.4) be trained as follows:

$$k_n = n, w_0 = 0, w_i = Y_i, \ c_i = X_i, \ i = 1, \cdots, n.$$

Consider radially symmetric kernel $K : \mathcal{R}^d \to \mathcal{R}$ which is defined by $K(x) = \phi(||x||)$. Network (25.4) can be rewritten

$$f_k(x) = G_n(x) = \frac{1}{n} \sum_{i=1}^{n} Y_i K_n(x - X_i).$$

We define RBF estimate of G by

$$G_n(x) = \frac{1}{n} \sum_{i=1}^{n} Y_i K_n(x - X_i)$$

where $K_n(x)$ is some square integrable kernel. We consider MISE of $G_n(x)$

$$Q = E \int (G(x) - G_n(x))^2 \, dx \tag{25.5}$$

where $\int$ is taken over $\mathcal{R}^d$. We are interested in deriving an optimal kernel K_n^* minimizing Q.

For the sake of simplicity in the remainder of the paper we only consider scalar case $d = 1$. In what follows we will use the elements of Fourier transform theory [9]. Denote by Φ_g Fourier transform of g, i.e.

$$\Phi_g(t) = \int g(x)e^{itx} dx$$

and thus inverse Fourier transform is given by

$$g(x) = \frac{1}{2\pi} \int \Phi_g(t)e^{-itx} dt.$$

The optimal form of K_n is given in Theorem 14.

Theorem 14. *The optimal kernel K_n^* minimizing (25.5) is defined by the equation*

$$\Phi(K_n^*) = \frac{n|\Phi_G|^2}{EY^2 + (n-1)|\Phi_G|^2}. \tag{25.6}$$

The optimal rate of MISE corresponding to the optimal kernel (25.6) is given by

$$Q_n^* = \frac{1}{2\pi} \int \frac{(EY^2 - |\Phi_G(t)|^2) |\Phi_G(t)|^2}{EY^2 + (n-1)|\Phi_G(t)|^2} dt.$$

Observe that

$$Q_n^* = \frac{EY^2}{2\pi} \int \frac{|\Phi_G(t)|^2}{EY^2 + (n-1)|\Phi_G(t)|^2} dt$$

$$- \frac{1}{2\pi} \int \frac{|\Phi_G(t)|^4}{EY^2 + (n-1)|\Phi_G(t)|^2} dt$$

$$\leq \frac{EY^2 K_n^*(0)}{n} + \frac{1}{2\pi(n-1)} \int |\Phi_G(t)|^2 dt. \tag{25.7}$$

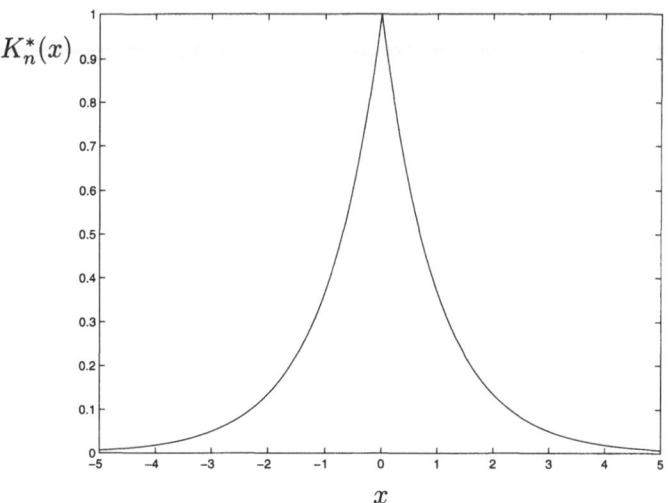

$K_n^*(x)$

x

Figure 25.4. Optimal RBF kernel

Kernel (25.6) is related to the superkernel of [3, 4]. Notice that for band-limited R and f the rate of MISE convergence with kernel (25.6) is

$$nQ_n^* \to \frac{1}{2\pi} \int_{-T}^{T} \left(EY^2 - |\Phi_G(t)|^2\right) dt \tag{25.8}$$

as $n \to \infty$, where T is the maximum of bandwidth of R and f. Therefore $Q_n^* = O(1/n)$. For other classes of R and f we will get different optimal kernels and rates.

To get an idea how the optimal kernel may look like consider the following examples

- $G(x) = e^{-x}$. It can be shown that

$$K_n^*(x) = \frac{n}{EY^2} \sqrt{\frac{\pi EY^2}{2EY^2 + 2(n-1)}} \exp\left(-\sqrt{1 + \frac{n-1}{EY^2}}|x|\right)$$

(see Figure 25.2) so by inequality (25.7) we have

$$\sqrt{n}Q_n^* \le \sqrt{\frac{\pi EY^2 n}{2EY^2 + 2(n-1)}} - \frac{\sqrt{n}}{2\pi(n-1)} \int |\Phi_G(t)|^2 dt \to \sqrt{\frac{\pi EY^2}{2}}$$

as $n \to \infty$.

- $G(x) = \frac{EY^2}{\pi(1+x^2)}$. One can show that

$$K_n^*(x) = \frac{n}{n-1} \frac{e^{-\pi x/2}}{1 - e^{-\pi x}} \sin \frac{x \log(n-1)}{2} + \frac{2n}{\pi} \sum_{k=1}^{\infty} \left(\frac{-1}{n-1}\right)^{k+1} \frac{k}{x^2 + 4k^2}$$

and

$$(n/\log n)Q_n^* \to (1/2\pi)$$

as $n \to \infty$.

We next consider polynomial and exponential classes of functions that is classes of R and f with tails of Φ_R and Φ_f decreasing either polynomially or exponentially. The rate of decrease affects the shape of optimal kernel and the optimal rate of convergence of MISE.

We say that Φ_G has **algebraic rate** of decrease of degree $p > 0$ if

$$\lim_{|t| \to \infty} |t|^p |\Phi_G| = \sqrt{K} > 0.$$

One can show that

$$n^{1-1/2p} Q_n^* \to \frac{1}{2\pi} \left(\frac{K}{EY^2} \right)^{1/2p} \int \frac{dx}{1 + |x|^{2p}}$$

as $n \to \infty$ for $p > 1/2$.

Φ_G has **exponential rate** of decrease of coefficient $\rho > 0$ if

$$|\Phi_G| \le A e^{-\rho|t|}$$

for some constant A and all t, and

$$\lim_{s \to \infty} \int_0^1 [1 + e^{2\rho s} |\Phi_G(st)|^2]^{-1} dt = 0.$$

It can be shown

$$\frac{n}{\log n} Q_n^* = \frac{1}{2\pi} \frac{n}{\log n} \int \frac{|\Phi_G(t)|^2}{EY^2 + (n-1)|\Phi_G(t)|^2} dt \to \frac{EY^2}{2\pi\rho}$$

as $n \to \infty$.

These results show that in exponential class of regressions and input densities we get parametric rate of convergence $O(1/n)$ for RBF networks with kernel defined by (25.6). It may be interesting to investigate how robust is K_n^* and the rate of convergence in (14) to changes in classes of functions.

25.3 MISE optimal RBF networks for unknown input density

In this section we will derive the optimal RBF regression estimate and the optimal rate of convergence in case when density f is estimated from $(X_1, Y_1), \cdots, (X_n, Y_n)$. Consider RBF regression estimate

$$R_n(x) = \frac{G_n(x)}{f_n(x)} = \frac{\frac{1}{n} \sum_{i=1}^n Y_i K_n(x - X_i)}{\frac{1}{n} \sum_{i=1}^n K_n(x - X_i)}.$$

This is normalized RBF network. It was introduced by Specht [18] and studied in [20] where its connection to the kernel regression estimate [5] was exploited.

Instead of working with MISE directly we will use the following relationship

$$EX \leq \epsilon + (M - \epsilon)\mathbf{P}\{X > \epsilon\}$$

which provides for bounded random variables the upper bound for MISE in terms of the probability of deviation provided that $X \leq M$. We will optimize the upper bound with respect to K_n. Let

$$Q = E \int |R_n(x) - R(x)|^2 f(x) dx. \tag{25.9}$$

The next theorem gives the form of the radial function minimizing the bound on Q and the corresponding upper bound on the rate of convergence.

Theorem 15. *The optimal kernel K_n^* minimizing upper bound on (25.9) is defined by the equation*

$$\Phi(K_n^*) = \frac{n[|\Phi_f|^2 + |\Phi_G|^2]}{(1 + EY^2) + (n-1)[|\Phi_f|^2 + |\Phi_G|^2]}. \tag{25.10}$$

The optimal rate of MISE corresponding to the optimal kernel (25.10) is given by

$$\bar{Q}_n^* = \frac{1}{2\pi} \int \frac{\left(EY^2 - [|\Phi_f(t)|^2 + |\Phi_G(t)|^2]\right) |\Phi_G(t)|^2}{(1 + EY^2) + (n-1)[|\Phi_f(t)|^2 + |\Phi_G(t)|^2]} dt.$$

The optimal radial function depends on density of X and on regression function or poster class probability G. It is clear that MISE rate of convergence of RBF net in nonlinear learning problem with band-limited f and G is $O(1/n)$. For classes of functions with algebraic or exponential rate of decrease we get similar rates as in the previous section. Observe that we achieve a parametric rate in intrinsically non-parametric estimation problem.

References

[1] D. S. Broomhead, D. Lowe, Multivariable functional interpolation and adaptive networks, *Complex Systems* 2 (1988) 321-323.

[2] K. B. Davis, Mean integrated error properties of density estimates, *Annals of Statistics* 5 (1977) 530-535.

[3] L. Devroye, A note on the usefulness of superkernels in density estimation, *Annals of Statistics* 20 (1993) 2037-2056.

[4] L. Devroye, *A Course in Density Estimation*, Birkhauser, Boston, 1987.

[5] L. Devroye, A. Krzyżak, An equivalence theorem for L_1 convergence of the kernel regression estimate, *J. of Statistical Planning and Inference* 23 (1989) 71-82.

[6] L. Devroye, L. Györfi and G. Lugosi, *Probabilistic Theory of Pattern Recognition*, Springer-Verlag, New York, 1996.

[7] F. Girosi, G. Anzellotti, Rates of converegence for radial basis functions and neural networks, in: R. J. Mammone (Ed.), *Artificial Neural Networks for Speech and Vision*, Chapman and Hall, London, 1993, 97-113.

[8] U. Grenander. *Abstract Inference.* Wiley, New York, 1981.

[9] T. Kawata, Fourier *Analysis in Probability Theory*, Academic Press, New York, 1972.

[10] A. Krzyżak, The rates of convergence of kernel regression estimates and classification rules, *IEEE Transactions on Information Theory* 32 (1986) 668-679.

[11] A. Krzyżak, T. Linder, G. Lugosi, Nonparametric estimation and classification using radial basis function nets and empirical risk minimization, *IEEE Transactions on Neural Networks* 7 (1996) 475–487.

[12] A. Krzyżak, T. Linder, Radial Basis Function Networks and Complexity Regularization in Function Learning, *IEEE Transactions on Neural Networks* 9 (1998) 247–256.

[13] D. F. McCaffrey, A. R. Gallant, Convergence rates for single hidden layer feedforward networks, *Neural Networks* 7 (1994) 147–158.

[14] J. Moody, J. Darken, Fast learning in networks of locally-tuned processing units, *Neural Computation* 1 (1989) 281-294.

[15] J. Park, I. W. Sandberg, Universal approximation using radial-basis-function networks, *Neural Computation* 3 (1991) 246-257.

[16] M. J. D. Powell, Radial basis functions for multivariable approximation: a review, in J. C. Mason and M. G. Cox (Eds.), *Algorithms for Approximation*, Oxford University Press, Oxford, 1987, 143–167.

[17] H.P. Schwefel, *Evolution and Optimum Seeking*, John Wiley, New York, 1995.

[18] D. F. Specht, Probabilistic neural networks, *Neural Networks*, 3 (1990) 109-118.

[19] G. S. Watson, M. R. Leadbetter, On the estimation of the probability density, I, *Annals of Mathematical Statistics* 34 (1963) 480-491.

[20] L. Xu, A. Krzyżak, A. L. Yuille, On Radial Basis Function Nets and Kernel Regression: Statistical Consistency, Convergence Rates and Receptive Field Size, *Neural Networks* 7 (1994) 609-628.

26

Instability in Nonlinear Estimation and Classification: Examples of a General Pattern

Steven P. Ellis[1]

Summary

"Instability" of a statistical operation, Φ, means that small changes in the data, D, can lead to relatively large changes in the data description, $\Phi(D)$. A "singularity" is a data set at which Φ is infinitely unstable to changes in the data, essentially a discontinuity of Φ. Near the set, S, of its singularities Φ will be unstable. A general topological theory gives conditions under which singularity will arise and gives lower bounds on the dimension, $\dim S$, of S. $\dim S$ is related to the probability of getting data near S. We describe this theory by showing how it manifests itself in several common forms of data analysis and state some of the mathematical results.

26.1 Introduction: Nonlinearity and Instability

In data analysis, we perform operations on data to extract features of the data. In symbols, we have a data summary, Φ, that takes data sets, D, in a data (or sample) space, $\mathcal{D}$, and maps them to decisions, classifications, estimates, descriptions, or features in a feature (or parameter) space, $\mathcal{F}$. (Φ might not be defined at literally every point of $\mathcal{D}$.) If either the data or $\Phi(\mathcal{D}) \subset \mathcal{F}$ is not a linear, i.e. vector, space (over $\mathbb{R}$ = reals, say) then the data summary must be nonlinear.

It is tempting to say that the feature space will be nonlinear when the structure that the data summary tries to detect in the data is "complex". In this chapter we look at the instability of a statistical operation when the

---

[1]Steven P. Ellis is with the New York State Psychiatric Institute and Columbia University (Email: ellis@neuron.cpmc.columbia.edu). The author would like to acknowledge the helpful suggestions from the Editors.

feature space is nonlinear. "Instability" of a statistical operation means that small changes in the data can lead to relatively large changes in the data description. Instability is a basic issue in applied mathematics (Hadamard [15, p. 38] and Isaacson and Keller [17, p. 22]). However, instability in data analysis apparently has not been investigated much as a general phenomenon (but see Hampel [16, Section 5.3], Belsley [1, Chapter 11], and Breiman [4]) so it is worth examining. Our approach can detect only the coarsest nonlinear phenomena, but it will have the advantages of being extremely general and not requiring distributional assumptions or large sample asymptotics. An interesting feature of this approach is that it is based on the *global* geometry of data analytic problems.

We will see that the topology of the feature extraction problem can force there to be instability. Since topological properties of a phenomenon depend on only the most basic features of the phenomenon, our topological results concerning instability in the interpretation of data even apply to cognition in living organisms like sea slugs or data analysts. The papers Ellis [5, 6, 7, 8] prove results in special cases. This chapter describes a general approach to the problem. (Ellis [10] discusses the same general ideas that this chapter does, but from a biological point of view.) Since the aspects of a data analysis problem that lead to instability are rather general, I conjecture that instability will be common in cases where one attempts to extract complex features from data.

The focus of this chapter is on examples (classification, plane fitting, factor analysis, and the location problem for spherical data), but formal results are presented, without proof, in Section 26.8.

26.2   Classification: The Presidential Election in Florida

26.2.1   The Florida vote

Classification or hypothesis testing offer mathematically trivial, but from a practical standpoint extremely important, examples of instability. Consider one example. The data consist of numbers of votes in Florida for Bush, Buchanan, Gore, Nader, etc. For simplicity suppose Bush and Gore are the only possibilities. Unfortunately, the data are corrupted in various ways. The objective is to classify Florida as a Bush or Gore state. As we saw from what transpired in Florida in the fall of 2000, even majority rule is not a precisely defined decision rule, but here we allow practically any decision rule. Ignoring the ultimate discreteness of the data, the data space, $\mathcal{D}$, is a simplex. It consists of all pairs, (b, g), of nonnegative numbers whose sum is no greater than the total number of registered voters in Florida. Here, b

and g are the numbers of votes for Bush and Gore, respectively. (Not all registered voters vote.)

The feature space, $\mathcal{F}$, consists of two points: { Bush, Gore }. So $\mathcal{F}$ is not linear. It has a "hole" or "gap" separating its points. The hole is zero dimensional. (We use informal language here. See Section 26.8 for a more precise formulation.) A nontrivial data summary (decision rule), Φ, will map $\mathcal{D}$ onto $\mathcal{F}$. Since $\mathcal{F}$ is not linear this means Φ must ultimately also be nonlinear.

The majority rule decision procedure is unstable near the line $b = g$: A small change in b or g would have made Gore the winner in Florida. Instability is very important as a practical matter. Similar instability occurs in other classification problems or hypothesis testing.

In order to get insight into instability in general, we analyze the Florida vote example. Let $\Phi : \mathcal{D} \to \mathcal{F}$ be a decision rule. Various versions of majority rule are examples of such data summaries, but we will place only the mildest restrictions on decision rules. We will see that the behavior of a data summary on small collections of simple data sets, call them "test patterns", can determine the global stability of the data summary. For this example consider a space $\mathcal{T} \subset \mathcal{D}$ consisting of just two test patterns, say $B = (b = 4 \text{ million}, g = 0)$ and $G = (b = 0, g = 4 \text{ million})$. Any reasonable election decision rule should map B to "Bush" and G to "Gore". Even though $\mathcal{T}$ is a small collection of easy to analyze data sets, the behavior of decision rule on $\mathcal{T}$ samples a wide variety, in fact all, of the possibilities in $\mathcal{F}$.

26.2.2 Singularity in Florida

Continuing to ignore the ultimate discreteness of the data, $\mathcal{D}$ is pathwise connected so we can connect the test patterns B and G by a curve. Assuming Φ assigns B to "Bush" and G to "Gore" there must be a point or points along the curve at which the decision rule makes a sudden transition from one candidate to the other. A data set at which a transition takes place is a "singularity" of the decision rule. More precisely, a singularity of Φ is a data set, D, at which the limit, $\lim_{D' \to D} \Phi(D')$, does not exist. So a singularity is like a discontinuity except that Φ does not have to be defined at a singularity. The set of all singularities is the "singular set," $\mathcal{S}$, of the data summary, Φ.

Data analysis is concerned with the overall pattern of data, not with fine details. So it is disconcerting to find that small changes in the data can have relatively big impact on the data description. But this happens in the vicinity of a singularity. In the estimation context, a nearby bad singularity introduces uncertainty in the estimate that should be reflected by a big standard error (SE). But if one's data are near a singularity, one might not be able to estimate the SE well. Even if one can estimate the SE well, no one wants an estimate with a big SE. In theory one will never get a singularity as a data set, but I stress that singularity is an issue because one might get a data set *near* a singularity (e.g., Florida election data or Ellis [9, Figure 2]).

I expect that singularity is important mainly in small samples. But "small" should probably be interpreted relative to model size and the Florida election debacle shows that singularity can be a problem in large data sets, too.

We observed above that along *any* curve joining the test patterns B and G there must be at least one singularity. Since this is true for any curve, it should be clear from drawing a picture of the situation that *if* Φ is continuous near B and G, then the singular set must be at least one-dimensional. (In practice any reasonable, general notion of dimension should do [Falconer [13]]. Denote dimension by "dim".) The qualification that Φ be continuous near B and G is important. Otherwise, one could get by with a single singularity at B or G. I.e., if Φ is not continuous near B and G then the singular set might be 0-dimensional.

26.2.3 Codimension

It is often better to express a lower bound on the dimension as an *upper* bound on the "codimension" of $\mathcal{S}$:

$$\text{codim } \mathcal{S} = \dim \mathcal{D} - \dim \mathcal{S}.$$

Singularities are bad. So a high dim $\mathcal{S}$ or low codim $\mathcal{S}$ is bad.

If there are c Presidential candidates, then $c = \dim \mathcal{D}$ and, providing Φ is continuous on $\mathcal{T}$, the lower bound on dim $\mathcal{S}$ will be $c - 1$. But the upper bound on codim $\mathcal{S}$ will always be $c - (c - 1) = 1$. I.e., the upper bound on codim $\mathcal{S}$ does not depend on number of candidates. The small size of the upper bound indicates that singular sets in classification are big. That is bad news. In Section 26.6 we will see that the codimension of a singular set is related in a simple way to the probability of getting data near the singular set. (Note that we could make one of the c "candidates" the decision "a draw", i.e. no winner. This maneuver does not eliminate singularities, but it might reduce post election rancor a little.)

26.2.4 Summary

The feature space, $\mathcal{F}$, has a "hole", i.e., it has structure. Because there is a set, $\mathcal{T}$, of test patterns on which behavior of data summary, Φ, is sufficiently rich, Φ "feels" the structure of $\mathcal{F}$. The data space lacks the kind of structure $\mathcal{F}$ has. This mismatch forces the existence of singularities. The structure of $\mathcal{F}$ and $\mathcal{T}$ yields upper bound on codim $\mathcal{S}$, at least if Φ is continuous at B and G. From the point of view of generalizing these principles, it is better to express the assumption that Φ is continuous at test patterns as "$\mathcal{S} \cap \mathcal{T}$

is 'small'". Then the conclusion takes this form.

$$\text{codim } \mathcal{S} \leqslant \text{dimension of "hole"} + 1 \ (= 0 + 1 = 1). \tag{26.1}$$

(This bound is tight. Majority rule achieves it.)

On the other hand, if Φ is not continuous in the vicinity of $\mathcal{T}$, then, without further assumptions, all we can say is

$$\dim \mathcal{S} \geq \text{dimension of } \mathcal{T} - \text{dimension of "hole"}. \tag{26.2}$$

Since $\mathcal{T}$ consists of a finite number (2) of points, $\dim \mathcal{T} = 0$. Hence, the right hand side of (26.2) is $0 - 0 = 0$.

26.3   Singularity in Plane Fitting

26.3.1   Structure of problem

In the Florida example singularity is a zero-dimensional phenomenon because it hinged on the zero-dimensional "hole" in $\mathcal{F}$. Now we look at example in which singularity is a one-dimensional phenomenon. A very common data analytic operation is fitting a plane to multivariate data. Let n = sample size, p = number of variables, k = dimension of plane to be fitted. Assume $n > p > k > 0$. Then $\mathcal{D} = \mathbb{R}^{pn}$.

Singularity is inherent in plane fitting (Ellis [5, 7, 8, 9, 11]). For simplicity, consider fitting a line ($k = 1$) to bivariate ($p = 2$) observations. We will not be interested in the offset of the line from origin. So we can always shift the line so that it passes through the origin. Thus, $\mathcal{F}$ = space of all lines through origin of $\mathbb{R}^2$. $\mathcal{F}$ is a circle in disguise so it has a one dimensional "hole".

As for $\mathcal{T}$ we can take it to be a simple closed curve (homeomorphic image of a circle) in $\mathcal{D}$ with the following two properties (Ellis [7, Proof of Proposition 4.1], [11]). First, in each $D \in \mathcal{T}$ all the observations lie exactly on a unique line, $L = L(D)$, passing through the origin. Second, L maps $\mathcal{T}$ onto $\mathcal{F}$. If Φ is a line fitting method it is reasonable to demand that Φ map a data set lying exactly on a unique line to that line. That means that the line-fitter Φ should map each $D \in \mathcal{T}$ to $L(D)$. (Actually, Φ only needs to be defined on dense subsets of $\mathcal{D}$ and $\mathcal{T}$.) Since L maps $\mathcal{T}$ onto $\mathcal{F}$, so does Φ. This construction generalizes to fitting planes of any dimension and $\mathcal{T}$ remains a simple closed curve.

26.3.2   Codimension of singular sets in plane fitting

Extrapolating from the classification case (26.1) one would expect that an upper bound on the codimension of the singular set, $\mathcal{S}$, of a plane fitting method, Φ, is

$$\text{codim } \mathcal{S} \leqslant (\text{dimension of "hole" in } \mathcal{F}) + 1 = 1 + 1 = 2. \tag{26.3}$$

Providing $S \cap T$ is "small". "$S \cap T$ is 'small' " means two things:

The restriction of Φ to $T \setminus S$ has a continuous extension to T (26.4)

("$\setminus$" indicates set theoretic subtraction) and

$$S \cap T = \varnothing \text{ or } \dim(S \cap T) < \dim T - \text{(dimension of "hole")}. \quad (26.5)$$

(If T is a simple closed curve as above, conditions (26.4) and (26.5) are redundant.) The popular plane fitting methods principal component analysis (Johnson and Wichern [18]) and least absolute deviation regression (Bloomfield [2]) satisfy conditions (26.4) and (26.5) and so their singular sets are large. (Ellis [7, Example 2.3] and Ellis [9].)

For least squares linear regression condition (26.5) fails and so does (26.3). However, our theory does apply to least squares. The idea is to replace the very small T described above by a larger T, viz., the set of *all* data sets that lie exactly on a unique plane (of the right dimension). With this choice condition (26.5) still fails for least squares but now

$$\dim T - \text{(dimension of "hole")} = nk + (k+1)(p-k) - 1. \quad (26.6)$$

Analogously to (26.2), for any plane fitting method satisfying (26.4) this is a lower bound on the dimension of the singular set (Ellis [7, Theorem 2.6]). Least squares achieves this lower bound (Ellis [7, Example 2.8])

26.4   Factor Analysis

We have seen singularity as a zero-dimensional phenomenon (classification) and as a one-dimensional phenomenon (plane fitting). Factor analysis (Johnson and Wichern [18]) has a kind of singularity that is apparently a two-dimensional phenomenon (Ellis [12]). I have not yet computed an upper bound on the codimension of the set of these singularities, but I expect it is 2.

26.5   Location Problem for Spherical Data

Let $x_1, \ldots, x_n$ be points on the p-sphere, $S^p \equiv \{x \in \mathbb{R}^{p+1} : |x| = 1\}$ ($p =$ positive integer). Consider the problem of measuring (assigning) location on the sphere of such data clouds. (Fisher *et al* [14] and Watson [22] are general references on directional data. Ellis [6] discusses existence of singularities, but not dimension, for this problem.) In this case the data space is the Cartesian product, $\mathcal{D} = (S^p)^n$, and the feature space is just the sphere $\mathcal{F} = S^p$. $\mathcal{F}$ clearly has a p-dimensional "hole". Take the test pattern space, T, to be the "diagonal"

$$T = \{(x_1, \ldots, x_n) \in \mathcal{D} : x_1 = \cdots = x_n\}.$$

If Φ is a measure of location, it is reasonable to suppose

$$\Phi(x, \ldots, x) = x, \ x \in S^p. \tag{26.7}$$

So Φ is a homeomorphism of $\mathcal{T}$ onto $\mathcal{F}$. Note that in this case condition (26.5) becomes

$$\Phi \text{ has no singularities in } \mathcal{T}. \tag{26.8}$$

On the basis of what we have seen so far ((26.1), (26.3)), one would expect that if (26.8) holds, then

$$\text{codim } \mathcal{S} \leqslant p + 1, \tag{26.9}$$

where $\mathcal{S}$ is the singular set of Φ. As a check, consider the Φ that takes $(x_1, \ldots, x_n) \in \mathcal{D}$ to x_1. This Φ satisfies (26.7) and (26.8), but this Φ has no singularities, contradicting (26.9)! What is wrong? The problem is that $\mathcal{D}$ has holes of the same dimension as the hole in $\mathcal{T}$ and $\mathcal{F}$. But with an additional innocuous assumption we can proceed. Assume

$$\Phi \text{ is symmetric in its arguments.} \tag{26.10}$$

With this assumption, (26.9) holds for Φ satisfying (26.8).

The "directional mean" is the measure of location that takes $(x_1, \ldots, x_n) \in \mathcal{D}$ to $\bar{x}/|\bar{x}|$, providing $\bar{x} \neq 0$, where $\bar{x}$ is the sample mean of $x_1, \ldots, x_n$ regarded as vectors in $\mathbb{R}^{p+1}$ and $|\cdot|$ is the Euclidean norm. The singularities of the directional mean are precisely data sets for which $\bar{x} = 0$. It is easy to see that the directional mean satisfies (26.8) and achieves the bound (26.9).

26.6   Codimension and Probability

By definition, a data summary is unstable in the vicinity of a singularity, so getting data near a singularity is bad. We have also suggested measuring the size of singular sets by their (co)dimension. It turns out that the smaller the codimension of $\mathcal{S}$ the higher is the probability of getting data near $\mathcal{S}$. More precisely, suppose the data are random with a nowhere vanishing continuous density. Then, roughly speaking, as $\epsilon \downarrow 0$, the probability of getting data within ϵ units of $\mathcal{S}$ goes to 0 at least as slowly as $\epsilon^{\text{codim } \mathcal{S}}$ (Ellis [7]). This is another reason why codimension is usually a better measure of size of $\mathcal{S}$ than dimension (subsection 26.2.3). See Ellis [11] for further examination of dimension and probability in the line fitting setting. Numerical analysts have also been interested in this issue (Blum et al [3, Chapters 11-13] and references therein).

26.7   Mitigating Singularity

So far we have only discussed the existence and magnitude of singularity problems. Here we say a few words about reducing the impact of singularity. There are two approaches.

First, one can try to choose a data analytic method, Φ, that is appropriate for the problem at hand but whose singular set lies in a region of $\mathcal{D}$ of low probability. This choice can be made before seeing the data. If one wants to choose Φ on the basis of the data this choice must be made cautiously. To see this, suppose one starts with a family, $\boldsymbol{\Phi} = \{\Phi_\alpha, \ \alpha \in A\}$, of procedures appropriate to the problem at hand. Given data, D, choose some member, $\Phi_{\alpha(D)}$, whose singular set appears remote or improbable with respect to D. Summarize the data by $\Phi^*(D) \equiv \Phi_{\alpha(D)}(D)$. At first glance it would appear that Φ^* has good singularity properties. However, by piecing together the procedures in $\boldsymbol{\Phi}$ one may inadvertently create singularities. For example, $\alpha(D)$, the index that selects the member of $\boldsymbol{\Phi}$ based on the data D, may be a piece-wise constant function of $D \in \mathcal{D}$. Φ^* may have singularities at some of the jumps in the function $D \mapsto \alpha(D)$.

A second approach to countering the singularity problem is through the use of "diagnostics". For example, suppose Φ has singular set $\mathcal{S}$. Let $\mathcal{C}$ denote the "cone over $\mathcal{S}$" defined as follows. Start with the Cartesian product, $\mathcal{S} \times [0, 1]$. The cone, $\mathcal{C}$, is the space we obtain by identifying set $\mathcal{S} \times \{0\}$ to a point, the "vertex", v. If $D \in \mathcal{D}$, let $\rho(D) = \mathrm{dist}(D, \mathcal{S})$, the distance from D to $\mathcal{S}$. Now define a new data analytic procedure, $\Phi^\#$ taking values in $\mathcal{C}$, as follows.

$$\Phi^\#(D) = \begin{cases} \left(\Phi(D), \frac{\rho(D)}{1+\rho(D)} \right), & \text{if } D \notin \mathcal{S}, \\ v, & \text{if } D \in \mathcal{S}. \end{cases}$$

The quantity $\rho(D)/[1 + \rho(D)]$ is a diagnostic for Φ at D. The augmented data summary $\Phi^\#$ is actually a continuous function on $\mathcal{D}$ that contains all the information in Φ (assuming $\Phi(D)$ is meaningless for $D \in \mathcal{S}$). However, such use of diagnostics is not always helpful. E.g., once someone has been elected President, his/her margin of victory is largely irrelevant.

26.8   Statement of Results

26.8.1   Two step process

The theory discussed here is an application of algebraic topology (e.g., Munkres [20]). In order to apply this theory to a problem, one must carry out two steps. One step is always the same: Apply the following theorem. Further hypotheses can be found in Subsection 26.8.2.

Theorem 16. *Let $0 \leqslant r \leqslant t = \dim \mathcal{T}$ be an integer. Suppose*

1. *The restriction of* Φ *to* $\mathcal{T} \setminus \mathcal{S}$ *can be continuously extended to* $\mathcal{T}$. *Let* Θ *be the continuous extension of that restriction.*

2. $\mathcal{S} \cap \mathcal{T} = \varnothing$ *or* $\dim(\mathcal{S} \cap \mathcal{T}) < t - r$.

Then

$$\Phi_* \big[H_r(\mathcal{D} \setminus \mathcal{S}) \big] \supset \Theta_* \big[H_r(\mathcal{T}) \big]. \tag{26.11}$$

(Any coefficient group is permissible, except if $\mathcal{T}$ *is non-orientable. In that case use* $\mathbb{Z}/(2\mathbb{Z})$.*)*

Note that hypotheses 1 and 2 are just conditions (26.4) and (26.5). This theorem is only interesting when

$$\Theta_* \colon H_r(\mathcal{T}) \to H_r(\mathcal{F}) \text{ is non-trivial}, \tag{26.12}$$

which happens when Θ "feels" r-dimensional "holes" in $\mathcal{F}$. In applying this theorem the idea is to choose $\mathcal{T}$ so the hypotheses of Theorem 16 and (26.12) are easy to check.

The second step takes (26.11) and (26.12) and goes on to prove $H^{d-r-1}(\mathcal{S}) \neq 0$, where $d > t$ is the dimension of $\mathcal{D}$. It then follows that $\mathcal{S}$ has dimension at least $d - r - 1$. (I.e., codim $\mathcal{S} \leqslant r + 1$.) How this is done varies from problem to problem. However, we have the following.

Corollary 1. *Suppose* $\mathcal{D}$ *is a manifold and* $H^{d-r}(\mathcal{D}) = 0$. *Then, under the hypotheses of theorem, codim* $\mathcal{S} \leqslant r + 1$.

$H^{d-r}(\mathcal{D}) = 0$ holds, e.g., if $\mathcal{D}$ is a sphere of dimension $> d - r > 0$. In the plane fitting problem (Section 26.3) this applies because we can restrict attention to spheres lying in $\mathcal{D}$. Doing so also gives information about where the singularities lie. This will not work for the classification example, but there the "second step" is still easy to carry out. Corollary 1 does not seem to help with the "second step" in the spherical location case either. For that example the "second step" requires (26.10) and some work.

26.8.2 Fine print.

Here are some additional technical hypotheses for Theorem 16 and Corollary 1. Assume $\Phi \colon \mathcal{D} \setminus \mathcal{S} \to \mathcal{F}$ is continuous, $\mathcal{T} \subset \mathcal{D}$ is a compact manifold, and $(\mathcal{T}, \mathcal{S})$ is a triangulable pair. For Corollary 1 I need to assume that $\mathcal{D}$ is a compact manifold and $(\mathcal{D}, \mathcal{S})$ is also a triangulable pair.

The assumption that $\mathcal{S}$ is triangulable is annoying because in the course of checking that assumption one would naturally find out the dimension of $\mathcal{S}$, rendering this general theory pointless! However:

1. In practice it is probably a safe assumption that $\mathcal{S}$ is triangulable (Munkres [19], van den Dries [21]).

2. I *think* versions of these results can be proved without any triangulability assumptions.

I also think the compactness assumptions on T and D might be dispensable.

References

[1] Belsley, D.A. (1991) *Conditioning Diagnostics: Collinearity and Weak Data in Regression.* John Wiley & Sons, New York.

[2] Bloomfield, P. and Steiger, W.L. (1983) *Least Absolute Deviations: Theory, Applications, and Algorithms.* BirkhaYser, Boston.

[3] Blum, L.; Cucker, F.; Shub, M.; and Smale, S. (1998) *Complexity and Real Computation.* Springer-Verlag, New York.

[4] Breiman, L. (1996) "Heuristics of instability and stabilization in model selection," *The Annals of Statistics* **24**, 2350-2383.

[5] Ellis, S.P. (1991a) "The singularities of fitting planes to data," *Annals of Statistics* **19**, 1661-1666.

[6] Ellis, S.P. (1991b) "Topological aspects of the location problem for directional and axial data," *International Statistical Review,* **59**, 389-394.

[7] Ellis, S.P. (1995) "Dimension of the singular sets of plane-fitters," *Annals of Statistics* **23**, 490-501.

[8] Ellis, S.P. (1996) "On the size of singular sets of plane-fitters," *Utilitas Mathematica* **49**, 233-242.

[9] Ellis, S.P. (1998) "Instability of least squares, least absolute deviation, and least median of squares linear regression" (with discussion), *Statistical Science* **13**, 337-350.

[10] Ellis, S.P. (2001) "Topology of ambiguity" in *Computational Neuroscience: Trends in Research 2001,* edited by James M. Bower, Publisher: Elsevier; –OR– *Neurocomputing* **38-40**, 1203-1208.

[11] Ellis, S.P. "Fitting a line to three or four points on a plane," (submitted)

[12] Ellis, S.P. "Instability of statistical factor analysis," (submitted)

[13] Falconer, K. (1990) *Fractal Geometry: Mathematical Foundations and Applications.* John Wiley & Sons, New York.

[14] Fisher, N.I.; Lewis, T.; and Embleton, B.J.J. (1987) *Statistical Analysis of Spherical Data.* Cambridge University Press, New York.

[15] Hadamard, J. (1923) *Lectures on Cauchy's Problem in Linear Partial Differential Equations.* Yale University Press, New Haven.

[16] Hampel, F.R. (1974) "The influence curve and its role in robust estimation," *Journal of the American Statistical Association* **69**, 383-393.

[17] Isaacson, E. and Keller, H.B. (1966) *Analysis of Numerical Methods.* Wiley, New York.

[18] Johnson, R.A. and Wichern, D.W. (1992) *Applied Multivariate Statistical Analysis, Third Edition.* Prentice Hall, Englewood Cliffs, NJ.

[19] Munkres, J.R. (1966) "Elementary Differential Topology (Rev. Ed.)" *Annals of Math. Studies* **54**, Princeton Univ. Press, Princeton.

[20] Munkres, J.R. (1984) *Elements of Algebraic Topology.* Benjamin/Cummings, Menlo Park.

[21] van den Dries, L. (1998) *Tame topology and o-minimal structures.* Cambridge Univ. Press, New York.

[22] Watson, G. (1982) "Directional data analysis," In *Encyclopedia of Statistical Sciences,* S. Kotz and N.L. Johnson, Eds., 376-381. Wiley, New York.

27

Model Complexity and Model Priors

Angelika van der Linde[1]

Summary

In hierarchical models it is often hard to specify a (hyper-)prior distribution for a model parameter at the highest level of the model. Several authors suggested to define a hyperprior as some decreasing function of "model complexity" or "generalized degrees of freedom." In this chapter the proposal is discussed from a conceptual point of view. In particular, "model complexity" is evaluated as a model discriminating transformation of the hyperparameter. Also, criteria for model choice of the form "goodness of fit - penalty" are related to non-informative (reference) priors for model parameters in representative examples.

27.1 Introduction

Consider an abstract hierarchical model where Y is the observed variable with density $p(y|\theta)$, and θ is the parameter of interest in inference. Let a prior on θ be defined by a density $p(\theta|\phi)$ with hyperparameter ϕ, which indicates the model for θ. For a Bayesian analysis additionally a prior on ϕ is required.

For example, in a regression setup one may be interested in $\theta = EY$, and ϕ may denote a scale factor of the prior covariance K, e.g. $\theta \sim N(0, \phi K)$. A focused prior on θ will describe a rather restricted model for θ, whereas a flat prior will correspond to a flexible or, intuitively, complex model. Thus model complexity reflects the strength of the prior (taking into account the likelihood, though). The classical definition of model complexity as "number of unknown parameters" = k, if $\theta \in \mathbb{R}^k$, may be considered as a special case corresponding to a non-informative prior on θ. In regression,

---

[1] Angelika van der Linde is with the Institut für Statistik, Universität Bremen.

if $\widehat{\theta}(y) = A(\phi)y$, the trace of the "hat matrix" $A(\phi)$, tr $A(\phi)$, is an agreed definition of model complexity.

Based on such a reasoning it is tempting to define the hyperprior $p(\phi)$ as a decreasing function of model complexity $c(\phi)$, say. For example,

$$p(\phi) \propto \exp(-\gamma c(\phi)), \tag{27.1}$$

with an additional scale factor γ, was suggested and tried by Holmes and Denison [6]. A similar proposal using degrees of freedom was made by Hodges and Sargent [5]. The intuitive justification for such a hyperprior mainly is parsimony: simple models are preferred. More formally, it is also argued (by Holmes and Denison [6]), that using (27.1) the posterior

$$p(\phi|y) \propto p(y|\phi)p(\phi) \quad \text{with} \quad p(y|\phi) = \int p(y|\theta)p(\theta|\phi)d\theta$$

on the log scale

$$\log p(\phi|y) \propto \log p(y|\phi) - \gamma c(\phi) \tag{27.2}$$

is decomposed into "goodness of fit - penalty," and thus has the form of many criteria used for model choice.

In this chapter several conceptual issues related to the hyperprior (27.1) are discussed. The first issue results from the question how generally the proposal (27.1) can be applied. There is no unique definition of model complexity beyond the regression setup, but there are several attempts (Akaike [1]; Efron [4]; Ye [9]; Spiegelhalter et al. [7]) yielding measures of model complexity which in general depend on either the data or the unknown parameter θ, and hence their use in hyperpriors is limited. More generally, $c(\phi)$ is to be evaluated as a transformation of ϕ which competes with other model discriminating parameterizations. This issue is addressed in Section 27.2 of the chapter. The second issue is the link between hyperpriors and penalties of goodness of fit, as suggested by (27.2). $p(y|\phi)$ does not assess any fit, but $p(y|\widehat{\theta}(y))$ does and is conventionally used. The appropriate formal link and its impact is the topic of Section 27.3. A third issue to be considered is an approach reverse to (27.1): instead of defining a hyperprior based on a measure of model complexity we derive non-informative (reference) priors for standard examples and investigate if, respectively how they are related to measures of model complexity. This amounts to asking whether (27.1) or model choice based on penalized goodness of fit is equivalent to using a non-informative (reference) hyperprior. Some findings are reported in Section 27.4. Finally, results are summarized and further discussed in Section 27.5 of the chapter.

27.2 Model discriminating transformations of the hyperparameter

27.2.1 Measures of model complexity

For now we consider ϕ to be fixed and omit it in our notation.

Measures of model complexity were derived as part of criteria for model choice that take the form "goodness of fit - penalty." Three most prominent examples are briefly reviewed. Throughout $\widetilde{y}$ will denote a future observation, and $\delta(p||q) = \int p(x) \log \frac{p(x)}{q(x)} dx$ will denote the Kullback-Leibler distance between densities p and q.

AIC

Akaike's information criterion [1] is based on an approximation of a true distribution $p^*(\widetilde{y})$ by a best parameterized model $p(\widetilde{y}|\theta^*)$ minimizing $\delta(p^*(\cdot)||p(\cdot|\theta^*))$ with respect to θ. The Kullback-Leibler distance then is to be minimized with respect to competing models p. As θ^* is unknown, the target evaluated at the maximum likelihood estimate $E_{\widetilde{Y}|p^*}(-\log p(\widetilde{y}|\widehat{\theta}_{ML}(y)))$ is used and estimated by $-\log p(y|\widehat{\theta}_{ML}(y)$. This estimate is corrected for bias with respect to $E_{Y|p^*}$ and the correction term interpreted as "model complexity,"

$$c_{AIC} = E_{Y|p^*} \left[E_{\widetilde{Y}|p^*} \left(-\log p \left(\widetilde{y}|\widehat{\theta}_{ML}(y) \right) \right) + \log p \left(y|\widehat{\theta}_{ML}(y) \right) \right]. \quad (27.3)$$

Based on a second order Taylor expansion an approximation can be derived

$$c_{AIC} \approx \text{tr} \left[I_{p^*}(\theta^*) \text{cov}_{Y|\theta^*} \left(\widehat{\theta}_{ML}(y) \right) \right],$$

where $I_{p^*}(\theta^*) = E_{Y|p^*}(-\frac{\partial^2 \log p(y|\theta)}{\partial \theta^2}|_{\theta=\theta^*})$. c_{AIC} thus depends on the unknown true distribution p^* or the unknown true parameter θ^*, and if estimated, depends on the data. For details see Burnham and Anderson [3], Ch.6.

Expected optimism

Efron's target with respect to model choice [4] is the expected loss $E_{\widetilde{Y}|\theta}(-\log p(\widetilde{y}|\widehat{\theta}(y))$ that occurs when fitting a future observation $\widetilde{y}$ using the estimate $\widehat{\theta}(y)$. Its estimate $-\log p(y|\widehat{\theta}(y))$ is "biased due to overfit," and the corresponding measure of model complexity is

$$c_E = E_{Y|\theta} \left[E_{\widetilde{Y}|\theta} \left(-\log p \left(\widetilde{y}|\widehat{\theta}(y) \right) + \log p \left(y|\widehat{\theta}(y) \right) \right) \right]. \quad (27.4)$$

Again this depends on θ. $\widehat{\theta}$ can be any estimator, possibly a Bayesian estimator.

DIC

The deviance information criterion (Spiegelhalter et al. [7]) is Bayesian in spirit and does not evaluate expectations over observables Y and $\widetilde{Y}$ but refers to the posterior expectation of θ. The target now is the expected loss $E_{\theta|y}E_{\widetilde{y}|\theta}(-2\log p(\widetilde{y}|\theta))$ which is estimated by $-2\log p(y|\bar{\theta})$ where $\bar{\theta} = E(\theta|y)$. The corresponding measure of model complexity is

$$c_{DIC} = 2E_{\theta|y}\left(-\log p(y|\theta) + \log p(y|\bar{\theta})\right),\qquad(27.5)$$

and an approximation is given by

$$c_{DIC} \approx \operatorname{tr}\left[I_{obs}(\bar{\theta})\operatorname{cov}(\theta|y)\right],$$

where $I_{obs}(\bar{\theta}) = \frac{\partial^2 \log p(y|\theta)}{\partial\theta^2}\big|_{\theta=\bar{\theta}}$ is the observed Fisher information matrix. DIC in general depends on the data.

Summarizing, all these criteria of model complexity capture the "difficulty in estimating θ" in comparisons of "theoretical residuals" $\log p(\cdot|\theta)$ to "empirical residuals" $\log p(\cdot|\widehat{\theta})$. Let

$$
\begin{aligned}
(*) = \quad &-\log p(\widetilde{y}|\widehat{\theta}(y)) &&+\log p(y|\widehat{\theta}(y))\\
&+\log p(\widetilde{y}|\theta) \quad -\log p(\widetilde{y}|\theta)\\
&\qquad\qquad +\log p(y|\theta) \quad -\log p(y|\theta).
\end{aligned}
$$

Then $E(*)$ including two expected residuals (for the actually observed y and the future observation $\widetilde{y}$) and a term of sampling expectation 0 is regarded as measure of model complexity. The criteria differ with respect to the expectation operator and the estimate of θ.

In the sequel we are explicit about dependence on ϕ.

In general the criteria depend on either unknown parameters or the data, and hence their use in defining a hyperprior distribution is limited. In a regression setup, however, where $Y \sim N(\theta, \Sigma)$, and $\widehat{\theta}(y) = A(\phi)y$,

$$c_E(\phi) = c_{DIC}(\phi) = \operatorname{tr}A(\phi),$$

and if in addition $p(\theta)$ becomes non-informative, i.e. either with variance tending to infinity or data asymptotically prevailing,

$$\operatorname{tr}A(\phi) \to c_{AIC}(\phi) = k \qquad \text{for } \theta \text{ in } \mathbb{R}^k.$$

It is this special case that motivated the proposal to specify $p(\phi)$ as decreasing function of some $c(\phi)$.

Subsequently we shall focus on $c_{DIC}(\phi)$ because it is derived within a Bayesian framework. From a Bayesian point of view it is more instructive to interpret model complexity in terms of the Kullback-Leibler distance

$$\delta_\phi(y) := \delta\left(p(\theta|\phi, y)\|p(\theta|\phi)\right) = E_{\theta|\phi,y}\left(\log\frac{p(\theta|\phi, y)}{p(\theta|\phi)}\right)\qquad(27.6)$$

of the posterior to prior distribution rather than in terms of residuals. Note that by Bayes' theorem

$$p(y|\phi)\frac{p(\theta|\phi,y)}{p(\theta|\phi)} = p(y|\theta) \qquad \text{for all } \phi \text{ and } \theta \qquad (27.7)$$

and hence

$$
\begin{aligned}
c_{DIC}(\phi) &= 2E_{\theta|\phi,y}\left(-\log\frac{p(y|\theta)}{p(y|\bar{\theta}_\phi)}\right) \\
&= 2E_{\theta|\phi,y}\left(-\log\frac{p(\theta|\phi,y)}{p(\theta|\phi)} + \log\frac{p(\bar{\theta}_\phi|\phi,y)}{p(\bar{\theta}_\phi|\phi)}\right) \qquad (27.8) \\
&= 2\left(-\delta\left(p(\theta|\phi,y)\|p(\theta|\phi)\right) + \log\frac{p(\bar{\theta}_\phi|\phi,y)}{p(\bar{\theta}_\phi|\phi)}\right) \\
&= 2\left(-\delta_\phi(y) + \hat{\delta}_\phi(y)\right)
\end{aligned}
$$

where $\hat{\delta}_\phi(y) = \log\frac{p(\bar{\theta}_\phi|\phi,y)}{p(\bar{\theta}_\phi|\phi)}$. That is, $c_{DIC}(\phi)$ is twice the difference between the Kullback-Leibler distance of the posterior to the prior distribution and its estimate.

27.2.2 Model discrimination

A standard model discriminating quantity is the Kullback-Leibler distance between the marginal densities corresponding to a model of interest (ϕ) and a reference model (ϕ_0),

$$\delta_{\phi,\phi_0} := \delta\left(p(y|\phi)\|p(y|\phi_0)\right).$$

We argue that the measure of model complexity $c_{DIC}(\phi)$ does have a structure similar to that of δ_{ϕ,ϕ_0} but is based on an internal reference $\bar{\theta}_\phi$ rather than an external reference ϕ_0. Using again (27.7) one obtains

$$\delta_{\phi,\phi_0} = E_{Y|\phi}\left[-\log\frac{p(\theta|\phi,y)}{p(\theta|\phi)} + \log\frac{p(\theta|\phi_0,y)}{p(\theta|\phi_0)}\right] \qquad (27.9)$$

(which does not depend on θ), in comparison to (27.8) resp. $E_{Y|\phi}c_{DIC}(\phi)$, where expectations are aligned. One might consider using δ_{ϕ,ϕ_0} instead of a $c(\phi)$ in order to define a prior distribution on ϕ. At least it provides some guidance to the choice of a model discriminating parameterization of a prior distribution on models.

27.3 Hyperpriors and penalties

In the decomposition (27.2) of the posterior distribution

$$\log p(\phi|y) \propto \log p(y|\phi) - \gamma c(\phi)$$

goodness of fit is described by $p(y|\phi)$. In contrast, in criteria of model choice usually goodness of fit is represented by $p(y|\widehat{\theta}(y))$, a term that actually measures fit due to an estimate. This motivates exploring (27.7) in order to relate priors on ϕ to penalties of goodness of fit in terms of $p(y|\widehat{\theta}(y))$. Equation (27.7) can be read in two ways.

(i) **From right to left:** Which hyperprior is induced by a penalty on $p(y|\widehat{\theta}(y))$? Application of the posterior expectation $E_{\theta|\phi,y}$ after taking logarithms yields $-\frac{1}{2}c_{DIC}$ on the rhs of (27.7),

$$\log p(y|\phi) + \delta_\phi(y) = E_{\theta|\phi,y} \log p(y|\theta) \approx \log p(y|\bar{\theta}_\phi) - \frac{1}{2}c_{DIC}(\phi)$$
(27.10)

(to be maximized). $\delta_\phi(y)$ on the lhs tends to be large if the data is influential, or if the prior on θ is weak and the model is complex.

(ii) **From left to right:** Which penalty on $p(y|\widehat{\theta}(y))$ is induced by a hyperprior? Rearranging terms in (27.10) yields

$$\log p(y|\phi) \approx \log p(y|\bar{\theta}_\phi) - \widehat{\delta}_\phi(y).$$
(27.11)

If a uniform hyperprior $p(\phi)$ is specified maximizing $p(y|\phi)$ with respect to ϕ amounts to maximum likelihood estimation of ϕ and hence to using the penalty $\widehat{\delta}_\phi(y)$ on $p(y|\widehat{\theta}(y))$. If, alternatively we specify a non-informative hyperprior, in particular, if we derive a reference hyperprior $\pi(\phi)$ (Bernardo and Smith [2], ch.5.4), does it turn out to be a(n inverse) function of some measure of model complexity ? A reference prior indicates whether and possibly to which extent $p(\phi) \propto \exp(-\gamma c(\phi))$ (or a criterion like DIC) can be regarded as corresponding to a non-informative hyperprior.

27.4   Reference priors on hyperparameters: two examples

A multivariate reference prior can be thought of as being sequentially built from (conditional) univariate Jeffreys' priors. Due to the choice of a sequence of one-dimensional parameters of increasing interest, the reference prior is not invariant under multidimensional one-to-one-transformations and depends on the chosen parameterization. If derived for hyperparameters ϕ, it is based on the marginal likelihood $p(y|\phi)$.

27.4.1   The Normal case

Assume $Y|\theta \sim N(\theta, \Sigma)$, $\Sigma > 0$ known, $\theta \sim N(0, \phi K)$, K known, such that marginally $Y|\phi \sim N(0, \Sigma + \phi K)$.

Model complexity

In this setup

$$\bar{\theta}_\phi = E(\theta|\phi, y) = (\Sigma^{-1} + (\phi K)^{-1})^{-1}\Sigma^{-1}y = A(\phi)y, \quad \text{say}$$

and

$$c_{DIC}(\phi) = \text{tr}A(\phi) = \text{tr}(I + \frac{1}{\phi}\Sigma K^{-1})^{-1} \tag{27.12}$$

is increasing in ϕ. Denoting the eigenvalues of ΣK^{-1} by κ_i

$$c_{DIC}(\phi) = \sum_{i=1}^{n} \frac{\phi}{\phi + \kappa_i} = \sum_{i=1}^{n} \alpha_i, \quad \text{say.} \tag{27.13}$$

Referring to (27.10) and (27.11) it is also of interest to evaluate

$$E_{Y|\phi}(\delta_\phi(y)) = E_{Y|\phi}[\delta(p(\theta|\phi, y)||p(\theta|\phi))] = \frac{1}{2}\sum_{i=1}^{n}\log\left(1 + \frac{\phi}{\kappa_i}\right) \tag{27.14}$$

which is again increasing in ϕ.

Model discrimination

If we choose $\phi_0 = 0$, a prior distribution concentrated on the prior mean of θ and hence being most informative about θ, we obtain

$$\begin{aligned}
\delta_{\phi\phi_0} &= \delta(p(y|\phi)||p(y|\phi_0)) \\
&= \frac{1}{2}\left[\sum_{i=1}^{n}(-\log(1 + \frac{\phi}{\kappa_i}) + (1 + \frac{\phi}{\kappa_i})) - n\right], \tag{27.15}
\end{aligned}$$

also increasing in ϕ. The discriminative parameter $\frac{\phi}{\kappa_i}$ resp. ϕ is the inverse smoothing parameter.

The reference prior

The reference prior for ϕ (and related parameters in Normal models with two variance components) were derived by van der Linde [8]. Here it is

$$\begin{aligned}
\pi(\phi) \propto (\text{tr}[(\Sigma + \phi K)^{-1}K]^2)^{\frac{1}{2}} &= \left(\sum_{i=1}^{n}\frac{1}{(\phi + \kappa_i)^2}\right)^{\frac{1}{2}} \\
&= \frac{1}{\phi}\left(\sum_{i=1}^{n}\alpha_i^2\right)^{\frac{1}{2}} \\
&= \left(\sum_{i=1}^{n}\frac{1}{\kappa_i}\frac{1}{(1 + \frac{\phi}{\kappa_i})^2}\right)^{\frac{1}{2}}. \tag{27.16}
\end{aligned}$$

It is decreasing in ϕ and thus assigns low probability to complex models. Comparison with (27.10) suggests that DIC in the Normal case may not correspond to using a reference prior. $\pi(\phi)$ is not explicitly a function of any measure of model complexity and not exactly inverse to such a measure, but in an involved way depending on similar terms as those measures.

27.4.2 One-parameter exponential families with conjugate prior (canonical parameterization)

As a prototype example consider

$$\text{independent } Y_i \sim B(1, \mu), \qquad Y = (Y_1 ... Y_n)', \qquad R := \sum_{i=1}^{n} Y_i$$

with canonical parameter

$$\theta = \text{logit } \mu = \log\left(\frac{\mu}{1-\mu}\right).$$

The conjugate distribution is

$$\mu \sim \text{Beta}(\alpha, \beta)$$

with hyperparameters $\phi = (\phi_1, \phi_2)$, $\phi_1 = \alpha$, $\phi_2 = \alpha + \beta$ corresponding to the canonical parameterization. For further interpretation also recall

$$m_\phi := E_\phi(\mu) = \frac{\alpha}{\alpha + \beta},$$

$$\text{var}_\phi(\mu) = m_\phi(1 - m_\phi)\rho \qquad \text{where} \qquad \rho = \frac{1}{\alpha + \beta + 1} \in (0, 1).$$

Model complexity

$c_{DIC}(\phi)$ can be calculated exactly (Spiegelhalter et al. [7]),

$$c_{DIC}(\phi) = 2n\psi(\alpha + \beta + n) - \log\left(e^{\psi(\beta+n-r)} + e^{\psi(\alpha+r)}\right), \qquad (27.17)$$

where ψ denotes the digamma function, $\psi(x) = (\log \Gamma(x))'$. In this case $c_{DIC}(\phi)$ depends on the data y resp. r, but it can be approximated by

$$c_{DIC}(\phi) \approx \frac{n}{\alpha + \beta + n} \qquad (27.18)$$

which, like the prior variance is a decreasing function of $\phi_2 = \alpha + \beta$.

Model discrimination

$\delta_{\phi\phi_0} = \delta(p(y|\phi)\|p(y|\phi_0))$ mainly depends on the ratio of the normalizing constants of the prior and posterior Beta-distribution.

$$\delta_{\phi\phi_0} = E_{Y|\phi} \log \frac{\omega(\overline{\phi}(y))}{\omega(\phi)} - \log \frac{\omega(\overline{\phi}_0(y))}{\omega(\phi_0)} \qquad (27.19)$$

where $\overline{\phi}(y) = (\alpha+r, \beta+n-r)$ parameterizes the posterior Beta-distribution of μ and

$$\omega(\phi) = \frac{\Gamma(\alpha)\Gamma(\beta)}{\Gamma(\alpha+\beta)}$$

denotes the normalizing constant. (The notation is extended in the obvious way to $\overline{\phi}(y)$, $\overline{\phi}_0(y)$.) (27.19) reminds of the ratio of posterior to prior variance approximating $c_{DIC}(\phi)$ (see (27.5)). Under the assumption of multivariate Normality of the likelihood and the prior one obtains

$$\begin{aligned}
c_{DIC}(\phi) &\approx \operatorname{tr}\left(\operatorname{cov}(\theta|\phi, y) I_{obs}(\overline{\theta}_\phi)\right) \\
&\approx \operatorname{tr}\left(\operatorname{cov}(\theta|\phi, y)\left[\operatorname{cov}(\theta|\phi, y)^{-1} - \operatorname{cov}(\theta|\phi)^{-1}\right]\right) \\
&= \operatorname{tr}\left(I - \operatorname{cov}(\theta|\phi, y)\operatorname{cov}(\theta|\phi)\right).
\end{aligned}$$

Reference priors

The findings reported here result from ongoing work with J.M.Bernardo (Valencia, Spain). A detailed paper will be published elsewhere.
The joint reference prior derived for the parameterization (m_ϕ, ϕ_2) is

$$\pi(m_\phi, \phi_2) \propto_{\text{approx}} \left(\frac{1}{m_\phi} \frac{1}{1-m_\phi} \frac{1}{\phi_2}\right)^{\frac{1}{2}} \frac{1}{1+\phi_2}. \tag{27.20}$$

The transformation of variables $(m_\phi, \phi_2) \to (m_\phi, \rho)$ yields

$$\pi(m_\phi, \rho) \propto_{\text{approx}} \operatorname{Beta}\left(m_\phi \Big| \frac{1}{2}, \frac{1}{2}\right) \operatorname{Beta}\left(\rho \Big| \frac{1}{2}, \frac{1}{2}\right).$$

Given a Bernoulli likelihood the $Beta(\frac{1}{2}, \frac{1}{2})$-distribution is the reference prior for the proportion. Its density is U-shaped, assigning higher probability to more extreme values. The same assignment occurs with respect to the dispersion factor ρ, indicating that the joint reference prior is not inverse to model complexity. Using yet another transformation of variables to (α, β), one obtains

$$\pi(\alpha, \beta) \propto_{\text{approx}} (\alpha\beta(\alpha+\beta)(\alpha+\beta+1))^{-\frac{1}{2}} \tag{27.21}$$

which varies along with, not inversely to $c_{DIC}(\phi)$. Hence in this case it is suggested that a prior of the form $p(\phi) \propto \exp(-\gamma c(\phi))$ is an informative prior.

27.5   Discussion

The proposal to define a hyperprior density as decreasing function of a measure of model complexity was investigated emphasizing three crucial points.

- A measure of model complexity is not always a suitable model discriminating transformation of a hyperparameter. In general, it depends on unknown parameters or the data. A competing transformation is the Kullback-Leibler distance between the densities of a model of interest and a (minimal) reference model. Furthermore it was pointed out that from a Bayesian point of view a measure of model complexity describes the distance of the posterior to the prior distribution of the parameter of interest in inference.

- Criteria of model choice of the form "goodness of fit - penalty" can be related to hyperprior distributions in various ways. If "goodness of fit" in terms of $\log p(y|\widehat{\theta}(y))$ is penalized, it is not obvious why the hyperprior in turn should act like a penalty on $\log p(y|\phi)$. It is of interest to understand whether such penalties induce uniform or non-informative hyperpriors.

- If non-informative hyperpriors are specified to be reference hyperpriors, their relation to quantities of model complexity can be examined at least in examples. Two examples were reported indicating that the reference hyperprior does depend on terms that also constitute measures of model complexity though not always in a way that the hyperprior assigns low probability to complex models. The issues are somewhat obscured by problems of mathematical tractability, parameterizations and approximations. The results also suggest that a hyperprior constructed as a decreasing function of a(n approximate) measure of model complexity may be informative.

Certainly further research is needed to provide more findings and links before any general conclusions can be drawn.

References

[1] Akaike, H. (1973). Information theory and an extension of the maximum likelihood principle. In *2nd Intl. Symp. on Information Theory*, (Eds. B. Petrow and F. Csaki), Akademiai Kiado, Budapest.

[2] Bernardo, J.M. and Smith, A.F.M. (1994). *Bayesian Theory*. Wiley: New York.

[3] Burnham, K.P. and Anderson, D.R. (1998). *Model Selection and Inference*. Springer: New York.

[4] Efron, B. (1986). How biased is the apparent error rate of a prediction rule ? *J. Amer. Statist. Ass.* **81**, 461-70.

[5] Hodges, J. and Sargent, D. (2001). Counting degrees of freedom in hierarchical and other richly parameterised models. *Biometrika* **88**, 367-379.

[6] Holmes, C.C. and Denison, D.G.T. (1999). Bayesian wavelet analysis with a model complexity prior. *Bayesian Statistics 6*, (Eds. J.M. Bernardo et al.), Oxford University Press, 769-776.

[7] Spiegelhalter, D., Best, N., Carlin, B., van der Linde, A. (2002). Bayesian measures of model complexity and fit. *J. Roy. Statist. Soc. B* **64** (3), 1-34 (without discussion).

[8] van der Linde, A. (2000). Reference priors for shrinkage and smoothing parameters, *J. Statist. Pl. Inf.* **90**, 245-274.

[9] Ye, J. (1988). On measuring and correcting the effects of data mining and model selection. *J. Amer. Statist. Ass.* **93**, 120-131.

28

A Strategy for Compression and Analysis of Very Large Remote Sensing Data Sets

Amy Braverman[1]

28.1 Introduction

NASA launched its first Earth Observing System (EOS) satellite, Terra, into polar orbit on December 18, 1999. Terra carries five instruments for studying Earth's climate systems over a six year period, and is producing vast quantities of data; more than what the user community is equipped to handle. In geoscience, traditional strategies for coping with this problem are two-fold. The first is to work only with spatio-temporal subsets. The second is to work with low-resolution summaries typically created by partitioning data for a specified time period into 1° latitude by 1° longitude regions, and summarizing each region by its mean and standard deviation. The first strategy fails to take advantage of the global nature of Terra data, and requires researchers to know ahead of time where interesting phenomena exist. The second strategy fails to capture multivariate structure, and may aggregate away important high-resolution features.

This chapter describes a method for summarizing these data in a way that approximately preserves high-resolution data structure while reducing data volume and maintaining global integrity of very large, remote sensing data sets. The method is under development for one of Terra's instruments, the Multi-angle Imaging SpectroRadiometer (MISR). The strategy is to partition data for each month into 1° latitude by 1° longitude spatial

---

[1]Amy Braverman is with the Jet Propulsion Laboratory and the California Institute of Technology. The author would like to thank the referee and editors for their careful and helpful comments. This work was performed at the Jet Propulsion Laboratory, California Institute of Technology, under a contract with the National Aeronautics and Space Administration.

cells, and summarize each cell with a set of clusters. Each cluster is represented by its centroid and the number of original data points it contains. The combination of cluster centroids and counts is a compressed version, or summary, of the original data. Researchers wishing to conduct global, exploratory analysis can do so using the compressed data with the understanding that results should be confirmed using appropriate portions of the original data.

The algorithm used to construct these summaries is a modification of the Entropy-constrained Vector Quantization algorithm (ECVQ) of Chou, Lookabaugh and Gray [1], and is described in Section 28.3. Section 28.2 describes the MISR data stream, and Section 28.4 provides an example analysis using compressed MISR data.

28.2   MISR Data

MISR (Diner et. al. [2]) is a set of nine cameras mounted underneath Terra looking down at Earth at nine view angles: 70.5°, 60.0°, 45.6°, and 26.1° aft; 0° (nadir), and 70.5°, 60.0°, 45.6°, and 26.1° forward along the spacecraft's north-south flight path. Each camera has four line arrays each of which is comprised of 1504 pixels across the field of view. The line arrays are each sensitive to one of four wavelengths: blue, green, red, and NIR (446, 558, 672 and 866 nanometers), and each pixel views a square ground footprint 275 × 275(meters)^2. Thus, one orbital swath on the daylight side of Earth tiles the view into disjoint, contiguous 275 meter spatial regions, and produces 36 nearly simultaneous radiance measurements for each one. Data for the nadir camera and the red bands in the other cameras are transmitted to Earth at full 275 meter resolution. Data for all other channels are averaged up to 1.1 km resolution on-board the spacecraft to limit data rate. The instrument does not take data as the satellite travels up the night side, so sequential orbits are geographically separated. After 16 days 233 overlapping orbits have completed covering the whole Earth. Every 234th orbit covers the same ground track as the first to within 20 kilometers.

Several steps are taken to process these radiance data. First, they are geometrically and radiometrically calibrated to create the so-called Level 1B2 product. Then Level 2 data are created by converting Level 1B2 data into geophysical quantities. Radiometric calibration is necessary to adjust for differences in camera sensitivities. Geometric calibration is necessary because there is a seven minute lag between the forward-most and aft-most views of the same scene. Geometric calibration aligns the observations so that 36 nearly simultaneous measurements (nine angles by four wavelengths) are associated with the latitude and longitude of each pixel center. The time lag does result in slight temporal inconsistencies, but these do not have serious scientific consequences given MISR's relatively coarse temporal sampling. MISR produces about 40 GB per day of Level 1B2 data.

Level 2 data sets are created from Level 1B2 by applying science algorithms. These are based on the photon scattering properties of various media when viewed at different wavelengths and from different angles. For example, measurements taken within a 17.6 kilometer area are used to derive aerosol type and amount by matching observed radiances with those predicted by various physical models. Other quantities such as cloud height, wind direction and speed, and surface properties are derived at other spatial resolutions, typically 1.1, 2.2, and 35.2 kilometers. Level 2 processing generates about 20 GB per day of derived geophysical data.

The third processing step creates Level 3 monthly summaries by partitioning observations according to their membership in cells of a 1° latitude by 1° longitude spatial grid, and summarizing by grid cell. At the time of this writing the intention is to routinely produce compressed Level 3 data products derived from a select set of Level 2 geophysical variables. However, the method for producing grid cell summaries is also applied to portions of the Level 1B2 radiance data for research and analysis purposes. One such application is the topic of Section 28.4.

28.3 Monte Carlo Extended ECVQ

The method used to create grid cell summaries is based on the Entropy-constrained Vector Quantization algorithm. ECVQ is a practical algorithm for summarizing data in a way that balances two competing goals: to remain faithful to the original data, and to achieve data compression. ECVQ was originally introduced as a means of estimating distortion-rate functions [1] of stochastic information sources. A distortion-rate function describes the trade-off between data compression and statistical information loss in quantizing a stream of signals for data storage. The same paradigm is applied here in quantizing a data set for the purpose of summarizing it. The data set is the information source, and data compression is measured by the reduction in entropy achieved by the summary. Statistical information loss is measured by the mean squared error between the original data and the summary.

ECVQ is an iterative algorithm that solves the following constrained optimization problem. Suppose one has a discrete probability distribution with support points $\{y_n\}_{n=1}^N$ and associated probabilities $\{\pi_n\}_{n=1}^N$. How does one combine support points into clusters so that (1) the mean squared error between the y_n's and the centroids of their clusters is minimized, while (2) ensuring that the entropy of the distribution defined by the resulting clusters does not exceed a prescribed level? If p_k is the mass associated with cluster k, $k = 1, 2, \ldots, K$, then the entropy of the distribution defined by the clusters is

$$h = -\sum_{k=1}^{K} p_k \log p_k.$$

If $q(y_n)$ denotes the centroid of the cluster to which y_n is assigned, then the mean squared error of the clustering is

$$\delta = \sum_{n=1}^{N} \|y_n - q(y_n)\|^2.$$

δ is also called distortion.

ECVQ finds a solution using the Lagrange multiplier λ and the objective function

$$L_\lambda = \sum_{n=1}^{N} \left\{ \|y_n - q(y_n)\|^2 + \lambda \left[-\log p(y_n) \right] \right\}, \tag{28.1}$$

where here $p(y_n)$ is understood to be the probability associated with cluster to which y_n is assigned. λ controls the level of compression. High values put a premium on the penalty $-\log N(k)/N$, cause summaries to collapse down to fewer, more highly concentrated clusters, and result in higher mean squared errors. Low values of λ usually result in greater numbers of clusters, higher entropies, and lower mean squared errors. Since entropy measures descriptive complexity of the cluster distribution, λ parameterizes the trade-off between distortion and complexity.

When ECVQ is used to summarize a data set, the y_n's are data vectors, and $\pi_n = 1/N$ for all n. In describing the algorithm it will be helpful to change notation slightly. Let $\alpha(y_n)$ be an integer indexing the cluster to which y_n is assigned, let $\beta(k)$ be the centroid of the cluster indexed by k, and let $N(\alpha(y_n))$ be the number of data vectors assigned to y_n's cluster. Equation 28.1 can be written

$$L_\lambda = \sum_{n=1}^{N} \left\{ \|y_n - \beta[\alpha(y_n)]\|^2 + \lambda \left[-\log \frac{N[\alpha(y_n)]}{N} \right] \right\}. \tag{28.2}$$

$-\log(N(k)/N)$ is positive and varies inversely with $N(k)$. Thus, even if $\|y_n - \beta(k_1)\|^2 > \|y_n - \beta(k_2)\|^2$, y_n could be assigned to cluster k_1 if the difference in the terms involving logarithms compensates. If $\lambda = 0$, L_λ is euclidian distance, and ECVQ is equivalent to the batch version of the K-means clustering procedure (MacQueen [3]).

Briefly, the ECVQ algorithm works as follows:

1. Fix the maximum number of clusters allowed, K, and the compression parameter, λ.

2. Arbitrarily assign the y_n's to the K clusters by specifying initial values for $\alpha(y_n)$. Compute centroids and counts of these clusters, and denote them $\beta(k)$ and $N(k)$ respectively, for $k = 1, 2, \ldots, K$.

3. Reassign each y_n to the cluster with the smallest loss:

$$\alpha(y_n) = \arg\min_k \left\{ \|y_n - \beta(k)\|^2 + \lambda \left[-\log \frac{N(k)}{N} \right] \right\}.$$

4. Update $\beta(k)$ and $N(k)$ for all k.

5. Eliminate any clusters for which $N(k) = 0$.

6. Repeat steps (3), (4) and (5) until convergence.

The ECVQ solution has the property that the $\beta(k)$'s are the means of the y_n's they represent, a property call self-consistency by Tarpey and Flury [4]. However, assignment of data points to clusters does not minimize mean squared error since some data points may not be assigned to the cluster with the nearest euclidian distance centroid.

The algorithm is guaranteed to converge in a finite number of steps, but not necessarily to either a local or global minimum [1]. However, the solution does improve on the starting point providing a sensible summary of the y_n's in the sense described by MacQueen: "The point of view taken in this application is *not* to find some unique, definitive grouping, but rather to simply aid the investigator in obtaining qualitative and quantitative understanding of large amounts of ... data by providing him with reasonably good similarity groups." ([3], page 288.)

To apply ECVQ to large quantities of geophysical data, two modifications are made. First, since the algorithm is $O(n^2)$ and cell populations are large, a sample of data points from the cell being summarized is selected. The sample size, M, is chosen as large as possible given computational constraints and in view of the total number of cells. ECVQ is applied to the sample, and an initial set of centroids, $\{\beta^*(k)\}_{k=1}^{K^*}$, obtained. This is the design step. Then each original data point in the cell is assigned to its nearest euclidian distance $\beta^*(k)$, empty clusters are deleted, and centroids and counts updated. This is the binning step. In other words, a preliminary set of representatives is determined from a sample, then the entire cell data set clustered using it. The ultimate set of centroids and counts thus reflects all the data, and also minimizes mean squared error. The resulting summary is denoted $\left\{ \tilde{\beta}(k), \tilde{N}(k) \right\}_{k=1}^{\tilde{K}}$. This modification constitutes the Extended ECVQ (EECVQ) procedure.

The second modification addresses the fact that EECVQ is sample dependent, and $\left\{ \tilde{\beta}(k), \tilde{N}(k) \right\}_{k=1}^{\tilde{K}}$ is subject to sampling variation. To account for this EECVQ is repeated a number of times, say S times, using a different random sample of size M in the design step each time. As with M, S is chosen as large as possible, but more often than not there is a trade-off to be made between S and M. The latter should be large enough to be representative of the data being summarized, and the former should be large enough to provide a diverse set of starting points for the ECVQ algorithm. When there is a conflict, we err on the side of large M. This produces S summaries of the cell data, each one having a mean squared error δ_s. The best summary is the one having the smallest δ: $s_{opt} = argmin_s\{\delta_s\}_{s=1}^{S}$, and $\left\{\tilde{\beta}_{s_{opt}}(k), \tilde{N}_{s_{opt}}(k)\right\}_{k=1}^{\tilde{K}_{s_{opt}}}$ is selected to represent the original data. $\delta_{s_{opt}}$ is reported as a goodness of fit measure, and the entropy of the best summary,

$$h_{s_{opt}} = -\sum_{k=1}^{\tilde{K}} \frac{\tilde{N}_{opt}(k)}{N} \log \frac{\tilde{N}_{opt}(k)}{N},$$

is reported as a measure of descriptive complexity of the underlying data. Average mean squared error over trials, $\bar{\delta} = S^{-1} \sum_{s=1}^{S} \delta_s$, is also reported as an overall figure of merit. This procedure that embeds EECVQ in a Monte Carlo simulation is called Monte Carlo Extended ECVQ (MCEECVQ).

Finally, a value of λ must be selected. Choosing λ for any cell in isolation amounts to deciding how much compression one wants to achieve in that cell beyond that assured by fixing the initial number of clusters, K, and how much mean squared error one is willing to tolerate. When summarizing many cells in concert, it is desirable that distortions be as equal as possible across cells so differences in summaries reflect differences in data they represent, not differences how well summaries fit those data. As a consequence of the trade-off between mean squared error and entropy, this tends to produce summaries with entropies that reflect concentrations of mass in underlying empirical distributions. Figure 28.1 illustrates a simple example. The top two panels show two data sets drawn from mixtures of bivariate normal distributions. The middle panels show those data summarized using five clusters in each case: $K = 5$, $\lambda = 0$. Data from the top panels are shown on plot floors, the positions of the spike show locations of cluster representatives, and spike heights show cluster counts. Bottom panels show how these data are summarized by ECVQ with $K = 5$ and $\lambda = .04$. In the $\lambda = .04$ regime, high density regions are reflected by fewer, more massive clusters. The sums of squared distances between points and their nearest cluster centroids are more nearly equal in the bottom panels than in the middle ones.

In practice one selects a value of K to limit the maximum size of the MCEECVQ output to $K \times B$, where B is the number of cells being summarized. This determines an overall level of mean squared error. Then, one selects λ to minimize the variance of $\bar{\delta}$'s across cells:

$$Var(\bar{\delta}) = \frac{1}{B} \sum_{b=1}^{B} (\bar{\delta}_b - \hat{\delta})^2,$$

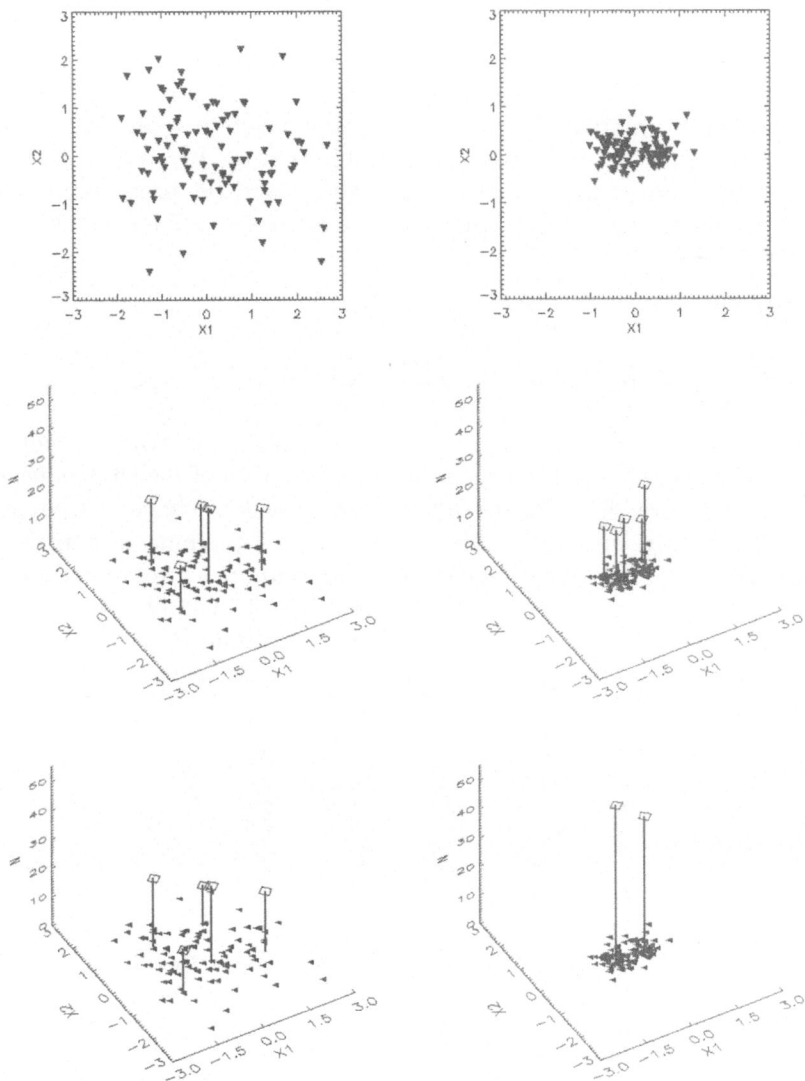

Figure 28.1. Top left: 100 observations from $f(x) = \pi_1 f_1(x) + \pi_2 f_2(x)$ where f_1 and f_2 are independent bivariate normals with $\mu_1 = (-.5, 0)$, $\mu_2 = (+.5, 0)$ and $\sigma_1 = \sigma_2 = 1$, and $\pi_1 = \pi_2 = .5$. Top right: same as top left except $\sigma_1 = \sigma_2 = .3$. Middle left and right: summaries of data with $K = 5$ and $\lambda = 0$. Bottom left and right: summaries of data with $K = 5$ and $\lambda = .04$.

where b indexes cell and $\hat{\bar{\delta}} = B^{-1} \sum_{b=1}^{B} \bar{\delta}_b$. This requires testing various values of λ beforehand, and can be computationally intensive. If necessary this can be done using a subset of cells. The procedure is to begin testing

λ values in the range $[0, 1]$. Zero is a natural lower bound, and one is the value of λ that makes one unit of distortion worth as much as one bit of information. The number of values tested is ideally large, but in practice it seems that nine intermediate values suffice. If the variation in the distortion across cells is minimized by $\lambda \in (0, 1)$, that value of λ is selected. If the variation in distortion across cells is minimized by $\lambda = 0$ or 1, the range is refined to either $[0, .1]$ or $[1, 2]$ respectively, and the procedure repeated with the new range. This continues until a distortion-minimizing λ is found in the interior of the range.

28.4   Application to MISR Data

To illustrate MCEECVQ, it is applied to MISR Level 1B2 data collected over the eastern United States on March 6, 2000. 15 of the 36 radiances are shown in Figure 28.2. The five panels shows the scene in three (red, green, blue) of the four spectral bands and from five of the nine view angles. All data used for this exercise have been averaged up to 1.1 km resolution. The data set partially depicted in Figure 28.2 has 491,044 observations, each representing a 1.1 km spatial region, and 38 columns: one for each view angle-spectral band combination, and latitude and longitude of the pixel center.

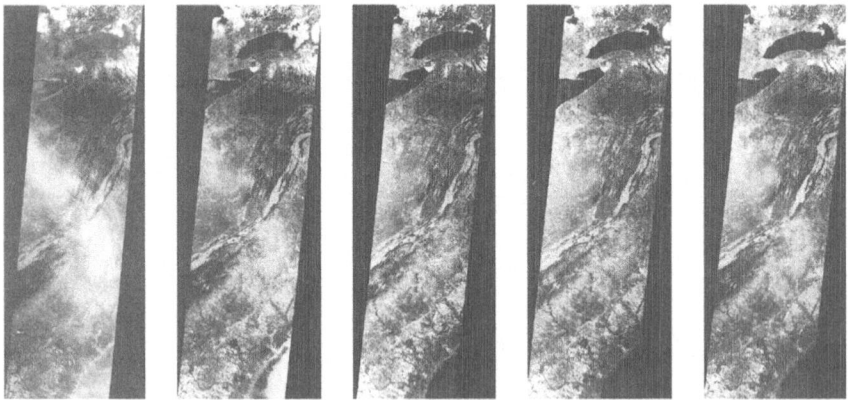

Figure 28.2. Left to right: 70.5° forward, 45.6° forward, nadir, 45.6° aft, and 70.5° aft.

A frequent objective in the analysis of remote sensing data is to classify pixels in a scene. In Figure 28.2 water, ice (in the far eastern end of Lake Erie), clouds (over central New York state), vegetated land of various types and terrain, and haze are all evident. MISR's multi-angle observations provide a novel kind of information useful for these classifications. For example, haze is more obvious in the oblique angles because these views represent longer paths through the atmosphere. The combination of 36 radiances is expected to provided better discriminatory power than single view-angle, multi-spectral data.

One way to classify the scene in Figure 28.2 would be to run a K-means cluster analysis on all 491,044 pixels. Even choosing a modest value for K of, say ten, is computationally intensive and time consuming. Instead, the following procedure was used. First, the data were partitioned into 84 $1° \times 1°$ strata, and each strata summarized using MCEECVQ. MCEECVQ was applied with $K = 10$, $\lambda = 2$, sample size $M = 200$, and $S = 50$ trials. These values were chosen to balance the need for representative samples in a relatively high-dimensional space with the need for sufficient number of random starting points for the algorithm. The samples used in the design step were first standardized using the grand means and variances for all 491,044 data points, and then projected into the space of the first ten principal components calculated from the grand correlation matrix. Ten principal components account for over 98 percent of the total variation in the data. MCEECVQ was applied to the transformed data, and representatives were re-transformed back to the original 36-dimensional data space before the binning step. $\lambda = 2$ was chosen after testing 15 values ($\lambda = 0, .5, 1, \ldots, 6.5, 7$) and determining $\lambda = 2$ minimized $Var(\bar{\delta})$ across the 84 strata in the reduced space. This produced between one and ten representatives and associated counts for each spatial cell.

The MCEECVQ output contained a total of 479 representatives and counts, and 84 associated values of $\delta_{s_{opt}}$, $h_{s_{opt}}$ and $\bar{\delta}$. $\delta_{s_{opt}}$'s are the mean squared errors between grid cell summaries and the data they summarize. They measure the lost statistical information due to quantization. They are shown, by cell, in the second panel of Figure 28.3 relative to the average squared magnitude of the cell data. That is, each cell is color-coded on the to reflect $\delta_{s_{opt}}/N^{-1} \sum \|y_i\|^2$, where the sum in the denominator is over all N pixels with centers inside the cell. The first (leftmost) panel of Figure 28.3 shows average distortion, $\bar{\delta}$, over the $S = 50$ repetitions of the design step, relative to $N^{-1} \sum \|y_i\|^2$. This can be thought of as a process performance measure, since it incorporates both summary accuracy and variability. The difference between the two panels is slight, indicating that most of the error in the summaries is due to quantization rather than algorithm instability. The center panel in Figure 28.3 shows entropy of the best summary in each cell. It is highest in the same cells where distortion is high indicating these cells are difficult to summarize well even when the summaries have relatively high information content. The final two panels in the figure are the numbers of clusters, $\tilde{K}$, and the number of original data vectors in each grid cell. Note that cells with the largest values of N are not necessarily those with the largest numbers of clusters.

Next, a weighted K-means analysis with $K = 10$ was applied to the 479 representatives. Here, the 479 cluster centroids serve as data

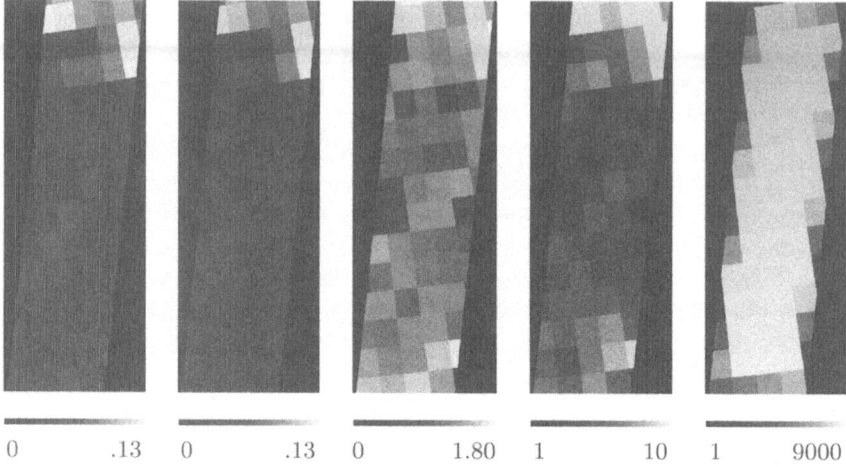

0   .13   0   .13   0   1.80   1   10   1   9000

Figure 28.3. Left to right: Average cell mean squared error (over S trials in the MCEECVQ simulation) as a proportion of the average squared norm of data points in the cell; best cell summary mean squared error as a proportion of the average squared norm of data points in the cell; entropy of the cell summary; number of clusters in the cell; summary; cell population.

points to be clustered. Combining representatives and counts to form $\left\{\tilde{\beta}_{s_{opt}}(k), \tilde{N}_{s_{opt}}(k)\right\}_{k=1}^{479}$, an initial set of ten representative is selected at random with probabilities $\tilde{N}_{s_{opt}}(k)/\sum_{k=1}^{479} \tilde{N}_{s_{opt}}(k)$. These serve as the initial "supercluster" centroids, and each of the 479 cluster means is then assigned to the supercluster with the nearest centroid. The following steps are iterated until convergence: (1) supercluster centroids are updated by computing the weighted averages of members with weights proportional to $\tilde{N}_{s_{opt}}(k)$, and (2) the 479 cluster means are reassigned to the nearest euclidian distance supercluster centroid. The weighted K-means procedure was repeated 50 times, and the solution with the smallest weighted mean squared error adopted. Weighted mean squared error is

$$\delta_{wtd} = \sum_{k=1}^{479} \|\tilde{\beta}_{s_{opt}}(k) - q(\tilde{\beta}_{s_{opt}}(k))\|^2 \frac{\tilde{N}_{s_{opt}}(k)}{\sum_{k=1}^{479} \tilde{N}_{s_{opt}}(k)},$$

where $q(\tilde{\beta}_{s_{opt}}(k))$ is the centroid of the cluster to which $\tilde{\beta}_{s_{opt}}(k)$ is assigned in this second-stage K-means analysis. The resulting $K = 10$ supercluster centroids are taken as the ten types used to classify the scene.

Finally, each of the original 491,044 data points is assigned an integer between one and ten indicating which of the ten superclusters it is nearest in euclidian distance. The map in Figure 28.4 shows the resulting classification of the scene. The classification identifies the band of haziness (in red) that is barely visible in the nadir image but more apparent in the 70° forward

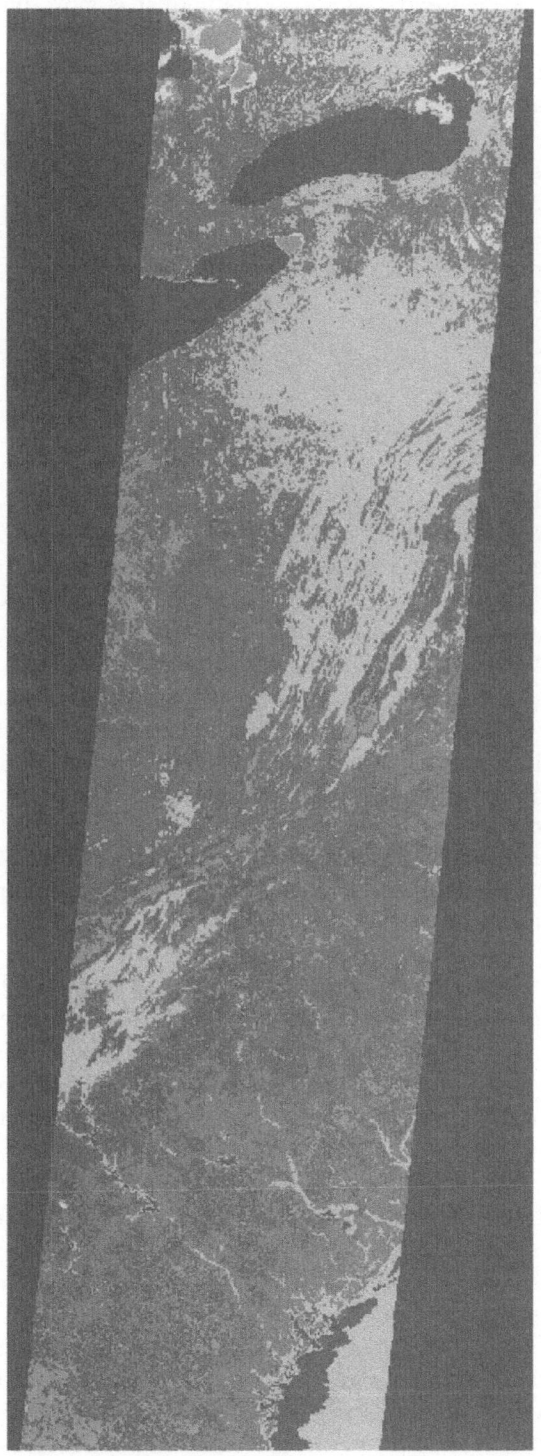

Figure 28.4. MISR thematic classification map of the Appalachian scene.

view in Figure 28.2. Figure 28.4 also distinguishes sun glint on water off the southeast coast and in Lake Ontario seen in the 45° forward view, ice at the extreme east end of Lake Erie, and ice and clouds over Lake Simcoe in the upper left corners of the images in Figure 28.2. The ability to differentiate between pixel types that are indistinguishable at a single nadir view highlights the principal behind multi-angle imaging. This K-means analysis is an example of a procedure scientists are interested in conducting on data of this type, but which may be impractical if not for a volume-reduced version of the data that approximately preserve high-dimensional relationships.

28.5 Summary and Conclusions

This chapter describes a randomized version of the ECVQ algorithm for creating compressed versions of large geophysical data sets. The technique is especially well suited to remote sensing data such as that obtained from MISR because the method is easily parallelized, and operates under a range of algorithm parameter settings. Thus, the trade-off between accuracy and computational speed can be customized. Remote sensing data are also especially appropriate for this technique because they are naturally stratified by geographic location, and have strong high-dimensional structure. Natural stratification makes it easier to understand the relationships between summaries of various spatial regions, and high-dimensional structure allows for dimension reduction.

The technique is demonstrated by partitioning a test MISR data set according to membership in 1° latitude by 1° longitude spatial regions, and compressing data in each region. The compressed data are a set of representative vectors and associated counts which can be thought of as multivariate histograms with variable numbers of bins, and bins with sizes and shapes that adapt to the shape of the data in high-dimensional space. The algorithm is applied to all regions using common values of algorithm parameters K and λ. K specifies the maximum number of representatives and is set to limit the size of the output to no more than K times the number of spatial regions. This determines the overall level of error between the summaries and their parent data. λ sets the level of compression over and above that resulting from the choice of K, and is selected so that the quality of the summaries is as uniform as possible. Compressed data are then used in place of original data in a cluster analysis to create a thematic map of the Appalachian region of the US as seen by MISR.

This exercise was performed on a relatively small amount of test data. Samples were used in the design step, but the full test data set is used for binning in each trial of the simulation. This requires $S+2$ passes through all the data: one to collect samples, one for each trial, and one to finally bin the data. In larger applications, it may not be possible to scan the full data set multiple times. Another version of the algorithm is under development to address this problem. Also, choosing λ requires testing multiple candidate values, 15 in this case. This necessitates running MCEECVQ 15 times on each 1° latitude by 1° data set. Again, the strategy may not be practical for larger applications. An alternative is to test for the best λ on a subset of 1° latitude by 1° regions. For example, in a global application those regions indexed by latitudes and longitudes evenly divisible by five or ten degrees could be used.

The method described here is on its way to becoming an operational algorithm for compressing a portion of MISR Level 2 geophysical data products on a monthly basis. To a large extent, its efficiency and effectiveness will depend on the data themselves. How well MCEECVQ scales up, and what further modifications are necessary remain to be seen.

References

[1] Chou, P.A., Lookabaugh, T., and Gray, R.M. (1989), "Entropy-constrained Vector Quantization," *IEEE Transactions on Acoustics, Speech, and Signal Processing*, **37**, 31-42.

[2] Diner, David J., Beckert, Jewel C., Reilly, Terrance H., Bruegge, Carol J., Conel, James E., Kahn, Ralph H., Martonchik, John V., Ackerman, Thomas P., Davies, Roger, Gerstl, Siegfried A. W., Gordon, Howard R., Muller, Jan-Peter, Myeni, Ranga B., Sellers, Piers J., Pinty, Bernard, and Verstrate, Michel M. (1998), "Multi-angle Imaging SpectroRadiometer (MISR) Instrument Description and Experiment Overview," *IEEE Transactions on Geoscience and Remote Sensing*, **36**, 4, 1072-1087.

[3] MacQueen, James B. (1967), "Some Methods for Classification and Analysis of Multivariate Observations," *Proceedings of the Fifth Berkeley Symposium on Mathematical Statistics and Probability*, **1**, 281-296.

[4] Tarpey, Thaddeus and Flury, Bernard (1996), "Self-Consistency: A Fundamental Concept in Statistics," *Statistical Science*, **11**, 3, 229-243.

29

Targeted Clustering of Nonlinearly Transformed Gaussians

Juan K. Lin[1]

29.1 Introduction

In this manuscript we present research which touches upon recent exciting developments in the fields of machine learning, theoretical computer science, and robust statistics. In particular, we focus on model-based nonlinear descriptions of local clustered data. Two main aspects are presented in this chapter. First we introduce a multivariate statistical model consisting of non-linear volume preserving transformations of the multivariate normal. Second, we describe a robust local fitting algorithm for extracting individual non-linearly transformed Gaussian clusters.

29.2 Nonlinear volume preserving transformations of the multivariate normal distribution

We begin by introducing an extension of the multivariate normal distribution, first presented in Lin & Dayan [3], which is characterized by a non-linear volume preserving transformation. Consider a multivariate probability density function in the exponential family of the form

$$f(\vec{x}; \vec{\theta}) \propto \exp[p(\vec{x}, \vec{\theta})], \qquad \vec{x} \in \Re^n \tag{29.1}$$

where $p(\vec{x}, \vec{\theta})$ is a polynomial in $\{x_1, \ldots, x_n\}$ with coefficients which are functions of the parameters $\vec{\theta}$.

A familiar example is obtained if we take $p(\vec{x}, \vec{\theta})$ to be a negative definite quadratic form. The random vector $\vec{X}$ in this case is multivariate normal.

---

[1]Juan K. Lin is Assistant Professor, Department of Statistics, Rutgers University, Piscataway, NJ 08854 (Email: jklin@stat.rutgers.edu).

Since $f(\vec{x}; \vec{\theta})$ is assumed to be a probability density function, care must be taken to insure that it is indeed normalizable. More specifically, an arbitrary polynomial $p(\vec{x}, \vec{\theta})$ in the exponent is not guaranteed to generate a proper normalizable p.d.f. For example, if the polynomial $p(\vec{x}, \vec{\theta})$ is an indefinite quadratic form, it would be impossible to normalize $f(\vec{x}; \vec{\theta})$ through the adjustment of a multiplicative proportionality constant.

In the field of smoothing and splines, often polynomials are fit to the log-density, in essence, directly estimating the polynomial $p(\vec{x}, \vec{\theta})$ in the exponent. The normalization issue is addressed by the construction of the polynomial $p(\vec{x}, \vec{\theta})$ using basis functions over finite regions, as delineated by knots, which have proper limiting behaviors in the large and small x limits. The basis functions must satisfy the equivalent of a normalizability condition.

Our method of addressing the normalization problem in models with p.d.f.'s of the form $f(\vec{x}; \vec{\theta}) \propto \exp[p(\vec{x}, \vec{\theta})]$ is to consider only invertible, volume-preserving transformations of a multivariate normal random vector with diagonal covariance, $\vec{\varphi} \sim N(0, D)$ of the form

$$
\begin{aligned}
\varphi_1 &= X_1 - c_1 \\
\varphi_2 &= X_2 - m_2(X_1) \\
&\vdots \\
\varphi_n &= X_n - m_n(X_1, \ldots, X_{n-1})
\end{aligned}
$$

where $m_i(X_1, \ldots, X_{i-1})$ is a polynomial in $\{X_1, \ldots, X_{i-1})\}$. The Jacobian of the transformation is strictly diagonal, with ones along the diagonal. Thus, the determinant of the Jacobian is one, and the transformation is volume-preserving by construction. We use the notation

$$
\mathcal{J} \cdot \vec{X} = \vec{\varphi} \tag{29.2}
$$

to represent the relationship between the random vectors $\vec{X}$ and $\vec{\varphi}$, where $\mathcal{J}$ is an invertible volume-preserving transformation of the form described above. Analytic solutions for the maximum likelihood estimators of the transformation parameters were derived in Lin and Dayan [3].

In two dimensions, it is known that *all* invertible polynomial transformations can be written as the so called "amalgamated product"

$$
\mathcal{A}_1 \cdot \mathcal{J}_1 \cdots \mathcal{A}_j \cdot \mathcal{J}_j \tag{29.3}
$$

of affine transformations $\mathcal{A}_1, \ldots, \mathcal{A}_j$ and invertible volume-preserving transformations $\mathcal{J}_1, \ldots, \mathcal{J}_j$ of the form described above. In higher dimensions, it is not known whether all invertible polynomial transformations are of this form.

We consider successively more complex polynomials for $m_i(X_1, \ldots, X_{i-1})$.

29.2.1   Quadratic Terms

Let $\vec{\varphi}$ be multivariate normal with diagonal covariance D. Consider the transformation model $\vec{\varphi} = K\vec{X} - \vec{c}$, where the strictly lower diagonal matrix K has components $[K]_{ij} = k_{ij}$ if $i > j$, and is zero otherwise. This model was first presented in Lin & Dayan [3].

These so called "Curved Gaussian Models" contain parameters in the transformation matrix K which effectively capture non-linear curvature information in the data. Unfortunately, these models do not contain multivariate normal distributions with arbitrary covariance structure.

29.2.2   Linear and Quadratic Terms

In this chapter, we consider more general volume preserving polynomial transformations of the multivariate normal distribution with diagonal covariance structure, as defined through the generating model

$$\vec{\varphi} = L\vec{X} + K\vec{X}^2 - \vec{c},$$

where $\vec{X}^2 = (X_1^2, \ldots, X_n^2)^T$, and both L and K are strictly lower diagonal matrices. Just as with the case where only quadratic terms are considered, solutions for the maximum likelihood estimators are still analytically determined from the sufficient statistics.

This transformation model is still of the form $\mathcal{J} \cdot \vec{X} = \vec{\varphi}$, though now all multivariate normal distributions with arbitrary covariance structure are contained as special cases. By letting K be the matrix of zeros and considering only the linear terms in L, the model simplifies to $\vec{\varphi} = L\vec{X} - \vec{c}$. Since $\vec{\varphi} \sim N(0, D)$, therefore $\vec{X} \sim N(L^{-1}\vec{c}, (L^T DL)^{-1})$. This can be seen to be an alternative transformation model of the multivariate normal with arbitrary mean and covariance. Instead of generating a multivariate normal with arbitrary mean and covariance structure through a dilation and subsequent rotation of a standard multivariate normal, this model generates the general multivariate normal structure through a dilation and subsequent shear of a standard multivariate normal. This corresponds to a LDL^T (lower-diagonal-upper) factorization of a symmetric matrix. There are $(n^2 - n)/2$ lower diagonal elements in L, which along with the n parameters in the diagonal matrix D, give the correct number of degrees of freedom contained in a covariance matrix. (see e.g. [7])

This more general model combines the linear shearing transformation which generates all multivariate normal distributions, with some higher order polynomial transformations which effectively capture non-linear aspects of the data.

29.3   Local Fitting Problem

Since there are analytic solutions for the maximum likelihood estimators of the parameters in the transformation model, the EM algorithm can be used for fitting mixtures of non-linearly transformed Gaussians. Here we introduce a local fitting algorithm for extracting individual non-linearly transformed Gaussians. The goal is to be able to extract out an individual cluster, perhaps based on prior location and spread information, without having to fit a mixture model to all of the data.

The algorithm is motivated by two main applications.

1. In gene expression data, there is interest in finding other genes belonging to the same gene function cluster. For example, given one gene which has been implicated in causing breast cancer, other genes in the same gene function cluster can potentially be targeted for cancer treatment.

2. In location based data such as personal GPS or cell phone call detail record (CDR), the data consists of an unknown number of location clusters at various length scales, connected by approximately uniformly distributed data along the roads. Fitting a mixture model with the EM algorithm to try to model all possible location clusters for an individual can easily run into local maxima. On the other hand, algorithms designed to extract individual clusters can be useful for finding important location clusters such as an individual's home, office, favorite restaurants, and shopping areas.

29.3.1   Individual Cluster Extraction Algorithm

The local cluster extraction algorithm is as follows. At time step t

1. Compute the sphered coordinate (non-linear extension of Mahanabolis distance)

$$\mathcal{D}_{(t-1)}(\vec{x}) = \vec{\varphi}_{(t-1)}^T D_{(t-1)}^{-1} \vec{\varphi}_{(t-1)} \qquad (29.4)$$

where

$$\varphi(\vec{t-1}) = L_{(t-1)}\vec{x} + K_{(t-1)}\vec{x}^2 - \vec{c}_{(t-1)}, \qquad (29.5)$$

and define the region of interest

$$S_{(t)} = \{\vec{x} : \mathcal{D}_{(t-1)}(\vec{x}) < \gamma\}. \qquad (29.6)$$

The region of interest $S_{(t)}$ essentially defines a weighting function which weights data inside the region by 1 and those outside by 0.

2. Find the maximum likelihood estimators for $L_{(t)}$, $K_{(t)}$, $D_{(t)}$ and $\vec{c}_{(t)}$ for the subset of the data in $S_{(t)}$.

These steps are iterated until either strict convergence is achieved, or a suitable convergence criteria is satisfied. Some numerics showing convergence, and robustness of the cluster finding algorithm to other clusters are shown in Figure 29.1. The successive best fit models to the expanding local regions of interest are represented in the figure as expanding curved ellipses. Since at each step, only the data within the curved regions of interest are being considered, the cluster extraction algorithm is robust to data outside of the converged region of interest. Explicitly, if no new data points are introduced in the new region of interest, the algorithm converges.

Intuitively, the local cluster extraction algorithm consists of local transformed Gaussian kernels, each of which are iteratively updated by:

1. Fitting a transformed Gaussian to the data within the local region of interest

2. Expanding the local region of interest in accordance with the fitted model

This process of expanding the observation in accordance to the model, and adjusting the model in response to updated observations is repeated until the model parameters converge with respect to the iterative dynamics.

The local cluster extraction algorithm described above captures a natural interaction between observation and prediction by iteratively using a model of the data within a localized region to predict the structure in an expanded region of interest. Instead of directly seeking a global model of the entire set of observed datapoints, we consider only the data within a local region of interest.

29.3.2   Choice of parameters and convergence issues

The parameters in the local inference algorithm consist of the initial region of interest $S_{(1)}$, and γ. Since the goal is to find individual Gaussian, or Gaussian-like clusters, care must be taken in choosing γ. Locally, Gaussians, or in fact any distribution, looks uniform. Thus, we wish successive iterations of the algorithm to expand the region of interest when the data within the current region of interest looks uniformly distributed.

From the calculation in the Appendix, a random vector uniformly distributed within a ball of radius r in $\Re^n$ has variance-covariance matrix $\text{cov}(\vec{X}) = \frac{r^2}{n+2}\mathcal{I}$. The parameter γ must be chosen so that the region of interest in the next iteration does not shrink in size. Since the sphering transformation in this case would effectively multiply the coordinates by $\frac{\sqrt{n+2}}{r}$, the critical value of $\gamma = \sqrt{n+2}$ results in a steady-state where the region of interest remains the same, $S_{(t)} = S_{(t-1)}$. In order for the region of interest to successfully capture an individual cluster, γ needs to be slightly larger than $\sqrt{n+2}$.

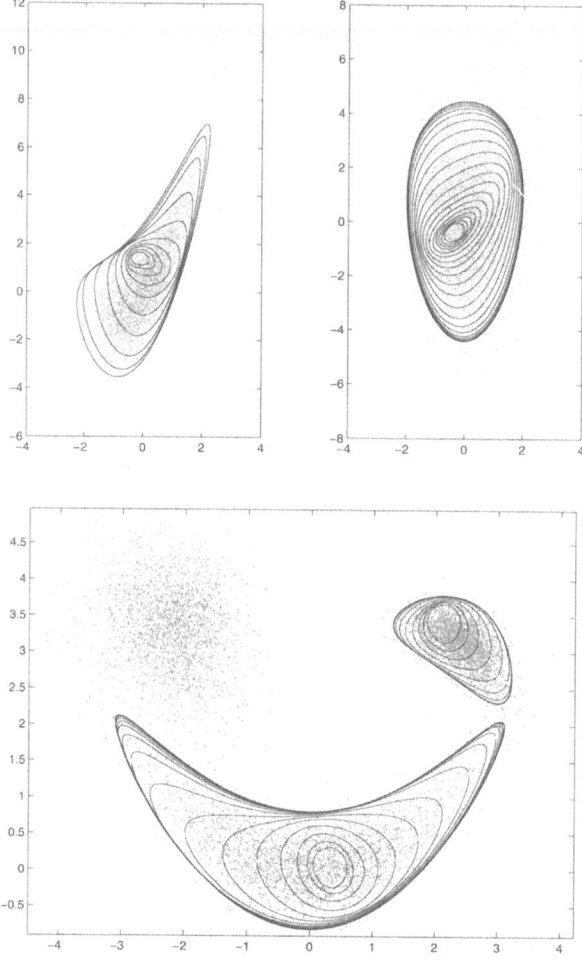

Figure 29.1. Numerical examples of the local cluster extraction algorithm. In all figures, the initial regions of interest are the smallest elliptical regions. The regions of interest expand in accordance with the data contained in the respective regions of interest. A smaller value of γ is used in figure in the top right, resulting in a more conservative inference based on local data and slower convergence than in the top left figure. The robustness of the algorithm to data from outlier clusters can be seen in the figure on the bottom.

For Gaussian clusters, the regions of interest tend asymptotically to a limit based on the covariance matrix of the Gaussian. In particular, the regions of interest are not always expanding. For example, if the current region of interest already encompasses all the datapoints belonging to a Gaussian cluster, the region could in fact shrink to better fit the cluster.

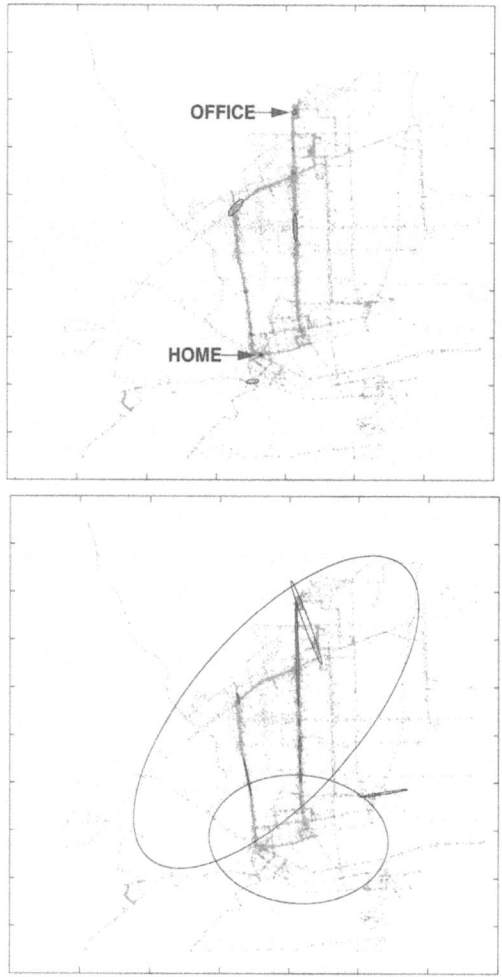

Figure 29.2. Analysis of an individual's GPS location data. Comparison of 10 randomly targeted clusters found with the local cluster extraction algorithm (top), with the EM algorithm after 200 iterations for a mixture of 10 Gaussians (bottom). The Gaussians have arbitrary covariance structure in both cases. Targeted cluster sampling (top) successfully found the individual's home and office location clusters as labeled. GPS data courtesy of Groundhogtech. (www.groundhogtech.com)

Precise calculation of the converged region of interest for Gaussian clusters involve solutions of transcendental equations.

The choice of γ should properly balance speed of convergence, and the separation of the Gaussian clusters. Smaller γ values correspond to conservative inference based on the data within the current region of interest, while larger γ values aggressively predict data outside of the region.

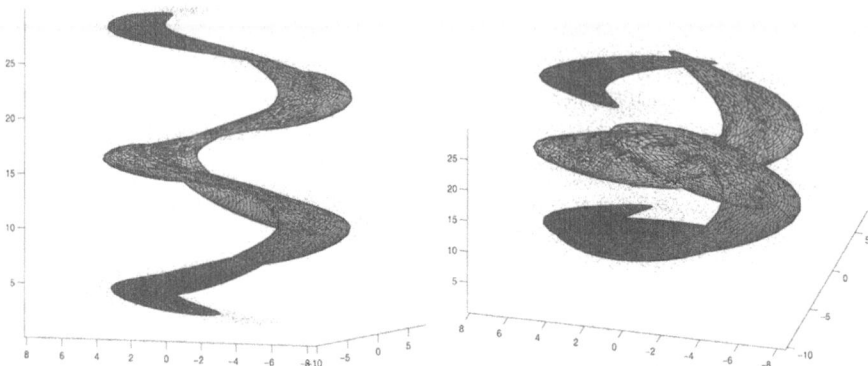

Figure 29.3. Two views of 10 (non-interacting) clusters found with the local cluster extraction algorithm for the artificially generated three dimensional noisy spiral data.

Larger γ values result in faster convergence and cluster extraction, however, smaller γ values can better separate out closely spaced clusters. (See Figure 29.1.)

Additional numerical results are shown in Figures 29.2 and 29.3. In Figure 29.2, the local cluster extraction algorithm and the EM algorithm are contrasted in the analysis of an individual's GPS location data. In Figure 29.3, the local cluster extraction algorithm effectively finds good local curved descriptions of the noisy spiral data.

29.4   Discussion

The non-linearly transformed Gaussian models described in this chapter is attractive from a few perspectives. First, the statistical model contains Gaussians with arbitrary covariance structure, as well as elements which capture non-linear aspects in the data. Second, there are analytic solutions for the maximum likelihood estimators, thus the models are easy to fit. And finally, since the non-linear volume preserving models are infinitely paramaterizable, the models can be custom tailored to scale better. For example, only quadratic transformation terms $\{k_{n1}, \ldots, k_{nn-1}\}$ in the least variance principal component can be included, thus resulting in a reasonably flexible transformed Gaussian model with parameters which scale linearly instead of quadratically with dimension.

For analyzing mixtures of transformed Gaussians, the local cluster extraction algorithm is introduced as an alternative to the EM algorithm. Much like for the split-and-merge EM algorithm (SMEM, Ueda et.al. [9]) which introduces local criteria for splitting a Gaussian, and merging two Gaussians, a future challenge would be to incorporate individual clusters found with the local cluster extraction algorithm to enhance mixture model fits of the entire data. In a sense, the cluster extraction algorithm can be viewed dynamically as an extreme case where all the Gaussian kernels are completely non-interacting, whereas the Gaussian kernels evolved according to EM algorithm dynamics are fully interacting. It will be very interesting to investigate intermediate algorithms where the Gaussian kernels have local or pairwise interaction.

Both the SMEM algorithm and the local cluster extraction algorithm bring up the issue of the quality of the local cluster fit. This in turn leads to the need for a good test for multivariate normality. The fitting of the transformed Gaussian models $\mathcal{J} \cdot \vec{X} = \vec{\varphi}$ can likely be turned into a useful test of multivariate normality.

There is much recent interest in capturing non-linear, lower dimensional structure in the data. Though the transformed Gaussian model is not as flexible as kernel PCA (Scholkopf et.al. [6]), Isomap (Tenenbaum et.al. [8]) or locally linear embedding (Roweis & Saul [5]), it does provide a probabilistic description of the data. An extension of the transformation to the amalgamated product (Eqn. 1.3) will add flexibility, and is currently being investigated.

Finally, there is a close relation between the local cluster extraction algorithm, Dasgupta's [1] provably correct algorithm for learning a mixture of Gaussians, and in the context of robust statistics, redescending M-estimators for location and scatter (Kent & Tyler [2]), and weighted likelihood estimating equations (Markatou, Basu, and Bruce G. Lindsay [4]). Future work will focus on a close investigation of the connection, as well as a Bayesian formulation of the local cluster extraction algorithm.

29.5 Appendix

Let $B_n(r)$ denote a ball of radius r in $\Re^n$, and let S_{n-1} denote the surface of the unit ball in $\Re^n$. Let w_n be the volume of B_n. Let the random vector $\vec{X} = (X_1, \ldots, X_n)^T$ be uniformly distributed inside a ball of radius r in $\Re^n$.

By symmetry, $\mathrm{cov}(X_i, X_j) = 0$ for $i \neq j$. Computation of $\mathrm{var}(X_i)$ proceeds as follows.

$$
\begin{aligned}
\mathrm{var}(X_i) &= \frac{1}{r^n w_n} \int_{\vec{x} \in B_n(r)} x_i^2 dx_1 \ldots dx_n \\
&= \frac{1}{r^n w_n n} \int_{\vec{x} \in B_n(r)} (x_1^2 + \cdots + x_n^2) dx_1 \ldots dx_n \\
&= \frac{1}{r^n w_n n} \int_{u \in S^{n-1}} \int_0^r r^2 r^{n-1} dr du
\end{aligned}
$$

$$= \frac{1}{r^n w_n n}(n w_n)\frac{r^{n+2}}{n+2}$$

$$= \frac{r^2}{n+2}$$

Thus the variance-covariance matrix of $\vec{X}$ is $\text{cov}(\vec{X}) = \frac{r^2}{n+2}\mathcal{I}$.

References

[1] Dasgupta, S. (1999) Learning mixtures of Gaussians. *IEEE Symposium on Foundations of Computer Science*

[2] Kent, J.T. & Tyler, D.E. (1991) Redescending M-Estimates of Multivariate Location and Scatter. *Annals of Statistics* Volume 19, Issue 4, 2102-2119.

[3] Lin, J.K. & Dayan, P. (1999) Curved Gaussian Models with Application to the Modeling of Foreign Exchange Rates. *Computational Finance* vol. **6**.

[4] Markatou, M., Basu, A. and Lindsay, B. (1998) Weighted Likelihood Equations With Bootstrap Root Search. *Journal of the American Statistical Association* Volume 93, Number 442, 740-750.

[5] Roweis, S.T. & Saul, L.K. (2000) Nonlinear Dimensionality Reduction by Locally Linear Embedding. *Science* Vol.290, 2323-2326.

[6] Schölkopf, B., Smola, A. & Müller, K.-R. (1998) Nonlinear Component Analysis as a Kernel Eigenvalue Problem. *Neural Computation* **10**, 1299-1319.

[7] Strang, G. (1997) Every Unit Matrix is a LULU. *Linear Algebra and its Applications*, **265**:165-172.

[8] Tenenbaum, J.B., de Silva, V. & Langford, J.C. (2000) A Global Geometric Framework for Nonlinear Dimensionality Reduction. *Science* vol.290, 2319-2322.

[9] Ueda, N., Nakano, R., Ghahramani, Z. & Hinton, G.E. (2000) SMEM Algorithm for Mixture Models. *Neural Computation* **12**, 2109-2128.

30

Unsupervised Learning of Curved Manifolds

Vin de Silva and Joshua B. Tenenbaum[1]

Summary

We describe a variant of the Isomap manifold learning algorithm [1], called 'C-Isomap'. Isomap was designed to learn non-linear mappings which are *isometric* embeddings of a flat, convex data set. C-Isomap is designed to recover mappings in the larger class of *conformal* embeddings, provided that the original sampling density is reasonably uniform. We compare the performance of both versions of Isomap and other algorithms for manifold learning (MDS, LLE, GTM) on a range of data sets.

30.1   Introduction

We consider the problem of manifold learning: recovering meaningful low-dimensional structures hidden in high-dimensional data. An example (Figure 30.4) might be a set of pixel images of an individual's face observed under different pose and lighting conditions; the task is to identify the underlying variables (angle of elevation, direction of light, etc.) given only the high-dimensional pixel image data [1].

Recently Tenenbaum et al. introduced the Isomap algorithm, which extends the classical techniques of principal components analysis (PCA) and multidimensional scaling (MDS) [2] to a class of nonlinear manifolds, those which are isometric to a convex domain of Euclidean space. This includes

---

[1]Vin de Silva is with the Department of Mathematics, Stanford University (Email: silva@math.stanford.edu). Joshua B. Tenenbaum is with the Department of Brain and Cognitive Sciences, MIT (Email: jbt@mit.edu). The authors gratefully acknowledge the support of the DARPA Human ID project, the Office of Naval Research, and the National Science Foundation (grant DMS-0101364). The authors also thank Sam Roweis for stimulating discussions, and Larry Saul for suggesting the conformal fishbowl example as a test case.

manifolds like the 'Swiss roll' (Figure 30.1(b)), but not those like the 'fish-bowl' (Figure 30.1(c)), which have intrinsic curvature. For nonlinear but intrinsically flat manifolds such as the Swiss roll, Isomap efficiently finds a globally optimal low-dimensional representation that can be proven to converge asymptotically to the true structure of the data.

Here we extend the Isomap approach to a class of intrinsically curved data sets that are conformally equivalent to Euclidean space. This allows us to learn the structure of manifolds like the fishbowl, as well as other more complex data manifolds where the conformal assumption may be approximately valid.

The chapter is organised as follows. In Section 30.2 we introduce a hier-archy of generative models for manifold learning and describe prototypical data sets representing each of the models. Sections 30.3 and 30.4 describe the Isomap and the conformal Isomap (C-Isomap) algorithms, respectively. In Section 30.5 we compare C-Isomap and several alternative algorithms on test data sets.

30.2   Generative models for manifold learning

In the spirit of [3], we can view the problem of manifold learning as trying to invert a generative model for a set of observations.

Let Y be a d-dimensional domain contained in the Euclidean space $\mathbf{R}^d$, and let $f : Y \to \mathbf{R}^N$ be a smooth embedding, for some $N > d$. Data points $\{y_i\} \subset Y$ are generated by some random process with probability density $\alpha = \alpha(y)$, and are mapped by f to give the *observed data*, $\{x_i = f(y_i)\} \subset \mathbf{R}^N$. We refer to Y as the *latent space* and to $\{y_i\}$ as the *latent data*.

The task is to reconstruct f and $\{y_i\}$ from the observed data $\{x_i\}$ alone. The answer can take various forms. The approach taken by Isomap is to present *reconstructed data*, $\{z_i\} \subset \mathbf{R}^{d'}$, for some d'. In a successful solution, $d' = d$ and the configurations $\{z_i\} \subset \mathbf{R}^{d'}$ and $\{y_i\} \subset \mathbf{R}^d$ are congruent in a suitable sense. A mapping $f' : \mathbf{R}^{d'} \to \mathbf{R}^N$ can be constructed from the pairs (z_i, x_i) by any reasonable interpolation procedure (such as radial basis functions [4]); this amounts to recovering f.

Before the general problem stated above becomes meaningful, we require some constraints on the mapping f and on the density α of the random sampling process. Table 30.1 lists three such sets of constraints, together with an algorithm in each case that is provably correct: MDS exactly re-covers the original configuration (see [8]), and the two Isomap algorithms converge to the right answer as the number of data points tends to infinity. Note that as we move from top to bottom in the table, assumptions about the mapping are relaxed but we require increasingly strong assumptions about the sampling density. In manifold learning, as in other areas of sta-

Table 30.1. Under different assumptions about the distribution of latent variables (first column) and the mapping from latent variables to observations (second column), different algorithms are appropriate.

DISTRIBUTION	MAPPING	ALGORITHM
arbitrary	linear isometry	classical MDS
convex, dense	isometry	Isomap
convex, uniformly dense	conformal embedding	conformal Isomap

tistical learning and pattern recognition [9], some inductive bias is always required in order to draw meaningful inferences from data.

The simplest case is when f is a linear isometry $\mathbf{R}^d \to \mathbf{R}^N$; for example, a set of 2-dimensional data mapped into a plane in $\mathbf{R}^3$ (Figure 30.1(a)). The latent distribution may be arbitrary. PCA recovers the d significant dimensions of the observed data. Classical MDS produces the same results, but requires only the Euclidean distance matrix as its input (not the actual coordinates of the embedding in $\mathbf{R}^N$).

The second case is where $f : Y \to \mathbf{R}^N$ is an isometric embedding, in the sense of Riemannian geometry. In other words, f preserves the length of (and angles between) infinitesimal vectors in Y. The image set $f(Y)$ is then an *intrinsically flat* submanifold of $\mathbf{R}^N$; our standard example is the Swiss roll (Figure 30.1(b)). When Y is a convex domain in $\mathbf{R}^d$, and provided the data points are sufficiently dense,[2] Isomap successfully recovers the approximate original structure of data sets generated in this way. The crux of the method is to estimate the global metric structure of the latent space using only local geometric information measured in the observation space $\mathbf{R}^N$, by computing shortest paths in a graph connecting only neighboring observations.

The intrinsically flat model becomes unsuitable when the data manifold has any non-negligible Gaussian curvature; for example, data lying on the surface of a sphere. Curvature in the data manifold may occur in two places: it may be intrinsic to the latent space Y, or it may be introduced by the mapping f. The first situation represents a hard problem; it is not even clear what form the output from a manifold learning algorithm should be expected to take. The Euclidean assumption in our generative model deliberately excludes this possibility. The second situation, as we now show, can be handled for a certain class of functions.

Specifically, our third model allows f to be a *conformal* embedding. At each point y in Y, angles between infinitesimal vectors are preserved by f,

---

[2]The data density required depends on certain geometric parameters of the embedding: the minimum radius of curvature, and the minimum branch separation [1].

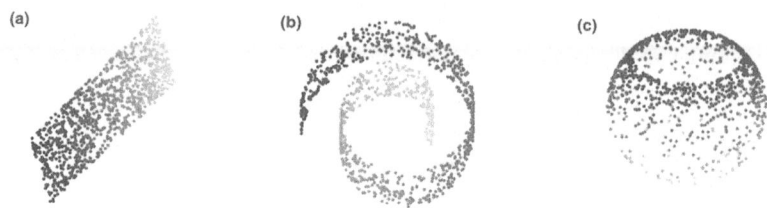

Figure 30.1. Prototypical datasets for the three generative models described in Table 30.1.

but their lengths are scaled by a factor $\phi = \phi(y) > 0$, which varies smoothly over Y. Compensating for this extra degree of freedom, we require that the original data be *uniformly* dense in Y.

The simplest example of a conformal map is the stereographic projection in $\mathbf{R}^3$ from the plane $z = 0$ to the unit sphere.[3] A large, centered disk in the plane maps to a 'fishbowl' under this map. Note that points which are uniformly sampled in the disk bunch up non-uniformly near the rim of fishbowl.

In the next two sections we describe Isomap and explain how it can be modified to deal with uniformly sampled, conformally embedded data. The key idea is to use the density of the observed data to estimate $\phi(y)$. This strategy is similar to the motivation behind conformal approaches to nonlinear ICA [5], but differs in that our goal is to reduce dimensionality rather than to find a better basis for observations in a space of the same dimensionality.

It is worth contrasting the *stereographic fishbowl* described above with a *uniform fishbowl* dataset, where points are sampled uniformly from the fishbowl surface (Figure 30.2, column 3). Here there is no bunching effect at the rim, and hence no way to estimate $\phi(y)$. Our new methods do not naturally extend to the uniform fishbowl; indeed it is not clear what the 'correct' answer should be in that case.

30.3   The Isomap algorithm

We briefly describe the standard Isomap procedure [1].

1. Determine a *neighbourhood graph* G of the observed data $\{x_i\}$ in a suitable way. For example, G might contain $x_i x_j$ iff x_j is one of the k nearest neighbours of x_i (and vice versa). Alternatively, G might contain the edge $x_i x_j$ iff $|x_i - x_j| < \epsilon$, for some ϵ.

---

[3]The Mercator projection is another well-known conformal map.

2. Compute shortest paths in the graph for all pairs of data points. Each edge $x_i x_j$ in the graph is weighted by its Euclidean length $|x_i - x_j|$, or by some other useful metric.

3. Apply MDS to the resulting shortest-path distance matrix D, to find the reconstructed data points $\{z_i\}$ in $\mathbf{R}^{d'}$.

The premise is that *local* metric information (in this case, lengths of edges $x_i x_j$ in the neighbourhood graph) is regarded as a trustworthy guide to the local metric structure in the original (latent) space. The shortest-paths computation then gives an estimate of the global metric structure, which can be fed into MDS to produce the required embedding.[4]

As an example, Isomap is effective in recovering the original rectangular structure of a Swiss roll data set with 2000 points, as illustrated in the first column of Figure 30.2. We used the k-nearest neighbours method with $k = 10$. Compare the Isomap output (row 3) with the output from MDS by itself (row 2). We will return to this example later.

The embedding dimension ($d' = 2$ in this case) is chosen at the end, by consulting a suitable 'residual variance' function, which measures how badly the MDS embedding preserves the distance matrix obtained during Step 2. Figure 30.3 shows the residual variance for Isomap and MDS embeddings in dimensions $d' = 1, 2, \ldots 8$. At $d' = 2$, the residual variance of the Isomap embedding has fallen virtually to zero, so there is little gain in choosing a higher value of d'. In a similar way, it is clear from the graph that MDS prefers a three-dimensional representation of the data. In general, one looks for an 'elbow' in the graph to indicate the preferred dimensionality [1].

When does Isomap succeed? The *recovered* shortest-paths distance matrix D needs to be a good approximation to the *original* Euclidean distance matrix for the points Y. When that is the case, it becomes trivial to read off the true dimensionality using the 'elbow' method, because the residual variance falls essentially to zero at the true dimension. The following convergence theorem is proved in [7]:

Theorem. *Let Y be sampled from a bounded convex region in $\mathbf{R}^d$, with respect to a density function $\alpha = \alpha(y)$. Let f be a C^2-smooth isometric embedding of that region in $\mathbf{R}^N$. Given $\lambda, \mu > 0$, for a suitable choice of neighborhood size parameter ϵ or k, we have*

$$1 - \lambda \leqslant \frac{\text{recovered distance}}{\text{original distance}} \leqslant 1 + \lambda$$

with probability at least $1 - \mu$, provided that the sample size is sufficiently large. [The formula is taken to hold for all pairs of points simultaneously.]

---

[4]Note that in the extreme case where G is a complete graph, meaning that all distances are regarded as trustworthy, the computation reduces to Step 3 alone, the MDS calculation. Thus Isomap may be viewed as a generalisation of MDS.

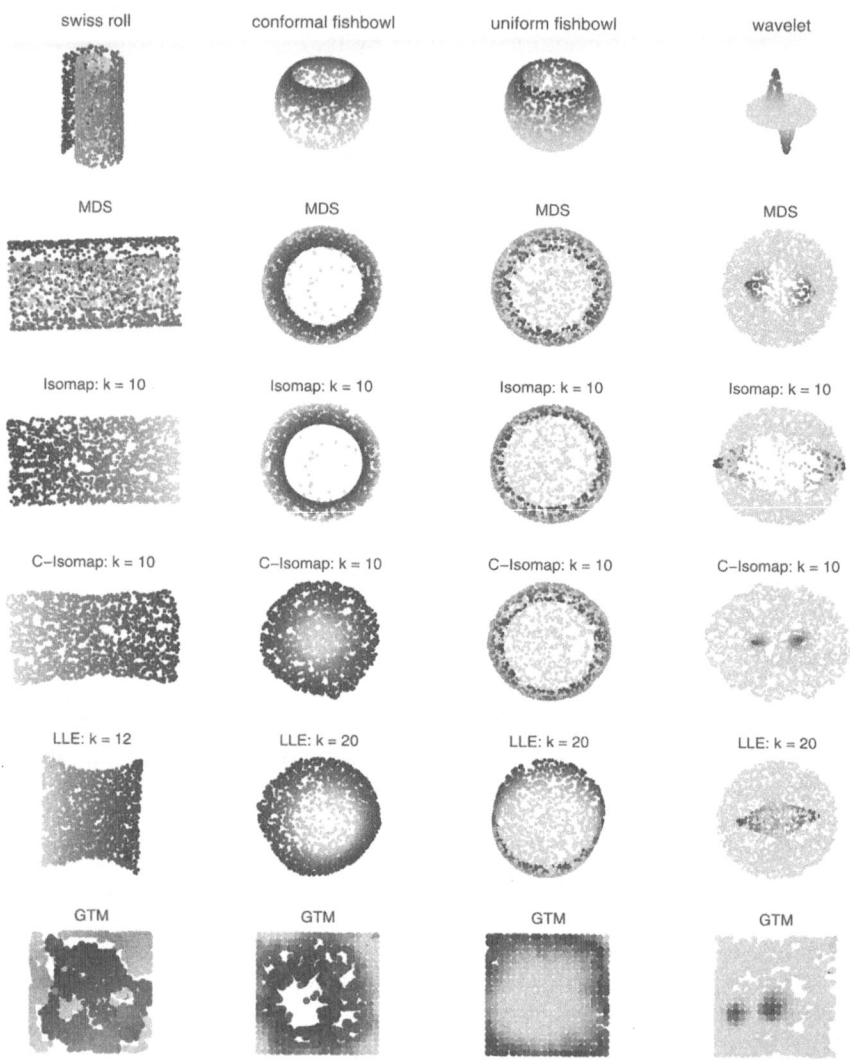

Figure 30.2. Applying MDS, Isomap, C-Isomap, LLE and GTM to four different data sets. The top row gives a 3-dimensional view of each data set; note that the Swiss roll is shown at a different angle than in Figure 30.1(b).

Explicit bounds for "sufficiently large" are given in [7] which depend on certain geometric parameters measuring the nonlinearity of the embedding. Specifically, if the embedding is highly curved, or if widely separated points in the domain are brought close to each other by f, then the task is more difficult and the required sample size increases.

The proof takes the following line. Since f is an isometric embedding, by definition it preserves the *infinitesimal* metric structure of Y perfectly.

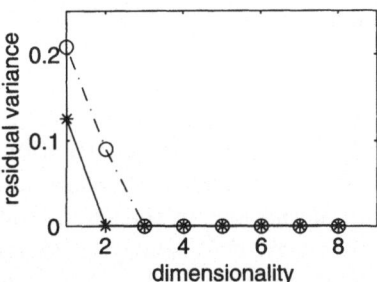

Figure 30.3. The residual variance plot shows that Isomap [unbroken line] detects the true 2-dimensional structure of the Swiss roll data, whereas MDS [dots-and-dashes] identifies the data as being 3-dimensional.

When the neighbourhood size is small, the *local* metric structure of the data is preserved approximately. The shortest-paths calculation effectively computes an approximation to geodesic distance in the original domain. For convex domains, geodesic distance is equal to Euclidean distance, which finishes the proof. Details are in [7].

30.4   The C-Isomap algorithm

C-Isomap is a simple variation on standard Isomap. Specifically, we use the k-neighbours method in Step 1, and replace Step 2 with the following:

2a. Compute shortest paths in the graph for all pairs of data points. Each edge $x_i x_j$ in the graph is weighted by $|x_i - x_j|/\sqrt{M(i)M(j)}$. Here $M(i)$ is the mean distance of x_i to its k nearest neighbours.

We can motivate Step 2a as follows. Assume that we have a very large number of uniformly sampled data points. In the latent space, the k nearest neighbours of a given point y_i occupy a d-dimensional disk of approximately a certain radius r that depends on d and on the sampling density, but not on y_i. The map f carries this approximately to a d-dimensional disk of radius $r\phi(y_i)$ in $\mathbf{R}^N$. The expected average distance of these k points from x_i is therefore proportional to $\phi(y_i)$. The computed quantity $M(i)$ is a fair approximation, so Step 2a has the effect of correcting for ϕ.

Using this kind of reasoning, one can prove a convergence theorem for C-Isomap:

Theorem. *Let Y be sampled uniformly from a bounded convex region in $\mathbf{R}^d$. Let f be a C^2-smooth conformal embedding of that region in $\mathbf{R}^N$. Given $\lambda, \mu > 0$, for a suitable choice of neighborhood size parameter k, we have*

$$1 - \lambda \leqslant \frac{\text{recovered distance}}{\text{original distance}} \leqslant 1 + \lambda$$

with probability at least $1 - \mu$, provided that the sample size is sufficiently large.

Explicit lower bounds for the sample size are much more difficult to formulate here; certainly we expect to require a larger sample than in regular Isomap to obtain good approximations. In situations where both Isomap and C-Isomap are applicable, it may be preferable to use Isomap, being less susceptible to local fluctuations in the sample density.

In general, the crude effect of C-Isomap is to magnify regions of the data where the point density is high, and to shrink regions where the point density is low. Whether or not the conformal model is valid in a given case, this tendency towards uniformity may still be useful for representing the large-scale structure of a data set. The 'wavelet' data set discussed later is a good illustration of this.

30.5   Comparative performance of C-Isomap

We present the results of tests using four toy data sets—Swiss roll, stereographic fishbowl, uniform fishbowl and 'wavelet'—and two more realistic data sets of face images.

C-Isomap is compared to four other algorithms: MDS, Isomap, Locally Linear Embedding (LLE) [6], and the Generative Topographic Mapping (GTM) [3]. LLE, like Isomap, computes a global low-dimensional embedding of the data from only local metric information, but differs in that the embedding aims to preserve only that local metric structure, rather than an estimate of the manifold's global metric structure, as in Isomap or C-Isomap. By respecting local constraints only, LLE can handle more curvature than Isomap, at the cost of sometimes producing less globally stable results. GTM uses the EM algorithm to fit a version of the generative model in Section 30.2, under the assumptions of a uniform square density in latent space and a soft smoothness prior on the nonlinear mapping. Assuming only a sufficiently smooth mapping is more general than our conformal assumption, but the strong constraint of a uniform square density and the use of a greedy optimization procedure prone to local minima problems frequently lead GTM to distorted representations of a data manifold.

For the toy data sets, each algorithm was used to obtain a 2-dimensional embedding of the points. Figure 30.2 summarises the results. Each column shows the results of applying the five algorithms on a particular data set. The data set itself is shown at the top, in a 3-dimensional representation. The Swiss roll data points are shaded to indicate one of the original rectangular coordinates. The shading on the other three data sets indicates the original z-varaiable.

The images shown are intended to be reasonably typical of the results a careful user might obtain; specifically the parameter k in Isomap, C-Isomap and LLE was tuned for good performance. For the GTM analysis of the two fishbowls and the wavelet, we used $20^2 = 400$ sample points on a grid; the same number of basis functions; relative width $\sigma = 1$; and weight regularisation factor $l = 2$. In all three cases the training process converged by the 50th iteration (often much sooner); we show the posterior mean coordinates after convergence. Some tuning was needed for the Swiss roll; we ran 200 iterations at $\sigma = 1.8$.

Results: Swiss roll, two fishbowls, and wavelet

The Swiss roll (column 1) was constructed by sampling 2000 points uniformly from a rectangle and mapping into $\mathbf{R}^3$ using an isometric spiral embedding. The 2-dimensional projection given by MDS fails to resolve the true non-linear structure of the data. Both Isomap and C-Isomap peform well. Isomap recovers the original rectangle, including its aspect ratio, with very little distortion. C-Isomap gives similar results but is noticeably slightly worse than Isomap, as we would expect for a dataset where Isomap's stronger metric assumptions hold. LLE successfully unfolds the manifold, but introduces substantial curvature in the sides of the rectangle. Note that LLE is unable to recover the aspect ratio of the rectangle, because it is constrained to give output with equal covariance in all directions. Finally, GTM performs poorly on the Swiss roll. After about 100 iterations, the output resembles that of classical MDS. This proves to be unstable, and by the 200th iteration the algorithm has converged to the hopelessly tangled form shown in the diagram. This solution is stable to at least 2000 iterations.

The stereographic fishbowl (column 2) was constructed by sampling 2000 points uniformly from a disk in the plane, and projecting the result stereographically onto a sphere. The MDS coordinates are just the 2-dimensional vertical projection of the data. There is an 'annulus of ambiguity' where the projection map is 2-to-1. As expected, regular Isomap does no better than MDS; however C-Isomap and LLE are both successful in flattening the fishbowl and recovering the original disk shape. GTM appears to be confused by the heavy point density around the rim; the output incorrectly indicates a ring-shaped structure.

The uniform fishbowl (column 3) was obtained by sampling 2000 points uniformly from the fishbowl surface itself. Again the first two dimensions of MDS give the vertical projection. Since the sampling density is uniform, Isomap and C-Isomap behave alike, giving a version of the MDS projection slightly widened at the rim. LLE is more successful at opening out the rim, but it remains partly turned in. GTM gives the best performace, flattening out the fishbowl. What is missing (and what GTM does not seek to provide) is any clear indication of the round shape of the data set.

The wavelet (column 4) was chosen as an example of a non-conformally generated data set on which C-Isomap performs well. 2000 points were sampled uniformly in a disk and translated in the z-direction by a function with one positive and one negative peak (specifically, we used the x-derivative of a Gaussian, scaled to exaggerate the peaks). The projection given by MDS is tilted; we see two dark 'horns' corresponding to the noise peaks, where the 2-dimensional representation is ambiguous. Regular Isomap and LLE show the same phenomenon, with Isomap doing worst. C-Isomap flattens the wavelet almost perfectly, with a slight stretching in the x-direction. GTM also flattens the data, but the results are comparatively distorted.

Face images

Two high-dimensional data sets of face images illustrate the kinds of real-world problems where Isomap and C-Isomap may succeed in recovering subtle low-dimensional structure. Both of these examples cause problems for classical approaches (PCA, MDS) and the other algorithms discussed above, but space does not permit a full comparison here.

In the first example, 698 images of a face were generated synthetically to cover a range of poses (with up-down and left-right parameters) and lighting directions (a single left-right parameter). Each image was rendered in 64-by-64 greyscale pixels; equivalently as 4096-dimensional vectors. Applied to the raw, unordered images, Isomap ($k = 6$) learns a 3-dimensional embedding of the data's intrinsic geometric structure. Each coordinate axis correlates strongly with one of the three original parameters: x-axis with left-right pose ($R = 0.99$); y-axis with up-down pose ($R = 0.90$); z-axis with lighting direction (slider position, $R = 0.92$). Figure 30.4 shows the x- and y-coordinates. More details of this example can be found in [1].

For the second example, we constructed a 2-dimensional family of images (Figure 30.5) in the following way. A rectangular photograph (of a human face) was overlaid digitally onto a uniform grey background of fixed size, using two parameters: 'position'—the image center was kept at a fixed height, and varied left-right—and 'zoom'—the photograph image ranging from large and high-contrast, to small and low-contrast (simulating the effect of distance). 2000 images were generated, with both parameters chosen randomly and independently within a prescribed interval.

Both algorithms recover a two-dimensional embedding, but the Isomap embedding suffers from much greater distortion. As a result, only the C-Isomap embedding faithfully represents the original generating parameters of position and zoom in its principal axes.

Why should C-Isomap work well here? Small changes in the zoom and position parameters can be measured by Euclidean distances in pixel space; however the effect is much smaller, for both parameters, when the image is zoomed out. In this situation, a conformal model is the natural approximation, with the conformal scale parameter ϕ depending on zoom.

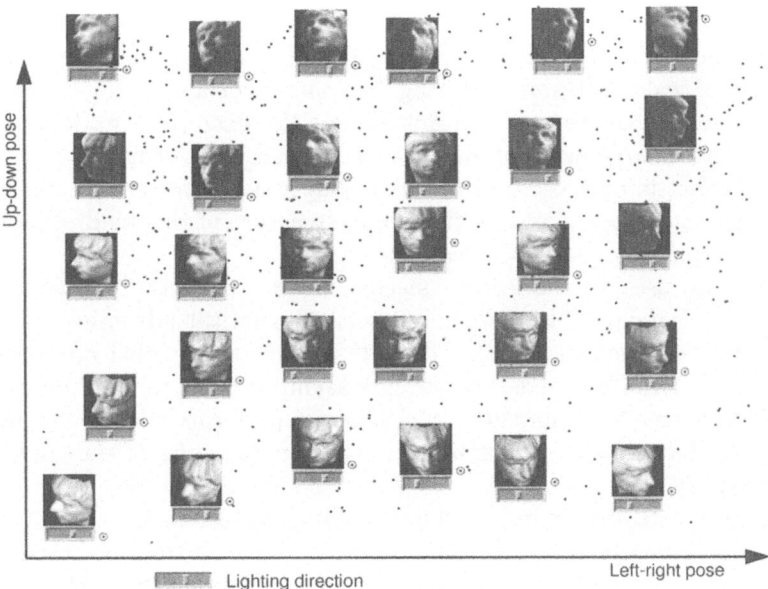

Figure 30.4. Isomap successfully recovers the three parameters of a family of face images, which vary by pose (left-right and up-down) and lighting direction. The first two Isomap coordinates (which correlate with the pose parameters) are shown here, together with sample images at various points.

[Reprinted with permission from Tenenbaum et al., *Science* 290, 2319 (2000), Figure 1A. ©2000 American Association for the Advancement of Science.]

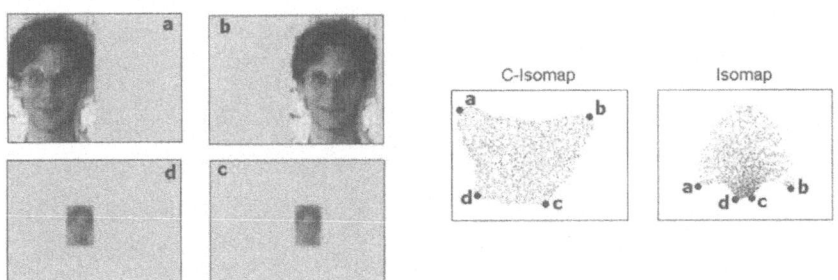

Figure 30.5. C-Isomap successfully finds the two independent degrees of freedom (left-right position and 'zoom', or apparent distance) in this data set of face images. Isomap finds a 2-dimensional embedding, but fails to separate the two variables. Four sample images are shown.

30.6   Conclusions

We have presented a simple variant of the Isomap manifold learning algorithm which addresses a recurrent difficulty in nonlinear data analysis: the problem of scale. Naturally occuring data manifolds are often generated by the action of some group of transformations (for example, rotations, translations and enlargements of an image). A given transformation may have a small effect in one region of the data, but a very large effect somewhere else. Any attempt to discover the global structure of this kind of data set must in some way deal with this issue.

C-Isomap uses an elementary mechanism to defuse the problem. As a variant of Isomap, C-Isomap inherits its theoretical advantages: it is a non-iterative, computationally efficient solution to a global optimisation problem, and can be proved to converge asymptotically to the right answer for a large class of manifolds. In practice it performs well on a range of data sets, including some outside the strict parameters of the conformal generative model.

We close with two open questions for future research. First, what kinds of natural data sets, other than those involving scaling transformations, are suitable domains for conformal dimensionality reduction? Second, are there ways to extend our approach to dimensionality reduction based on geometric invariants to more general classes of curved manifolds, beyond those that are conformally equivalent to Euclidean space? Positive answers could represent important steps towards increasing the practical applicability of nonlinear dimensionality reduction methods.

References

[1] J. B. Tenenbaum, V. de Silva and J. C. Langford. *Science* **290**, 2319 (2000).

[2] K. V. Mardia, J. T. Kent and J. M. Bibby. *Multivariate Analysis*, (Academic Press, London, 1979).

[3] C. M. Bishop, M. Svensén, and C. K. I. Williams. *Neural Computation* **10**, 215 (1998).

[4] D. Beymer, T. Poggio, *Science* **272**, 1905 (1996).

[5] A. Hyvärinen and P. Pajunen (1998). Nonlinear Independent Component Analysis: Existence and Uniqueness Results. *Neural Networks* **12(3)**, 423 (1999).

[6] S. Roweis and L. Saul. *Science* **290**, 2323 (2000).

[7] M. Bernstein, V. de Silva, J. C. Langford, and J. B. Tenenbaum. Preprint dated (12/20/2000) available at: http://isomap.stanford.edu/BdSLT.pdf

[8] T. F. Cox and M. A. A. Cox, *Multidimensional Scaling*, (Chapman & Hall, London, 1994).

[9] T. M. Mitchell, *Machine Learning*, (McGraw-Hill, New York, 1997).

31

ANOVA DDP Models: A Review

Maria De Iorio, Peter Müller,
Gary L. Rosner, and Steven N. MacEachern[1]

Summary

This paper briefly describes the results and methodology presented in De Iorio *et al.* [1] and Müller *et al.* [6]. We consider dependent non-parametric models for a set of related random probability distributions and propose a model which describes dependence across random distributions in an ANOVA type fashion. We define a probability model in such a way that marginally each random measure follows a Dirichlet Process and use the dependent Dirichlet process (DDP; MacEachern [3]) to define ANOVA type dependence across the related random measures.

31.1 Introduction

In this article we summarize two related manuscripts describing the construction and application of ANOVA DDP models for inference about related random functions. (See De Iorio *et al.* [1] and Müller *et al.* [6] for a full discussion.) We consider dependent non-parametric models for a set of related random probability distributions or functions. We propose a model which describes dependence across random distributions in an ANOVA type fashion. Specifically, assume that random distributions F_x are indexed by a p-dimensional vector $x = (x_1, \ldots, x_p)$ of categorical covariates. For example, in a clinical trial F_{x_1,x_2} could be the random effects

---

[1]Maria De Iorio is with the Department of Statistics, University of Oxford. Peter Müller Gary L. Rosner is with the Department of Biostatistics, The University of Texas M. D. Anderson Cancer Center, Houston, TX. Steven N. MacEachern is with the Department of Statistics, The Ohio State University, Columbus, OH.

distribution for patients treated at levels x_1 and x_2 of two drugs. We define a non-parametric probability model for F_x in such a way that marginally for each x the random measure F_x follows a Dirichlet Process $DP(M, F_{0x})$ with total mass parameter M and base measure F_{0x} (Fergurson [2]). We introduce dependence for F_x across x using the dependent Dirichlet process (DDP) as defined by MacEachern [3]. The random measures F_x are almost surely discrete with the point masses generated marginally from the base measure F_{0x}. MacEachern [3] introduces dependence across random measures generated marginally by a DP by imposing dependence in the distribution of these point masses, maintaining the base measure as the marginal distribution. We use the DDP to define ANOVA type dependence across related random measures by assuming ANOVA models for these point masses. The resulting probability model defines an overall average effect and offsets for each level of the categorical covariates. If desired this can be generalized to include interaction effects. We describe how the ANOVA DDP can be easily embedded in a more complex hierarchical model and we present a full posterior analysis in a Bayesian framework.

We use the proposed models to analyze data from a clinical trial for anti-cancer agents. We define a semi-parametric model for blood counts over time with a non-parametric random effects distribution for patient-specific random effects vectors. Marginally, for each treatment combination we use a DP model for the random effects distribution. Across different treatments the non-parametric random effects distributions are linked by introducing dependence using the proposed DDP ANOVA model.

In section 2 we develop the basic model as a dependent DP model. Section 3 illustrates the proposed model on longitudinal data with non-parametric random effects distributions.

31.2   The ANOVA DDP

Our starting point is the Dirichlet Process, described through Sethuraman's representation (Sethuraman [8]). The random distribution F can be described by the following expression

$$F(y) = \sum_{h=1}^{\infty} p_h I(\theta_h \leqslant y)$$

where $\sum_{h=1}^{\infty} p_h = 1$ and $I(\cdot)$ denotes the indicator function. The θ_h are usually referred to as locations and p_h is referred to as the mass assigned to location h. The distribution of F is then defined by placing a distribution on the θ_h and the p_h. These distributions are governed by the parameter of the Dirichlet process, a non-null, finite measure α. The measure α is often described by its mass, M, and its shape F_0, which is a distribution function. The θ_h and p_h are mutually independent, where $\theta_1, \theta_2, \ldots$, are independent and identically distributed draws from F_0 and the p_h follow a rescaled beta distribution. More precisely, let $v_1, v_2, \ldots,$ be a sequence of draws from the Beta$(1, M)$ random variable then set $p_h = v_h \prod_{j=1}^{h-1}(1 - v_j)$.

MacEachern [3] generalizes the Dirichlet process to DDP, defining a probability model for a collection of non-parametric distributions, the realizations of which are dependent. For the remainder of this section we will consider only the *single-p DDP model* which is a special case of the DDP model. Consider a set of random distributions $\{F_x, x \in \mathcal{X}\}$, where $\mathcal{X}$ is any covariate space. The collection of random distribution is then specified as follows:

$$F_x(y) = \sum_{h=1}^{\infty} p_h I(\theta_{hx} \leqslant y)$$

for each $x \in \mathcal{X}$, where $\sum_{h=1}^{\infty} p_h = 1$. Hence the main idea is that, in the presence of a covariate, the locations, θ_h, of the Dirichlet process can be replaced by the sample path of a stochastic process, denoted $\theta_{h\mathcal{X}}$. This sample path provides the location at each value of the covariate and therefore the degree of dependence among the random distributions, $\{F_x, x \in \mathcal{X}\}$ is governed by the level of the covariate x. The p_h are once again random variables of a single dimension (defined as in the case of the simple DP model) and the vector of p_h and the collection of stochastic processes, $\theta_{\mathcal{X}}$, are mutually independent. The marginal distribution, F_x follows a Dirichlet process with mass M and base measure F_{0x}, $F_x \sim \mathrm{DP}(M, F_{0x})$, for each $x \in \mathcal{X}$. This simple model can be generalized to more complex settings. See MacEachern [3] for a more detailed and mathematical description. In conclusion, the key idea behind the DDP is to introduce dependence across the measure F_x by assuming the distribution of the locations θ_{hx} are dependent across different levels of x, yet still independent across h.

We consider an extension of the dependent Dirichlet process to multiple categorical covariates and in the next section we will show how to use it as a component in more complex hierarchical Bayesian models. To achieve the desired generalization to multiple categorical covariates, we will use an ANOVA-type probability model for the locations θ_{hx}.

Assume $\mathcal{F} = \{F_x, x \in X\}$ is an array of random distributions, indexed by a categorical covariate x. For simplicity of explanation, assume for the moment that $x = (v, w)$ is a bidimensional vector of categorical covariates with $v \in \mathcal{V} = \{1, \ldots, V\}$ and $w \in \mathcal{W} = \{1, \ldots, W\}$. The covariates (v, w) could be, for example and as we shall see later, the levels of two treatments in a clinical trial, and the distributions F_x might be sampling distribution for recorded measurements on each patient or random effects distribution.

The main result in De Iorio *et al.* [1] is to build a model for the random distributions F_x which allows us to build an ANOVA type dependence structure. For example, we want random distributions F_x and $F_{x'}$ for $x = (v_1, w_1)$ and $x' = (v_1, w_2)$ to share a common main effect due to the common factor v_1. To achieve this it is natural to consider the DDP framework, imposing the following structure on the locations θ_{hx} (for each $x \in V \times W$)

$$\theta_{hx} = m_h + A_{hv} + B_{hw} \qquad (31.1)$$

with $m_h \overset{iid}{\sim} p_m^o(m_h)$, $A_{vh} \overset{iid}{\sim} p_{Av}^o(A_{vh})$, and $B_{wh} \overset{iid}{\sim} p_{Bw}^o(B_{wh})$, with independence across h, v and w. We refer to the joint probability model on $\mathcal{F}$ as $\{F_x, \, x \in X\} \sim$ ANOVA DDP(M, p^o). m_h can be interpreted as the "overall mean" and A_v and B_w are the "main effects" for covariate levels v and w. In this way standard ANOVA concepts are extended to a class of random distribution and it is also straightforward to include interaction effects. The model is parameterized by the total mass parameter M and the base measure p^o on the ANOVA effects in (31.1). Marginally, for each $x = (v, w)$, the random distribution F_x follows a DP with mass M and base measure F_x^o given by the convolution of p_m^o, p_{Av}^o and p_{Bw}^o. Therefore the dependence among the random distribution F_x is determined by the covariance structure of the point masses θ_{xh} across x.

In such a way we have defined an "array" of random distributions F_x by associating a different random distribution to each combination of covariates such that the realizations of these distributions are dependent.

This model can be utilized to account for any p-dimensional vector of categorical covariates and/or to impose constraints on some of the offsets. In fact, a crucial feature of the ANOVA DDP model is that model specification and computation are dimension independent.

One source of practical problems in the implementation of the proposed model is related to identifiability concerns. Therefore, as in standard ANOVA models, we might need to introduce some identifiability constraints. For example, we may impose any of the standard constraints such as $A_1 = B_1 \equiv 0$.

31.3   Meta-analysis over related studies

In this section we show the use of the ANOVA DDP model as component of a more complex hierarchical model using a real data example. We consider two hematologic studies. The data records white blood cell counts (WBC) over time for each of the patients in 2 studies. Denote with y_{it} the measured response on day t for patient i, recorded on a log scale of thousands/microliter, i.e. $y_{it} = \log(\text{WBC}/1000)$. The profiles of white blood cell counts over time look similar for most patients. See [7] for a detailed description of the data. Figure 31.1 shows some randomly selected patients from the 2 studies. There is an initial base line, followed by a sudden decline when chemotherapy starts, and a slow, S-shaped recovery back to approximately the base-line after the treatment ends. In this case, profiles can be

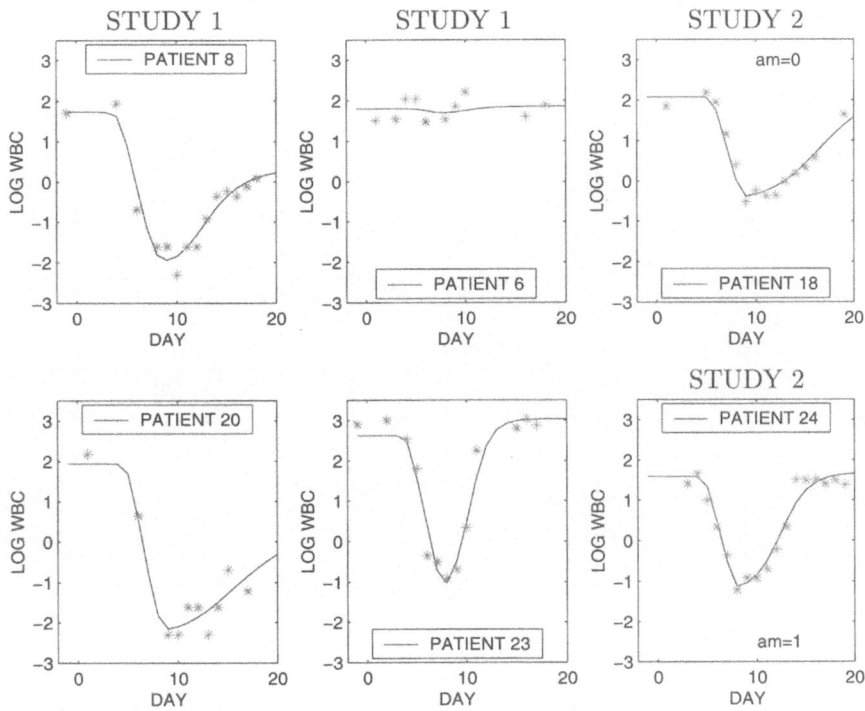

Figure 31.1. Some patients from the two hematologic studies. The data are represented by the crosses. The solid lines represent the fit obtained using the ANOVA DDP model.

reasonably well approximated with a piecewise linear-linear-logistic regression, using a 7-dimensional parameter vector $\theta = (z_1, z_2, z_3, \tau_1, \tau_2, \beta_0, \beta_1)$. (See [7] for more details). The nonlinear regression parameters might differ significantly across patients. So we introduce a patient specific random effects vector θ_i. Conditional on θ_i, we assume a nonlinear regression using the piecewise linear-linear-logistic regression model,

$$y_{it} = f_{\theta_i}(t) + \epsilon_{it} \tag{31.2}$$

$$\epsilon_{it} \sim N(0, \sigma^2) \tag{31.3}$$

where

$$f_\theta(t) = \begin{cases} z_1 & t < \tau_1 \\ rz_1 + (1-r)g(\theta, t) & \tau_1 \leqslant t < \tau_2 \\ g(\theta, t) & t \geq \tau 2 \end{cases}$$

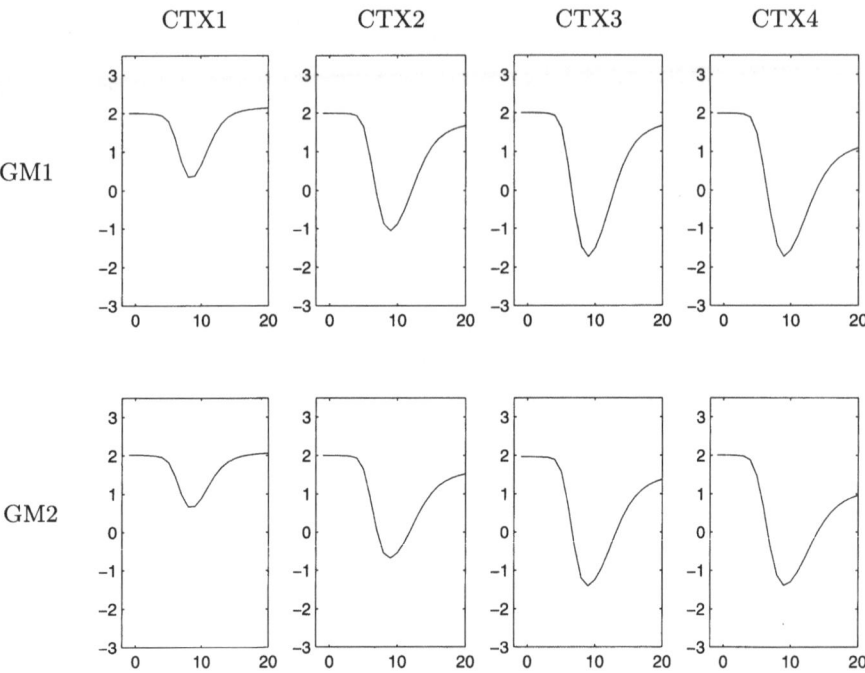

Figure 31.2. Predicted WBC response curves for hypothetical future patients in study 1. The dashed lines give the 95% prediction interval. CTX1 = 1.5, CTX2 = 3.0, CTX3 = 4.5, CTX4 = 6.0, GM1 = 5, GM2 = 10.

and

$$ r = \frac{\tau_2 - t}{\tau_2 - \tau_1} $$

$$ g(\theta, t) = z_2 + \frac{z_3}{1 + (1 + \exp\{-\beta_0 - \beta_1(t - \tau_2)\})} $$

The model is then completed with a random effects model G_x. The random effects distribution G_x should depend on the treatment levels x. In the first study patients are given two treatments, the actual anti-cancer agent cyclophosphamide (CTX), and a second drug (GM-CSF) which is given to mitigate some adverse side effects of the chemotherapy. Each patient receives a combination of both treatments, with CTX being given in one of four possible dose levels (1.5, 3, 4.5 or 6 grams per square meters of body surface area) and only two levels for GM-CSF (5 or 10 micrograms per kilogram of body weight). The objective of the second study, built on the experience of the first, was to determine if adding amofostine (AM) would reduce the hematologic side effects of aggressive chemotherapy. In the second study patients receive the same level of CTX (=3) and GM-CSF (=5), but only some of them are given amofostine (AM=1). Therefore to each patient we can associate a 4-dimensional vector of categorical covariates $x = $ (CTX,GM-CSF,AM,STUDY), where AM is equal to 0 (absence of amofostine) or 1 (presence of amofostine). As mentioned earlier, our aim is to introduce dependence among the patients random effects distributions, with the dependence determined by the level of the covariates.

Let $x_i = (v_i, w_i)$ denote the treatment for patient i. We assume

$$\theta_i | x_i = x \sim G_x(\theta) \tag{31.4}$$

$$G_x(\theta) = \int \text{N}(\theta \mid \mu, S) dF_x(\mu) \tag{31.5}$$

$$\{F_x, \, x \in X\} \sim \text{ANOVA DDP}(M, p^o) \tag{31.6}$$

In words, we convolve the discrete measure F_x with a Normal Kernel to remove discritness. Still if two patients have the same covariates vector then they will have the same random effects distribution. If they differ only in one covariate level then their random effects distributions will be similar, although not identical. It can be shown that the model specified in eq. (31.4)-(31.6) can be reformulated as a mixture of ANOVA models. It follows from this equivalence that any Markov chain Monte Carlo (MCMC) scheme for DP mixture models can be used for posterior simulation in DDP ANOVA models of the type (31.5). Using Normal priors for the base measure p^o, and the additional Normal kernel leads to a straightforward Gibbs sampling scheme. See, for example, [4]

Figure 31.2 shows posterior predictive inference for hypothetical future patients in study 1 for each combination of treatments CTX and GM-CSF. Given the level of GM-CSF, we see that higher doses of CTX depress the WBC and push the lowest count level reached down. On the other hand, for a given level of CTX, higher doses of GM-CSF seem to increase the white blood cell counts. This is to be expected as this drug is known to stimulate the production of white blood cells by the bone marrow. Figure 31.3 shows the posterior predictive mean response curves for hypothetical new patients in study 2. The presence of amifostine (AM=1) does not seem to have a relevant effect on WBC. We refer to Müller *et al.* [6] for a thorough discussion of this example.

References

[1] Maria De Iorio, Peter Müller, Gary L. Rosner, and Steven N. MaEach-ern. An ANOVA model for Dependent Random Measures. Technical report, 2002.

[2] Thomas S. Ferguson. A Bayesian analysis of some nonparametric problems. *Annals of Statistics*, 1(2):209–230, April 1973.

[3] S.N. MacEachern. Dependent nonparametric processes. *Journal of the American Statistical Association*, 2001.

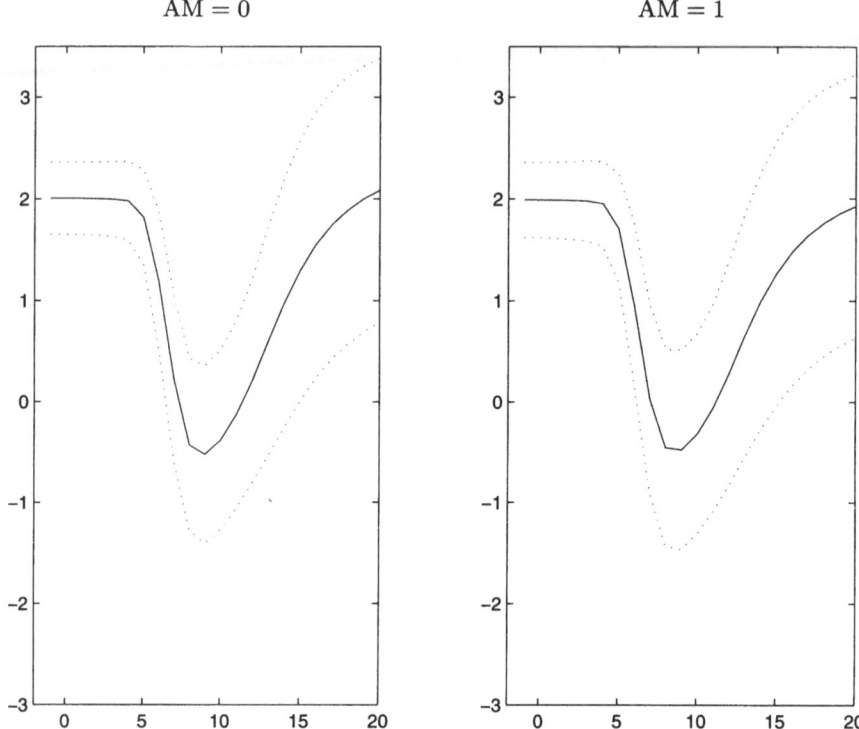

Figure 31.3. Predicted WBC response curves for hypothetical future patients in study 2. The dashed lines give the 95% prediction interval. AM=1 denotes presecnce of amifostine, while AM=0 denotes its absence

[4] S.N. MacEachern and P. Müller. Estimating mixture of Dirichlet process models. *Journal of Computational and Graphical Statistics*, 7:223–239, 1998.

[5] Steven N. MacEachern and Peter Müller. Efficient MCMC schemes for robust model extensions using encompassing dirichlet process mixture models. In Fabrizio Ruggeri and David Rios Insua, editors, *Robust Bayesian Analysis*. New York, NY, USA, 2000.

[6] Peter Müller, Maria De Iorio, and Gary L. Rosner. ANOVA Mixture Models for Non-Parametric Bayesian Inference over Related Studies. Technical report, 2002.

[7] Peter Müller and Gary Rosner. Semiparametric pk/pd models. In Dipak Dey, Peter Müller, and Debajyoti Sinha, editors, *Practical Nonparametric and Semiparametric Bayesian Statistics*, pages 323–337. New York, NY, USA, 1998.

[8] J. Sethuraman. A constructive defnition of the Dirichlet process prior. *Statistica Sinica*, 2:639–650, 1994.

Lecture Notes in Statistics

For information about Volumes 1 to 117, please contact Springer-Verlag

118: Radford M. Neal, Bayesian Learning for Neural Networks. xv, 183 pp., 1996.

119: Masanao Aoki and Arthur M. Havenner, Applications of Computer Aided Time Series Modeling. ix, 329 pp., 1997.

120: Maia Berkane, Latent Variable Modeling and Applications to Causality. vi, 288 pp., 1997.

121: Constantine Gatsonis, James S. Hodges, Robert E. Kass, Robert McCulloch, Peter Rossi, and Nozer D. Singpurwalla (Editors), Case Studies in Bayesian Statistics, Volume III. xvi, 487 pp., 1997.

122: Timothy G. Gregoire, David R. Brillinger, Peter J. Diggle, Estelle Russek-Cohen, William G. Warren, and Russell D. Wolfinger (Editors), Modeling Longitudinal and Spatially Correlated Data. x, 402 pp., 1997.

123: D. Y. Lin and T. R. Fleming (Editors), Proceedings of the First Seattle Symposium in Biostatistics: Survival Analysis. xiii, 308 pp., 1997.

124: Christine H. Müller, Robust Planning and Analysis of Experiments. x, 234 pp., 1997.

125: Valerii V. Fedorov and Peter Hackl, Model-Oriented Design of Experiments. viii, 117 pp., 1997.

126: Geert Verbeke and Geert Molenberghs, Linear Mixed Models in Practice: A SAS-Oriented Approach. xiii, 306 pp., 1997.

127: Harald Niederreiter, Peter Hellekalek, Gerhard Larcher, and Peter Zinterhof (Editors), Monte Carlo and Quasi-Monte Carlo Methods 1996. xii, 448 pp., 1997.

128: L. Accardi and C.C. Heyde (Editors), Probability Towards 2000. x, 356 pp., 1998.

129: Wolfgang Härdle, Gerard Kerkyacharian, Dominique Picard, and Alexander Tsybakov, Wavelets, Approximation, and Statistical Applications. xvi, 265 pp., 1998.

130: Bo-Cheng Wei, Exponential Family Nonlinear Models. ix, 240 pp., 1998.

131: Joel L. Horowitz, Semiparametric Methods in Econometrics. ix, 204 pp., 1998.

132: Douglas Nychka, Walter W. Piegorsch, and Lawrence H. Cox (Editors), Case Studies in Environmental Statistics. viii, 200 pp., 1998.

133: Dipak Dey, Peter Müller, and Debajyoti Sinha (Editors), Practical Nonparametric and Semiparametric Bayesian Statistics. xv, 408 pp., 1998.

134: Yu. A. Kutoyants, Statistical Inference For Spatial Poisson Processes. vii, 284 pp., 1998.

135: Christian P. Robert, Discretization and MCMC Convergence Assessment. x, 192 pp., 1998.

136: Gregory C. Reinsel, Raja P. Velu, Multivariate Reduced-Rank Regression. xiii, 272 pp., 1998.

137: V. Seshadri, The Inverse Gaussian Distribution: Statistical Theory and Applications. xii, 360 pp., 1998.

138: Peter Hellekalek and Gerhard Larcher (Editors), Random and Quasi-Random Point Sets. xi, 352 pp., 1998.

139: Roger B. Nelsen, An Introduction to Copulas. xi, 232 pp., 1999.

140: Constantine Gatsonis, Robert E. Kass, Bradley Carlin, Alicia Carriquiry, Andrew Gelman, Isabella Verdinelli, and Mike West (Editors), Case Studies in Bayesian Statistics, Volume IV. xvi, 456 pp., 1999.

141: Peter Müller and Brani Vidakovic (Editors), Bayesian Inference in Wavelet Based Models. xiii, 394 pp., 1999.

142: György Terdik, Bilinear Stochastic Models and Related Problems of Nonlinear Time Series Analysis: A Frequency Domain Approach. xi, 258 pp., 1999.

143: Russell Barton, Graphical Methods for the Design of Experiments. x, 208 pp., 1999.

144: L. Mark Berliner, Douglas Nychka, and Timothy Hoar (Editors), Case Studies in Statistics and the Atmospheric Sciences. x, 208 pp., 2000.

145: James H. Matis and Thomas R. Kiffe, Stochastic Population Models. viii, 220 pp., 2000.

146: Wim Schoutens, Stochastic Processes and Orthogonal Polynomials. xiv, 163 pp., 2000.

147: Jürgen Franke, Wolfgang Härdle, and Gerhard Stahl, Measuring Risk in Complex Stochastic Systems. xvi, 272 pp., 2000.

148: S.E. Ahmed and Nancy Reid, Empirical Bayes and Likelihood Inference. x, 200 pp., 2000.

149: D. Bosq, Linear Processes in Function Spaces: Theory and Applications. xv, 296 pp., 2000.

150: Tadeusz Caliński and Sanpei Kageyama, Block Designs: A Randomization Approach, Volume I: Analysis. ix, 313 pp., 2000.

151: Håkan Andersson and Tom Britton, Stochastic Epidemic Models and Their Statistical Analysis. ix, 152 pp., 2000.

152: David Ríos Insua and Fabrizio Ruggeri, Robust Bayesian Analysis. xiii, 435 pp., 2000.

153: Parimal Mukhopadhyay, Topics in Survey Sampling. x, 303 pp., 2000.

154: Regina Kaiser and Agustín Maravall, Measuring Business Cycles in Economic Time Series. vi, 190 pp., 2000.

155: Leon Willenborg and Ton de Waal, Elements of Statistical Disclosure Control. xvii, 289 pp., 2000.

156: Gordon Willmot and X. Sheldon Lin, Lundberg Approximations for Compound Distributions with Insurance Applications. xi, 272 pp., 2000.

157: Anne Boomsma, Marijtje A.J. van Duijn, and Tom A.B. Snijders (Editors), Essays on Item Response Theory. xv, 448 pp., 2000.

158: Dominique Ladiray and Benoît Quenneville, Seasonal Adjustment with the X-11 Method. xxii, 220 pp., 2001.

159: Marc Moore (Editor), Spatial Statistics: Methodological Aspects and Some Applications. xvi, 282 pp., 2001.

160: Tomasz Rychlik, Projecting Statistical Functionals. viii, 184 pp., 2001.

161: Maarten Jansen, Noise Reduction by Wavelet Thresholding. xxii, 224 pp., 2001.

162: Constantine Gatsonis, Bradley Carlin, Alicia Carriquiry, Andrew Gelman, Robert E. Kass Isabella Verdinelli, and Mike West (Editors), Case Studies in Bayesian Statistics, Volume V. xiv, 448 pp., 2001.

163: Erkki P. Liski, Nripes K. Mandal, Kirti R. Shah, and Bikas K. Sinha, Topics in Optimal Design. xii, 164 pp., 2002.

164: Peter Goos, The Optimal Design of Blocked and Split-Plot Experiments. xiv, 244 pp., 2002.

165: Karl Mosler, Multivariate Dispersion, Central Regions and Depth: The Lift Zonoid Approach. xii, 280 pp., 2002.

166: Hira L. Koul, Weighted Empirical Processes in Dynamic Nonlinear Models, Second Edition. xiii, 425 pp., 2002.

167: Constantine Gatsonis, Alicia Carriquiry, Andrew Gelman, David Higdon, Robert E. Kass, Donna Pauler, and Isabella Verdinelli (Editors), Case Studies in Bayesian Statistics, Volume VI. xiv, 376 pp., 2002.

168: Susanne Rässler, Statistical Matching: A Frequentist Theory, Practical Applications and Alternative Bayesian Approaches. xviii, 238 pp., 2002.

169: Yu. I. Ingster and I.A. Suslina, Nonparametric Goodness-of-Fit Testing Under Gaussian Models. xiv, 453 pp., 2003.

170: Tadeusz Caliński and Sanpei Kageyama, Block Designs: A Randomization Approach, Volume II: Design. xii, 351 pp., 2003.

171: D.D. Denison, M.H. Hansen, C.C. Holmes, B. Mallick, and B. Yu (Editors) Nonlinear Estimation and Classification. viii, 474 pp., 2003.